Hans B. Kief
Helmut A. Roschiwal

CNC Handbook

Translated by Jefferson B. Hood

New York Chicago San Francisco Lisbon London Madrid
Mexico City Milan New Delhi San Juan Seoul
Singapore Sydney Toronto

The McGraw-Hill Companies

1 2 3 4 5 6 7 8 9 0 CTP/CTP 1 8 7 6 5 4 3 2

ISBN 978-0-07-179948-5
MHID 0-07-179948-6

Sponsoring Editor
Michael Penn

Editing Supervisor
Stephen M. Smith

Production Supervisor
Richard C. Ruzycka

Acquisitions Coordinator
Bridget L. Thoreson

Project Manager
Harleen Chopra, Cenveo Publisher Services

Copy Editor
James K. Madru

Proofreader
Paul Sobel

Indexer
Edwin Durbin

Art Director, Cover
Jeff Weeks

Composition
Cenveo Publisher Services

This book was previously published as *CNC-Handbuch 2011/2012* by Carl Hanser Verlag GmbH & Co. KG, München, Germany, copyright © 2011.

This book is printed on acid-free paper.

Contents

Preface

Since 1976 the German *CNC Handbook* has kept pace with the rapid transition from NC to CNC and the emergence of new technologies. Each new development has been described as it appeared. As a result the range of topics covered by the book has been expanded more and more. In order to meet the needs of readers with various levels of existing knowledge, the introductory chapters have been tightened up, the contents of the chapters on punching/nibbling, laser technology, and generative manufacturing methods have been updated, and the subject of energy efficiency has been added.

NC/CNC: Both terms are used in industry and in the literature. Today, up-to-date products are all CNC systems, CNC hardware, or CNC machines. It is still common, however, to speak of NC technology, NC programs, and NC programming—as we have done in this book.

CNC machines and mechatronics: Machine tools are in themselves already highly developed mechanical systems. CNC machines, flexible manufacturing systems, and robots are based on the close interaction of mechanical, electronic, and IT elements. As such, they are typical examples of highly complex mechatronic systems. In addition, there are process measurement technologies, CAD/CAM programming, and data network technology. Computers and CNC systems are command centers for automated production processes. They contain intelligent functions in the form of software that costs almost as much as the hardware.

People: When it comes to digitally controlled manufacturing systems, human beings—from mechatronics technicians to managers—are still essential to perform important and responsible tasks. It is also vital for them to know the functions of the individual mechatronic components and to understand how they interact in a complete functioning system.

Thanks to our contributors: It is no longer possible today for any one author to master all of the specialized areas related to CNC in detail and to describe them at a basic level as well as at a highly sophisticated technical level. We therefore would like to thank all of our contributors for the excellent support they have provided in developing new chapters and illustrations. It is our goal not only to give our readers an overall survey of the wide field of digital manufacturing technology, but also to provide them with the necessary fundamental knowledge.

We also thank all of our **reviewers and critics** for their suggestions. These have enabled us to make continual improvements in each new edition.

Hans B. Kief

Helmut A. Roschiwal

About the Authors

Hans B. Kief is acknowledged as one of the world's leading experts in the fields of CNC manufacturing and computer-controlled flexible manufacturing systems (FMS). He has worked as a consultant in the manufacturing industry, having developed CNC reliability software for the aircraft and automobile industries, and has traveled widely in his professional capacity through the United States and Europe. Mr. Kief has a degree in electrical engineering, founded the German NC Society, and has held memberships in the Society of Manufacturing Engineers (SME), the Computer and Automated Systems Association (CASA), and the American NC Society.

Helmut A. Roschiwal began his career as an apprentice to a toolmaker. He went on to earn his mechanical engineering degree and start his own mechanical engineering product development company: Roschiwal+Partner offers development services and design engineering to customer specifications (especially in the area of CNC machine tools) and has subsidiaries in Germany and Romania.

Part 1

Introduction to CNC Technology

1. Historical Development of Numerical Control Production

A look back on the introduction and development of numerical control (NC) technology shows that it was not just technical aspects that played an important role. Correct and incorrect management decisions, the beginning of globalization, and especially the Japanese challenge also were major factors influencing the overall changes in the market and the manufacturing field as a whole.

1.1 Germany after World War II

1945–1948: All manufacturing facilities in Germany were destroyed or unusable and in some cases dismantled and sent abroad as war reparations. Production was completely crippled.

The inner cities had been devastated and were largely uninhabitable; millions of tons of rubble blocked the streets and other transportation infrastructure. Electricity, gas, and water supplies were barely functioning, and with a few minor exceptions, industrial production was impossible.

1948 (currency reform)–1955: The machine tool and manufacturing industry was rebuilt, primarily on the basis of previously existing concepts. During the war and shortly thereafter, the development of new machine concepts was not possible.

Most machines were designed for manual operation, but there was a lack of experienced, skilled workers. The few machines that were still available were busy producing urgently needed mass-produced articles.

The demand for goods was practically unlimited. The existing machines operated in two or three shifts. New jobs were created, but there were not enough workers. Over 2 million German men had been killed in the war, and over 6 million were wounded, sick, or still held as prisoners of war (POWs).

The solution was "guest workers." They came from all the countries of Western Europe. There was plenty of work to be done.

The goal: To rebuild the destroyed cities, factories, bridges, houses, streets, and other infrastructure and to provide urgently needed transportation capacity. To do this, all kinds of machines and vehicles were needed, especially construction machinery, cranes, backhoes, and trucks.

The main focus of industrial manufacturing was on **mass production** using manual production machinery, transfer lines, and mechanical automatic machines. The life cycle of the products being manufactured was at least 10 years; there was no need for rapid production changeovers.

The cumulative result of this enormous demand, clever policies, and energetic workers was the German **"economic miracle."**

1.2 Rebuilding the German Machine Tool Industry

Thanks to the conditions just described, within a few years **(by about 1960–1970), Germany** had the newest stocks of machinery of any developed nation: The average age of the machines was five to six years. But there were too few of them, and the statistics were lagging behind. Furthermore, some of the new machines were still using prewar technology.

During this period (approximately 1960–1975), American machines were about 15 to 17 years old. Renewal was brought about through the use of NC machines (i.e., turning, milling, machining centers) in the automotive and aerospace industries. The NC technology developed in the United States was implemented much more quickly there than it was in Europe. Many projects were subsidized by the government, for example, in defense industries.

American manufacturers of NC machines were very successful in selling their machines worldwide, but they neglected to systematically pursue further development.

This led to a continuous rise in imports of inexpensive Japanese machines.

The rapid series of improvements in numerical controls had major effects on all types of machinery and demanded new, adapted designs. These were not forthcoming, which led to the bankruptcy of many American manufacturers.

At the beginning of the 1970s, Japan made large investments in machine tool production. These were simple, inexpensive NC machines, but they were constructed according to the latest principles. Soon Japan was able to supply machines off the shelf and for heretofore unbelievably low prices. These machines were built according to a different set of requirements: series-produced standard machines without any major modifications, reliable, with standard NC, and no choice of controller, but inexpensive.

While German manufacturers continued selling to their traditional European market, the Japanese based their strategy from the very beginning on the global market, with a focus on the United States and later also on Europe. Customer-specific modifications were systematically refused.

By the middle of the 1980s, Japan had caught up with Germany in its share of the global market. One sign of German manufacturers' diminishing competitiveness was the constantly increasing percentage of imports: From 1973 to 1981, imports increased by 11.9 to 33.3 percent, and, by 1991, to 41.2 percent.

1.3 Worldwide Changes

In many **developed nations,** obsolescent machines were being used for production even 10 to 15 years after the end of World War II. While at first these were entirely sufficient, as competition and the pressure to reduce costs increased and buying practices changed, it became essential to modernize the machinery in many manufacturing firms.

Moreover, the **1970s** saw the beginning of a worldwide trend toward a **buyer's market;** that is, quicker product changes and shorter life cycles became the rule for almost all products.

The result: A shift away from mass production and toward **smaller lot sizes**. Instead of rigid mass production on automatic machines and transfer lines, increasing use was made of more flexible NC machines. However, the increasing complexity of products owing to greater use of computer-aided design (CAD) systems necessitated the deployment of up-to-date machine tools with seamlessly integrated data for quicker NC programming.

New potential users of NC machines appeared, such as

- The German defense industry for tanks, armored fighting vehicles, transporters, etc.
- The German aerospace industry, with manufacturing under license of the Starfighter, Phantom, helicopters, and weapons and later with the Airbus, MRCA Tornado, Alpha Jet, and Dornier Do 27 programs.

The aerospace industries in France (i.e., Dassault, Aerospatiale, and Snecma), the United Kingdom (i.e., Hawker and British Aerospace), and the United States (i.e., Boeing, McDonnell Douglas, Fairchild, Lockheed, Sikorsky, etc.) also were looking for new machine concepts. What was wanted were high-precision machines that could be reconfigured quickly, new machine sizes (i.e., surface milling machines and large drilling machines), and machining centers.

Small and medium-sized subcontracting firms were a large potential market that was as yet undeveloped.

1.4 Typical New NC Machines

From **1968** onward, the reborn German aerospace industry and the German automotive industry provided a major boost to the country's machine tool industry:

- Large-surface milling machines and machining centers with a high degree of automation
- Three- and five-axis milling machines with simultaneous interpolation in all axes
- Gantry-type milling machines for large milling widths with up to eight parallel main spindles
- Electron-beam welding machines, flexible manufacturing cells, and a very high degree of automation in workpiece and tool handling, as well as machining
- High-speed cutting machines for tool and die making
- New programming and machining strategies (e.g., APT, CAD, CAD/CAM) brought in large contracts for many European manufacturers.

In the course of just a few years (1970–1980), Germany became the world's largest exporter of machine tools.

At first, any number of new features were simply added to the old "tried and true" machine concepts without modernizing their basic design.

The result: Too many parts.
Machines too heavy.
Machines take too long to build.
Excessively complex designs.
Machines too expensive.

And
Commissioning takes too long.
Machines break down too often.
There is excessive downtime.

Consequently, these machines were not cost-effective for "normal" industries. The urgently needed breakthrough into the general mechanical engineering industry became possible only when reworked, less expensive concepts became available.

1.5 The Japanese Influence

The Japanese standard NC machines, which in the meantime had undergone continuous improvement, made ever-greater inroads into the German market. The initial "psychological resistance" on the part of purchasers against Japanese products was broken down by their low prices, constantly improving quality, and reports of generally positive experiences with them.

These machines were built in large volumes, had unusually short delivery times, and featured very reliable numerical controllers (e.g., FANUC, Mitsubishi, Okuma, Mazak, etc.). What's more, the Japanese firms provided service and support on a grand scale. Soon more and more German machine manufacturers were adding Japanese controllers to their machines and used the international availability of service and support, for example, from FANUC, to sell machines worldwide.

1.6 The Crisis in Germany

After the boom years from 1985 to 1990, **from 1992 onward** the German machine tool industry was faced with its most serious crisis since World War II. By 1994, production had fallen by almost 50 percent and the number of people employed by 30 percent. At that point the machine manufacturers' structural and financial difficulties took on particularly serious dimensions.

This rapid decline was the cumulative result of a number of problems. The German machine tool industry entered into a crisis for similar reasons to those of the American industry in the 1980s. Instead of marshalling their resources to take on the Japanese competition, German manufacturers tried to fend them off by means of price reductions—a tactic that could not succeed in the long run. Furthermore, German manufacturers worked against each other instead of working with each other to develop new ideas to defend against the ever-stronger Japanese competition. Some good approaches might have been uniform tool holders and changing systems, uniform pallet changers, and coordinated table heights. Such an approach, for example, would have made it much simpler, cheaper, and thus more attractive to introduce flexible manufacturing systems by combining machines from different manufacturers. But there was no money to develop new, less expensive machines.

The competitive atmosphere made it impossible to find common, coordinated, complementary strategic solutions—the kinds of solutions that many major users were looking for.

The result: The contribution margins, which had shrunk to less than 5 percent, did not allow for any large-scale future-oriented development projects. Many German machine manufacturers had either no strategic concept or no money to realize it. Instead, almost all of them tried to escape "upward" to the field of special and custom machines. But this niche policy had no prospects for success because the special machines were too expensive, and the production of standard machines did not use enough of the manufacturing capacity. Moreover, potential buyers demanded extensive, detailed plans from multiple manufacturers without bearing the associated costs.

Many highly respected manufacturers headed into bankruptcy or were taken over by competitors in subsequent years.

1.7 Causes and Effects

German managers asked the question quite openly: What is it that the Japanese are doing better than we German machine manufacturers, who have become so accustomed to success? Was it their lower prices owing to lower production costs? Or their better technical ideas? Or their delivery time?

But that was only part of it! What was much more serious was the fact that the Japanese had better business ideas, produced in larger quantities, and had a **global market strategy!** German manufacturers were looking for **purchasers** for special machines, whereas Japanese manufacturers were looking for **markets** for standard machines!

The Japanese machines were of high quality and made do with about 30 percent fewer mechanical parts. Purchasers were impressed by their advantages, advantages that became even more pronounced as time went by.

Even hard-core German traditionalists bought more and more Asian products. For the price of one German "super-special custom machine" with a long delivery time, it was possible to buy two or three Japanese standard machines off the shelf. Truly convincing!

It was only toward the **end of the 1980s/beginning of the 1990s** that the surviving German machine manufacturers finally understood that they had to build "different" machines in order to once again attract buyers and be successful. The niches for the German special machine manufacturers had become too small.

For many German manufacturers, the solution was mergers—often forced on them by banks. Today, many of these manufacturers have become competitive once again and use as a selling point the fact that they have reduced the number of parts in their modernized machines by 30 to 35 percent. These companies have finally comprehended that neither outmoded ideas nor "technical overkill" with a niche strategy would put them on the right track. But purchasers also have accepted the fact that German machines have comparable specifications to those of Japanese machines without a large number of customer-specific special functions.

Not to be underestimated is the role played by the new, high-performance, dialog-oriented **NC programming systems,** which were available both as a programming stations and also directly on the machines.

The revival of the German machine tool industry also was helped by **new technological processes** and completely new types of machine, such as high-speed cutting, high-power lasers for welding and cutting, additive manufacturing methods such as rapid prototyping systems, and machines for hard machining of metals and ceramics. Most recently, machines with parallel kinematics (i.e., tripods and hexapods) have become ready for market introduction, but potential purchasers are still holding back. However, universal machines for complete machining in a single setup are attracting more and more interest.

Furthermore, the use of new, highly dynamic drives has been making these machines faster and faster.

1.8 Flexible Manufacturing Systems

In the 1970s, large American companies such as Caterpillar, Cummings Diesel, General Electric, and a number of machine manufacturers (e.g., Cincinnati Milacron, Kearney & Trecker, Sundstrand, etc.) started designing and installing the first **flexible manufacturing systems** (FMSs). These consist of several **self-replacing** (identical) or **complementary** (different) NC machines and a common workpiece transport and control system. Such systems can be used profitably to produce either single parts on an order-by-order basis or in small to medium-sized lots. In special cases, FMSs also can be used for large-scale series production.

During this time, the first FMSs were tested successfully in Japan and marketed internationally. Visitors came from all over the world and were astounded by the unmanned production in unlit factory shops.

In Germany, demand for FMSs was at first very guarded. The main factor in this reluctance to buy was the extensive engineering required, that is, the customer-specific planning and dimensioning of such systems at the customer's location and the very involved time, unit cost, and investment calculations generally required by the customers. All this led to high costs and prices. It was only when the fantastic visions of "factories without workers" gave way to "manufacturing with reduced personnel" based on affordable manufacturing concepts that German users started to take more interest in such systems.

In 1974, the Getriebe Bauer Company of Esslingen, Germany, installed one of the first FMSs in Germany. It was composed of nine identical machining centers (made by BURR) with Bosch/Bendix controllers, a recirculating pallet system for automatic workpiece transport, and pallet transfer stations at each machine. The decisive factor here was that at that time, the first NC systems using program memory instead of punched-tape readers had become available. In the following years, Bauer expanded this system to include 12 machines and in 1988 upgraded them with higher-performance computer numerical control (CNC) systems. More than 20 years of two- and three-shift operations have met the user's technical and economic expectations—and then some! Finally, it was possible to produce goods to order—to reduce warehouse stocks and still deliver on short notice.

After the first positive reports, other FMSs followed in many other manufacturing firms.

In **Japan, the United States, and Europe,** FMSs designed according to the latest state of the art are being installed at a steady rate. The positive experience gained with these systems and their cost-effectiveness has led to better FMS-compatible machines that are unproblematic to combine and operate. The integration of robots for tool and workpiece handling also has led to improved system concepts. For early detection of planning errors, powerful simulation and production planning systems (PPSs) were developed.

FMSs have been used in **Germany** for many years now, and the trend is increasing. Because of the unavoidably rapid intervals between updates and modifications of workpieces, flexible manufacturing systems are often more cost-effective than rigid, inflexible transfer lines, even in mass production.

1.9 Situation and Outlook

Manufacturing technology and automation are undergoing continuous worldwide development with new ideas and concepts. Today, the primary role is played by **NC machines and integrated robots,** which are available in a wide variety of designs and combinations for all types of applications. Highly dynamic linear drives, position-measuring systems with extremely high resolution and precision, and fundamentally new machine concepts have made NC machines the dominant type of manufacturing system—and not just for machining.

Also with regard to **robots,** Japanese industry initially responded faster than all other developed nations, developing economical standard robots that subsequently served as a basis for gaining valuable experience worldwide. They won acceptance quickly almost everywhere.

Today there is a great demand for robots worldwide. All developed nations have built up their own production and offer special robots for manufacturing, handling, and assembly.

Computer-integrated manufacturing systems (**CIM systems**) as they were originally conceived are now outdated. The plan was to use interlinked computers to implement factories without workers and to control all the processes fully automatically, from purchasing to manufacturing and assembly. But experience soon showed that implementation of these ideas was too expensive and therefore unrealistic. Rather, users came to the realization that processes should be automated only when to do so would reduce the price of products while maintaining high quality. CIM quickly developed a bad reputation—a reputation that it did not deserve!

So a new term was coined—**digital manufacturing**—to express almost the same concept. This has attracted renewed interest based on the following proposition: **Automate? Yes, but not at any price!** Otherwise, profitability will remain a mere pipe dream.

In contrast, FMSs are in greater demand today than ever before. The much-praised flexibility of such systems no longer resides in centrally controlled machines with an extremely high degree of automation and a host computer but rather in a decentralized, clearly organized arrangement of the system components. Specially adapted workpiece transport systems and, in particular, a seamlessly integrated data network are essential, however. With reliable monitoring systems for tool breakage, measuring probes with automatic measurement logs, corrective interventions and fault messages, monitoring systems for torque and feed force, and absolutely error-free data transmission, **temporary unmanned operation** of these systems is possible. This great potential for greater efficiency was quickly recognized and exploited.

And one should not forget very high performance industrial computers and well-tested software for (almost) all requirements. Much of the more **recently developed software** has further **reduction of times** as its main goal.

Current technical literature shows that the speed of production is becoming more and more important. **Wasted time** is coming under increasingly close scrutiny when looking for efficiency improvements. It is not just shorter and shorter tool changes or faster rapid traverses that are being demanded, but also more intelligent machining itself. To achieve this, existing NC programs have to be reworked and finely adjusted. Today's **simulation systems** are perfectly suited for this, both technically and economically, and they make it possible to save millions of euros per year.

In this area, the largest users of NC machines have exerted their influence on the software producers in order to bring about practice-oriented solutions.

Simulations also can be used to satisfy requirements for quicker **market introduction** of new products—with time savings of more than 50 percent in some cases. Production lines for new products often have to be planned and ordered at a point in time when the final form, size, and machining of the parts have not been finalized. With support from production planning systems (PPSs) and simulation systems, today it is possible to determine and plan with great precision the complete scope of all investments required for production. It is also no problem to incorporate later product modifications into such machinery.

Simultaneous engineering methods provide a perfect supplement to this accelerated pace.

Finally, it is important to mention **rapid prototyping development (RPD)** manufacturing methods, which in the meantime have become an established practice. Depending on the specific tasks, various methods and processes are available to create physical test workpieces from CAD models using special numerically controlled machines. Especially the tool and die industry is already making very heavy use of these options. Most machines for **rapid prototyping manufacturing** use laser beams as a universal tool with completely new manufacturing methods. Many German and other European manufacturers are active in this field, with great success.

1.10 Conclusions

In the course of about 40 years, NC technology has brought about major transformations not only in the machine tools themselves but also in entire manufacturing plants, the people who work there, and beyond. Manufacturers and users have learned not to strive for 100 percent automation on a purely theoretical basis but rather, when performing manufacturing and automation analyses, to include all the departments involved in the process. Only in this manner can one achieve technically and economically viable solutions. Working together, machine and controller manufacturers have developed technically sophisticated manufacturing concepts at market-oriented prices and have managed to contain the initial Japanese success.

Today's manufacturing industry would not be possible without such high-performance components as computers, new machine concepts, automatic transport and handling systems, reliable controllers, and intelligent monitoring systems. And that is certainly not all!

In order to exploit the benefits of this technology profitably, highly educated and trained personnel is absolutely necessary—all the way from the management suite to the shop floor. Only such personnel will be able to plan, implement, operate, and maintain these complex systems competently.

Since 2008, the topics of environmental protection and energy efficiency have been attracting more and more public discussion. Implementing these requirements also will have an effect on the mechanical engineering and manufacturing technology industries. Machine tool and controller manufacturers are already considering how they can optimize auxiliary drives, work processes, and NC parts programs for greater energy efficiency.

In the international machine tool industry, the trend is toward the shifting of markets to Asia since business is still booming in countries such as China and India.

2. Milestones in the Development of NC

The idea of controlling devices by means of stored commands, as used in today's NC machines, can be traced back to the fourteenth century. It began with chimes controlled by spiked cylinders.

1808 Joseph M. Jacquard uses perforated metal cards for the automatic control of mechanical looms.

These were the first exchangeable data-storage media for controlling machines.

1863 M. Fourneaux patents the player piano, which becomes world famous under the name *Pianola*, in which a paper strip with a width of approximately 30 cm contains perforations for controlling a supply of compressed air, which, in turn, operates the keyboard mechanism. This method undergoes further developments, and eventually it becomes possible to control the tone, the intensity of the attack, and the advance speed of the paper roll.

This represents the introduction of paper as a data-storage medium and controller of auxiliary functions.

1938 During his doctoral work at the Massachusetts Institute of Technology (MIT), Claude E. Shannon comes to the conclusion that rapid calculation and transmission of data are possible only in binary form using Boolean algebra and that electronic switches are the only realistic components for such applications.

This lays the foundations for today's computers, including numerical controls.

1946 Dr. John W. Mauchly and Dr. J. Presper Eckert supply the first digital electronic computer, the ENIAC, to the U.S. Army.

This creates the basis for electronic data processing.

1949 to 1952 John Parsons and MIT are contracted by the U.S. Air Force to develop "a system for machine tools, to directly control the position of spindles by means of the output from a computing machine and to produce a workpiece as proof of the proper functioning of the system."

Parsons outlined four basic approaches to this idea:
1. The calculated positions along a path are stored on punch cards.
2. The punch cards are read automatically by the machine.
3. The positions that are read are output continuously, and additional intermediate values are calculated internally, so that
4. Servomotors can control the motions of the axes.

This machine was intended to produce the increasingly complex integral parts required by the aircraft industry. Some of these workpieces could be described precisely using relatively little mathematical data but were very difficult to manufacture manually. The connection between computers and NC was specified from the very start.

1952 The first numerically controlled machine tool—a Cincinnati hydrotel with a vertical spindle—goes into operation at MIT. The control system is constructed using vacuum tubes and provides for simultaneous motion along three axes (three-dimensional [3D] linear interpolation), receiving its data via binary-coded punched tape.

1954 Having bought the rights to Parsons' patent, Bendix builds the first industrially produced NC system. It continues to make use of vacuum tubes.

1957 The U.S. Air Force installs the first NC milling machines in its own workshops.

1958 The first symbolic programming language, APT, is introduced in connection with the IBM 704 computer.

1960 NC systems with transistor technology replace controls using relays and vacuum tubes.

1965 Automatic tool changing increases the degree of automation.

1968 Integrated circuit (IC) technology makes controls more compact and reliable.

1969 First direct numerical control (DNC) installations in the United States using Sundstrand's Omnicontrol and IBM computers.

1970 Automatic pallet changing.

1972 The first NC systems with built-in, series-production minicomputers open up a new generation of powerful computerized NCs (CNCs), which, in turn, is replaced very quickly by microprocessor CNC systems.

1976 Microprocessors revolutionize CNC technology.

1978 Flexible manufacturing systems (FMSs) are implemented.

1979 The first examples of CAD/computer-aided manufacturing (CAM) integration are developed.

1980 Programming tools integrated into CNC systems ignite an ideological conflict over the pros and cons of manual data input controls.

1984 Powerful CNC systems with graphical programming tools set new standards for programming "on the shop floor."

1986 Standardized interfaces open the way . . .

1987 Toward the ideal of an automated manufacturing plant based on compatibility in the exchange of data: CIM.

1990 Digital interfaces between NC and drives improve accuracy and control characteristics of the NC axes and the main spindle.

1992 "Open" CNC systems allow customer-specific modifications, operations, and functions.

1993 First standardized use of linear drives in machining centers.

1994 The combination of CAD, CAM, and CNC in a single process is made complete through the use of nonuniform rational-basis spline (NURBS) as an interpolation method in CNC systems.

1996 Digital drive control, fine interpolation with resolutions in the submicron range (<0.001 µm/<0.04 µin), and feed rates up to 100 m/min (325 ft/min).

1998 Hexapods and multifunctional machines sufficiently developed for industrial use.

2000 CNC systems and programmable logic controllers (PLCs) with Internet interfaces allow worldwide exchange of data and intelligent fault diagnosis/troubleshooting.

2002 The first highly integrated, universally configurable IPC CNC systems, including data memory, PLC, serial real-time communication system (SERCOS) digital drive interface, and Process Field Bus (PROFIBUS) interfaces on a PC plug-in card.

2003 Electronic compensation for mechanical, thermal, and measurement-related sources of error.

2004 Increasing importance of external dynamic process simulation of NC programs on the personal computer (PC), including depiction of machines, clamping fixtures, and workpieces for the purposes of error detection and program optimization.

2005 CNC systems with nano and pico interpolation improve workpiece surfaces and accuracy.

2007 Teleservice: Support for personnel via telephone or data link during commissioning, fault diagnosis, maintenance, and repair of machines and systems.

2008 Special safety systems are developed in order to meet more stringent requirements as regards safety for people, machines, and tools. Requirements such as "safe motion," "safe processing of peripheral signals," and "safe communication" are realized via the CNC and drive systems.

There is no need to develop costly additional software or wiring.

2010 The introduction of multicore processors in CNC systems pushes performance to an even higher level. Functions that previously had to be precalculated now can be integrated into the controller (e.g., in spline interpolation).

2011 With the new feature "Energy Analysis," modern CNCs measure the energy consumption of the machine tool over a defined time period. With the help of data analysis and specially adapted control of all auxiliary devices, energy efficiency of machines can be improved.

2012 Designing of the so called "Hybrid Machines" is in process. These machines will be able to execute up to five different machining operations; such as turning, milling, grinding, gear cutting, and hardening; in one claming and with higher quality.

3. What Are NC and CNC?

NC is the abbreviation for "numerical control," that is, controlling by means of number input. CNC systems are computer-based numerical control systems for open- and closed-loop control of machine tools. The most important basic concepts are described in the following.

3.1 The Path to NC

The primary task of control systems used in production machinery is to ensure that sequences of motions are repeated quickly and with great precision, thus creating mass-produced articles of uniform quality without human intervention. Depending on the **control components** used, these can involve **mechanical, electrical, electronic, pneumatic, or hydraulic controls.**

In order to process a workpiece, a machine tool needs "information." Before the introduction of NC technology, the "path information" was either entered manually by the machine operator or using mechanical aids such as templates or cam discs. Changes in the sequence of operations or changes to different products therefore involved relatively long downtimes for the **changeover of the machines and controls.** Adjustable cams and cam rails were used in combination with limit switches to terminate motions at precisely defined positions. The precise adjustment of these limiting cams was very time consuming. Added to this were the time required for manual changes to tooling; for specifying the spindle speeds and feed rates, the time required for clamping the workpiece, and fine-tuning the setup of the machine; and for exchanging the program. The overall scope of control of these program controls was very limited in part because of the small number of switching steps that was possible.

The flexibility needed for frequent changeovers could not be achieved in a cost-effective manner with these machines.

The meant that a new **control concept** had to be developed that would meet the following requirements:

- The maximum possible scope of control as regards program length and motions
- No manual assistance through intervention in the machining process
- Sequential programs that could be saved and exchanged and corrected quickly
- No cams or limit switches for the different adjusting paths
- Simultaneous 3D motions of multiple axes that could be precisely defined for the machining of complex shapes and surfaces
- Quick changing of tools, including feed rates and speeds
- If necessary, automatic changing of the workpieces being machined.

The goal was **control systems** that could be converted quickly and without error between various machining tasks. The objective was to control the relative motion between tool and workpiece using the dimensions from the workpiece drawing. High-resolution position-measuring systems with measurement data that could be evaluated electronically would allow precise relative motion between the machine and tool.

Such control systems thus would function through the input of numbers, that is, numerically. This defined the basic idea of **numerical control systems**.

It should be possible to use additional numerals to program the feed rate, the spindle speed, and the tool number. It should be possible to use additional **on/off commands** (M functions) to activate automatic tool changing and to switch the coolant on and off.

The **NC program** that is used to control the machine comprises all the numerical values arranged step by step according to the processing sequence.

From NC to CNC

The first NC systems were constructed using relays and were either "hard-wired programmed" or simply "hard-wired." The first electronic function modules followed in rapid succession: vacuum tubes, transistors, and integrated circuits (ICs). But it was only with the use of microelectronics and microprocessors that control systems became more economical, more reliable, and more powerful.

In order to process workpieces, in addition to the path and switching information, CNC systems have to constantly process additional numerical values, for example, to compensate for various miller diameters and tool lengths or clamping tolerances. Thanks to their high processing speed, they are able to execute a complete range of administrative, display, and control functions in a timely manner. Notwithstanding this, it is also possible to input the next parts program on the machine using graphical-dynamic support.

3.2 Hardware *(Figures 3.1 and 3.2)*

The electronics of modern CNC systems make use of **microprocessors, ICs,** and possibly special modules for the servo control circuits. In addition to this, the systems can contain electronic **data memory** for a number of programs, subprograms, and many compensation values:

- **Read-only memory (ROM)** and **EPROM** components usually contain the unchanging parts of the CNC operating system, as well as frequently used fixed cycles and routines.
- **FEPROMs** are used to save data that can only be determined during commissioning and that must be impossible to lose and, if necessary, is modifiable, such as machine parameters, special cycles, and subprograms.
- **Random-access memory (SD-RAM, DDR-RAM)** with expandable capacities is primarily used for NC-programs, subprograms, and compensation values.

The graphical displays and dynamic simulations likewise require a great deal of computing power and memory. For this reason, most controllers use additional special **customer-designed very large scale integration (VLSI) chips.** These are highly integrated microelectronic modules that are designed according to specific customer requirements and are produced in large quantities. In this manner, it is possible to obtain compact dimensions, control units with high reliability and speed, and minimal maintenance requirements in subsequent operation.

All the electronic modules are located on one or more printed circuit boards (PCBs), which are plugged into a

① = DRAM module

② = PCIU bus controller

③ = EPLD module—erasable programmable logic device

④ = Gigabit receiver/transmitter—used for controlling liquid-crystal displays (LCDs)

⑤ = DRAM bank, max. 1 GB

Figure 3.1: Examples of large-scale integrated microelectronic modules.

Figure 3.2: Plug-in electronic module for an industrial PC (IPC). Drive, CNC, and PLC functions are grouped on one card. The front contains TCP/IP, PROFIBUS DP, and SERCOS interfaces.

module rack and are connected to each other via an internal bus connection (→ *Figure 3.2*). In order to prevent the CNC system from reacting in an undesirable manner, the electronics are built into a **metal case** that provides electrostatic and electromagnetic **shielding.** The case also should prevent contamination by oil and dirt because even the finest metal particles would jeopardize the proper functioning of the system if they were to become deposited on the circuit boards.

Therefore, circulating air must not be used to **cool** the interior of the case, even if it passes through a filter. Filters become clogged, thus rendering the cooling system ineffective. If the heat is not dissipated from the surface of the case adequately, the only acceptable solution to the problem is an active **cooling unit.** This increases the allowable ambient temperatures to a range of +10 to +45°C (+50 to +113°F). The air humidity should not exceed 95 percent. Even at lower humidity levels, it is often necessary for the user to guard against excessive accumulation of condensed water, which also can lead to malfunctions and damage.

3.3 Software

CNC systems require an **operating system,** which also may be described as **control software** or **system software**. It essentially comprises two parts:

- The standard software, and
- The machine-specific software

The **standard software**, for example, for data input, display, interfaces, or spreadsheet management, in some cases can be handled using standard commercial computers. The **machine-specific software** has to be designed especially for the machine type being controlled because there are often major differences in the kinematics and operating characteristics of different machines. One advantage of CNC systems is that modifications and adaptations can be made without intervention in the CNC hardware.

The operating system has a decisive influence on the overall performance of the machine. The software also manages a number of processes that run continuously in the background: monitoring and fault diagnostics, acquisition of machine data, and data interfaces. This also includes the CNC-integrated programming system with graphical simulation of the machining process and processing of the compensation values. It is also possible to adapt to machine-specific variants via the software. For example, these variations may consist of the number of axes, parameter values for the servo drives, various tool magazines and tool changers, software limit switches, or the attachment of tool-monitoring devices. These **machine parameter values** only need to be entered once during commissioning and are stored permanently (→ *Part 2: Functions of Numerical Control Systems*).

Modern CNC systems also feature an **integrated programming language** similar to BASIC, PASCAL, or C++. This programming language can be used to implement special functions without changing the fundamental CNC program. This gives the machine manufacturer the opportunity to incorporate specific know-how into the control system and to offer solutions for special manufacturing problems. The manufacturer even can adapt the screen graphics to create user help systems.

3.4 Control Methods

As regards their original development, a distinction can be made between four different levels of performance for the control method:

Point-to-point control systems (*Figure 3.3*)**.** These systems operate only in positioning mode. All the programmed axes

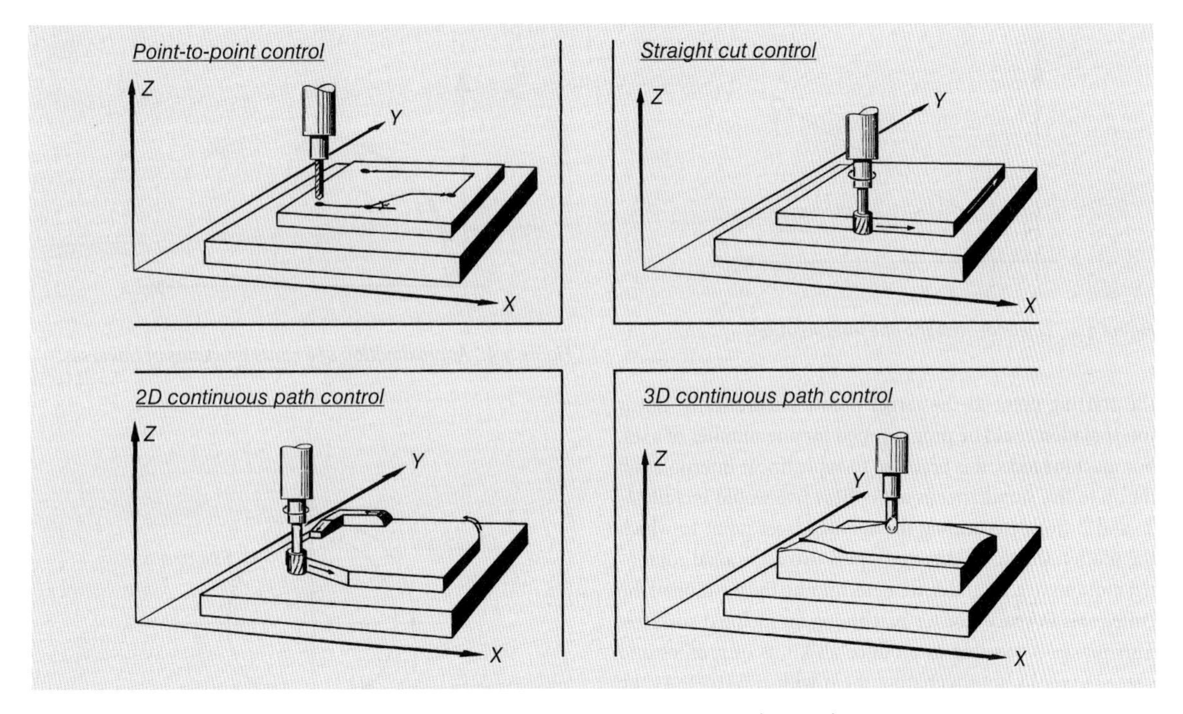

Figure 3.3: Development of NC technology from point-to-point to 3D continuous path control.

always start simultaneously at rapid traverse speed and continue motion until each axis has reached its target position. No tools are engaged while the positioning is underway. Machining does not begin until all the NC axes have reached their programmed positions. **Examples:** Drilling machines, punching machines, and feed motions on cutoff machines.

Straight-cut control systems. Motion in these systems can take place along the individual axes one after the other at a programmable feed rate; the tool can be engaged during this time. The travel motion is always parallel to the axis, and the feed rates have to be programmable. These severe technological limitations, and the fact that straight cut control is only slightly less costly than path control systems, means that it is in almost all cases a less attractive alternative. **Examples:** Feed control for drilling machines and workpiece handling.

Continuous-path control systems. These systems can "interpolate" the motions of two or more NC axes; that is, they can execute their motions in an exact ratio to each other. This motion is coordinated by an **interpolator,** which calculates the points along the path between the start and end points in a block-by-block manner. The NC axes do not stop at the programmed end point, however, but rather continue to move without interruption to the end point of the subsequent path section. The feed rates of the axes are controlled continuously in such a way that the designated cutting speed is maintained. This is called **three-dimensional continuous-path control,** or **3D control** for short. It allows tool motions to be executed in a plane or in space. **Examples:** Milling machines, lathes, erosion machines, machining centers—all types of machines, in fact.

Linear interpolation (*Figures 3.4 and 3.5*). In this method, the tool moves linearly, that is, in a straight line from

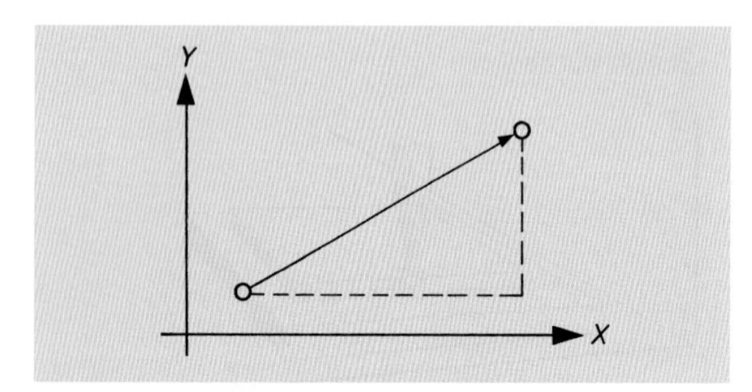

Figure 3.4: Linear or straight-line interpolation.

the starting point to the target point. Theoretically, linear interpolation could be programmed for any number of axes. For machine tools, it is practical to apply linear interpolation for up to five simultaneous axes; these five would include *X*, *Y*, and *Z* (for determining the spatial target point to which the motion should be directed), plus two additional rotational motions (e.g., *A* and *B*, for determining the position of the cutter axis in space), or for machining on inclined surfaces. This makes it possible to produce all kinds of profiles and space curves by approximating them with traverses. As the individual points of support become closer to each other, that is, as the tolerance width becomes narrower, the approximation to the specified profile becomes more accurate. However, a greater number of points also require a greater amount of data to be processed per unit time. This affects the processor speed required by the control system.

Circular interpolation (*Figure 3.6*). From a purely theoretical standpoint, it should be possible to approximate all kinds of paths by means of linear interpolation with traverses. The use of circular and parabolic interpolation reduces the amount of input data, resulting in easier programming and more accurate approximation of these paths. Circular interpolation is limited to the principal planes *XY*, *XZ*, and *YZ*. Circular interpolation is programmed using various methods depending on the type of control being used. These methods include programming in quadrants, in a full circle, by specifying the arc center, or by programming the arc end points and radius (→ *Part 5, Chapter 1*).

There is also parabolic, spline, and nano/pico interpolation (→ *Part 2*).

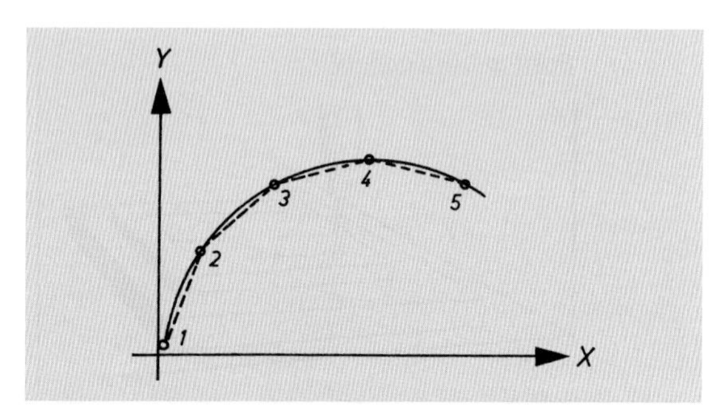

Figure 3.5: Approximation of a curve by means of traverses.

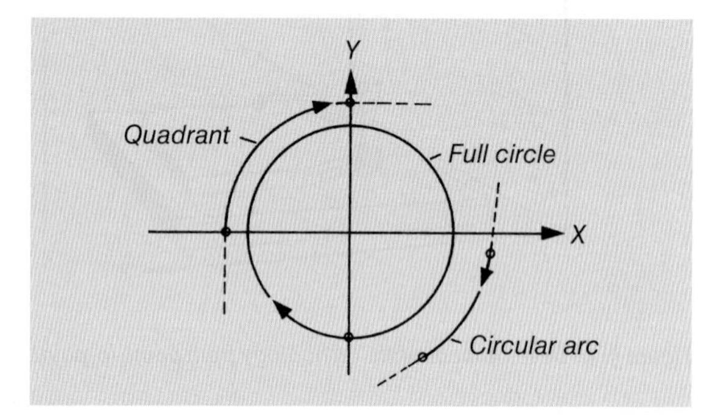

Figure 3.6: Circular interpolation.

3.5 NC Axes *(Figure 3.7)*

Depending on the specific machine, the **coordinate axes** can be designed as translational or rotary axes. Translational axes generally are perpendicular to each other, thus allowing movement to every point in the working space. Two additional rotary and swivel axes allow the machining of inclined workpiece surfaces or tracking of the cutter axis.

For axes to be controlled numerically, each NC requires

- **An electronic path-measuring system**
- **A servo drive**

The task of the CNC system is to compare the **target positions** predefined via the CNC system with the **actual position values** returned by the path-measuring system. If these values differ, the CNC system also must issue a control signal to the axis drives, which then will compensate

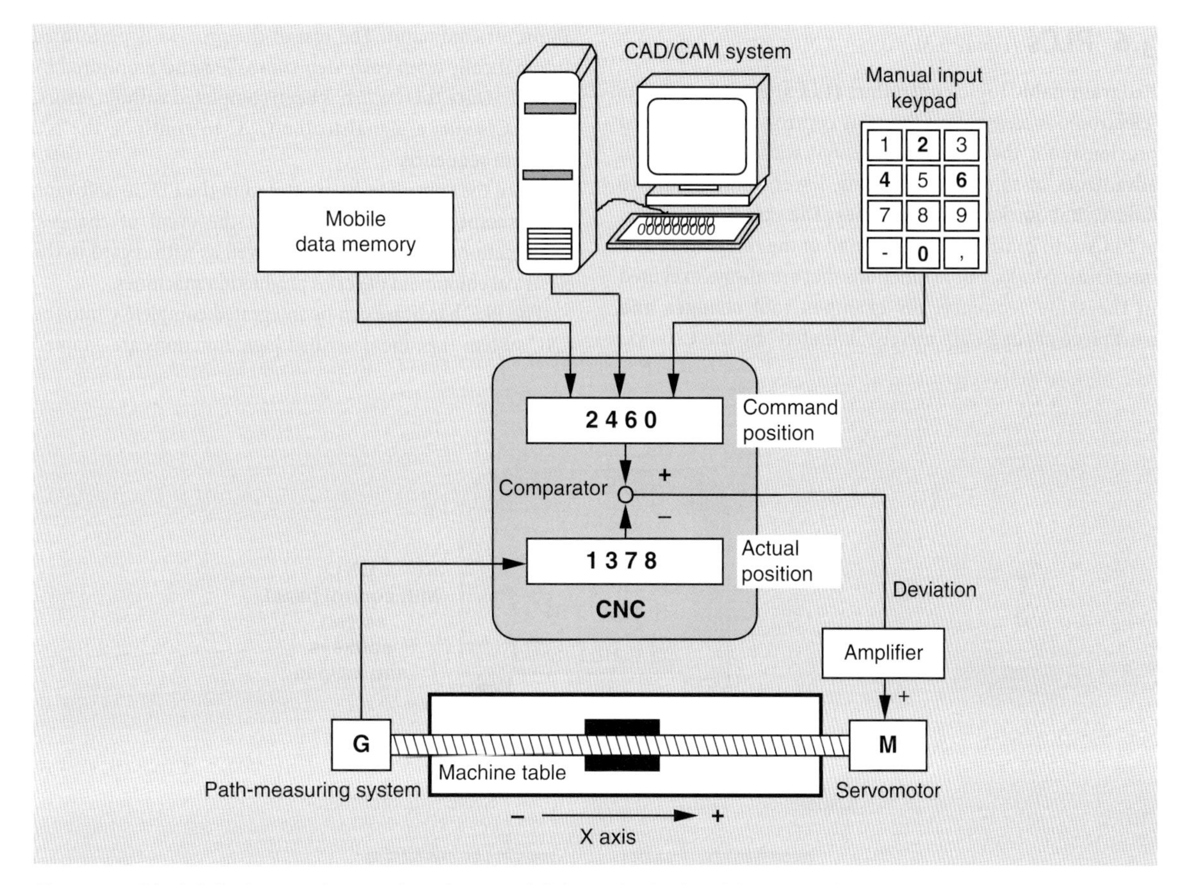

Figure 3.7: Principle for input and processing of geometric information in closed-loop control.

for the deviation (*Figure 3.7*). This is called **closed-loop control.** Continuous-path controls output new position values on an ongoing basis, and the controlled axes must follow these values. Continuous-path motions thus are achieved.

On lathes, the **main spindle** is also designed as an **NC axis** when rotating tools are used for drilling or milling. With drilling and milling machines, it is also possible to design the spindles as NC axes in order to allow programming of the functions "spindle orientation" and "helical interpolation."

Machining centers are usually equipped with **numerically controlled rotary tables.**

Increasingly, movement to the individual **positions in the workpiece magazine** is also being performed in the same manner as with NC axes. The use of position-measuring systems means that there is no need for other, costly coding devices for detection of the magazine position numbers or the tools. This makes the entire process for finding and exchanging the tools much faster.

The axis of the machine is designated according to the rules of the Cartesian coordinate system:

- Translational axes with the address letters ***X, Y, Z***
- In addition, parallel axes with ***U, V, W***
- Rotational or swivel axes ***A, B, C***
- (→ *Part 3, Chapter 1*)

3.6 PLCs *(Figure 3.8)*

Programmable logic controllers (PLCs) are in a way the "electronic successors" of the relay controls that were used previously for the same purpose, but with the additional advantages of smaller dimensions, lower vulnerability to failure, and quicker response times. The essential function of PLCs is to control and monitor all of the integration and interlocking tasks. Some functions that are always executed in the same sequence, for example, tool changes and workpiece changes, are merely "initiated" by the CNC via an on/off command. The rest of the process is performed automatically, with each step controlled and monitored by the PLC. Once this cycle has been completed without errors, the PLC sends a signal to the CNC to continue the NC program sequence.

All of the control tasks are stored in the PLC as software. This is especially advantageous in the event of changes, modifications, and expansions, as well as with regard to the electrical equipment of series production machines.

The PLC hardware can be integrated completely into the CNC system (i.e., the described logic functions are adopted

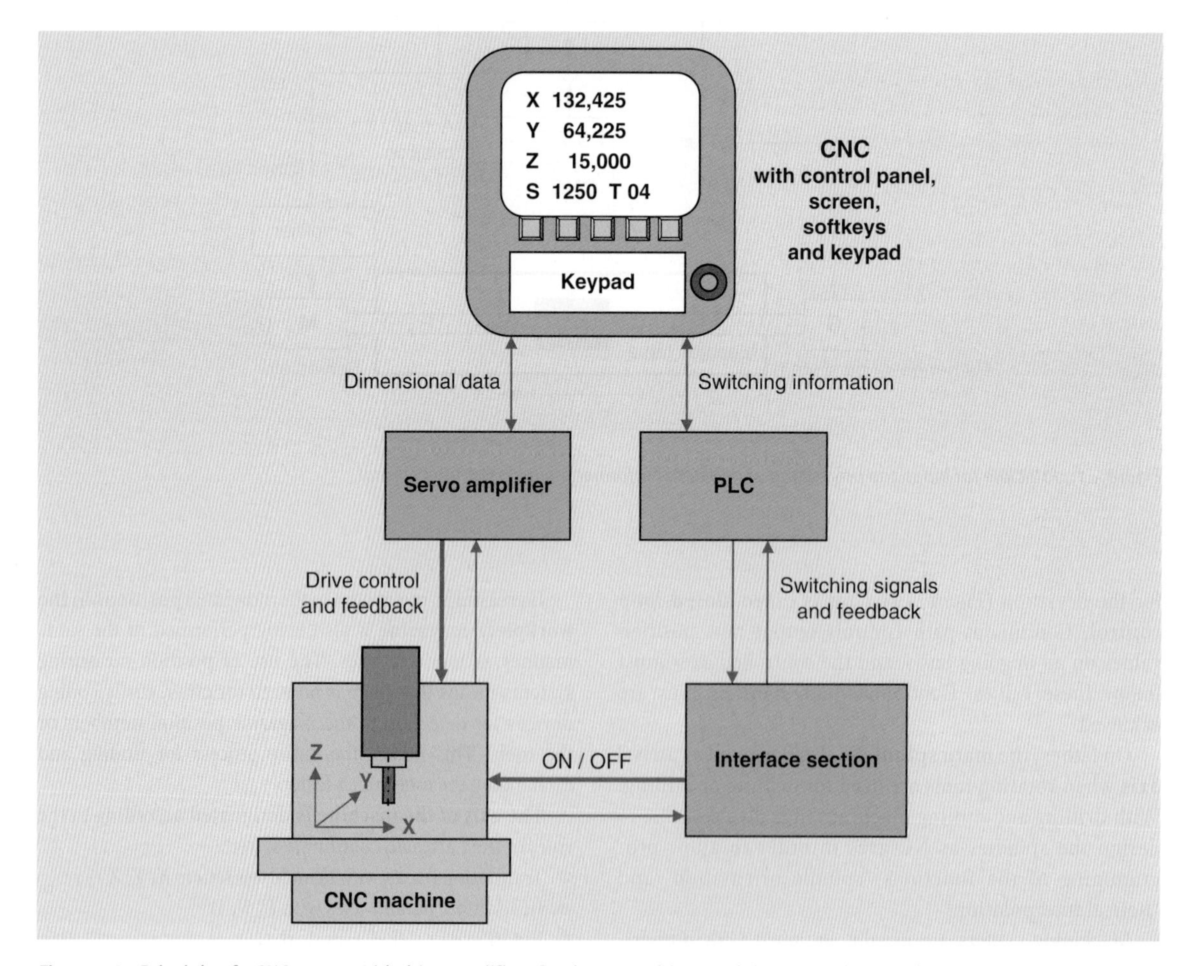

Figure 3.8: Principle of a CNC system with drive amplifiers for the servo drives and the external PLC with interface section.

by the CNC processor), and the control signals are output to the interface section. In the case of more complex machines, the manufacturer generally prefers to use a separate PLC. This has the advantage that the machine manufacturer can create and test the PLC program independent of the CNC system. This means that many functions can be put into operation even before the CNC system is delivered (→ *Part 2, Chapter 4: PLCs: Programmable Logic Controllers*).

In state-of-the-art machine tools, PLCs are also used to monitor the process itself. In higher-priority program sections, various CNC data are monitored, for example, in order to detect tool breakage within a fraction of a second and to quickly move the tool away from the workpiece in the right direction before further damage can occur.

3.7 Interface Section *(Figure 3.8)*

A distinction is made between an **interface section** and an **interface controller.**

An **interface section** comprises the electrical cabinet, including all the fuses, motor circuit breakers, transformers, contactors for auxiliary drives, amplifiers for high-performance drives, and terminals. The interface section is used, for example, to switch on the coolant supply, the chip removal system, or the auxiliary drives of the tool changer or workpiece changer.

In the case of CNC machines with a *separate* PLC, the PLC is likewise built into the interface section.

The task of the **interface controller** is to decode, interpret, execute logic operations, and control machine-specific sequences of functions. In today's CNC systems, this is the main task of the PLC. The necessary hardware generally is integrated into the CNC circuit board.

3.8 Computers and NC

From the beginning, computers and CNC systems were developed and put into use parallel to each other. Their fields of use expanded very quickly.

Computers in CNC

Today, microcomputers are the central element of every CNC system. Because the hardware components of industrial computers (IPCs) have become standardized, mass-produced products, the development focus for the computer user has been shifted to the software.

This also has given **CNC systems** a number of **significant advantages:**

- Software can be copied quickly, inexpensively, and without errors for use with subsequent machines.
- Software does not wear out and does not require repairs (i.e., it is low-maintenance).
- Software is relatively easy to work with (i.e., easy to improve, modify, and exchange if necessary) without having to change the hardware or "wiring."
- Software can be subdivided into function modules and recombined if necessary, which provides a variety of benefits.
- The series-produced hardware of IPCs and the computer operating system offer all the necessary, standardized interfaces for connecting peripheral devices.
- Quick analysis and display of faults is possible.
- The system can be modified remotely (teleservice).
- Low power consumption is the rule (Energy efficiency).

Thanks to computer technology, it has been possible to incorporate more and more CNC functions in smaller and smaller spaces. Using software, it has been possible to improve the integrated system monitoring and fault diagnostics of controllers to such an extent that malfunctions are eliminated.

The most important effect, however, was increasing performance while simultaneously reducing costs and prices.

The miniaturization of controllers also has resulted in outwardly visible benefits: The previously bulky electrical cabinets became smaller; in some cases they have turned into simple "switch boxes" built onto the machines.

Computers for NC Programming

Computers have long been used in the programming of NC machines. There have been considerable reductions in the expense of programming work and the time required to program geometrically complex workpieces and 3D surfaces. The computer handles the difficult and therefore time-consuming geometric calculations of points of intersection, transitions, contour segments, phases, curvatures, and shapes. Today, the NC programmer only needs to enter the required data from the drawing or import it directly from the CAD system. There is no need for auxiliary calculations.

Because the required computers have become smaller, more powerful, faster, and more economical, it also has become possible to integrate machine-specific NC programming systems directly into the CNC system. This gave rise to CNC systems with interactive, graphically based programming performed at the machine. These manual data input controls for **shop-floor programming (SFP)** provide excellent programming tools and graphical displays. Clearly, the best programming aids available today are color graphics and interactive programs. However, these require NC programming software that guides their user in a manner that is logical, understandable, and well suited for use on the shop floor.

Simulation software has been very helpful in this regard. Here, each CNC machine, including its workpiece, clamping device, and tools, as well as all its programmed motions, is shown three-dimensionally on a computer screen. The machining process can be viewed from any angle; by depicting the workpiece transparently, it is also possible to observe the machining process that takes place inside the workpiece. Any collisions that occur between the workpiece, the tool, the body of the machine, or the clamping device are clearly indicated by means of an alarm signal and color coding. The programming error can be corrected immediately, and the results can be verified.

Computers for Automation *(Figure 3.9)*

Computer technology also has had a major **impact on the design of CNC machines.** CNC-controlled loading and unloading stations for workpieces increase the **degree of automation and the flexibility** of machines. Exchangeable tool cassettes make it possible to prepare complete sets of tools externally for rapid exchange. The integration of robots makes tool changing faster and more flexible. The systematic use of new control and drive technologies allows a reduction of about 25 to 30 percent in the number of parts and the price of machines while simultaneously boosting performance.

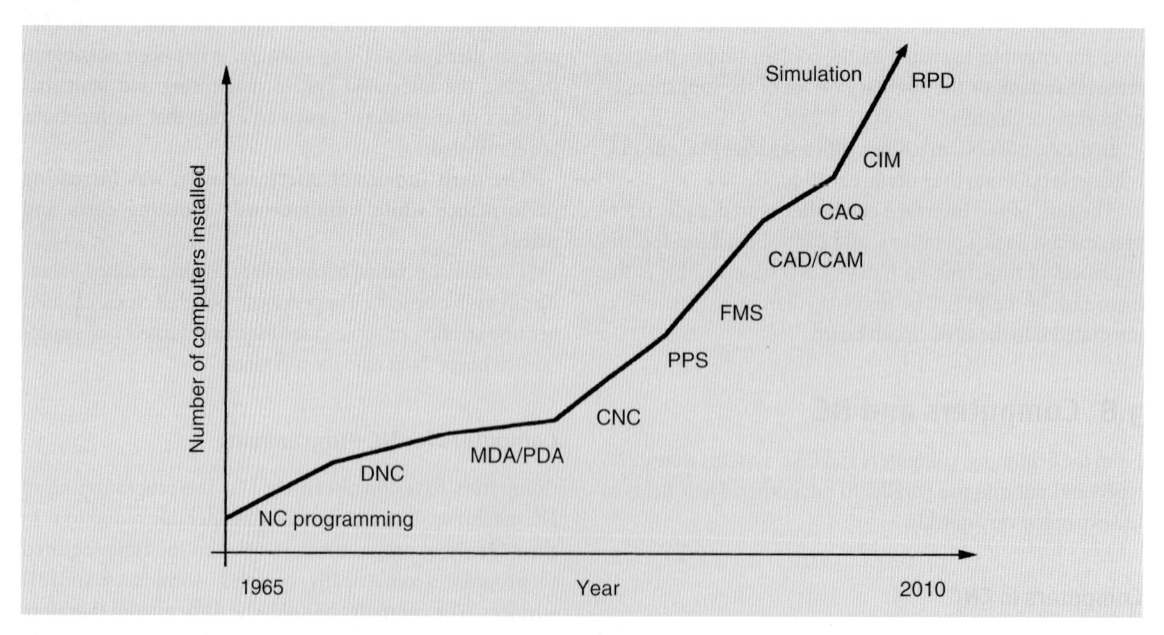

Figure 3.9: NC manufacturing and automated data processing meant that more and more computer systems were used to perform a wide range of tasks in the manufacturing sector. The trend is continuing.

An increasing number of specific tasks in manufacturing operations are being assigned to networked computer systems in order to make production as cost-effective as possible. These **computer-integrated manufacturing (CIM) systems** are already being used profitably in a wide range of configurations.

Combining multiple CNC machines into a **flexible manufacturing system (FMS)** could only be implemented when the task of overall production control for such a system could be assigned to one or more computers. Not only does this include **DNC systems** for the automatic transfer of NC programs to the CNC systems, but it is also necessary to have all workpieces, tools, and data available at the right machine at the right time. The movement of all parts through the manufacturing system also must be controlled and monitored, and the entire process must be documented. When minor problems occur in the system, emergency strategies should allow operation to continue so that there is no loss of valuable production time.

The combination of CAD, NC programming, and CNC manufacturing can be called **digital manufacturing.** No other alternatives are in sight, given the demand for faster conversion of CAD data into prototypes and for quicker introduction of new products. Computers are now being used more and more for planning, preparation, transport, measuring, testing, monitoring, installation, and adjustment.

3.9 NC Programs and Programming *(Figures 3.10)*

In order to machine workpieces using CNC machines, the user creates workpiece-specific **NC programs.** Depending on the organization or the complexity of the workpieces, this can take place either during production planning or directly at the machine.

An NC program contains all the information pertaining to the motion along the axes (**path information**) and for activating the **M-functions.** This information is combined in a step-by-step manner in the sequence required for machining. The information is saved on a machine-readable data-storage medium for input into the CNC system. If there is a direct data link to the computer (DNC), then the NC program is transmitted to the connected CNC systems directly via cable or wireless.

The **standardization of program structure** was an essential prerequisite for the introduction of NC machines. At a very early stage in NC development, it was agreed that the international standard code would be based on an International Standards Organization (ISO) recommendation, which was subsequently incorporated in **DIN 66 025.** This provided the greatest possible uniformity in program structure for all NC machines, making it possible to perform programming in an external, machine-independent manner with any programming system. Special adaptation to a particular CNC machine is the task of the **postprocessor**. This converter program can be implemented either in the CAD computer or in a downstream programming system (CAM).

Standardized **program structure** is described in *Part 5, Chapter1.*

Programming systems (CAD/CAM) (*Figure 3.11*) make programming easier, avoiding lengthy, time-consuming auxiliary calculations. The subsequent graphical-dynamic simulation of the machining process on the screen means that the programmer can be sure that there are no programming errors. If workpiece design is performed on a CAD system, then it is possible to import the workpiece data thus generated from the CAD computer to a CAM system and to use them directly for NC programming.

3.10 Data Input

NC programs are input into CNC systems using the same devices and data-storage media that we are familiar with from PCs:

- An **ASCII keyboard** for manual data input and for making corrections (editing)
- **Electronic data-storage media** and the associated interfaces (USB 2)
- **Directly** from a computer (DNC) via cables or wireless with suitable data interfaces
- **Keyboard**

Almost all of today's CNC systems allow quick, computer-based parts programming or program corrections at the machine. ASCII keyboards are the type most often used worldwide for interactive CNC systems and PCs.

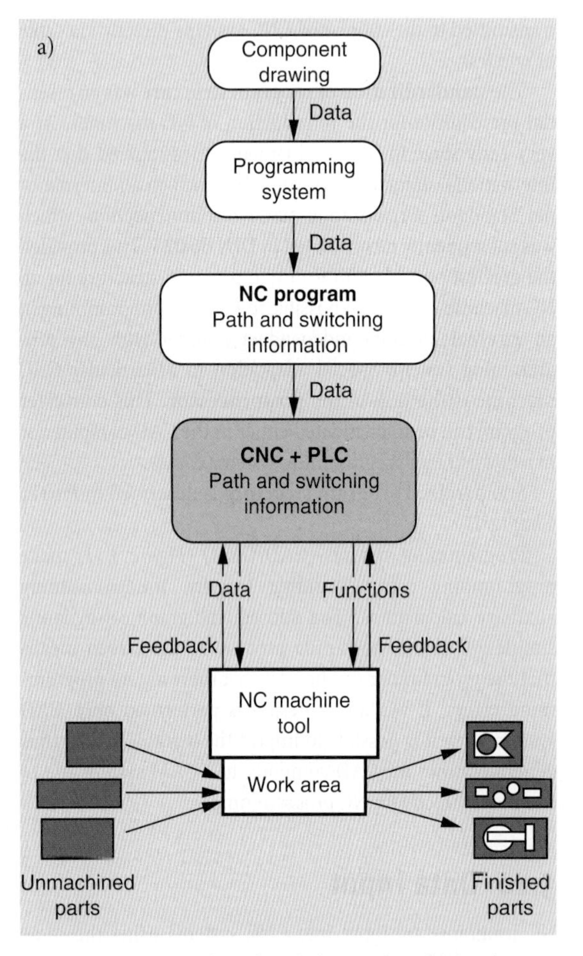

Figure 3.10a: Conversion of workpiece and machining data into path and switching information for an NC machine.

b)

```
N1000
N5 G90 F600 S3600 T08 M06
N10 M3
N15 G00 X-20 Y-20 M8
N20 G00 Z-5
N25 G41
N30 G01 X0 Y0
N35 G01 X0 Y30
N40 G02 X14.374 Y33 I7.5 J0
N45 G03 X18.965 Y30 I4.581 J2
N50 G01 X33.860
N55 G01 X44.5 Y40
N60 G01 X50
N65 G02 X65 Y25 I7.5 J-7.5
N70 G03 X49 Y5.932 I9.226 J-23.989
N75 G01 X0 Y0
N80 G40
N85 G00 X-25 Y-25
N90 G0 Z100 M9
N95 M5
N100 M30
```

Figure 3.10b: NC program (example).

Electronic Data-Storage Media

These are used first of all for quick saving and transport of NC programs and any compensation values in a form that is readable by the CNC system. A secondary requirement is to be able to read the data out and save them again, including any program corrections that have been made.

PC cards have been supplanted by **USB sticks** and **USB hard drives** (*Figure 3.12*). Almost all CNCs and PCs today have USB 2.0 ports and the necessary driver software.

USB sticks are a kind of flash ROM storage medium that retains the stored data when the power is switched off, but which can be deleted selectively. Their **storage capacity** has grown explosively: There are hardly any devices left on the market with less than 2 GB, and 256-GB sticks are already available.

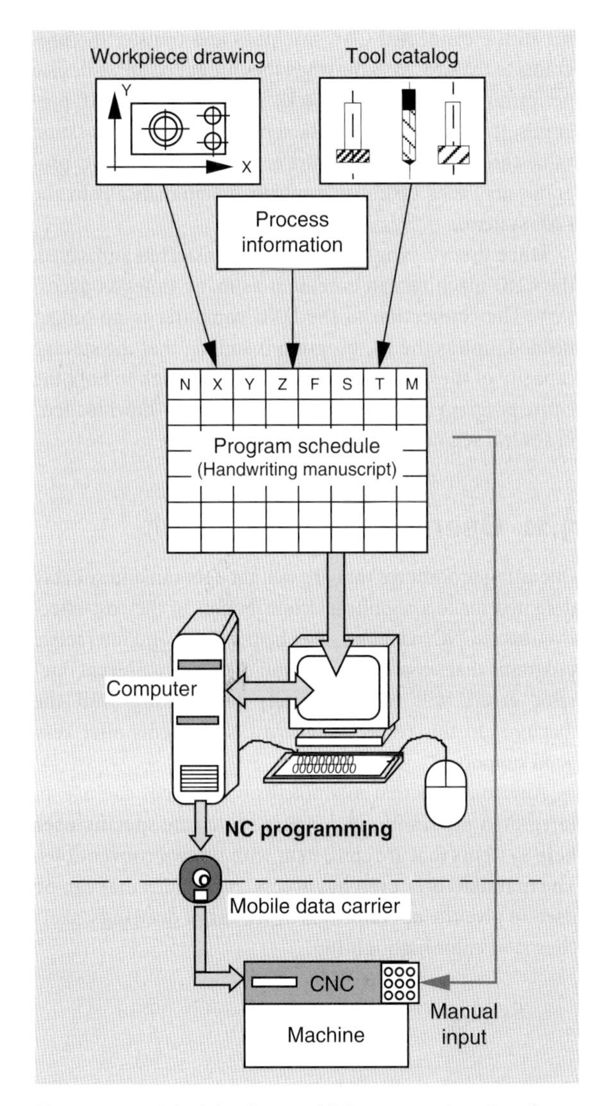

Figure 3.11a: Principle of manual NC programming: Step-by-step input of machine motions. Path and switching information are brought together step by step in the NC program and transferred to a mobile, machine-readable data carrier. Or the individual data sets are typed into the CNC system manually.

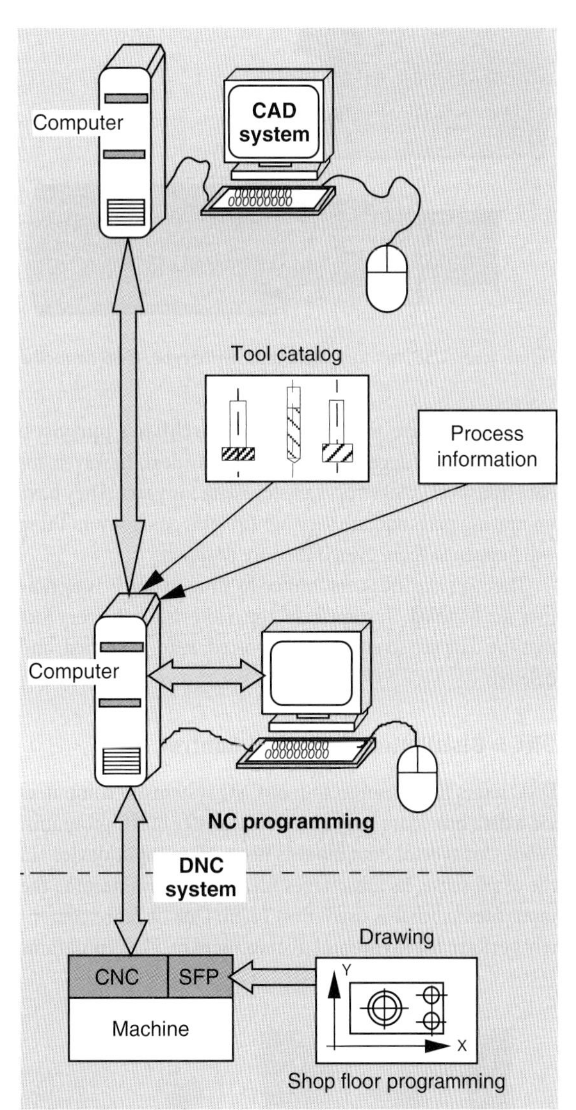

Figure 3.11b: CAD/CAM principle: Inputting the geometries of unfinished parts and workpieces and automatically generating machine motions based on them. The workpieces are designed on the CAD system, and the CAD data are used in the NC programming system to generate the machine motions for machining the parts.

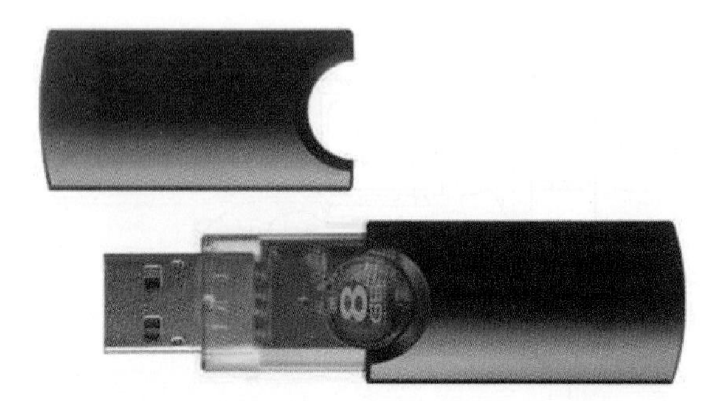

Figure 3.12: USB memory stick, read/write type, 8 GB capacity.

USB sticks are too expensive for **archiving purposes.** Their physical ruggedness makes them ideal, however, for data transport. USB sticks can store data for years. They have no moving parts and are very lightweight; generally, nothing will happen to them even if they are dropped.

Their service life is not limited by time so much as by how they are handled. In some case, the main risk is the very fact that the data are so easy to transported, read, modified, and deleted.

DNC = Distributed Numerical Control

DNC refers to the connection of CNC systems to computers via a data link (data bus) for direct transfer of NC programs. While this type of data input is not a "data-input device" in the usual sense, its advantages mean that it has become the input method that is used most often. One or more computers perform the saving and management of all NC programs for all the connected CNC machines and transfer the programs to the CNC systems when they are called up while also performing various safety checks. In addition to the NC programs, the necessary tool data, tool lives, and compensation values are transferred (→ *Part 6, Chapter 1*). The NC programs are called up either manually or automatically by the CNC system.

Once the NC program has been transferred and saved, the CNC machine can execute it as many times as necessary. The connection to the DNC computer is no longer needed, unless the NC program is so long that the storage capacity of the CNC system is not large enough to hold the entire program. In such a case, the program is "downloaded" in several segments.

3.11 Operation *(Figures 3.13 and 3.14)*

Operating procedures that are well thought out and suitably designed make a significant contribution to the cost-effectiveness of CNC machines. User help systems and interactive programs that support the operator and help to prevent operator errors will promote **safety in working** with the machine. In this respect, modern CNC systems offer very good support.

Any company that manufactures machines can use an integrated programming language to create specific **user help systems** that are consistent with its own conventions. The manufacturer even has access to graphics that can be used to give the operator tips by means of drawings, sketches, and color highlighting.

Figure 3.13: Siemens 840D with control panel for machine and CNC, servo amplifier, and PLC.

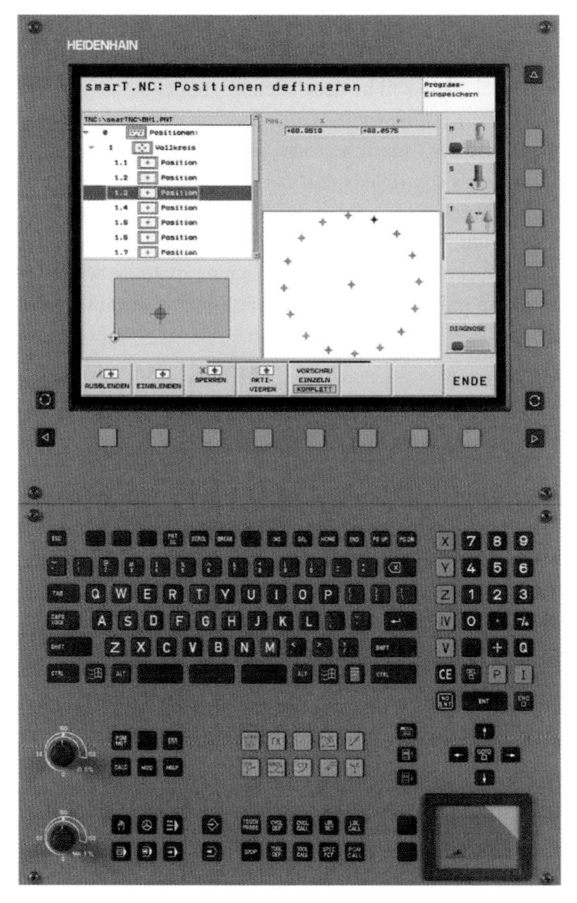

Figure 3.14: Example control panel for a standard 3–5 axis CNC system (Heidenhain).

The overall operating procedure should be as compatible as possible with the **operator's usual habits** and should not be too complicated. A simple, logical operating procedure has a significant effect on the subsequent productivity and profitability of the machine. Well-thought-out *man-machine interfaces* (MMIs) are now considered very desirable.

Interactive operating procedures speed up data input and reduce errors. Even highly automated machines occasionally require manual intervention. Feed rates, speeds, and auxiliary functions often need to be corrected, or interrupted programs have to be resumed. Sometimes data can be determined or optimized only after machining an initial test workpiece and performing analysis of clamping errors, vibrational characteristics, chip formation, or surface finish. In such cases, the person setting up the machine will be thankful if the appropriate correction options and the actions to be taken are explained on the CNC screen.

When a **tool breaks** or another fault occurs, it is not sufficient to simply switch off the machine. The operator must be able to intervene quickly and safely, stop the machine, move the tool away from the workpiece safely, insert a new tool, move back a few steps, and enter the new compensation values. The operator then must approach the point where the break occurred in such a way that the new tool enters the interrupted cut gently, with the CNC system putting the new compensation values into effect from that point onward. Without good operating prompting, downtimes will be much longer than necessary.

3.12 Summary

Owing to their numerous capabilities, CNC systems have been entrusted with a wide range of tasks that contribute to the fully automated, technically sophisticated operation of individual machines and flexible manufacturing systems. These include

- NC programming on the machine during ongoing operation
- Management of extensive tool tables with up to 50 data sets per tool
- Automatic generation and output of lists of tool differences when a new NC program is input
- Automatic communication with external computers, for example, computers belonging to measuring machines, DNC systems, and production planning systems (PPSs)
- Workpiece management in flexible manufacturing cells (FMCs) and flexible manufacturing systems (FMSs), with logging of production quantities, rejects, rework, and irregularities, or to specify processing priorities
- Integration of software for production data acquisition and machine data acquisition (PDA/MDA); for fault diagnostics; for maintenance, servicing, and troubleshooting; and for a large number of graphical depictions to support the operator

These functions will be explained in the following chapters.

In the course of just a few years, CNC systems have changed the design and handling of machines to such an extent that in many cases the machines can no longer be guided manually. Empirically speaking, the operator only has to intervene to make corrections in the event of a fault. Modern CNC machines contain many safety and monitoring devices to protect humans and machines from improper operating procedures and malfunctions.

Although all CNC systems operate using the same principle, different machine types also require different, specially adapted control systems with a large number of special functions. This is achieved by means of the operating software of the CNC system, even if the CNC hardware remains identical. This will be described in greater detail in *Part 2* and in *Part 3, Chapter 1.*

Today, CNC machine tools are equipped with multiple numerically controlled main axes, simple auxiliary axes, and a large number of M-functions. What's more, some types of machining, such as laser cutting or high-speed milling, require very rapid and precise machine motions, which, in turn, can only be achieved using fast, highly dynamic servo drives.

What Are NC and CNC?

Important points to remember:

1. NC stands for "numerical control," that is, **controlling with numbers.** When used in connection with machine tools, this term refers to the direct input of the **dimensions of the workpiece being shaped.**
2. Microprocessors are used in the construction of modern numerical controls. These controls are used in **CNC,** that is, **computer numerical control.**
3. Microprocessors offer such **high processing speeds** that a single processor is sufficient to control multiple machine axes simultaneously and with great precision.
4. **Continuous-path controls** are the most universally applicable and most commonly used controls. The number of axes that can be controlled simultaneously can be expanded in a modular fashion.
5. Continuous-path controls, of course, also can be used (programmed) as **point-to-point controls** and **straight-cut controls.**
6. The capabilities of a CNC system are completely contained within its **CNC operating program.** This program can be copied, maintained, and modified easily and without errors.
7. To a great extent, all CNC manufacturers make use of commercially available microelectronic components. However, they develop different **software** in order to tailor certain control functions for specific machines and customers.
8. The most powerful CNC systems offer an integrated programming language based on PASCAL or BASIC in order to provide customer-specific functions. The machine manufacturer can use these programming languages to supplement the CNC software with its own **know-how.**
9. To input NC programs into the CNC system, portable **electronic data-storage media** or direct input via data interfaces (**DNC**) is used today.
10. Standardized interfaces (e.g., Ethernet) are being used increasingly as **data interfaces.**
11. CNC machines are **user-programmable machines;** that is, the motion sequences of the individual axes can be specified by means of exchangeable NC programs.
12. CNC machines generally involve a combination of **translational and rotary axes.** Each axis is equipped with a **measuring system** that can be evaluated electronically and a **servo drive.**
13. In CNC machines, the **technological functions** are also programmable; these include the feed rate (*F*), spindle speed (*S*), tool number (*T*), and auxiliary functions (*M*).
14. Sequences that remain the same, such as the automatic changing of tools or workpieces, are determined by the machine. They are called up by means of an **M-function** (M00 to M99) and then proceed to run without operator intervention. Signals are output to the actuators via a PLC and appropriate amplifiers.

Part 2

Functions of Numerical Control Systems

1. Implementation of Dimensional Data

The main new feature in CNC-machines is the automation of all machine motions based on input of digital information for the axis and all machine functions. It is performed separately for each of the various motion directions of a machine, which are also called the *axes*. To position the tool in a plane, two translational axes are required, and for positioning in space, three translational axes are required. These are normally positioned perpendicularly to each other, thus forming a Cartesian coordinate system (*Figure 1.1*).

1.1 Introduction

The characteristic functional element of computer numerical control (CNC) machines is a **position-control loop—a combination of feed drives and position measurement.** Other characteristic components are automatic tool changing and workpiece changing. The CNC affects the machine configuration, including the machine frame, guides, and main drives.

The numerically controlled machine tool has a fully automatic sequence for all the functions that are necessary to execute a machining operation on a workpiece. The information for this sequence is saved in digital form, which means that is has to be converted into machine functions. A distinction has been made here between **dimensional data that define the motions of the machine and switching information that triggers fixed machine functions.**

If it is also necessary for the tool to engage with the workpiece in a freely determined direction, then two additional rotational axes are necessary. In large machines or machines in which a number of tools can operate simultaneously, there will be even more axes, some of which may be parallel axes.

Table 1.1 provides an example of how many axes are necessary or typical in various types of machine tools and which machine functions are automated.

1.2 Axis Designations *(Figure 1.1)*

The designation of coordinate axes and directions of motions in numerically controlled machine tools is defined in **DIN 66217.** This standard is related to International Standard Organization (ISO) **Recommendation R 841.** Both these are based on the *three-finger rule*, in which the right hand is used to define the directions of the main axes *X*, *Y*, and *Z*, which are perpendicular to each other:

- The thumb represents the *X* axis, the index finger the *Y* axis, and the middle finger the *Z* axis (*Figure 1.2*).
- The tips of the fingers point in the positive direction.

In order to use this rule to define the axes of a CNC machine, one imagines that they are sticking one's middle finger into the tool holder of the spindle. This is the *Z* axis, and the finger points away from the workpiece in the retraction direction of the spindle.

Then you turn your hand in such a way that your thumb points in the direction of motion of the longest axis: This is then the *X* axis, which is generally horizontal.

This also automatically determines the *Y* axis: The index finger is pointing in the positive direction.

All the additional axes are based on these three basic or main axes:

- *A*, *B*, and *C* are rotary or swivel axes, with *X*, *Y*, or *Z* as the central axes; that is, *A* rotates about *X*, *B* rotates about *Y*, and *C* rotates about *Z*.
- The positive direction of rotation of the rotary axes corresponds to clockwise rotation viewed in the direction of the positive axis direction; this is also known as the *corkscrew rule:* During "screwing in," the direction of the tip and the rotary motion are positive.
- *U*, *V*, and *W* are parallel axes to the three main axes *X*, *Y*, and *Z*.
- *P*, *Q*, and *R* are additional axes that do not necessarily have to be parallel to the main axes. *R* is used primarily in drilling cycles as an address for the reference plane of the workpiece, that is, where the *Z* axis switches over from rapid traverse to the feed rate (*R* = reference surface).

Machine type	Axes	Tool Changing	Workpiece Changing	Special Functions
Drilling machine	3	m/a	m	Special drilling cycles. Circuit boards with HSC.
Milling machine	3–5	m/a	m/a	Gantry and parallel axes, HSC, tool comp.
Turning machine 1 Turning machine 2 Turning machine 3	2 2 × 2 bis 3–8	a a a	m m/a a	Graphical programming, cycles, spindle orientation, driven tools, autom, reclamping, multiple slide machines.
Machining center	4–5	a	a	Tool management, cartridges, horiz./vertical head, pallet changer.
Grinding machine 1 Grinding machine 2	3 5 + 3 +n	m a	m a	Dressing cycles, oscillating axes. Multiple grinding supports. Automatic grinding disk and tool change.
Nibbling machine 1 Nibbling machine 2	2 5	m/a a	m/a a	Nibble function, tool changing, rotating tools, intermeshing, multiple tools.
Laser machine	3–5	m	m/a	Controlling beam power, high feed rate.
Hobbing machine	5+	m/a	m/a	Hobbing module, parameter programming.
Wire EDM machine	2–5	Wire m/a	m/a	Moving back on the original path.
Manufacturing cell	6 synchronous 3 asynchronous	a with manag. monitoring, exchange	a w/workpiece detection, progressive call-up	DNC interface, MDA/PDA functions, sensor connections, graphical fault diagnosis, pallet pool.
Flexible manufacturing system	as desired	a	a	List of tool differences, closed data circuit, pallet feeder.

Table 1.1: Various Requirements of Various Machine Tools on the Scope of Their Automation (m = Manual; a = Automatic)

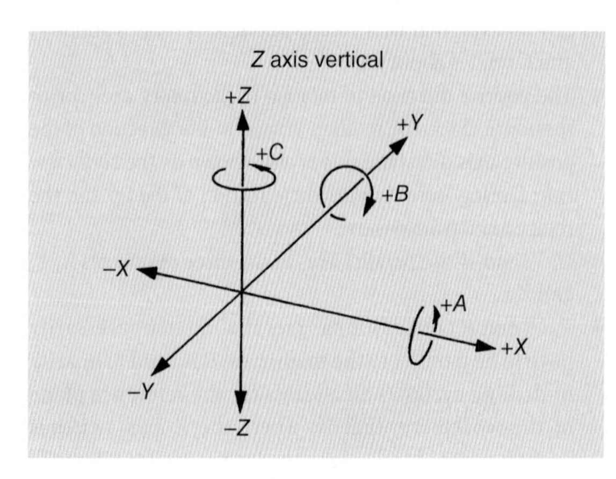

Figure 1.1: Cartesian coordinate system.

One frequently also sees designations such as *X*1/*X*2 or *Y*1/*Y*2. These involve movable gantries or cross-beams, so-called gantry axes (*Figure 1.3*). Because these axes have two widely separated guide tracks (on each side), they require two separate drives in order to move precisely in parallel even with varying loads. These axes thus are not independent axes with mutually independent motions but rather axes that move together and are also programmed under the same address: *X* or *Y*.

When defining the positive axis direction, it is assumed that the tool is always moving and the workpiece is always stationary. In this case, the positive axis directions are designated in the same way as the positive motion directions: +*X*, +*Y*, +*Z*, +*A*, or +*C*. If, however, the

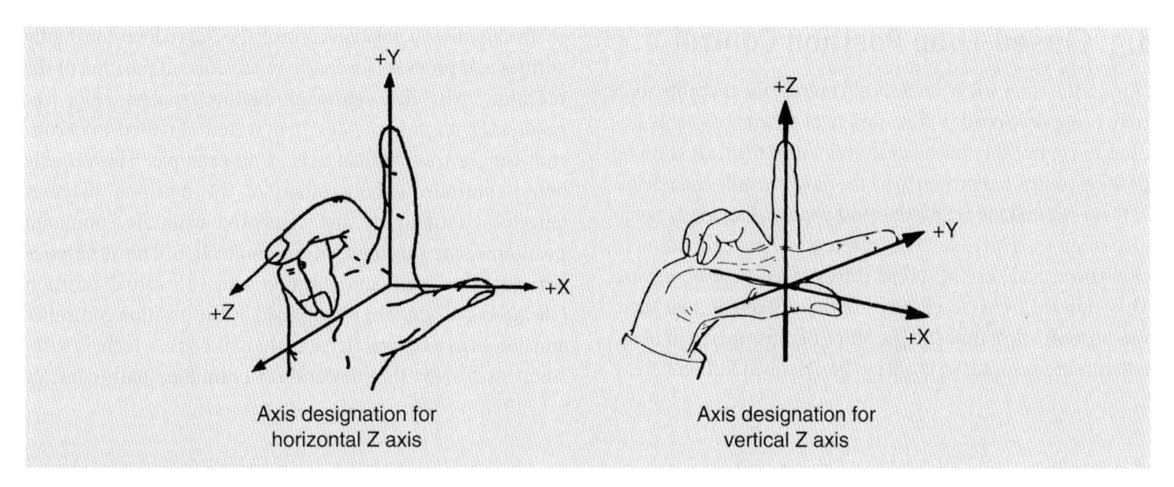

Figure 1.2: Use of the three-finger rule with the right hand.

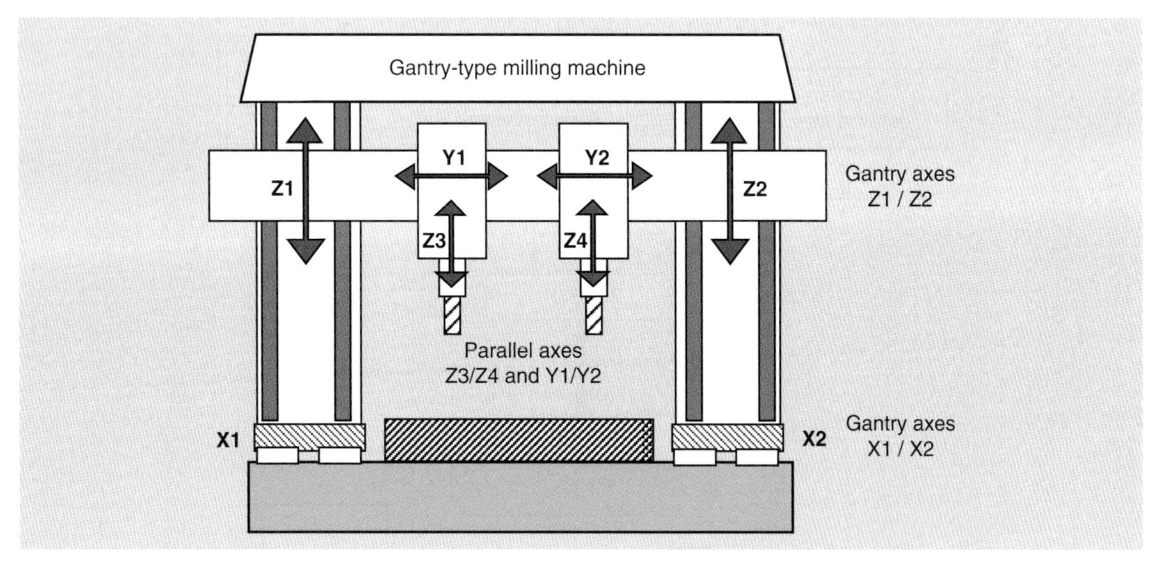

Figure 1.3: Difference between gantry axes and parallel axes.

workpiece is moved, such as with coordinate tables and rotary tables, then the direction of motion and the axis direction are inverted. If the table moves to the right, then the tool executes a relative motion to the left. In this case, the actual axis direction is specified, but the address is marked with an apostrophe: +X', +Y', +Z', + A', +B', or +C'.

Specifying the axis direction in this manner has the advantage that the programmer can create the programs independent of the structural design of the machine. The desired relative motion between the tool and workpiece is always executed in the correct direction regardless of the machine's configuration.

1.3 Closed-Loop Position Control

During the time when numerical controls for machine tools were being developed, a wide variety of different systems was used to control the machines' travel. Since then, closed-loop position control has proven to be the most versatile and reliable of these systems and is now the most popular. Previously open-loop control systems often also were used owing to their simplicity. These made use of special drives called *stepping motors.* They differed from closed-loop control systems in that there was no feedback of the actual position. Because this technology is hardly used nowadays, it will not be described further here.

In closed-loop position control, the closed loop constantly verifies and provides feedback on the current position of the machine axis, thus ensuring highly reliable, error-free motion. *Figure 1.4* provides an overview of closed-loop control using a translational axis as an example. The variable being controlled—the position of the machine slide—is detected continuously and compared with the command position from the higher-level controller. The difference between the command position and the actual position (*closed-loop deviation*) is amplified by the position controller and output as a control signal to the axis drive, which, in turn, compensates for this deviation. Continuous path controls

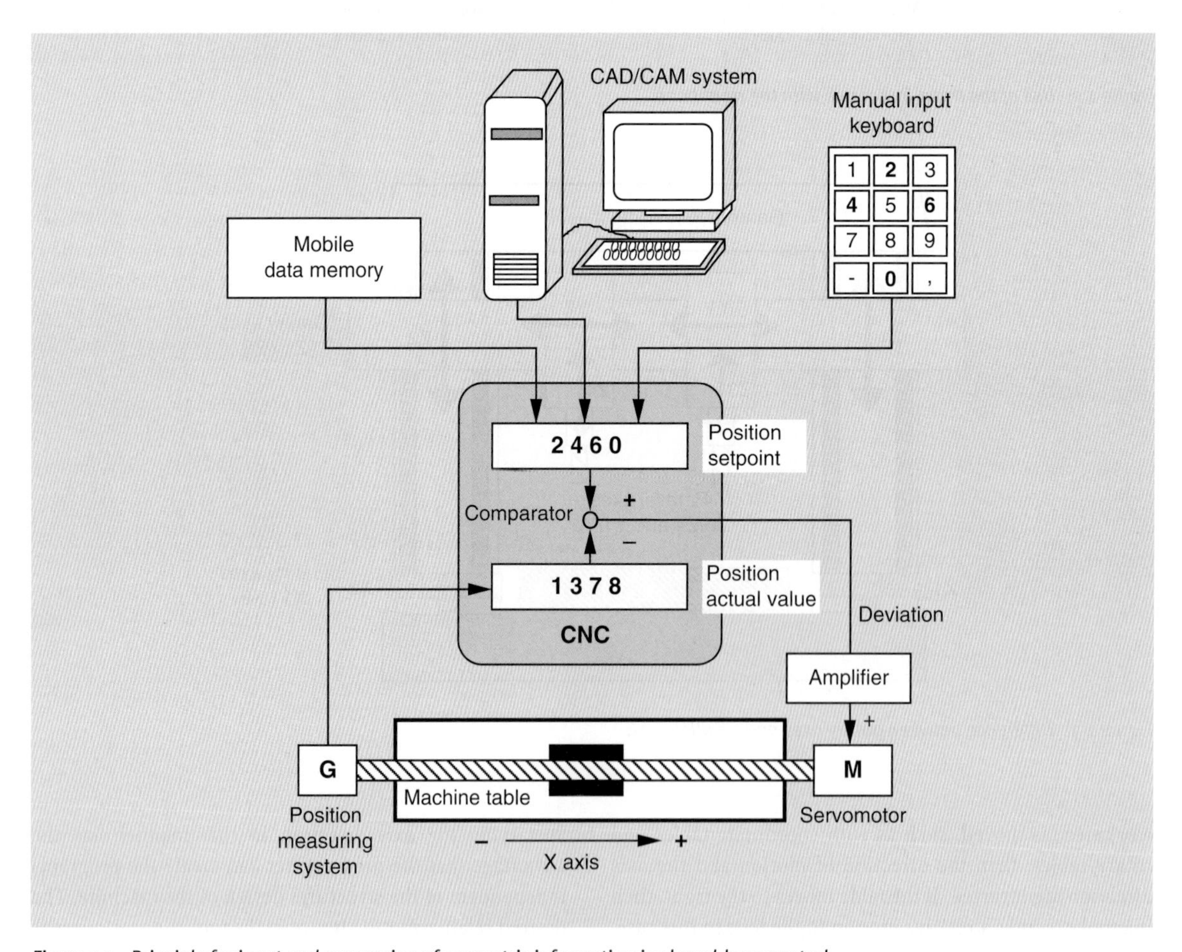

Figure 1.4: Principle for input and processing of geometric information in closed-loop control.

output new position values on an ongoing basis, and the controlled axes must follow these values. This means that it is possible to achieve continuous-path motions. **Thus each CNC axis requires the following:**

1. An electronic path-measuring system
2. A drive that is subject to open- or closed-loop control.

An important feedback characteristic for the position feedback control of a feed drive is the *amplification factor* (proportional amplification for feedback control). The position feedback controller is designed as a position controller. One important characteristic of the position controller is a persistent control deviation. The persistent control deviation of a closed-loop position controller, that is, the difference between the set-position value and the actual position value, is proportional to the current velocity of the motion and is referred to as the **contouring error** or **following error:**

$$X_s = \frac{V}{K_V}$$

where X_S = contouring error, in millimeters
K_V = amplification factor, in meters per minute per millimeter
V = velocity, in meters per minute

The magnitude of the following error for a certain traversing speed thus is determined by the amplification factor that can be achieved. The amplification factor thus is a measure of the machining precision that can be achieved for the dynamic characteristics of a feed drive.

This kind of closed-loop control, however, is a system that is subject to oscillations; if the amplification factor is too high, then there may be excessive oscillations in the closed-loop control. Because this may result in extremely negative effects on the quality of the workpiece being created, amplification of the control loop is also limited here. In order to increase the K_V value and thus reduce the following error, the simple closed-loop position control is expanded to include lower-level control circuits for motor speed and motor current, as shown in *Figure 1.5*. Here, the closed-loop deviation is the input variable for the lower-level speed-control loop. The speed-control loop gives the amplified speed deviation to the lower-level current controller as an input variable. The proportional-integral (PI) response of speed controllers and current controllers makes it possible to correct for even the smallest control deviations with no persistent deviation between the command value and the actual value.

The amplification factor that can be achieved for a feed drive is also affected by the design of the mechanical elements that are involved.

- As with every other oscillatory system, the masses being moved should be as small as possible, and the driving elements should be as stiff as possible.
- Any nonlinearities in the system, such as friction and play, should be as low as possible.

The various design potentials that arise here will be described later.

Friction and play in the area between the path-measuring system and the slide are also the source of *reversal errors*. The reversal error is the distance between the two actual positions that are created when a command position is moved to from opposite directions. It causes position deviations and therefore should be kept as small as possible. It largely can be compensated for by means of special corrections that are saved in the CNC system.

But the location of the displacement measurement, that is, the direct or indirect displacement measurement, also has a very large effect. *Figure 1.6* shows the difference.

1.4 Position Measurement

When it comes to position measurement of numerically controlled machines, different measuring devices are available for the various levels of precision and machine concepts. The machine manufacturer selects the suitable position-measuring devices for the CNC machines that it builds according to a number of criteria: the machine concept, type of drives, travel, traversing speed, precision and resolution (measurement steps), simple installation, and cost.

Position-measuring devices have the advantage that they can measure the position and the travel distance of a CNC axis precisely, with no time delay, and reliably, and they can constantly report it back to the closed-loop

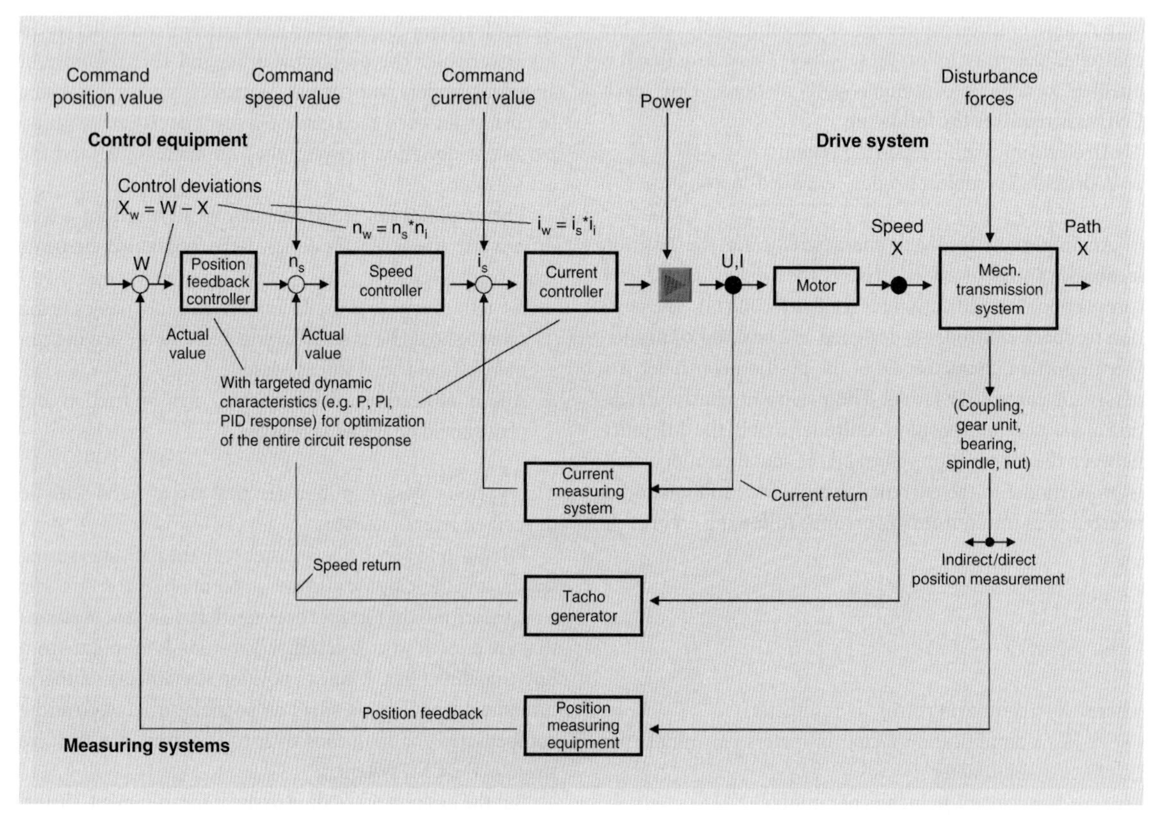

Figure 1.5: Schematic depiction of closed-loop control.

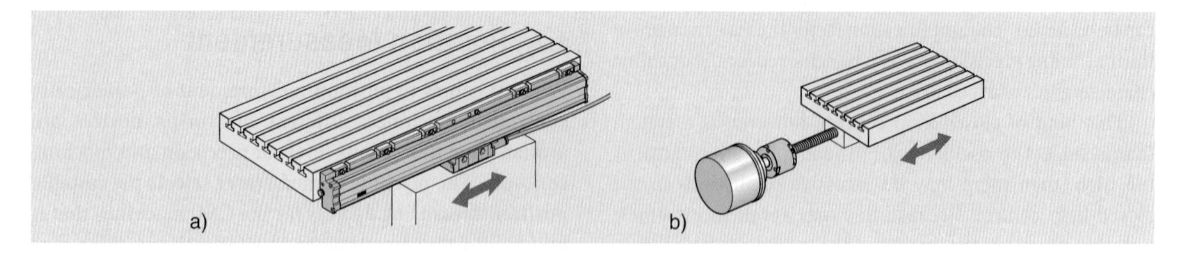

Figure 1.6: Principle of displacement measurement (a) with a length-measuring device, (b) with a spindle/nut/rotary encoder.

position control. In the case of direct drives and synchronous motors, the measured value is also used to detect the rotor position.

When it comes to the closed-loop speed or velocity control, the actual value is generated via differentiation of the position value. For highly dynamic drives, double differentiation of the position value can be used to obtain the actual acceleration value. Control cycle times of as little as 25 μs can be achieved (at present, the typical times are 50 or 62.5 μs). This also has to be supported by the position-measuring devices.

A **measuring device** comprises the entire measuring chain, for example, a **measuring scale with sensing head,** interpolation electronics, and a counter. The electronics and the counter can be integrated into the CNC system.

In length measurement using a spindle and rotary encoder, the measuring device also includes a spindle and nut that are used to traverse the machine element. A transmission gear unit and couplings also may be necessary. In the spindle/rotary encoder concept, accuracy may be compromised further by effects of the mechanical components that are not compensated for, for example, the longitudinal expansion of spindles owing to the effects of temperature. *Figure 1.7* may serve as an example.

All position-measuring devices for CNC machines are based on **material measures with a periodic pitch.** The pitch period and associated system accuracy components are important characteristics.

The photoelectric sensing principle provides the finest pitch periods and thus also the highest resolutions. Thus, for example, the pitch periods of Heidenhain measuring devices are typically 20 µm, and for more demanding requirements, variants with pitch periods of 8, 4, and even 0.512 µm are available. Other sensor principles, such as those used in magnetic, capacitive, or inductive measuring devices, are by their very nature unable to provide such precise pitches and are not suitable if high accuracy is required.

With **measuring spindles,** the pitch period corresponds to a spindle rotation (e.g., 5 or 10 mm). With rack and pinion systems, the pitch period corresponds to the path covered by one rotation of the pinion (e.g., 100 mm). Each rotation is subdivided into a number of periodic measurement steps by a **rotary encoder**, thus determining the position.

The material measure is of decisive importance for the precision of the measuring device. Measuring-scale measuring devices have a material measure in the form of an electronically scannable scale with a pitch. In spindle/rotary encoder systems, the drive spindle should be viewed as a material measure. Laser interferometers use the wavelength of the laser light as an immaterial material measure. For numerically controlled machines, a **measurement step** of 1 µm or less is generally required. This means that the pitch period of the material measure generally has to be subdivided again. With many position-measuring devices, an "electronic interpolation" of the individual pitches is performed even when the material measure is scanned **so that the signal period is smaller than the pitch period of the material measure.**

Thus the measurement steps that can be detected are finer than the scale graduation. In the case of measuring devices with a serial interface, it is necessary to perform interpolation of the sensing signals already in the measuring device.

In this case, it is common practice to interpolate high enough that such measuring devices can be used for all

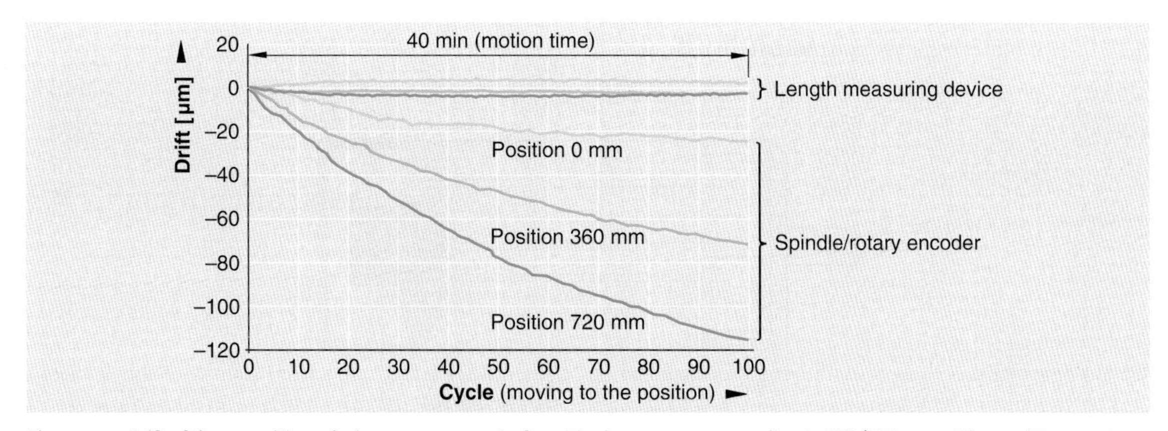

Figure 1.7: Drift of three positions during measurement of positioning accuracy according to ISO/DIS 230-3. The position sensing via spindle/rotary encoder demonstrates a considerable drift of the positions as a result of the thermal growth of the spindle.

applications, including direct-drive technology. Typical subdivision factors range from 1,024 to 16,384.

The new, digitally controlled servomotors represent an additional variant. Most of these motors have a position encoder for the rotor. In connection with the drive spindle, this generally can also be used for displacement measurement and position feedback control (*Figure 1.8*).

With such drives, the question arises as to whether an additional path-measurement device should be used, or whether it is sufficient to execute measurement using the feed spindle that is present already together with the drive rotary encoder. Here it is necessary to take into account the additional error sources of the mechanical transmission elements and the effects that cannot be compensated for, such as uneven elongation of the spindles owing to the effects of temperature.

Position Measurement via Linear Scales

Length-measuring devices, that is, linear scales for "direct" measurement of the relative motion between fixed and moving machine components, generally use a grating or graduations on substrates made of glass or steel. The sensing head, which moves close to the measuring scale, but without coming into contact with it, uses optoelectronic effect with electronic postprocessing to generate individual electronic pulses that can be clearly distinguished from each other. These are added to or subtracted from an electronic counter depending on the direction of motion. Depending on the design of the system, measurement steps between 0.1 μm and 1 nm (= 1/1000 μm) can be detected. The individual counter pulses are called *increments*, and the measuring process is correspondingly called *incremental position measurement.*

Position Measurement Via Thread Spindles and Rotary Encoders *(Figure 1.8)*

Position measurement via rotary encoders has been used since the beginnings of CNC technology because it works even with a simple rotary converter. This converts the rotations of the feed spindle into corresponding counter pulses for the travel. While the relative motions are detected directly by the measuring-scale measuring devices, with the mechanical material measures spindle/nut and rack/pinion, it is necessary to integrate a rotary angle encoder. The position value is determined only by converting the torsional angle into the corresponding linear dimension.

In contrast to direct position measurement using measuring scales, this measurement principle is also generally called indirect position measurement (→Figure 1.6).

Incremental and Absolute Measuring Processes

Measuring processes generally are subdivided into incremental and absolute principles. Furthermore, incremental measuring devices can be implemented as "absolute"

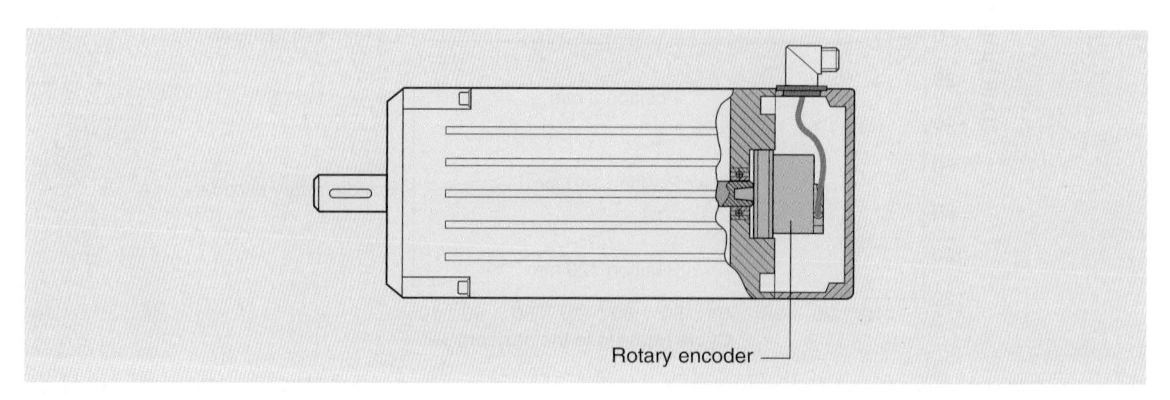

Figure 1.8: Motor for digital position and speed control.

devices by buffering the measurement values and backup batteries. These are also called **pseudo-absolute measuring devices.**

Incremental Measuring Devices

Incremental position measurement involves a measuring device that measures the positive or negative **increase** in the path that has been traveled over. This is done by dividing the entire distance into a large number of equally sized **measurement steps** (increments) and counting them depending on the direction. When the machine axis moves, an electronic counter reflects the sum of the counter pulses corresponding to the travel and the specific position of the axis. *Figure 1.9* shows a rotary encoder of this type.

This counter is always set to zero at the time of switch-on, regardless of the position in which the machine slide is located at that time.

To reproduce the absolute **machine zero point** of an axis after switch-on, photoelectric scales and rotary encoders have so-called **reference marks** that generate a pulse when they are moved over. This pulse is assigned precisely to a defined position value (*Figures 1.10 and 1.11*). Depending on the measuring device, the reference marks can have a number of different designs. For example, there are measuring devices with a single reference mark, reference marks that are applied cyclically (e.g., every 50 mm), and distance-coded reference marks. With reference marks that are applied cyclically, the position assignment in the machine also has to be determined by means of an additional contact (e.g., a cam or proximity switch). In contrast, distance-coded reference marks are designed so that every second mark is in a fixed grid pitch, whereas the ones in between are at defined varying distances. Even if only two neighboring reference marks are moved over, it is possible to restore the absolute position value by means of a mathematical logarithm.

Figure 1.9: Rotary encoder with stator coupling for stiff shaft connection.

With all these methods it is possible to "set" the counter to the correct position value. Because the zero point of the counter can be selected freely at any time, a **zero offset** in the CNC system does not present any problem.

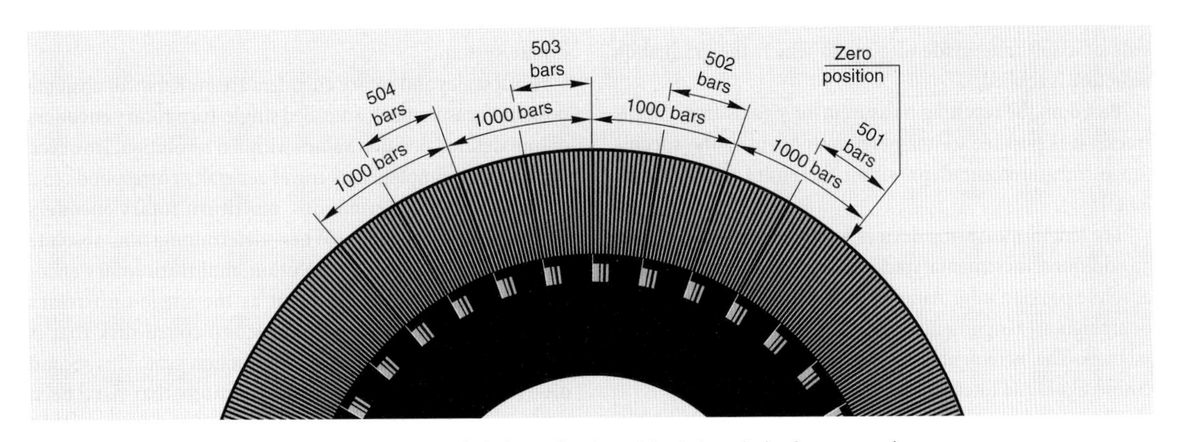

Figure 1.10: Schematic depiction of an incremental circle graduation with pitch-coded reference marks.

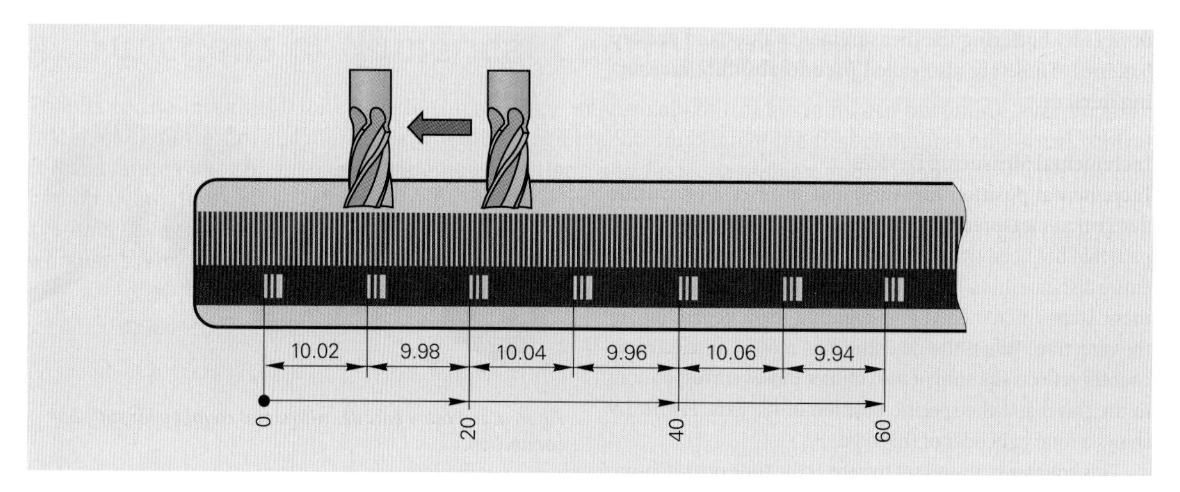

Figure 1.11: Incremental scale with pitch-coded reference marks (pseudo-absolute).

Absolute Measuring Devices

In contrast to the incremental measuring method, with **absolute position-measuring devices,** after the supply voltage is switched on again, the **position** of the machine axis is available immediately as an absolute value without first moving to a reference mark. Each position of the travel distance has a defined measurement value assigned to it. This measurement value is fixed and unambiguous.

Instead of a scale with a single graduation, the absolute measuring process uses a coded scale with a number of tracks, each with different pitch periods, or serially coded tracks and a suitable sensing head. The scale generally consists of a substrate made of glass to which the graduation structure is applied.

Based on the resulting measurement signals from all the tracks, it is possible to unambiguously identify the specific range (measurement step) within the entire measurement length.

For length-measuring devices with serially coded tracks, an additional incremental track is generally also applied. The absolute position information is determined by means of an unambiguous assignment of the serial code to the incremental track. The incremental track, in turn, is detected as a sine/cosine signal and interpolated up accordingly.

The precision of the measuring device depends on the precision of the finest track. For example, with this method, it is possible to create encapsulated length-measuring devices with measuring lengths of up to 3 m. With a smallest pitch of 20 μm, these devices can achieve absolute values with a 10-nm resolution and a system accuracy of ±3 or ±5 μm/m (Heidenhain LC 182/192 and LC 481/491, respectively).

The position value is output in dual code to allow easier calculation in the downstream electronics. The maximum traversing speeds for such length-measuring devices can be up to 180 m/min.

For **zero offset**, the CNC system has to constantly convert the absolute position values into the offset coordinate system.

Both scales and rotary encoders are available for absolute position measurement. So-called **single-turn rotary encoders** output defined position values within a revolution. However, because the complete axis travel generally corresponds to a number of spindle revolutions, **multiturn rotary encoders** generally are necessary. When determining the absolute position within a single revolution, multiturn rotary encoders use the same arrangement as the single-turn rotary encoder, but they also have a second coded disk that is connected to the first disk via a reduction gear. This second disk registers the number of revolutions that have been performed. The combination of signals from the two disks can be used to determine unambiguously the absolute position value over the entire travel distance. Current designs allow the coding of up to 4,096 spindle revolutions. This eliminates the need for limit switches or reference-point switches.

The main reason for wanting absolute measuring devices is to avoid problems that might occur after an interruption. It may be desirable (or essential) to avoid having to move to a reference point. For example, absolute measuring devices are necessary when robots are used in welding lines for car bodies. It is difficult, time-consuming, and in fact almost impossible to move all the robots to the reference point in five or six axes without damaging the car body. This situation is similar to that of assembly robots.

CNC machines are also being integrated into more and more interlinked systems, or the CNC machines themselves are so complex that moving to a reference point is not possible. Furthermore, when direct-drive technology with synchronous motors is used, it is necessary to control the motor windings correctly. Here, too, the absolute value of the rotor position is used.

Pseudo-Absolute Measuring Devices

Incremental measuring devices can be backed up either by the CNC system or in the measuring device itself in such a way that the counter value is actively retained. Here a backup battery allows the pulse generator and counter to remain operational for a limited time after a power failure. Every motion is detected, and the absolute value of the axis position is retained.

When the mains voltage is restored, the counter transmits the current value to the CNC system, and work can continue without having to move the reference point.

Cyclically Absolute Measuring Devices

In a like manner to systems with a battery backup, combinations of the various measuring methods are used here. For **single-turn rotary encoders** or single-track resolvers (→*Figure 1.9*), absolute position information is only available within a single revolution and thus only within a single spindle revolution. Besides this, an additional electronic counter counts the number of encoder revolutions starting at the reference point. This allows unambiguous determination of the absolute position. If this counter is cleared in the event of a power failure, then there will no longer be any unambiguous relationship, and it will be necessary to move to the machine zero point. This disadvantage could be avoided by using a second absolute encoder (M**ultiturn Systems**).

Types of CNC Measuring Devices

Various types of measuring devices are available for position measurement on machine tools. These devices have to stand up to harsh operating conditions. Brief descriptions of the major types are provided below.

Incremental Rotary Encoders and Angle Measuring Devices (*Figure 1.12*)

Incremental rotary encoders are the type most often used for position measurement. They are installed either via shaft couplings on the rotational axes being measured or else integrated into the drive motor. Various designs are available in a number of standardized sizes with housing diameters from 36 to 100 mm.

There can be from 5 to 10,000 graduations per revolution. The output signals can be rectangular signals with a digital HTL (high threshold logic) or TTL (transistor-transistor logic) level or sinusoidal signals with one V_{pp}.

A group of **incremental angle-measuring devices** is also available for use with precision rotary or swivel axes. These represent a significantly higher level of precision than incremental rotary encoders: better than ±10 angular seconds (= 0.003°). This can be achieved only through the use of high-precision bearings and the appropriate photoelectric sensor principles. These devices are equipped with up to 180,000 graduations, and the sinusoidal signals that are provided also can be interpolated up in the CNC system. With levels of precision of up to ±0.2 angular seconds, applications in high-precision machines also can be implemented.

Direction detection is important in incremental measuring systems, that is, distinguishing between motions in the positive and negative directions. This is done by scanning the grating of the scale by means of two gratings that are offset from each other by a quarter of the grating pitch. This results in two pulse sequences that are offset from one another by 90°. Based on these sequences, the direction of motion can be detected via an evaluating logic. Even more precise electronic interpolation can be achieved if additional scanning gratings are employed. *Figure 1.12* illustrates this type of scanning.

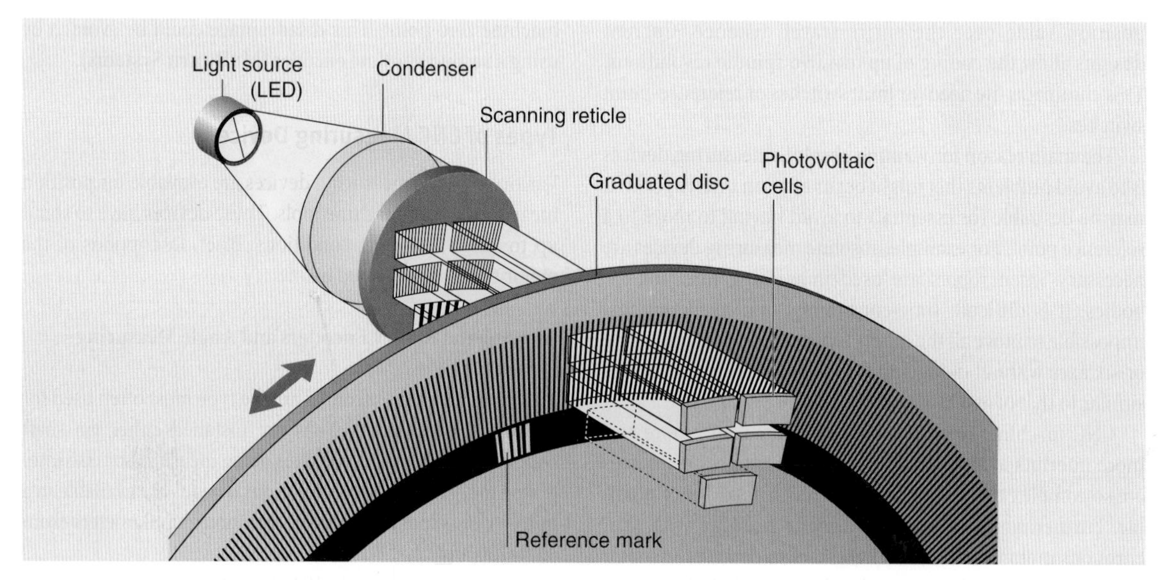

Figure 1.12: Principle of the photoelectric scanning of a rotary encoder disc with circular arc and reference track using the through-illumination process.

Absolute Rotary Encoders and Angle-Measuring Devices

With these devices, the angular position is derived directly from the graduation pattern of the graduated disk and converted into serial data that can be used by the CNC system. After switch-on, the absolute position value is generated from the pattern of the code tracks (*Figure 1.13*) or from a track with serial coding and is available immediately.

Absolute rotary encoders are available in the same mechanical designs as incremental rotary encoders. A major area of application relates to installation versions of motors for feed axes, which typically are designed as synchronous motors (→ *Figure 1.8*).

Motor control requires measurement of the absolute position value within a single revolution. Multiturn system designs have a mechanical material measurement for the spindle/rotary encoder that also provides an absolute position for the length. Typical resolutions of such systems range from 8,192 measurement steps per revolution (= 13 bits) to several million measurement steps per revolution (e.g., 2^{25} = 25 bits).

Like incremental angle measuring devices, **absolute angle-measuring devices** are used in precision rotary and swivel axes. Both their high precision and their high resolution are especially required when they are used to control torque motors. Such measuring devices are available with resolutions of more that 500 million measurement steps per revolution (2^{29} = 29 bits = 536,870,912 distinct position values per 360°).

The mechanical designs have integrated hollow shafts up to 100 mm, thus allowing almost any application.

Incremental Length-Measuring Devices

These measuring devices are likewise available in a number of measuring lengths, levels of precision, and designs, such as

- **Encapsulated,** in the form of a glass scale with a measuring length of up to 3 m for use on standard CNC machines, with measurement steps from 0.1 to 10 µm and maximum traversing speeds of 60 to 120 m/min. For measuring lengths of greater than 3 m, a one-piece steel measuring tape clamped in a special metal profile is used.
- **Open,** for higher levels of precision, with measurement steps from 1 nm (= 0.001 µm) to 0.1 µm and traversing speeds of up to 240 m/min (*Figures 1.14 and 1.15*).

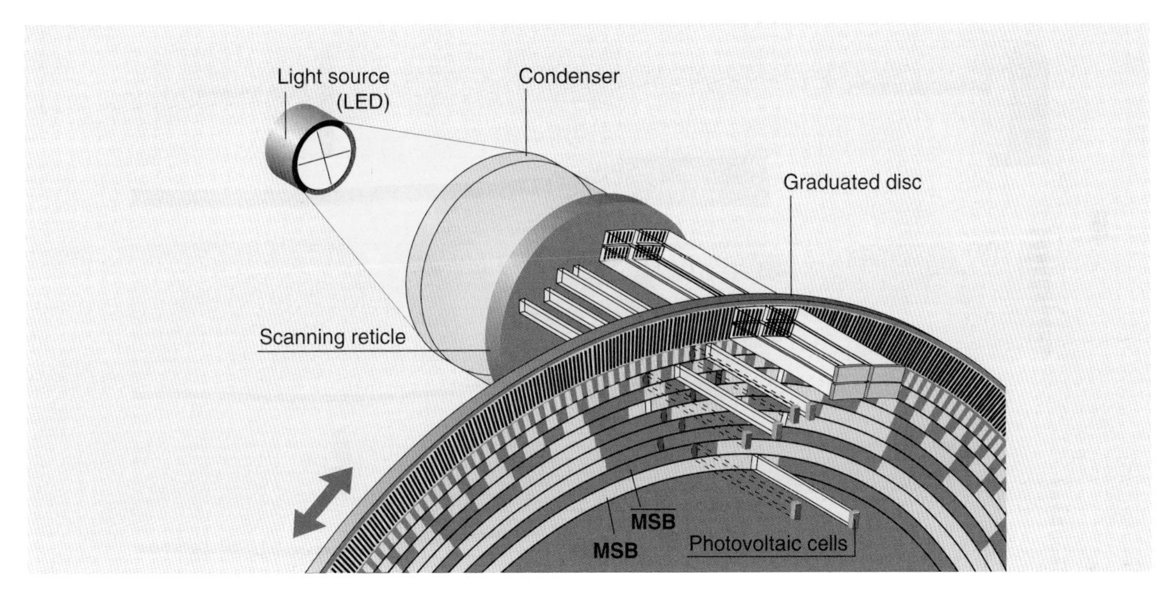

Figure 1.13: Principle of the photoelectric scanning of absolute rotary encoders.

These systems operate on a principle similar to that of incremental rotary encoders, for example, with photoelectric scanning of a grating consisting of opaque bars and translucent gaps of the same width. This grating is applied to either a glass scale or a reflective steel tape (*Figures 1.16 and 1.17*).

Depending on the manufacturer, larger scale lengths are put together from a number of individual pieces or delivered as a kit with a one-piece metal measuring tape. The light passing through (or reflected from) the grating reaches a scanning grating with the same graduations, after which it is collected by one or more photovoltaic cells. The relative motion between the scale and the scanning unit causes the bars and the gaps of the scale to be aligned alternately with those of the scanning grating. The photovoltaic cells convert the periodically changing light flux into sinusoidal electrical signals. These signals are used by the downstream electronics to generate direction-dependent counter pulses.

For use on machine tools, encapsulated path-measurement devices are recommended as a protection against fouling. A signal-monitoring function reports if the scale becomes too heavily fouled, allowing the machine to be switched off in good time before incorrect measurements are made.

Absolute Length-Measuring Devices

Encapsulated absolute length-measuring devices are externally identical to incremental systems, but in contrast, there is a limited variety of types. Scale lengths from 140 mm to about 3 m, provided in 100- or 200-mm steps, are offered. The greatest challenge with absolute measuring devices is the development of effective scanning systems. The simplest code is the **dual code** (*Figure 1.18a*). However, if the information is read at a particular position using photoelectric cells that are arranged in a row, then the corresponding code value is obtained directly. Assuming a period of 20 μm in the track with the most finely spaced interval, 17 tracks are needed for a measuring length of 1 m. In addition to the large number of tracks, scanning would present serious problems because multiple tracks change at the same time when the position is changed. A better system for scanning code tracks is the **Gray code** (*Figure 1.18b*) because this is a single-step code in which only one track changes when the position is changed. The Gray code can be converted into the dual code by means of a simple logic. But here, too, the number of tracks needed for the measuring length may be too large.

There is another measuring method that provides the position using only two graduation tracks. In this method,

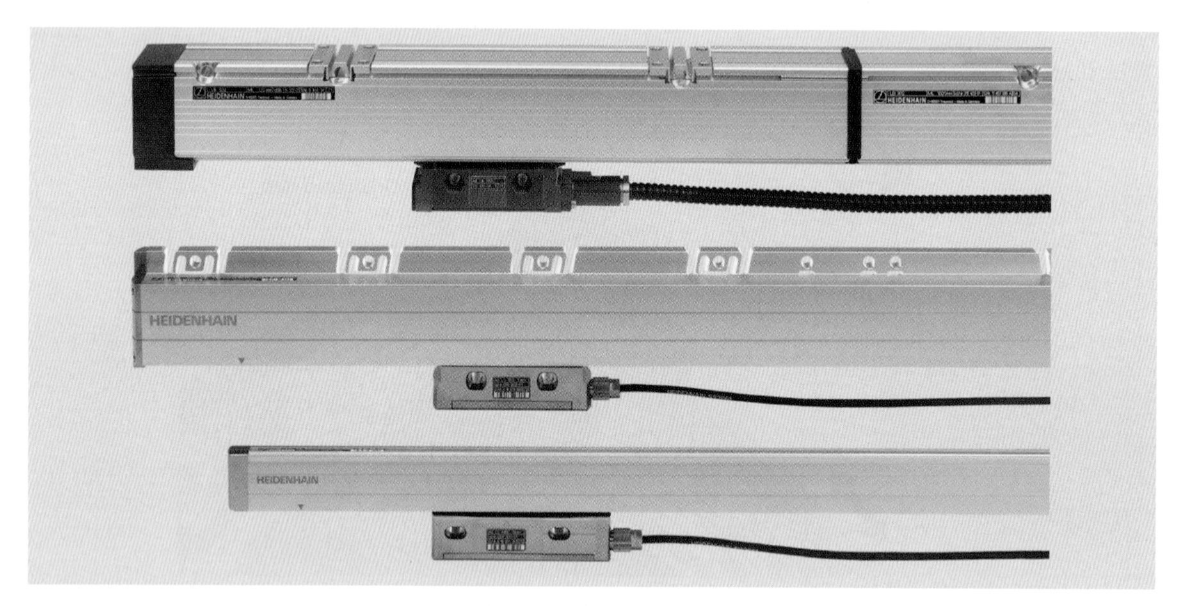

Figure 1.14: Encapsulated photoelectric length-measuring devices for installation on machine tools.

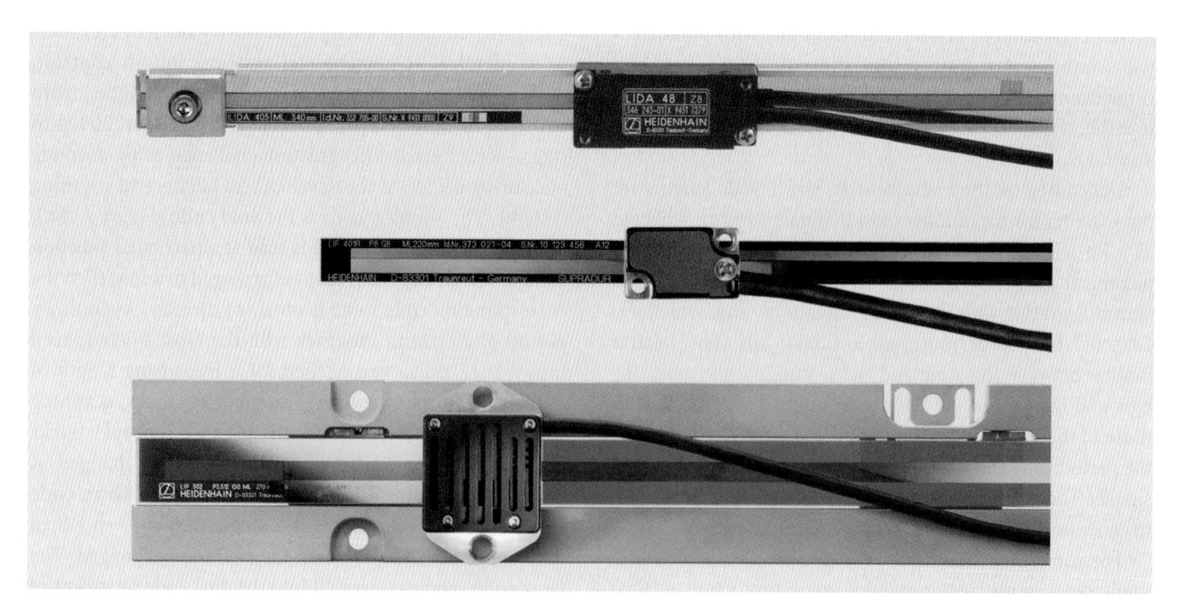

Figure 1.15: Open incremental lenght-measuring devices.

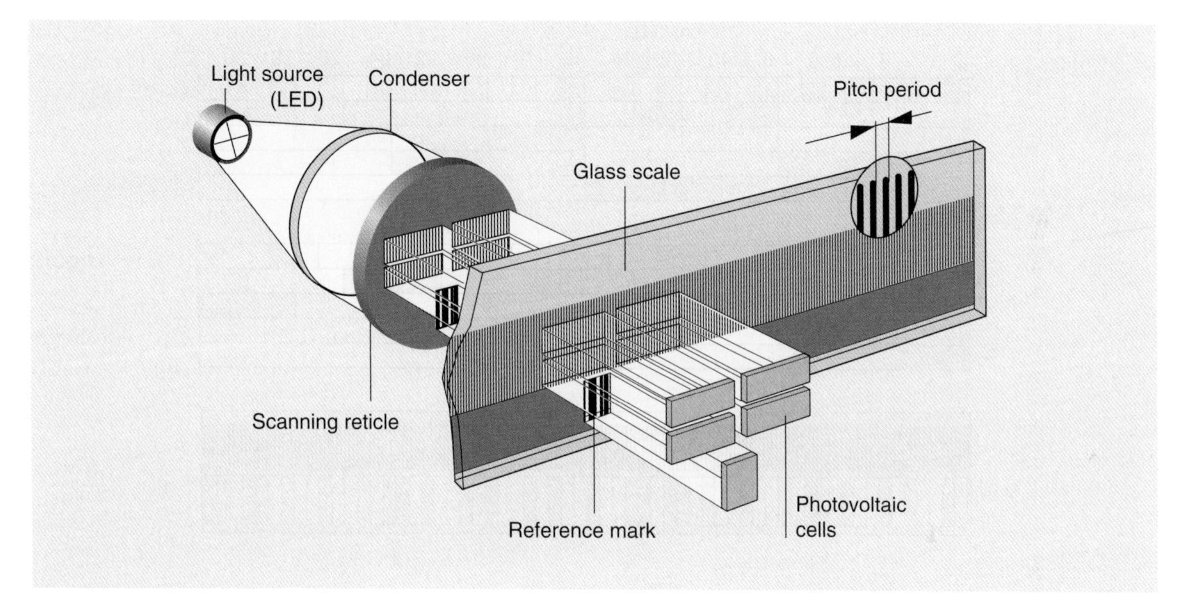

Figure 1.16: Principle of the photoelectric scanning of a glass scale using the through-illumination process.

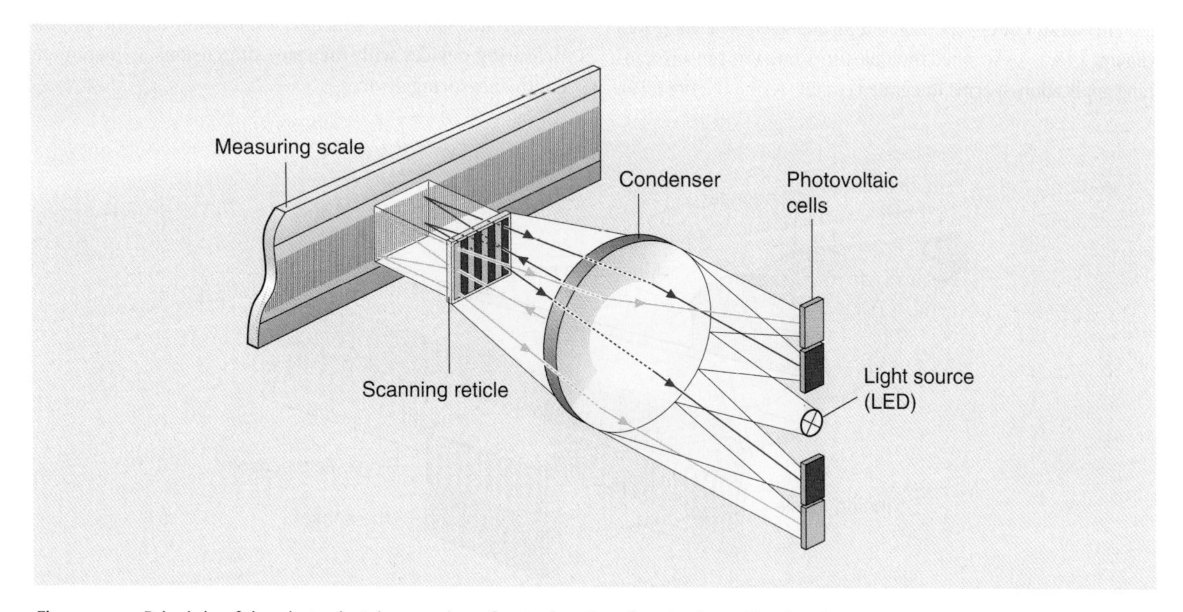

Figure 1.17: Principle of the photoelectric scanning of a steel scale using the front-illumination process.

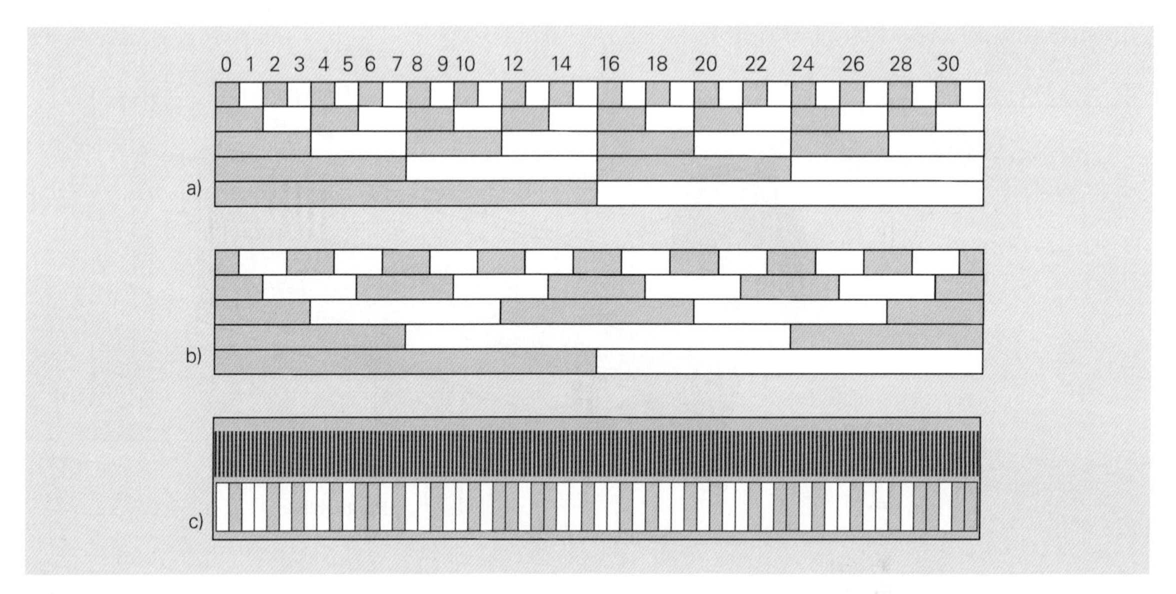

Figure 1.18: Code types: (a) dual code; (b) Gray code; (c) serial code.

in addition to the incremental track with the fine graduation (e.g., 20 μm), a second spur with a serial code is used.

The **serial code**—also called a *pseudo-random code* (PRC) (*Figure 1.18c*)—is scanned multiple times by a line sensor/scanning application-specific integrated circuit (ASIC) (*Figure 1.19*), and the absolute position can be assigned unambiguously to a signal period of the fine track.

With this scanning principle, it is possible to build length-measuring devices with the same dimensions as increment length-measuring devices.

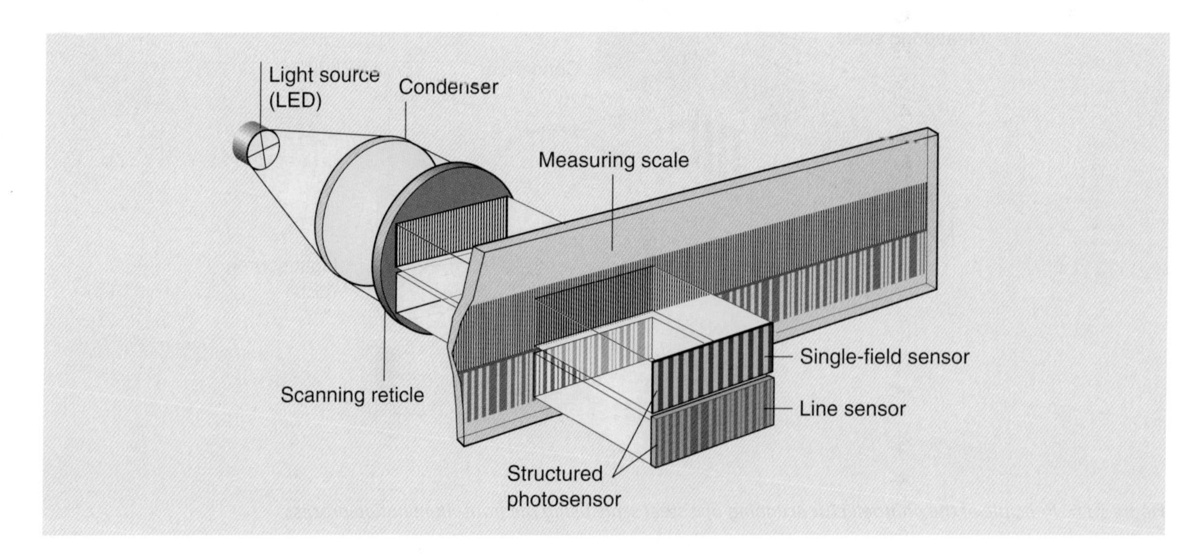

Figure 1.19: Principle of the photoelectric scanning of an absolute length scale with two tracks.

These measuring devices are likewise available in a number of measuring lengths, levels of precision, and designs: encapsulated, in the form of a glass scale with a measuring length of up to 3 m for use on standard CNC machines, with measurement steps up to 10 nm, and with serial interfaces appropriate for the control system.

Depending on their specific design, these length-measuring devices can be used for traversing speeds of up to 180 m/min.

Integrated Position Measurement in Linear Guides *(→Figure 1.20)*

This measurement principle is based on the AMR effect (Anisotropic Magnetic Resistance). In other words, the ohmic resistance of thin ferromagnetic layers is changed by external magnetic fields.

The material measure is bonded into the guide rail and sealed with a cover. The signal period is 200 μm.

The read head can be incremental (AMS) or absolute (AMSABS), with a protection classification up to IP68. An incremental system for long travel distances is also available (AMSA 3L), which theoretically can be of infinite length.

The output signal can be digital (SSI, Hyperface, TTL, RS) or analog (1Vpp).

Integrated Position Measurement

When a high-precision scale and a monorail guideway are brought together, the result is an integrated measuring system that can be installed directly without any assembly or adjustment work. This results in cost savings in the design, manufacture, and maintenance of the products. Schneeberger supplies quality integrated systems that can be installed right away. This reduces the degree of complexity when building machine axes with direct position-measurement systems (*Figure 1.21*).

The Magnetoresistive Measurement Principle

Here, the sensor is based on a specially adapted magnetoresistive measurement method. When there is a relative motion between the sensor and the material measure, the change in the field strength leads to an easily measured change in electrical resistance. An electric circuit in the form of a bridge means that disturbances owing to temperature, superimposed magnetic fields, displacement, and aging are minimal. In this case, the sensing head is in

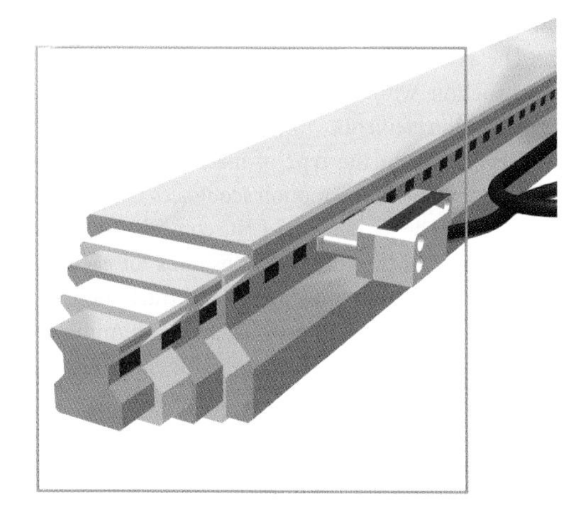

Figure 1.20: Integrated position-measuring system with the depiction of various command variables. (Image courtesy of Schneeberger.)

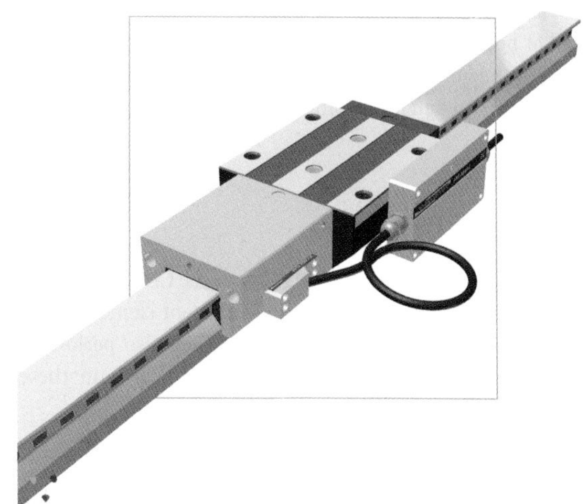

Figure 1.21: Recirculating roller guide with integrated measuring carriage, read head, and evaluation electronics. (Image courtesy of Schneeberger.)

contact, thus ensuring that the sensor's function will not be disrupted by any particles. The scanning process is efficient enough that it is possible to exchange all the read heads on all the rails without having to perform any adjustment work.

Close-to-Process Measurement of the Position

Good thermal coupling of the scale with the machine bed is possible, on the one hand, by means of the large-area connection of the guide rail with the integrated measuring tape, as well as a result of the fixed screw connection between the guide rail and the machine bed. As a consequence, the temperature changes in the machine bed are transferred directly to the scale. The good thermal coupling of the material measure to the guide rail, and thus also to the machine bed, ensures that such systems require no zero points and no temperature sensors to achieve very good process stability.

Thermal Expansion Like Steel

The magnetic material measure is inserted into a groove in the profiled rail guide. By using a specially adapted ferromagnetic material, it is ensured that the longitudinal expansion of the scale owing to thermal effects is identical to that of the steel rails.

The material measure is securely fastened to the guide rail at both ends and precisely follows the thermal expansion of the guide rail. Thus no temperature compensation is necessary when processing components that are made of steel.

1.5 Feed Drives

The mechanical energy needed to move the CNC axes is provided by **feed drives**. They are an important element in the position-control loops for these axes. They also perform a wide variety of transport and adjustment tasks in these machines.

The major components of a feed drive are as follows:

- The motor
- The drive controller, consisting of a power unit and a controller
- The axis mechanism and position-measuring system

The **motor** is an energy converter that supplies the mechanical energy required for motion and for maintaining position. In addition to the electrically active section, the motor includes additional components such as a holding brake and a position encoder. There is generally also a mechanical coupling to the drive shaft, which may, if desired, also feature integrated overload protection.

The motor is controlled by means of a **drive controller** (*Figure 1.22*). This device combines the functions of a controller and a power unit in a single component. In modern drive controllers, the current, speed, and position are controlled **digitally,** that is, by means of microprocessors. This allows greater precision and faster reaction times than with analog controllers.

Digital controllers also have a large number of additional application-specific functions, monitoring and diagnostic options, and communication interfaces.

In machine tools, feed drives generally are operated with modular drive controllers. These consist of a **supply module,** which rectifies the three-phase mains voltage and provides power to the **drive controller** via the **direct-voltage intermediate circuit.**

The **power-control element** of a modular drive controller consists of an inverter that takes current from the direct-current (dc) intermediate circuit and supplies three-phase current with a continuously adjustable frequency.

The **axis mechanism** essentially comprises the slide/axis construction with the guide system and the mechanical transmission elements.

Depending on the type of machine and the machining task, generation of the motion sequences may involve one or more axes. The higher-level CNC controls the axis motions by determining the command positions for each individual axis. The feed drives must be able to track these command positions as precisely as possible and with minimum delay. At the same time, the effects of interference must be kept as small as possible.

This means that feed drives have to meet the following **principal requirements:**

- High power density (high torque with a small size)
- Large range of speed control (>1:10,000)
- High overload capacity
- High positioning accuracy and repeatability
- Low mass moment of inertia

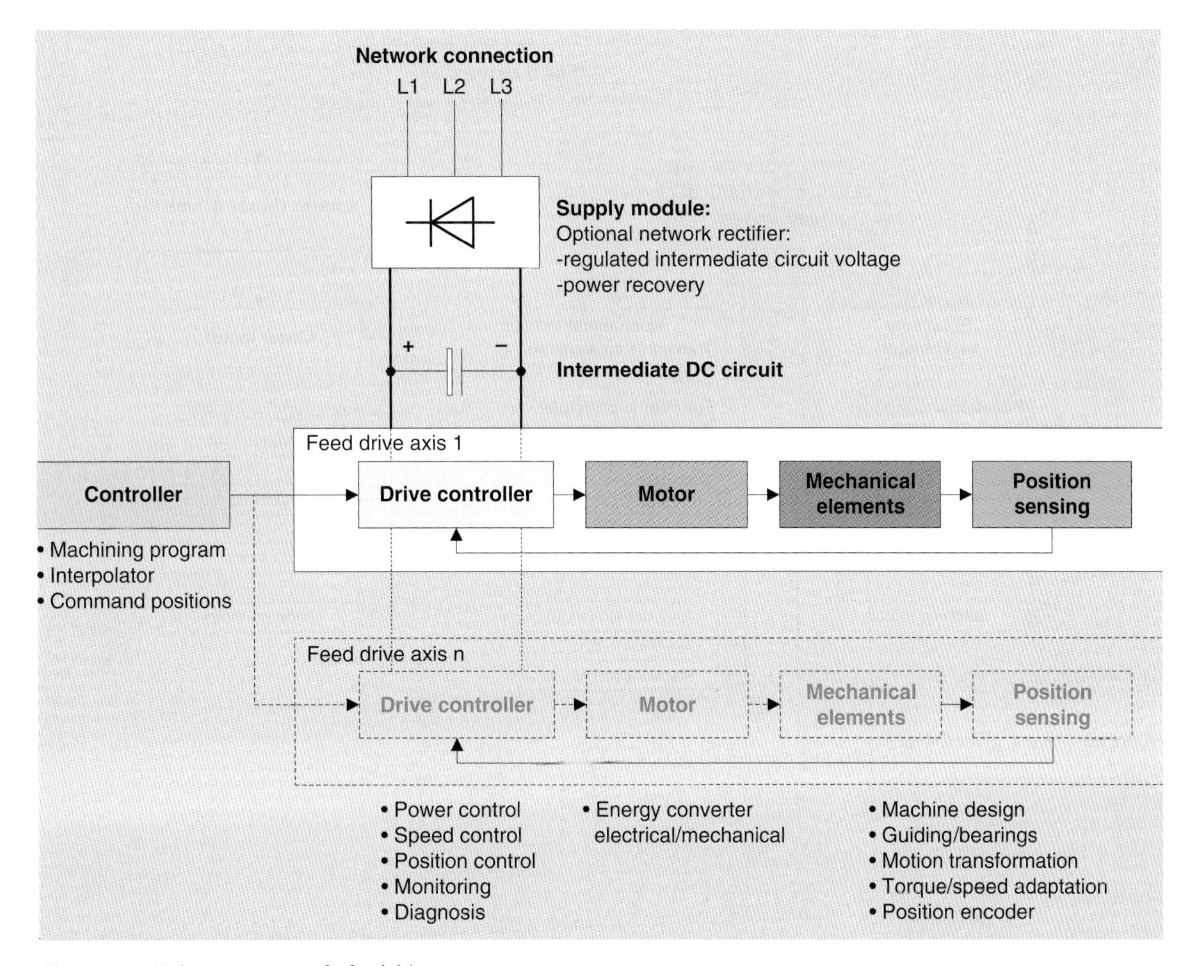

Figure 1.22: Main components of a feed drive.

Today's machine manufacturers and users also expect a number of additional features:

- Application-specific functions
- Simple commissioning and fault diagnostics
- Monitoring and safety functions
- Open, standardized interfaces
- No need for maintenance; high protection classification
- Low heating and high efficiency
- Low operating noise
- Small footprint
- Low costs

Types of Feed Drives

Feed drives provide all the necessary axis motions for the manufacturing process. *Figure 1.23* shows the various ways of implementing a translational feed motion.

Electromechanical Feed Drives

Today, most feed drives in machine tools are comprised of a rotational servomotor coupled with a mechanical gear unit such as a ball screw, which uses spindles and nuts to convert the rotational motion of the motor into a translational slide motion (*Figure 1.24*). In order to achieve optimal acceleration

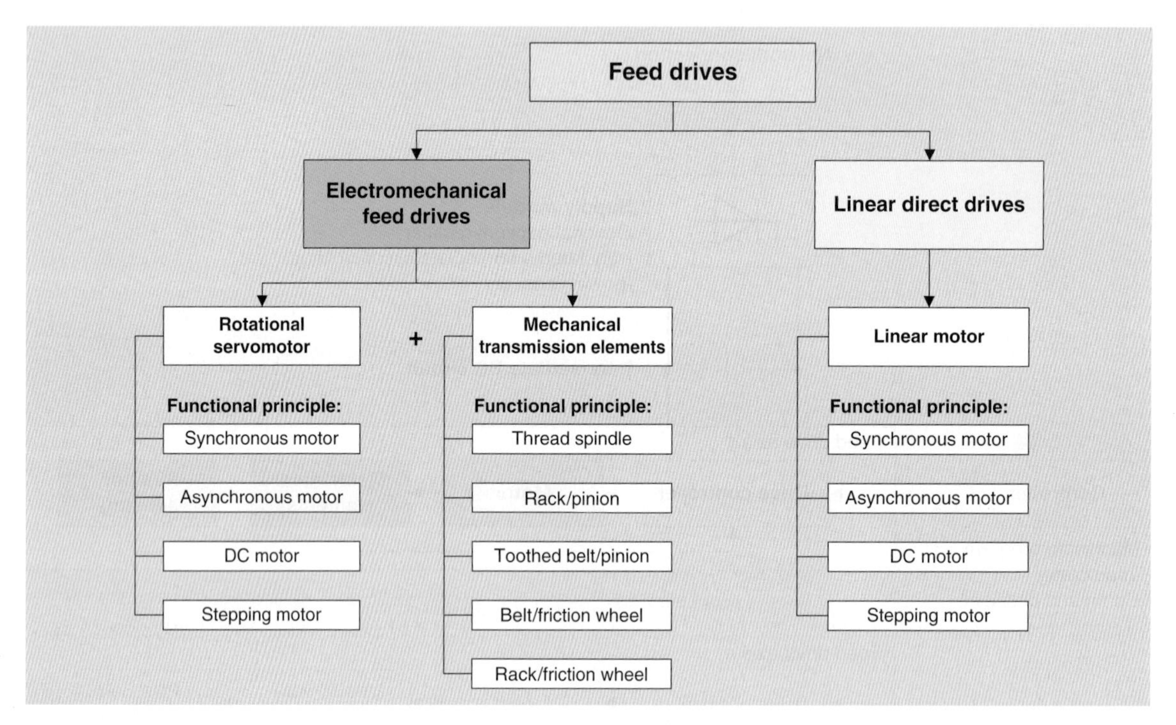

Figure 1.23: Types of feed drives.

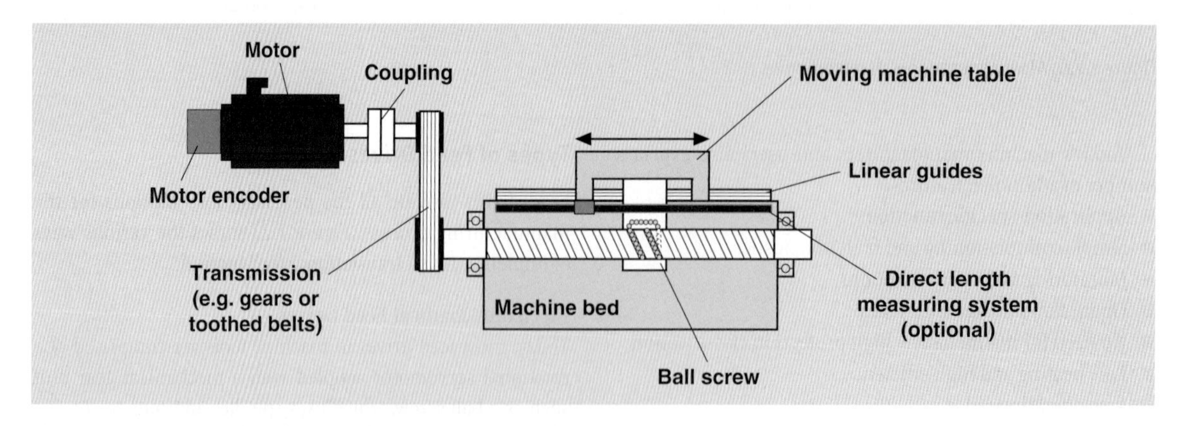

Figure 1.24: Structure of an electromechanical feed drive with ball screw.

or velocity of the mass that is being moved linearly, a gear unit is often placed between the motor and the thread spindle.

The slide position is detected by means of a direct length-measuring system. The angular position of the rotor is also determined via a rotational **motor encoder.** If a lower level of positioning accuracy or repeatability is required, the slide position can be determined by the motor encoder alone.

Relatively high feed forces can be achieved with ball screws owing to their ability to transfer power via spindles and nuts, as well as additional upstream transmission elements, such as toothed belts and gears. The acceleration capability of a ball screw is for the most part independent of the mass that is being moved linearly and is determined primarily by the pitch of the spindle and the moment of inertia of the motor and spindle. With today's highly dynamic ball screws, traversing speeds of up to 90 m/min can be achieved.

The control range of feed drives with ball screws is determined by the natural frequency of the mechanical system. The elasticities in the drive train, together with the masses being moved, lead to mechanical natural frequencies. In practice, this allows **amplification factors** of up to 5 m/min/mm (→ "Control of Feed Drives" below). The mechanical natural frequencies can be increased by using a larger spindle diameter. However, the moment of inertia of the spindle increases as the fourth power of the spindle diameter, thus limiting the dynamic performance that can be achieved.

When configuring ball screws as feed axes in highly dynamic machine tools, the optimal combination of velocity, acceleration, precision and service life is determined by several parameters. Decisive factors include the spindle pitch, the gear ratio between the motor and the thread spindle, and the ability to use various motors. Other limiting factors include the critical speed, the bending strength, the mass moment of inertia, and the position-dependent stiffness of the ball screw (*Figure 1.25*).

The integration of a ball screw into the machine does not pose any difficulties for the design engineer—this system has been proven in service over a number of years and is now regarded as the standard.

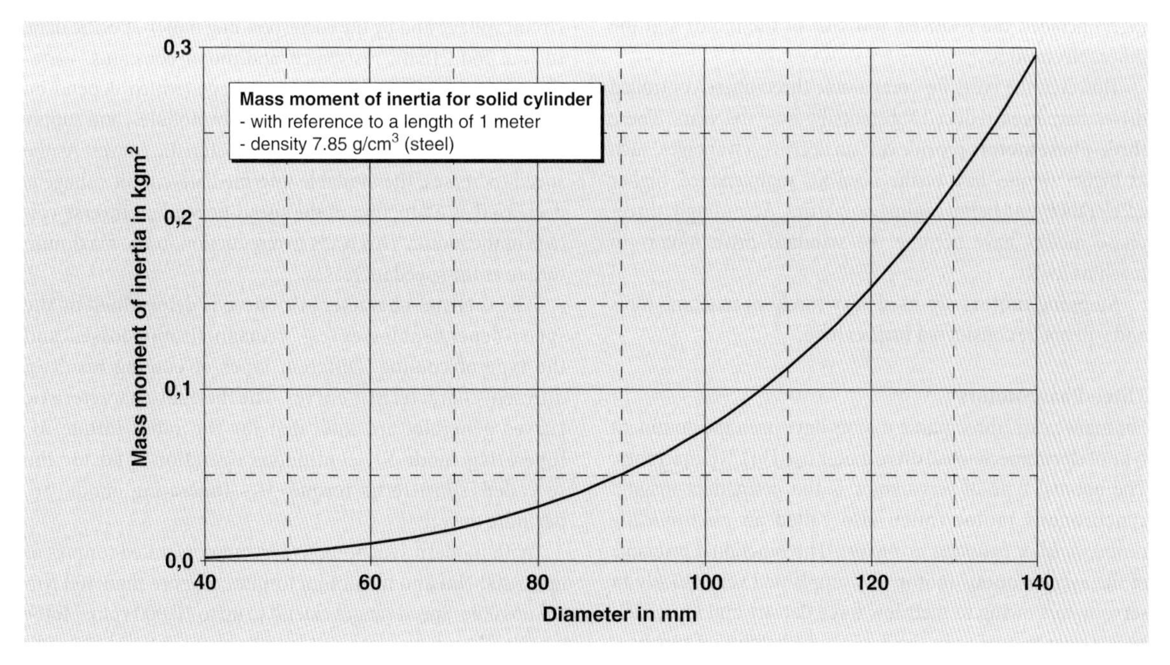

Figure 1.25: Mass moment of inertia of a ball screw as a function of the diameter.

The mechanical transmission elements in the drive train of a ball screw are components that are subject to wear. If axis collisions occur, then there is generally a risk of damage. This means that in practice there will be a corresponding amount of downtime depending on the maintenance and repair requirements.

In the case of highly dynamic ball screws or very long travel distances, it is generally necessary to use additional coolants to deal with thermal expansion.

Motors for Electromechanical Feed Drives

Servomotors have become the most popular type of feed drive in CNC machines and in the field of mechanical engineering as a whole. Functionally, rotating servomotors can be subdivided as follows:

- Direct-current (dc) motors
- Alternating-current (ac) and three-phase motors (synchronous/asynchronous)
- Stepping motors

In the past, **dc motors** had a number of advantages that made them preferable to other types of motors. These advantages included the relatively low cost of the motor and the power electronics.

However, the costs for comparable three-phase controlled drives have been reduced substantially over the years. These **three-phase motors** provide certain decisive advantages, such as higher torque, lack of the need for maintenance, higher acceleration, and better cooling properties. Accordingly, three-phase motors have become the standard drive motors in machine tools.

Stepping motors are used very rarely in machine tools and will not be considered further here.

Three-Phase Motors

For many years, three-phase motors have been the dominant type of electromechanical drive motor used in CNC machines. The optimal type of servomotor is the permanent-magnet **synchronous motor** (often also called an *electronically commutated* or *brushless dc motor*). The functional principle of the **asynchronous motor** was supplanted very quickly in servomotors owing to their low force density and low efficiency. Other factors were the easy controllability of synchronous motors and the lower cost of magnets.

The stator of the permanent-magnet synchronous motor has a three-phase winding, whereas the rotor has permanent magnets (*Figure 1.26*). Modern magnet materials allow high power density and thus high acceleration.

The main characteristics of synchronous motors are an equal rotational frequency and speed of the rotor and stator rotational fields (synchronism). In order to produce a constant torque, it is necessary to have constant synchronism between the rotational field frequency and the rotor speed. This is done by having the rotor position determined by the **motor encoder.** Based on the rotor position, the controller calculates and specifies the electrical angles of the stator currents. Depending on the characteristics of the dc motor, this is referred to as *electronic commutation*. The strength of the current is determined by the demand for torque. Speed is changed by altering the rotational frequency of the stator field.

The speed/torque characteristics of a permanent-magnet synchronous motor are provided in the form of characteristic curves. A distinction is made here between thermal and voltage-dependent characteristic curves (*Figure 1.27*).

The behavior of the voltage-dependent characteristic curves is determined by the strength of the intermediate-circuit voltage and by the corresponding motor-specific data, such as inductivity, resistance, and motor constants. Variations in the intermediate-circuit voltage (which depend on various drive controllers, power-supply modules, and supply voltages) may result in different characteristic curves. As the speed increases, the available intermediate-circuit voltage is decreased as a function of the speed-dependent inverse voltage of the motor. This leads to a reduction in the maximum torque as the speed increases.

The thermal-characteristic curve is determined by the speed-dependent losses (e.g., remagnetization losses) and the type of cooling. Different types of cooling result in different characteristic curves. The thermal-characteristic curves generally are indicated for the rated torque M_n (operating mode S1, continuous operation) and for the so-called short-term torque M_{kb} (operating mode S6, periodic duty).

With modern synchronous servomotors, rated torques of up to 200 Nm and maximum torques of more than 400 Nm are possible. Speed ranges extend to up to 10,000 rpm. Motor types with various sizes and lengths allow optimal tailoring for any individual application. The motor parameters of

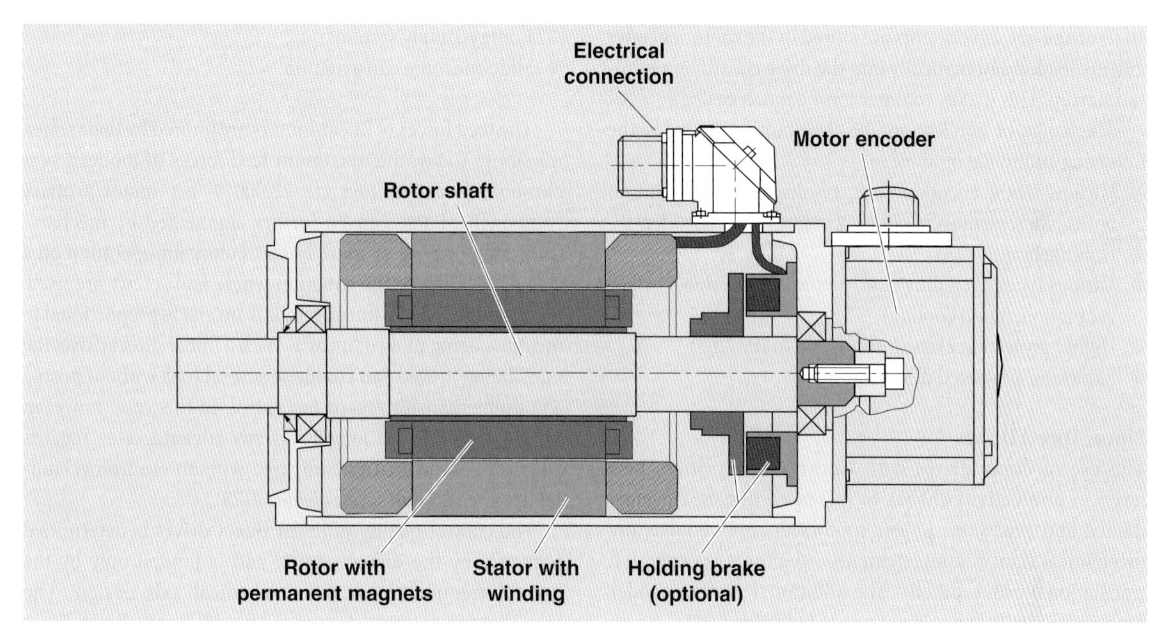

Figure 1.26: Structure of a permanent-magnet three-phase servomotor (synchronous motor).

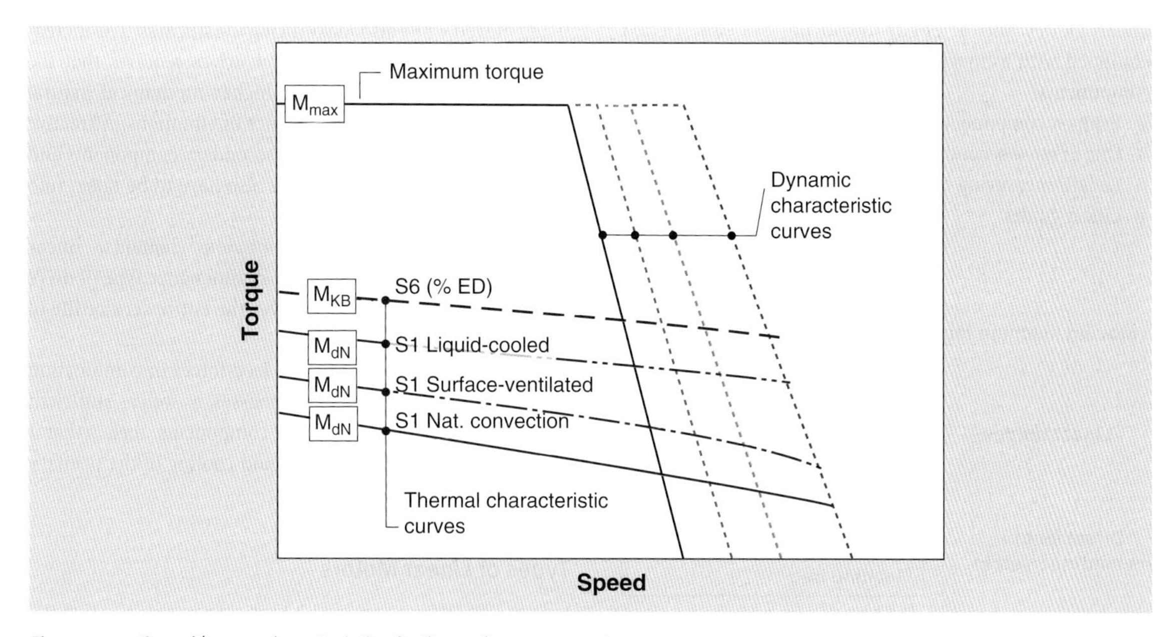

Figure 1.27: Speed/torque characteristic of a three-phase servomotor.

state-of-the-art servomotors are stored in the motor encoder and are loaded automatically into the drive controller on initial startup. This makes commissioning much easier.

The range of applications of servomotors includes the following additional options:

- Various motor encoders (e.g., resolvers, high-resolution optical incremental encoders, absolute-value encoders)
- Holding brake
- Various types of cooling (e.g., natural convection, surface ventilation, liquid cooling)
- Higher protection classifications (up to IP67/68)
- Explosion-protected designs

Linear Direct Drives

The use of direct drives with **linear motors** offers new levels of productivity thanks to increased dynamic performance and precision. Linear motors do not require any mechanical transmission elements—the force is generated translationally and directly. The additional motor encoder required for ball screws is not needed (*Figure 1.28*).

For machine tools, the linear motor generally is employed in the form of a kit motor. The primary and secondary components are delivered separately and are installed in the machine by the machine manufacturer, together with linear guides and a length-measuring system. An axis equipped with a linear motor generally has the following components:

- Primary component with three phase winding
- One or more secondary-component segments
- Length-measuring system
- Linear guides

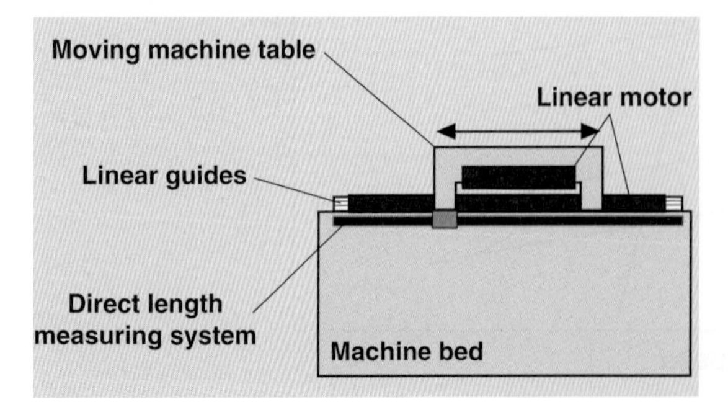

Figure 1.28: Structure of a linear direct drive with a linear motor.

- Energy-supply systems
- Slide/machine construction

The feed forces of linear motors are limited by their transmissions. Today, the maximum feed forces of modern synchronous linear motors are 22,000 N per motor/primary component. Force can be further augmented by mechanically coupling one or more linear motors for operation on a single axis. Unlike the situation where ball screws are used, the acceleration of a linear motor is inversely proportional to the mass being moved linearly. With a linear direct drive, the mass being moved can comprise several tons without posing any problems with regard to control. In this case, however, the linear motor no longer has any advantages as regards dynamic characteristics compared with an electromechanical drive with a ball screw (*Figure 1.29*).

The control quality of linear direct drives is determined primarily by the digital control and is limited only by the natural frequencies of the mechanical axis design. The amplification factors achieved today with linear direct drives in machine tools are typically from 20 to 30 m/min per mm.

Machine construction is subject to a number of requirements based on the design and characteristics of linear drives. For example, it is necessary to minimize the masses being moved while also maximizing the stiffness. The increased control range of linear direct drives requires that the machine structure has to have higher mechanical natural frequencies to avoid the occurrence of vibrations. Attractive forces between the primary and secondary components and attraction to ferromagnetic chips also have to be taken into account.

Without mechanical transmission elements, linear drives represent a low-wear, low-maintenance type of drive that maintains high precision over the entire service life of the machine.

The requirement for extreme feed forces in combination with minimum dimensions and masses, as well as sufficient thermal decoupling of the motor components, means that it is generally necessary to have liquid cooling of the primary component.

Types of Linear Motors

In principle, linear motors can be used to implement all the same types of functions as rotational motors (*Figure 1.30*).

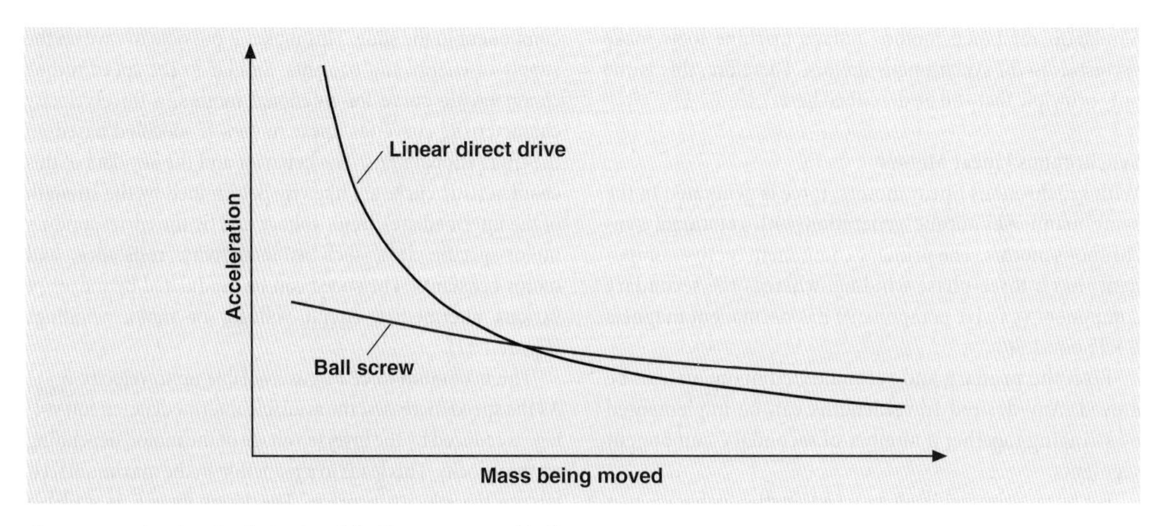

Figure 1.29: Acceleration behavior of the linear motor and ball screw.

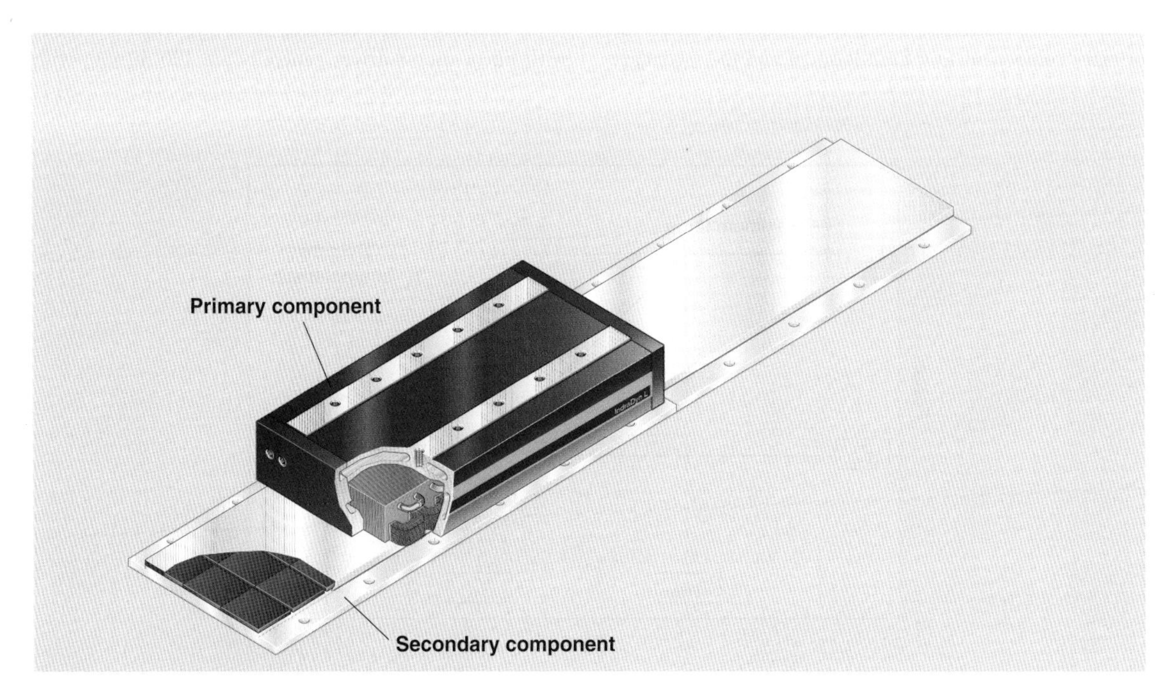

Figure 1.30: Structure of a synchronous linear motor.

Similar to rotational motors, today's machine tools make exclusive use of synchronous motors. Therefore, this is the only principle that will be described here.

Synchronous Linear Motors

With synchronous linear motors, force is generated in the same manner as torque generation with rotational synchronous motors. The primary component (active component) has a three-phase winding, whereas the secondary component (passive component) has permanent magnets (→ *Figure 1.30*).

Both the primary and secondary components can be moved. Any desired travel lengths can be implemented by stringing together a number of secondary-component segments.

For linear drives with longer travel lengths, higher magnet prices result in significantly higher costs compared with electromechanical feed drives. For travel lengths of up to 1 m—typical for most machine tools (e.g., machining centers)—the cost difference between linear drives and ball screws with a rotational servomotor is still quite small.

For very short travel lengths, such as the transverse axis of many turning machines, it may be advisable to assign the primary component to a fixed location and the secondary component to the slide. This makes it possible to save on the supply of energy and coolants. Similar to the speed/torque characteristic curve for rotational motors, a force/velocity characteristic curve for linear motors is specified based on the translational units. The behavior and the key data of this characteristic curve are likewise determined by the strength of the intermediate-circuit voltage and by the corresponding motor-specific data, such as inductivity, resistance, and motor constants. The speed can be modulated by means of various intermediate-circuit voltages or motor windings (*Figure 1.31*).

The maximum force F_{max} is available up to velocity $v_{F_{max}}$. As the speed increases, the available intermediate-circuit voltage is reduced by the inverse voltage of the motor, depending on the velocity. This leads to a reduction in the maximum feed force as the velocity increases. The design force F_n is available up to the maximum speed of the motor without any speed-related reductions.

The modular systems of modern synchronous linear motors offer a wide range of design options based on the sizes and lengths of their various primary components and their various secondary segments.

In machine tools, the feed rates implemented today with linear motor drives generally are 120 m/min, with

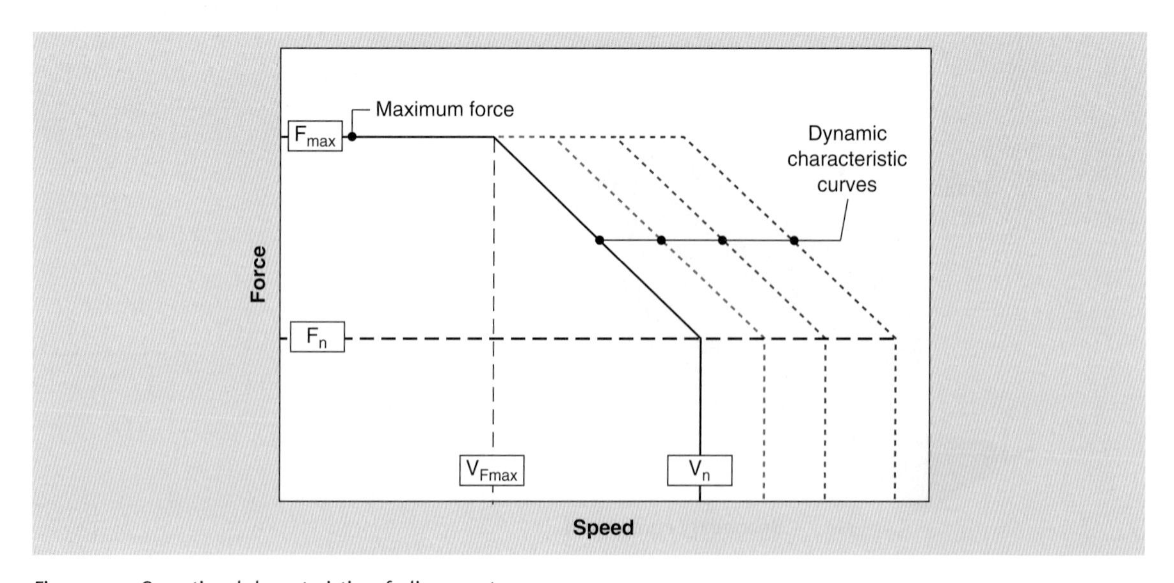

Figure 1.31: Operational characteristics of a linear motor.

accelerations of 10 to 20 m/s^2. The dynamic performance that can be achieved today is limited by the mechanical machine elements. For handling applications with linear motors, velocities of up to 300 m/min and accelerations of up to 80 m/s^2 can be achieved.

Appropriate cooling and encapsulation measures ensure that motors will function without any thermal signature and with a high protection classification and operational classification, even under harsh environmental conditions.

Advantages and Disadvantages of Linear Drives

The elimination of additional transmission elements produces a number of advantages:

- Freedom from wear and thus long service life
- No backlash, no elasticity of the drive train, large dynamic and static stiffness
- Low overall mass and small number of components
- In connection with digital controllers the possibility of achieving high control qualities with a large amplification factor (This means that a smaller following error and good positioning accuracy can be achieved even at high travel speeds.)
- High acceleration capability

Major disadvantages are lower efficiency and greater power loss. This leads to greater heating of the linear motor and results in additional costs for cooling systems.

Connecting the Drives to the CNC System

Numerical control of machine axes generally is position feedback control. An interpolator in the CNC calculates command positions for each machine axis cyclically, that is, at even, short time intervals. Each CNC axis follows these command values. This allows not only precise control of individual axes but also exact 2D or 3D continuous path control of multiple axes. Precise coordination of the axes depends on the precision of the interpolated command positions, the measuring accuracy, and the times of actual value measurement and processing.

In **analog drives,** the position control takes place in the CNC system. The CNC system merely transmits command velocities to the drive via a ±10-V interface. In this case, the CNC system has to have a great deal of computing power just to perform position control of the axes. Accordingly, the maximum number of axes that can be interpolated is limited.

In **digital drives,** the entire position control process is performed directly in the drive. This includes a lower-level velocity and current control loop, a large number of basic functions, and fine interpolation with extremely short cycle times. This arrangement allows much greater accuracy and higher velocities compared with position control in the CNC system. There is also a smaller load placed on the CNC system because only command positions have to be transmitted to the drive. Thus it is possible to synchronize a virtually unlimited number of drives. These advantages can only be realized if a suitable digital interface to the CNC system is employed (e.g., a **SERCOS interface**).

Digital drives operate cyclically; that is, all the command values and actual values must be updated with all the drives in each interpolation cycle of the CNC system (*Figure 1.32*).

Digital drives are adapted for various machines and control systems by entering parameter values via the CNC control panel or an external commissioning PC. Today's modern digital drives provide other additional functions to reduce the load on the CNC system. For example, standard procedures such as referencing are simply issued to the drive as a command by the CNC system. The drive executes the function independently and reports its status back to the CNC system. In addition, extensive **diagnostic functions** are available as standard, such as the so-called oscilloscope function using any desired actual values and command values on up to four channels. This can be used to save costs for external measuring equipment.

1.6 Summary

As in many other types of machines, in numerically controlled machine tools the working motions have to be controlled with great precision on the basis of digital information. This is done by means of position-control loops consisting of a position-measuring device and a servo drive. The position-measuring devices constantly measure the position of the element being moved. The position is compared with the specified command position, and the motion of the feed drive is corrected accordingly.

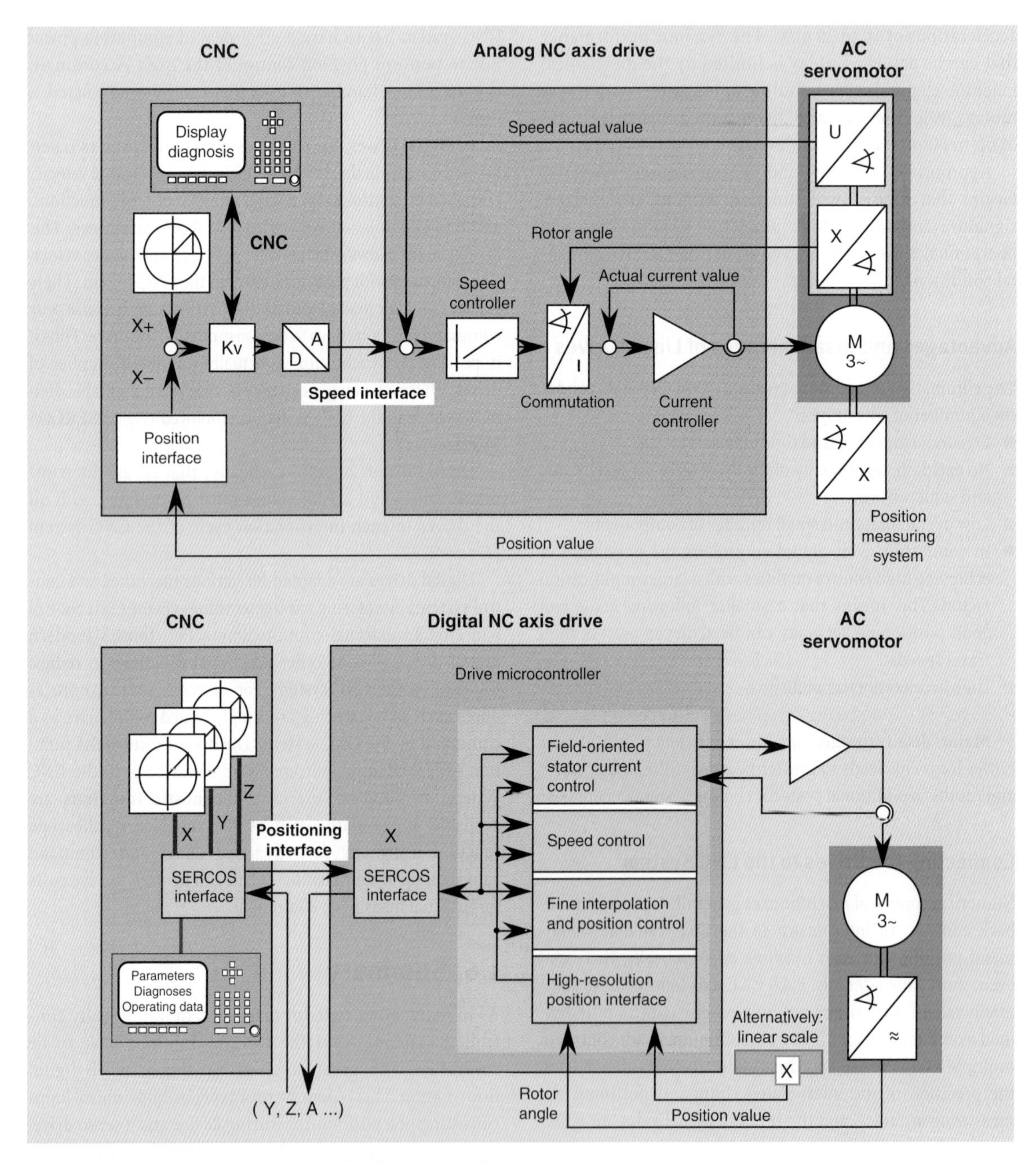

Figure 1.32: Schematic comparison of analog/digital CNC axis drives.

This places demanding requirements on the position-measuring devices and feed drives. For example, position-measuring devices have to be as resistant as possible to physical shocks, vibrations, accelerations, and electrical effects. For closed-loop position control, they must have a measuring accuracy in the range from 0.001 mm to several nanometers. They have to be able to measure rotary motions with a resolution of 0.001° with an accuracy of a few angular seconds. They have to be able to measure accurately even at velocities of up to 200 m/min and at several thousand revolutions per minute. They have to be able to withstand accelerations of multiple *g*-forces.

The **feed drives** that are used primarily today consist of a permanent-magnet three-phase motor with a ball screw. An optimal combination of maximum speed, acceleration, precision, and service life is based on the following parameters:

- Spindle pitch
- Gear ratio between motor and spindle
- Critical speed
- Mass moment of inertia
- Position-dependent stiffness and bending strength of the ball screw

The control range of **feed drives with ball screws** is determined by the natural frequency of the mechanical system. A maximum amplification factor of approximately 5 m/min/mm can be achieved.

The **amplification factor** is a measure of the machining precision and dynamic performance that can be achieved with a feed drive and indicates the velocity in meters per minute at which a CNC axis can move before a following error (also called a *lag*) of 1 mm is reached.

The amplification factor is determined by a number of variables and system characteristics, such as

- Stiffness of the mechanical components (bearings, gears)
- Control dynamics (motor data, masses, moments of inertia)
- Nonlinearities of the system (friction, backlash)
- Direct or indirect displacement measurement

Greater dynamic performance and precision can be achieved with **linear drives.** To use this type of drive, the machine must be designed to meet the following requirements:

- Minimization of the masses being moved
- High stiffness and high mechanical natural frequency of the machine structure
- Consideration of the attractive forces between the primary and secondary components and attraction to ferromagnetic chips

The control range of **linear direct drives** is determined solely by the digital control. A maximum amplification factor of up to 30 m/min/mm can be achieved.

Implementation of Dimensional Data

Important points to remember:

1. In the numerical control of machines, it is the specification of the dimensional data in digital (numerical) form and automatic execution of these data in the machine that are the main new features.
2. For this purpose, the machine has a position-control loop for each direction of motion (axis).
3. Turning machines have at least two axes, whereas most other machine tools have at least three axes. With five axes, a tool can be applied at any point on a workpiece in any desired direction.
4. Each of the axes is controlled by a position-control loop. Here, a position-measuring device continuously measures the actual position. This is compared with the command value derived from the dimensional data and is controlled in such a way as to reduce the positional deviation of the feed drive.
5. When an axis is in constant motion, a following error arises that is proportional to the velocity of the motion.
6. The amplification of control is determined by the amplification factor (K_v factor). This indicates the velocity in millimeters per minute at which the axis can move before a following error (also called a *lag*) occurs.
7. If the amplification factor is too great, oscillations occur in the position-control loop that can compromise the quality of the workpiece being machined.
8. In modern drive controllers, the current, speed, and position are controlled **digitally**, that is, by means of microprocessors. This allows greater precision and faster reaction times than with analog controllers.
9. **Stiffness** refers to the torque with which the drive holds a position that has been achieved despite the effects of counterforces.
10. **Dynamic performance** refers to the behavior of a drive when it has to execute changes in velocity as quickly as possible.
11. Position measurement is now only performed using digital graduated scales with optical scanning, either in the form of rotary encoders or, for greater precision, length-measuring or angle-measuring devices.
12. For rotary encoders it is necessary to take into account the additional dimensional deviations owing to mechanical material measures via spindles/nuts or racks/pinions, as well as temperature effects that cannot be compensated for.
13. Absolute position-measuring devices are nonresetting on power failure and can detect the absolute position immediately after a voltage interruption.
14. Feed drives should meet the following requirements:
    - Large range of speed control (1:10,000)
    - Good dynamic behavior
    - Stable speed response
    - Freedom from maintenance
    - High amplification factor
    - Compact dimensions
15. Three-phase synchronous motors and linear direct drives with digital closed-loop control are best suited for meeting these requirements.

2. Implementation of Switching Information

The automatic triggering of fixed machine functions, for example, tool changing, also was common among conventional automated machine tools. In numerically controlled machine tools, refinement of these switching machine functions and extensive integration of them into the automated work sequence made it possible to perform complete machining of workpieces in at most two operations or even a single operation. This was in any case the objective of the creation of a new type of machine, the machining center.

With so many types of machine tools and their diverse variants, the result has been a very confusing multitude of designs and functional principles for these additional machine functions. Therefore, we will only be able to describe here the most important functions—those which appear in many types of machine. These functions are automatic tool changing, automatic workpiece changing, and speed changing.

In order to machine a workpiece, most CNC machines require the use of a number of tools in a sequence defined by the work plan. In order to extend the automatic work sequence to cover the complete machining, automated tool changing is necessary.

Even long before the appearance of CNC machines, **tool turrets** were used, especially in drilling and turning machines. In these systems, each tool has a fixed position. After each operation, the turret automatically moves onward by one position, thus bringing the next tool to the working position.

The first CNC machines naturally made use of this proven technology for the automation of tool changing. It soon became clear, however, that the limited number of tools that could be accommodated in a turret meant that other tool-changing solutions were required. The Milwaukee-Matic from Kearney & Trecker, the first machining center developed for use in normal mechanical engineering applications, was presented for the first time in 1960. As the forerunner of all subsequent machining centers, it had a tool magazine with a double gripper and coded tools and even automatic workpiece changing with pallets. Because the company had registered these innovations, this newly developed technology was closed off for many years, but it nevertheless led to the development of many other solutions.

The decisive criteria for evaluating a tool-changing system are as follows:

- The number of tools available
- Limitations with regard to the dimensions and weight of the tools being used
- The tool-changing time, generally indicated as the *chip-to-chip time*
- The costs of any additional tool holders or tool blocks that may be required

2.1 Tool Changing

Tool Changing on Turning Machines

In turning machines, many tools are not specific to a particular shape or dimension. Thus the number of tools required is also smaller. This means that tool turrets are still the predominant solution here even today. The number of tools in a turret has been increased to 18. It also has proved possible to equip turning machines with two or even three turrets.

In contemporary turrets, the swivel motion is automatically clockwise or counterclockwise in order to bring the next tool into the working position over as short a path as possible. Inclining the axis of the turret by 45° or 90° produces various types of turret, such as star, crown, or disk turrets. In addition to turning tools and centrally located drilling tools, today all types of **driven tools** can be accommodated in the turret of a turning machine (*Figure 2.1*).

Although tool turrets do have a limited capacity, they are also more economical because they do not require any additional tool-changing grippers. Because the tools always remain in their holders, dimensional deviations on the workpiece

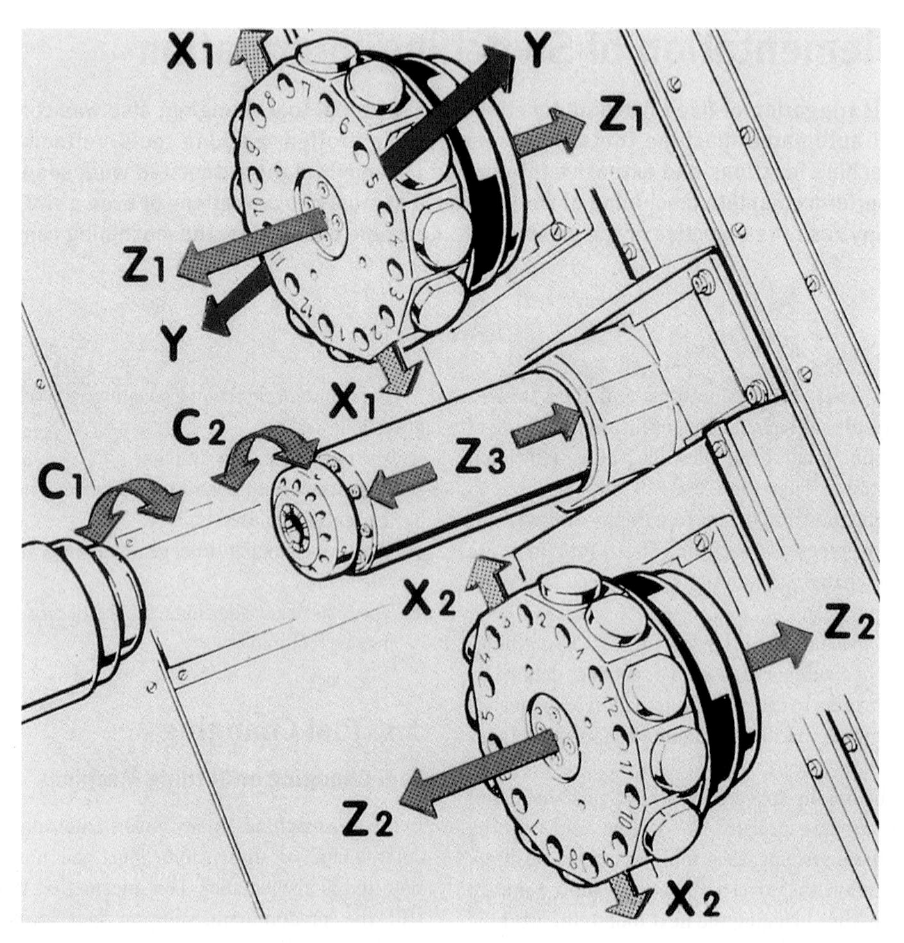

Figure 2.1: Turning machine with main spindle, counterspindle, and two turret heads, of which one has a Y axis for machining of cylinder-barrel surfaces.

caused by tool changing are avoided. What is more, tool changing is performed more quickly because there is no travel distance to the changing position. The turret generally can be swiveled after a short retraction.

Tool magazines are encountered less often; their main advantage is the lower risk of collision. The principle of using magazines has received an added boost from the development of tool systems that provide for only the tool head and insert (and not the entire tool block) to be interchanged. Such designs make it possible to store a large number of tool inserts in a small space and to keep them available over longer periods.

Tool Changing in Milling Machines and Machining Centers

In these machines, the number of tools required is generally significantly greater than with turning machines because many tools, such as drills, countersinks, reamers, and tapping tools, are not specific to a particular shape or dimension. Therefore, the use of various types of **magazines** has become predominant in these applications. The machine, tool magazine, and tool-changing system represent a single design element. A distinction is made between **chain, disk,**

circular, or cassette-type magazines depending on the storage element (*Figures 2.2 and 2.3*).

Solutions exist in which more than 100 tools are present in a machine's magazine. However, such large numbers of tools cannot be accommodated in magazines with a simple one-dimensional arrangement, such as circular magazines with a chain-type or annular configuration. One reason for this is that it takes too long to move to and locate a new tool. Circular magazines are limited first of all by their dimensions. Such magazines have been designed with up to 32 tools. **Chain magazines** are also subject to a limit owing to the weight of the chain and tools and the resulting drive power required for the chain.

For large numbers of tools, one solution was to equip machines with two magazines installed to the left and right of the machine base. There were even designs for tool-changing systems that had four magazine disks offset by 90° on a rotary table that could be brought to the working position in sequence (*Figure 2.4*). Another approach was to arrange the tools in a 2D system, for example, by placing them in two or three concentric rings or on a number of discs located coaxially one above another. Linear magazines featured the same principle; here, the tools were placed or suspended next to each other in a number of rows. Chain magazines were equipped with two independently driven chains. In all these cases, motion in the second dimension also was possible.

Figure 2.2: Circular magazine as a built-on unit. (Image courtesy of Miksch.)

However, **cassette-type magazines** have proved to be the preferred solution for storing a large number of tools. Four to six such cassette-type magazines can be incorporated into a single machine. Each of these magazines can house 20 to 30 tools in a rectangular field. The cassette-type magazines are loaded outside the machine and therefore can be exchanged and reconfigured while the machine is still in operation. This makes it possible to quickly prepare new tools and to remove tools that are worn or no longer needed. They are transferred to the machine by a floor-mounted robot that covers the entire surface of the magazines and makes the tools that have been called up available for the gripper in a changing position and also picks them up again there.

Various systems also have been developed for changing the tools between the magazine and the spindle. These basically have to perform the following set of functions—possibly in a different sequence, depending on the system:

- Moving the spindle and magazine to the changing position
- Presenting the new tool
- Removing the old tool from the main spindle
- Delivering the old tool to the magazine
- Removing the new tool from the magazine
- Inserting the new tool into the main spindle
- Moving the main spindle and magazine back to the working position

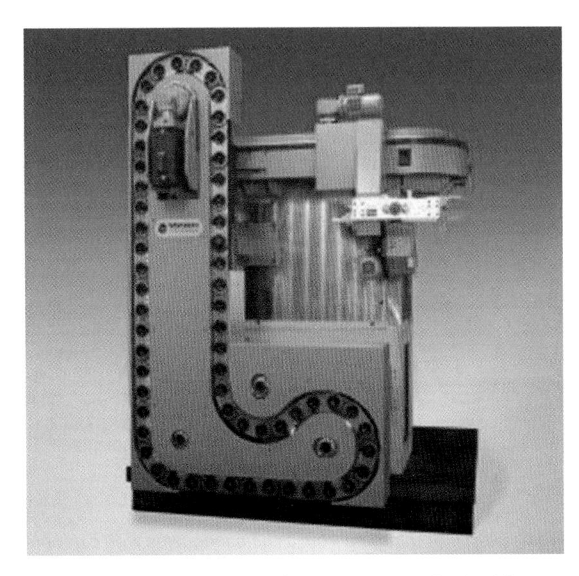

Figure 2.3: Chain magazine. (Image courtesy of Miksch.)

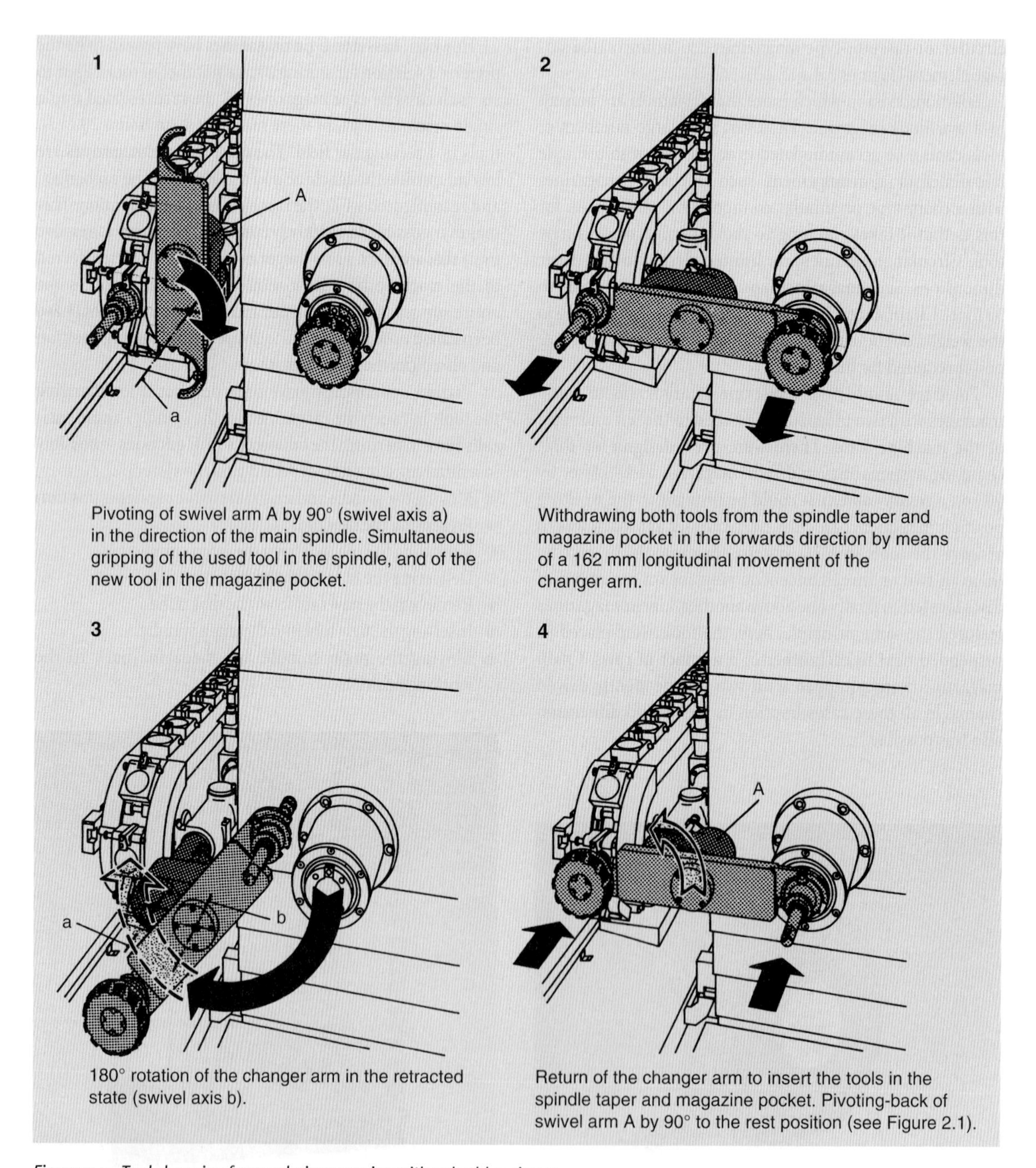

1
A
a
Pivoting of swivel arm A by 90° (swivel axis a) in the direction of the main spindle. Simultaneous gripping of the used tool in the spindle, and of the new tool in the magazine pocket.
2
Withdrawing both tools from the spindle taper and magazine pocket in the forwards direction by means of a 162 mm longitudinal movement of the changer arm.
3
a
b
180° rotation of the changer arm in the retracted state (swivel axis b).
4
A
Return of the changer arm to insert the tools in the spindle taper and magazine pocket. Pivoting-back of swivel arm A by 90° to the rest position (see Figure 2.1).

Figure 2.4: Tool changing from a chain magazine with a double gripper.

The sequence for this **changing process** is defined in the control system. Its duration is an important factor for the cost-effectiveness of the machine. It is indicated by the so-called chip-to-chip time.

The simplest method is to exchange the tools directly from the magazine to the spindle, for example, by moving the main spindle to the storage location of the tool using the normal CNC axes and setting the old tool down there or placing a new one in its mount (the pickup method). This requires a great deal of time, however.

Systems using a single- or double-armed gripper operate faster. Of these, **double grippers** are the fastest because the old and new tools are exchanged in a single operation, and the next tool is found and made ready during the machining time, that is, while the machine is working.

All magazines have limits when it comes to the dimensions and weights of the tools that can be stored and handled. This means that especially large or heavy tools may have to be exchanged manually. In exceptional cases, in addition to their normal magazine, machining centers sometimes are equipped with changing devices for such special tools, for example, large surfacing heads or drill heads.

Various **coding methods** are available for **tool identification.** Each of these has its own specific advantages and disadvantages. The following distinctions are made:

1. **Location coding,** in which the magazine locations are numbered, and the part program indicates not the tool number but rather the location number. Therefore, after use, each tool has to be placed back in its fixed location in the magazine.

Advantages:

- Use of standard commercial tools or tool holders
- Searching for locations via the shortest path
- No dynamic problems related to location detection owing to sufficiently long coding cams or an electronic location detection system (As a result, high search velocities are possible.)
- Oversized tools can be placed as desired; the adjacent positions are not used, and no collisions will occur.

Disadvantages:

- During setup for a new program, it is necessary to place all the tools in the magazine exactly as the programmer has defined them in the program.
- When producing the parts being machined as part of a product mix, problems may occur if different tools occupy the same locations in the programs for various workpieces.
- Configuring the magazine with alternate tools is problematic and may require the use of "tricks" in the CNC system.

2. Tool coding. In the past, this was implemented mechanically, for example, by means of cams or the like. Today, electronic memory chips are used. The systems available for this are described in more detail in *Part 4, Chapter 6.*

Advantages:

- The tools can be arranged in the magazine as desired.
- The tool number is programmed.
- Free placement of the tools in the magazine is allowed.
- The locations of the tools can be exchanged during the changing process.
- Electronic coding can include not only the tool number but also technological data, such as the exact tool diameter or the remaining tool life.

Disadvantages:

- The need for expensive tool holders with coding devices or memory chips.
- Each magazine requires a scanning or reading station.
- Reading may require the magazine to use a lower search speed.
- Coded tool holders cannot be used on all machines.
- Longer search times are necessary because the shortest path is not known.

3. **Random tool access.** In this system the operator places each tool in any desired location in the magazine and enters the tool number in the CNC system, which then performs all the further data management. This means that the tools can occupy a new location in the magazine after every changing operation because the CNC system records the new location. It is the tool number that is programmed, and the CNC looks for the location of the tool based on its accounting system. This method is seeing increasing use because it exploits the advantages described earlier while avoiding the disadvantages.

Advantages:

- Use of uncoded or electronically coded tools
- Use of the magazine's reliable location coding
- Programming of the tool number in the program
- Searching via the shortest path
- Short changing times thanks to a double gripper that exchanges the magazine and spindle locations for two tools simultaneously

Random tool access requires a CNC system that incorporates the necessary software. This has to be able to do the following:

- To establish the correct data assignments every time there is a tool change and to save the data permanently
- When an electronic coding system is used, to provide appropriate data interfaces for the read/write device of the data block and the tool data calculator
- To support manual tool changing by bringing the tool in question to a pickup station and displaying the tool number for checking
- Reserving fixed locations for oversized tools and keeping the adjacent locations free

What has been said here for the turrets of turning machines and for the magazines of drilling machines, milling machines, and machining centers is naturally also applicable for the application of such tool-changing systems to other numerically controlled machines.

2.2 Automatic Workpiece Changing

Automatic workpiece changing is an additional step in the direction of automated manufacturing. It makes it possible to avoid nonproductive time for the clamping and unloading of workpieces. The benefit is that the productivity of the machine tool is increased and is decoupled from human intervention—possibly even leading to unmanned production. Apart from being able to store a large number of machining programs in the CNC system, under certain conditions, it is even possible to implement unmanned, demand-driven machining of various parts of a group of related parts in one-off or small-series production. Moreover, automatic tool changing is an essential requirement for the integration of CNC machines into flexible manufacturing systems.

For a long time now, automatic tool changing has been comprehensively implemented into conventional automatic lathes that can operate starting with bar stock. These are even equipped with a bar magazine for automatic feeding of new bars and thus can operate for long periods without human operators.

In CNC machines, various systems are used for the automation of workpiece changing, in particular in regard to the type of clamping:

- Production from bar stock
- The pickup method
- Changing of the unclamped workpiece by a loading device, for example, a standard commercial industrial robot (*Figure 2.5*)
- Changing of the workpiece that is clamped by means of a clamping device, for example, a pallet

The decision in favor of one of these systems depends on the nature and size of the workpiece and the type of machining.

Production from bar stock is only practicable for relatively small workpieces that are machined from solid materials. It is thus used mostly with turning machines, although it is also employed with machining centers and milling/turning centers. Because the workpiece has to be cut off or sawed off at the end of the machining process, a removal device or robot is often used to pick it up, especially if it has to be passed on to further machining steps. In turning, it is disadvantageous if production from bar stock results in a major reduction of the permissible speed of the spindle. In combination with the performance of modern cutting materials, this would result in a significant increase in the machining time. This has led to a different mode of operation with bar stock in which the CNC program cuts the material to the correct length before machining, only after which the trimmed part is clamped.

In the **pickup method,** the clamping device picks up the workpiece directly from a deposit surface and presents it directly for machining. It is therefore used primarily in turning machines because the chuck can serve excellently as a gripper. Because the workpiece is fed in an unclamped state, it has to lie on a horizontal surface and therefore can be gripped and clamped only from above. In turning machines with a vertically suspended spindle, this is no problem. The spindle has to be able to move from a working position to the

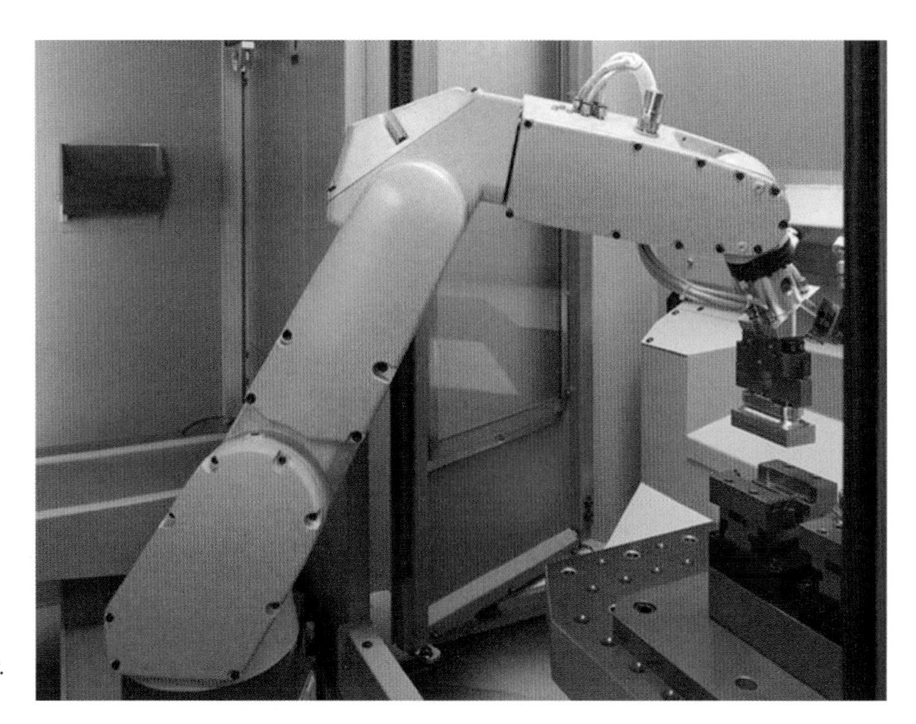

Figure 2.5: Workpiece changing using a standard commercial industrial robot. (Image courtesy of Chiron.)

receiving position and, after the machining, possibly to a different set-down position. This is very easy to implement if this motion is executed by the *X* axis of the CNC system. A simple conveyor is sufficient for feeding and removing the workpieces. Such systems are only suitable for handling large parts that lie in a stable position and generally have a round shape.

The **changing of the unclamped workpiece by a loading device** assumes that the clamping operation is reproducible in a simple manner and that the turned part is between centers or in chucks, unless a milling part has been properly aligned, mounted, and clamped in a vice or clamping fixture. Here, too, a distinction is made between systems with one or two grippers. Just as with tool changing, systems with two grippers have a significantly shorter changing time. Systems with a single gripper first have to move the finished part away, only after which they can retrieve the new part and bring it to the system. Systems with two grippers, on the other hand, can exchange the finished part for the new part directly.

Such systems are used for the handling of shafts, as well as for chuck components of turning and grinding machines, especially when these machines are combined with other machines via a special transport system. These generally involve gantry systems that introduce the workpieces to the working space from above. Industrial robots are often used when various complementary CNC machines form a flexible machining system, and the robot can tend a number of machines in succession.

Because automatic workpiece identification is not possible in the unclamped state, with small lot sizes or for one-off production, it has to be possible to follow the program instructions in another manner, for example, by tracking the movement of each workpiece in the system's host computer.

When changing a workpiece that has to be clamped on a clamping device, a pallet is always used. As a result, **pallet changing** has practically become the standard type of workpiece changing for milling machines and machining centers (*Figure 2.6*).

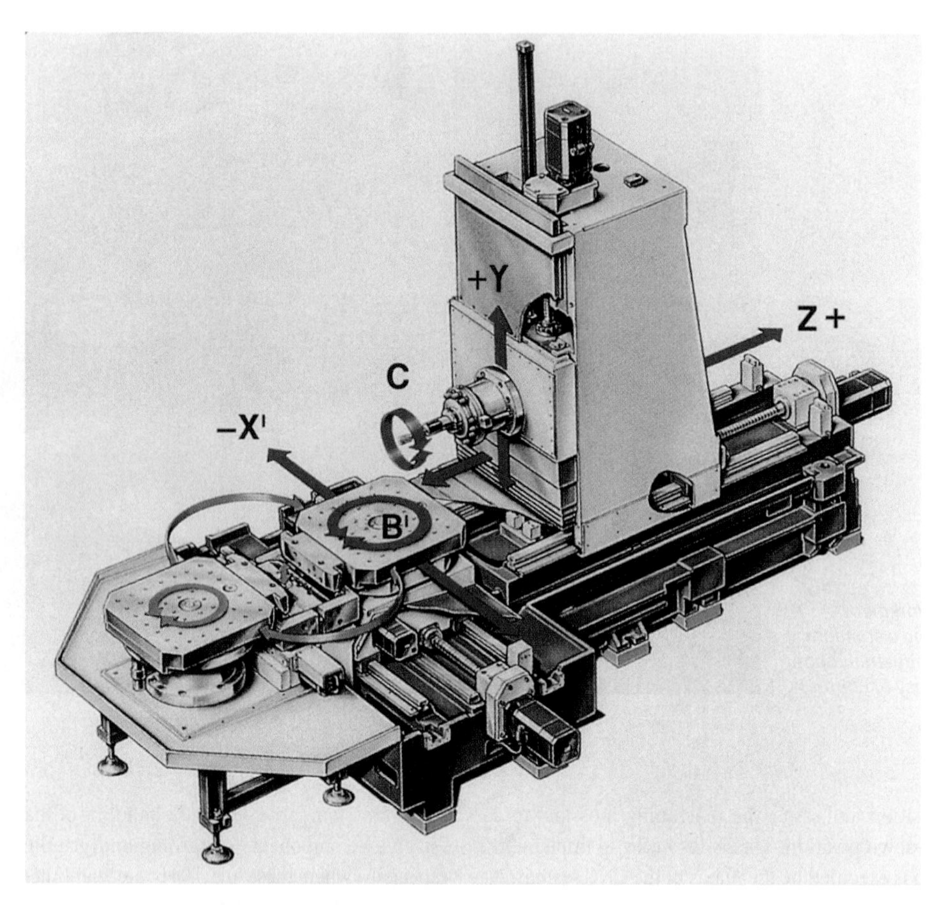

Figure 2.6: Machining center with pallet changer. (Image courtesy of Heckert.)

Pallets are tool carriers whose undersides serve the purpose of precise location and clamping of the surfaces and functional elements suitable for the table of the machining center. But even with carousel turning machines, entire faceplates are exchanged as pallets. The machines are equipped with tool-changing equipment in which the pallet with the workpiece being machined is presented automatically, and a new pallet with an unprocessed workpiece is brought from the waiting station to the working space of the machine. As a result, the clamped workpieces can be exchanged in only a few seconds. Clamping and unclamping then take place during the productive time outside the machine's working space.

In the case of **flexible manufacturing cells** or systems, additional **pallet pools** or linking systems automatically supply new pallets with workpieces to the machines and transport processed pallets to the central clamping station. These thus allow automatic workpiece changing within a manufacturing time that can be as long as possible. With individual machines, a pallet-storage unit offers a number of specific advantages relative to an interlinked system. When four to eight pallets per shift are required, a **pallet pool** is a cost-effective alternative that allows unmanned production over several hours. The price advantage is particularly evident when operation is begun with a single machine, and additional machines have to be added later. In such cases, the expense for the interlinked system only has to be paid when the second machine is acquired.

The pallet pool is controlled via the machine CNC and thus is very simple to program and operate. The linear chain linking requires a separate control unit to process the complex transport tasks. The priorities are also defined here, the individual devices are assigned to specific machines, and the load is simulated.

With two machining centers, the footprint is smaller than with a linear transport system. This is particularly noticeable in machines with an integrated pallet pool for smaller workpieces.

The availability of pallet pools can never be achieved using linear chain linking because the mechanical effort, the range of functions, and thus also the causes of faults are much more extensive. The decisive disadvantage of pallet pools is the fact that they cannot be expanded. It is not possible to expand the system by adding machine tools, washing and measuring machines, automatic tool-delivery systems, and additional clamping locations.

In most cases, the pallets are equipped with coding devices for tracking and motoring. These enable automatic identification of the pallets and thus of the workpieces that are clamped on them and activate the corresponding CNC program in the CNC memory. When used in flexible manufacturing cells, the coding device has to be settable and readable automatically in order, for example, to specify the workpiece number, machine number, and any order that has to be observed for a sequence of machining operations on multiple CNC machines. In these coding devices it is also necessary to be able to determine after machining which machine of the flexible machining system the pallet was located in for machining. This should make it easier to determine which machine or tools are faulty in the event of machining errors, exceeded tolerances, or rejected parts.

2.3 Changing the Speed

The speed of the main drive also has to be changed repeatedly during machining of the workpiece. In turning machines, for any given cutting speed, every change in diameter requires a different speed of the main spindle, which for machines with a recirculating tools means that a different speed is needed at least every time there is a tool change. For this reason, automatic work sequences also require automation of the speed setting.

In conventional machine tools driven by a normal three-phase asynchronous motor, the different speeds were produced primarily by means of stepped toothed gear units and less often via stepless transmissions. But positive-fit sliding-gear units were not very suitable for automation. More suitable were clutch transmissions that could be shifted under load. Accordingly, these were often used in turning machines. These transmissions were so large and heavy, however, that they could not be used for driving spindles mounted in heads, such as in boring mills or machining centers. Hydraulic drives were used frequently in such cases.

This problem became much easier to solve with the advent of steplessly adjustable electric motors. For this reason, these motors gained very rapid acceptance despite their initially very high prices. However, the adjustment range was still inadequate, and a downstream stepped transmission often was necessary to produce the required speed range. If the stepped transmission is designed as a positive-locking gear unit, that is, as a sliding-gear unit or with claw couplings, then the drive has to be stopped to allow shifting. This was done by means of a part program stored in the controller. This is why the preferred solution was to switch such stepped transmissions using electromagnetic multiple-disk clutches. This could be done while the drive was still running.

In the meantime, however, the adjustment ranges of steplessly adjustable electric motors have been developed to such a point that downstream stepped transmissions are now largely unnecessary. For more information on this, → *Part 2, Chapter 3*.

2.4 Summary

The automatic triggering of switching machine functions during automatic work sequences makes it possible to a large extent to perform complete machining of workpieces on a single machine. The most important of these functions are automatic tool changing, automatic workpiece changing, and speed changing.

Automatic tool changing makes it possible to use a number of separate tools when machining a workpiece. Various designs are used depending on the number of tools required. Turning machines require a smaller number of different tools. Here the tool turret is predominantly in the form of a star, crown, or disk turret. Driven tools also can be used here to perform drilling or milling operations.

Milling machines and machining centers require more tools because many tools are specific to a particular dimension or at least a particular shape. As a result, magazines for up to 100 tools have appeared, as well as a very wide variety of tool-changing devices.

A distinction is made between **chain, disk, and circular-type magazines, which are generally limited to 32 tools, and cassette-type magazines,** in which the tools are arranged in a plane.

The simplest and easiest type of tool changing is the pickup method, although this has the disadvantage of long changing times. Systems with a single or especially a double gripper are faster.

Identification of the tools has a great effect on the execution of the tool-changing operation and thus also on the magazine. A distinction is made here among tool coding, tool location coding, and random tool access.

Automatic tool changing extends the automatic work sequence to cover production of multiple workpieces, for example, all the workpieces in a particular lot or even different workpieces of a group of related parts. Production from bar stock is familiar from turning machines. Relatively simple workpieces often can be exchanged without clamping devices using the pickup method or by means of a loading robot. If the workpiece needs a special clamping device, then it can be exchanged in the clamped state using a pallet.

Automatic speed changing is always necessary in turning machines owing to the wide range of turning diameters that may be required. For machines with recirculating tools and automatic tool changing, automatic speed changing, of course, may not be necessary. Today speed changing is implemented mainly using steplessly adjustable electric motors. Additional stepped toothed-gear units are now only necessary if a very large range of speeds is required.

Implementation of Switching Information

Important points to remember:

1. In order to expand an automatic work sequence to cover various operations on a single workpiece, a number of switching functions also have to be automated. These are in particular
   - Tool changing
   - Workpiece changing
   - Changing the speed
2. The decisive criteria for evaluating a tool-changing system are as follows:
   - The number of tools available
   - Limitations with regard to the dimensions and weight of the tools being used
   - The tool-changing time, generally indicated as the *chip-to-chip time*
   - The costs of any additional tool holders or tool blocks that may be required
3. In turning machines, automatic tool changing generally involves a tool turret in the form of a star, crown, or disk turret.
4. In milling machines and machining centers, automatic tool changing can be implemented using a very wide variety of tool-changing devices and magazines.
5. A distinction is made between the following types of magazines:
   - Magazines in which the tools are arranged linearly, such as chain, disk, and circular-type magazines
   - Magazines in which the tools are arranged in a plane, especially cassette-type magazines with up to 100 tools
6. The identification of tools is also important. It has a major effect not only on searching for tools in the magazine and on the changing operation but also on the amount of work required for changeover of the machine. The following distinctions are made:
   - Tool coding
   - Location coding
   - Random tool access
7. Automatic tool changing expands the automatic work sequence to include the machining of multiple similar workpieces or even different workpieces. Various options exist for workpiece changing:
   - Production from bar stock
   - The pickup method
   - Exchanging in the clamped state with a pallet
8. Automatic speed changing is always necessary in turning machines. For milling machines and machining centers with tool changing, it may not be necessary. Automatic speed changing is implemented using steplessly adjustable electric motors.

3. Functions of Numerical Control Systems

Thanks to the integration of computer technology, numerical control systems are becoming ever smaller, faster, more powerful, and more user-friendly. Since the introduction of the first computer numerical control (CNC) systems in 1975, more and more new functions and tasks have been added, especially with the aim of improving the degree of automation and the reliability of CNC machines. Computer-based control systems have extensive functions that increase the productivity of the machines they control and the people who use them.

3.1 Definition

CNC refers to numerical control systems containing one or more **microprocessors** for the execution of control functions. The external characteristics of a CNC are a screen and a keyboard (*Figure 3.1*). The operating system of the controller, also called the **CNC software,** comprises all the necessary functions, such as interpolation, position feedback control, velocity control, displays and editors, data storage, and data processing. Also required is an **interface program** to the machine being controlled. This program is created by the machine manufacturer and is integrated into the interface programmable logic controller (PLC). All the machine-related links and interlocks for special functions are defined there, such as those for tool changing, workpiece changing, and axis limitations.

Workpiece-specific control of the machine motions during machining is performed according to the **part programs.** These are created by the machine user and are **not** a part of the CNC software.

3.2 Basic Functions of CNC

The typical task of numerical control systems, that is, to precisely control the relative motion of the tool and workpiece in a machine tool, is being augmented by more and more tasks and functions. While some of these run in the "background," for example, to monitor safety, others demand attention and occasional intervention on the part of the operator. For this reason, it is important for the controller to be clearly organized and easy to operate. All the more so because CNC has turned simple control systems that could understand simple numerical input into complex data-processing process control computers with completely new functions. These will now be introduced and explained briefly.

Today, a typical **basic CNC system** includes the following:

- A large color **graphics monitor** (→ *Figure 3.1*) for display, programming simulation, operation, and diagnostic functions
- **Operator prompting with interactive dialogs** and the ability to switch between at least two languages
- A **program memory** for a number of part programs, compensation values, tool data, zero-point tables, and cycles
- A **bus-coupled or integrated PLC** with a high processing speed for controlling the switching functions
- **Programmable software end limits** for the NC axes, thus replacing mechanical limit switches and their associated wiring
- **Manufacturing data acquisition (MDA)** and **production data acquisition (PDA)** and an automatic logbook for documentation of operator errors, fault messages, function sequences, warnings, and manual interventions

On top of this are **new functions** that make the machines more precise, more reliable, and more user-friendly, such as

- **Temperature error compensation** for machine inaccuracies owing to thermal effects
- **Random tool access** to accelerate the search and changing operations
- **Tool breakage and tool life monitoring** for automatic operation
- Automatic reading of **tool data** into the compensation-value memory

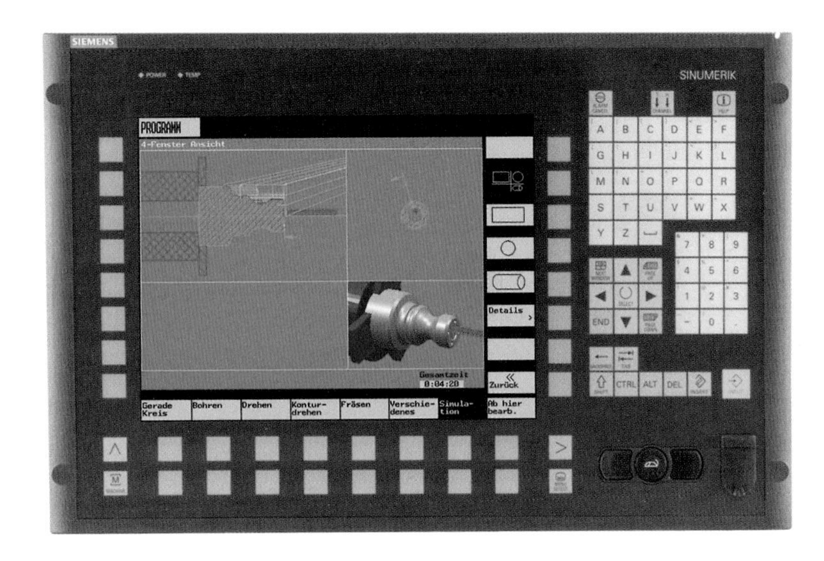

Figure 3.1: Four-window view of an operator panel.

- Simultaneous control of synchronous **main axes** and asynchronous **auxiliary axes** with no wait times
- Inputting **machine parameter values** via keyboard instead of tedious adjustment work during commissioning—and much more

The numerical control system performs a large number of **additional functions and tasks** in the automatic production sequence. These functions are now assumed as a matter of course.

Some of these special functions will now be enumerated and explained. Control systems from different manufacturers, however, may have different names for the same functions or may have different operational sequences or performance ranges.

Asynchronous Axes

These are auxiliary or secondary axes that are not interpolated with and move independently of the main axes (e.g., tool or workpiece handling devices in a machine).

Automatic Return to the Contour *(Figure 3.2)*

After a broken tool or emergency stop during machining, the tool has to reenter the interrupted program before the point where the break occurred, thus resuming the machining without making any marks on the workpiece. The new tool compensation values also have to be taken into account here.

Block Delete, Skip Block

When a program is processed, blocks that have a slash in front of the block number (/N147 X . . . Y . . .) can be executed or skipped depending on the switch setting. This can be used to activate programmed probing cycles or a machine stop. If this function is switched off, then all additional sections are executed without these interruptions.

Block Number Search

A time-saving option after a program interrupt to let the program run through quickly to a selected block number without any machine motion so that on reentry into the program the correct tool is available together with all compensation values, the correct feed velocity, and the right spindle speed.

Compensation Values

These are current tool data that have been input and saved for each tool located in the machine (e.g., diameter, length, and

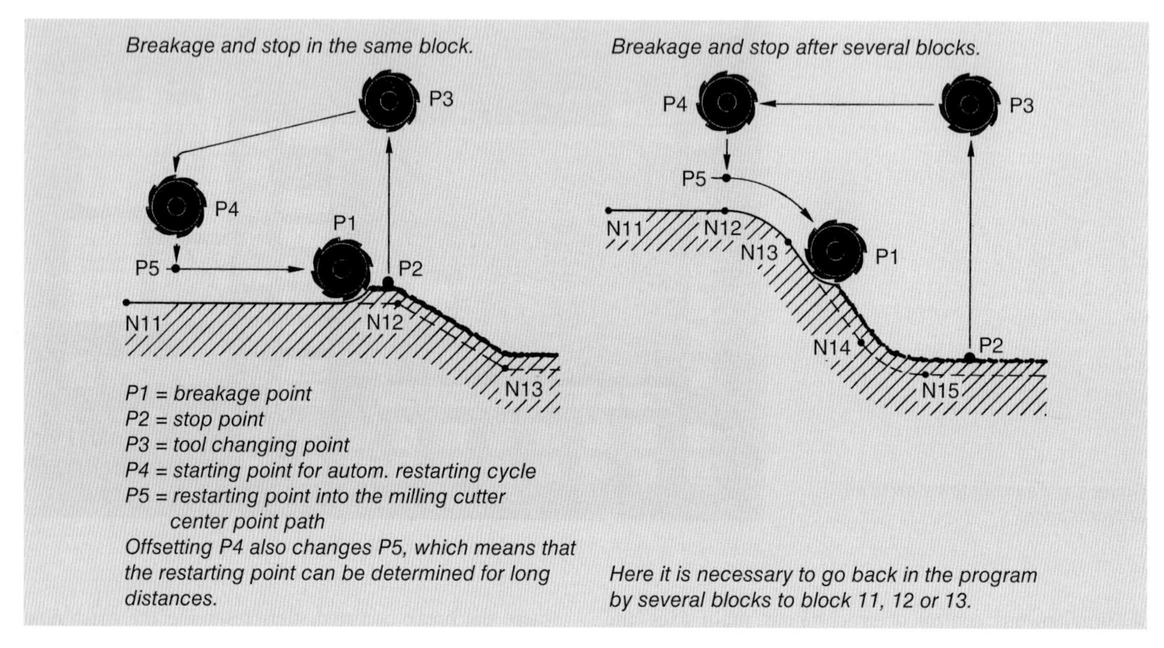

Breakage and stop in the same block.
P3
P4
P1
P5
P2
N11
N12
N13
P1 = breakage point
P2 = stop point
P3 = tool changing point
P4 = starting point for autom. restarting cycle
P5 = restarting point into the milling cutter center point path
Offsetting P4 also changes P5, which means that the restarting point can be determined for long distances.
Breakage and stop after several blocks.
P4
P3
P5
N11
N12
N13
P1
N14
P2
N15
Here it is necessary to go back in the program by several blocks to block 11, 12 or 13.

Figure 3.2: Restarting to the contour, automatic restarting cycle after milling cutter breakage.

radius). These data are taken into account during program execution. Compensation values also include error compensations, zero-point offsets, clamping tolerances, and wear values.

Cutting the Tool Free

At the end of a machining operation, the feed stops for a programmable time while the spindle continues to turn, only after which the tool is retracted.

Data Interfaces *(Figure 3.3)*

These are the interfaces for connecting CNC systems to higher-level computers to allow data exchange or execution of remote-control functions. Such interfaces are also required for automatic workpiece-identification and tool-identification functions.

Deactivating Axes

The selective stopping of some or all CNC axes allows quick testing of a CNC program for programming errors without moving the axes in question. To save time, it is also possible to deactivate tool changing, pallet changing, coolants, and the main spindle.

Diagnostic Software

This refers to functions that can be activated permanently or via a program to monitor machine and control performance for the purpose of automatic documentation of errors and their causes. The CNC system facilitates this by depicting the measured values on the screen in the form of curves, diagrams, or digital data. All data also can be output via an interface.

In addition to error diagnosis, controller manufacturers also offer special diagnostic software that supports the user in optimizing part programs. This makes it possible to reduce the processing time (cycle time) significantly.

Example:
If the PLC detects a broken tool, tool-changing motions can be performed using an asynchronous subprogram. Once

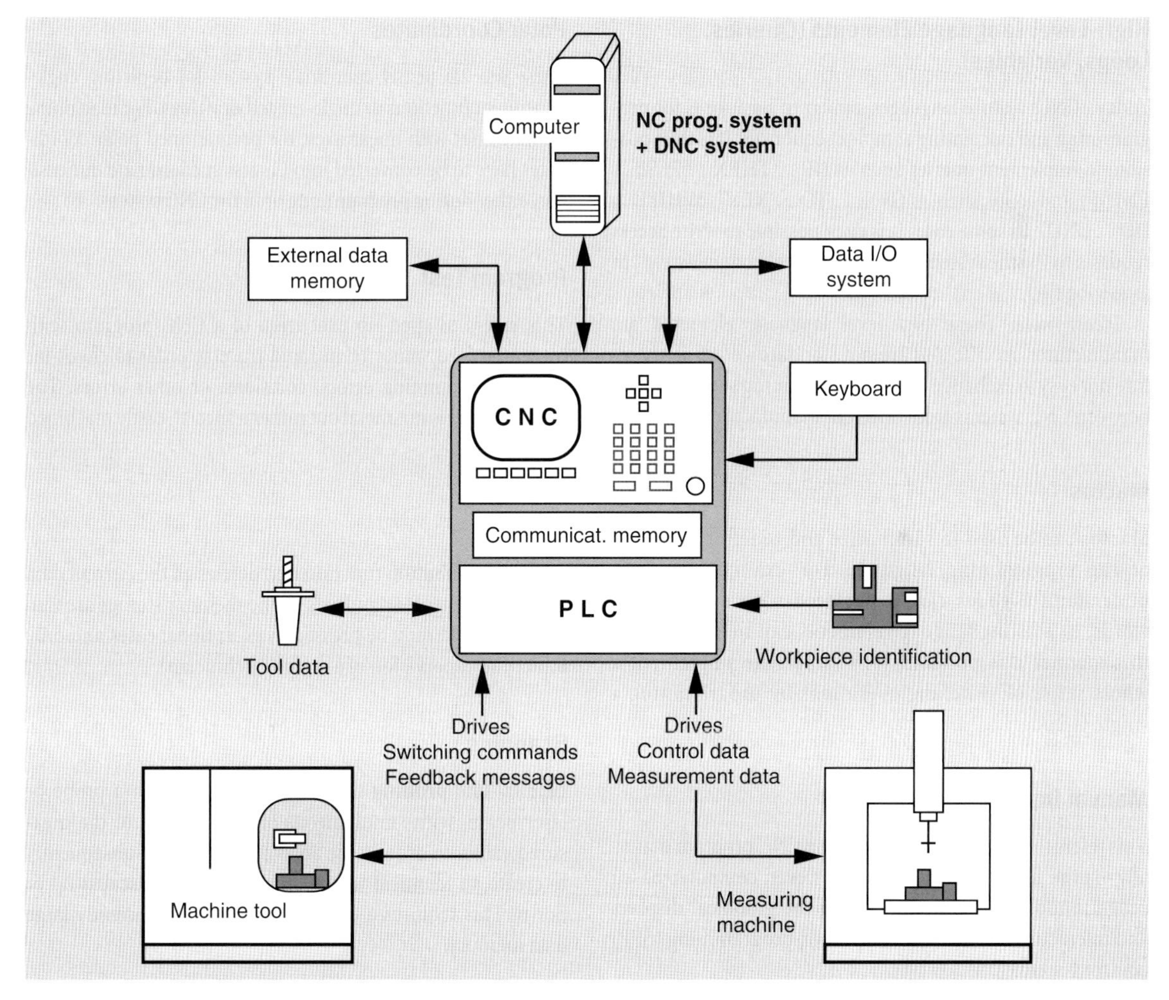

Figure 3.3: Data interfaces of a CNC/PLC for transmission of various production-related data.

there, the damaged tool is exchanged for a new one, and machining is continued from the last position.

Driven Tools

This term refers to tools such as drills or milling cutters that are used in turning machines to perform machining operations on stationary workpieces. As such, they require their own drive. To do this, the main spindle has to have continuous path control (*C* axis).

Energy Efficiency

Some current CNC systems have programs for analyzing energy consumption. For example, the switching times of the supply modules can be used to determine and record their energy consumption. This can help the machine manufacturer in designing supply modules so that they are rated properly for a specific application. The user has the opportunity to optimize work sequences and part programs in order to avoid wastage of energy. This especially pays off in large-scale series production.

High-Level Language Elements (Queries, Loops, Variables)

Today's CNCs feature languages similar to Basic or C for programming and calculating complex sequences. It is thus possible to implement queries such as (IF . . THEN . . ELSE . . ENDIF) and loops such as (FOR . . TO . . NEXT, WHILE . . DO . . END). In some cases it is even possible to write access requests to the file system in this high-level language (e.g., to save a log file).

Please note: These high-level language elements are manufacturer-specific and are not standardized. It is therefore not easy to exchange programs containing such elements between CNC systems from different manufacturers.

Macros

Macros can be used to summarize and redefine elements of the programming language. For example, hard to understand G-codes can be replaced with easy-to-read words, or existing language elements can be overlaid. In this manner it is possible, for example, to implement a whole series of switching operations by programming a single G-code.

Manual Input

This refers to typing in and correcting a CNC program manually via the CNC system's keyboard or even computer-based programming on the machine using the graphical displays and interactive dialogs of an shop-floor-programming (SFP) control system.

Mirror Imaging, Rotating, Shifting

The programmed dimensional data can be mirrored onto a specified axis, rotated, and/or shifted by a defined distance. This makes it easier to program parts whose geometry repeats, for example.

Offset

This refers to electronic compensation for clamping tolerances in the workpiece or tools in order to eliminate the need for mechanical alignment or adjustment.

Polar Coordinates

These are 2D or 3D coordinate system for depicting angle-dependent functions or angle-related drawings. For machining on machines with linear axes, the programmed polar coordinates have to be converted into Cartesian coordinate dimensions, either during programming or in the CNC system.

Program Test

This refers to sped-up execution of a CNC program with increased feed values or in rapid traverse order to check for major programming errors, collisions, or other errors. The material used is not metal but rather a special, easily machined plastic.

Reset

This is a pushbutton function that deletes all the current data in the main memory and sets all the memory storage areas to ZERO (should only be pressed in exceptional cases and only under the supervision of a service technician).

Scanning

Line-by-line **probing** of a shaped surface with a probe or laser beam, with simultaneous ongoing saving of the measured values in order to be able to use the data subsequently to create an identical workpiece or one that is scaled up or down. This requires a CNC system with a sufficiently large data memory.

Setting the Position (Position setup)

Using a dial gauge or other aid, the operator aligns the spindle center point with a fixed point on the workpiece or the fixture and sets the axis position to the values indicated in the drawing or the CNC program.

Simulation

This is graphical depiction of the machining operation (travel paths of the tools) and the final workpiece, taking into account the tool-compensation data and the geometry of the unfinished part. Depending on the type of control system, it may be possible to simulate the entire work sequence and display it as a three-layer view or as a solid. By calculating the complete program, it is possible to detect error sources and estimate the machining time in advance. The simulation is executed directly on the machine with the CNC system (→ *Part 5, Chapter 4: Manufacturing Simulation*).

Slope

This is a settable acceleration and braking response of CNC axes in order to prevent impacts and wear on the mechanical elements. It is important to set all the axes to the same value in order to prevent path deviations.

Synchronous Axes

These refer to all CNC axes of a machine whose movements are interpolated and coordinated simultaneously. As a rule, this includes all the main axes of a machine (antonym: asynchronous axes).

Subprograms

These include permanently stored programs such as whole patterns, drilling cycles, tapping cycles, and milling cycles, including the necessary data (parameter value). These programs can be called up and executed as often as desired (also called *parameterizable subprograms*).

For a detailed description of this function, please consult the documentation for the specific CNC system in question.

3.3 Special Functions of CNC

As a general rule, the performance and expandability of a CNC system are defined by the manufacturer during the initial concept and development of the system. In addition, new CNC concepts have an **open software interface** to the CNC system software. This gives the machine manufacturer and user the capability to integrate special functions or their own know-how at a later point. For this purpose, the CNC system includes special programming software that can be used to integrate special solutions of this kind. With this software, it is even possible to access the control system's graphics in order to visually depict user help systems, selection menus, or dynamic simulations. It is no problem, for example, to take a CNC system for machine tools and reconfigure it as a master controller for a pallet-transport system. What is more, this makes it possible for the machine manufacturer to test newly developed machines at an early stage without having to inform the CNC system manufacturer.

Let us now take a look at some of the special solutions that can be implemented with modern CNC systems.

Axis Change *(Figure 3.4)*

Axis change makes it possible to convert CNC programs created for milling machines with a vertical spindle to use them on machines with a horizontal spindle and an angle head placed in front (exchanging the *Y* and *Z* axes).

Working-Area Limit

The enabled working area of a CNC machine can be limited temporarily by programming the upper and lower limit values for each axis. If a displacement that lies outside these "software limits" is entered, then an error signal is generated, and the machine stops immediately.

Example:
N1 G25 X100 Y255 Z70 $
= lower-limit values for the *X*, *Y*, and *Z* axes
N2 G26 X440 Y321 Z129 $
= upper-limit values for the *X*, *Y*, and *Z* axes

Asynchronous Subprograms

It is possible to define a (small) subprogram in the CNC system to interrupt the normal execution and perform special functions. This subprogram can be triggered via the PLC or another channel.

Example 1:
Two work units of a machine have an overlapping working space. If one of them has to move into the working space of the other one, then it can use an asynchronous subprogram to interrupt its work and move out of the way. It then can continue from its last position once the first unit has exited its working space.

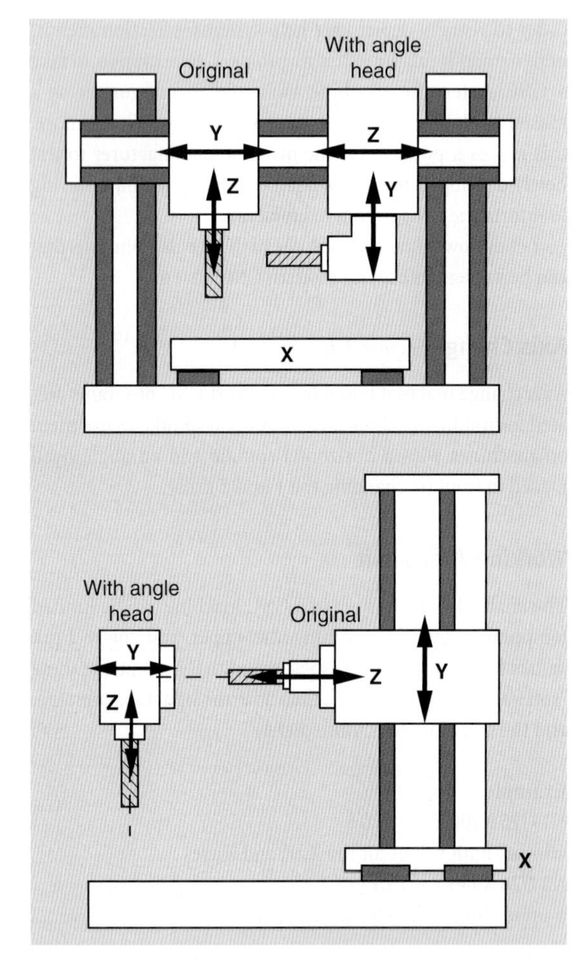

Figure 3.4: Exchanging axes. When the angle head is placed in front, it is possible to exchange the Y and Z axes in order to use NC programs that were created for a vertical Z axis.

Example 2:
If the PLC detects a broken tool, tool-changing motions can be performed using an asynchronous subprogram. Once there, the damaged tool is exchanged for a new one, and machining is resumed from the last position.

Automatic Tool Length Measurement *(Figure 3.5)*

After a tool is inserted, a measuring cycle is executed. This is done by moving to a measuring probe to determine and

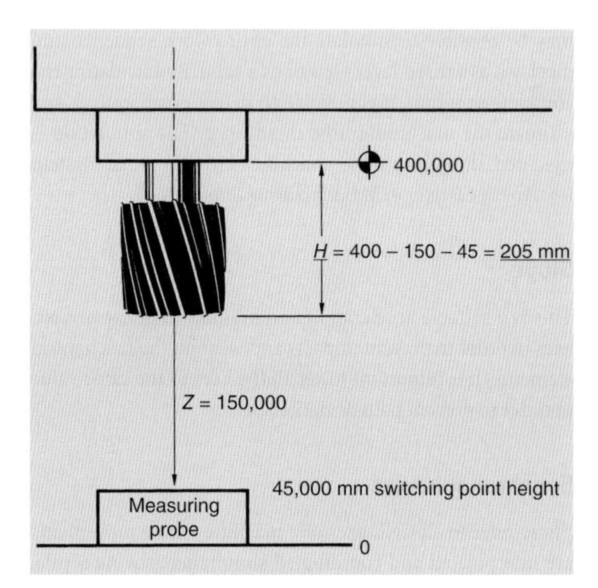

Figure 3.5: Automatic tool length measurement by means of an encapsulated switching measuring probe.

save the absolute tool length. This is especially advantageous and economical in terms of time for multiple-spindle milling machines.

Automatic System Diagnoses, NC Analyzers

These refer to special software for switching the CNC screen over to oscilloscope operation. This can be used as a convenient method for testing NC programs and finding any errors that may be present in the programming or during processing.

The NC analyzer answers, among other things, the following questions:

- Where in the program did the error appear for the first time?
- What effects does the error have on the program?
- What are the effects on other variables or program sections?
- How important is the error for the program?

Block Cycle Time *(Figures 3.6 and 3.7)*

In order to achieve high surface quality and contour accuracy, the CNC system has to execute the CNC program very

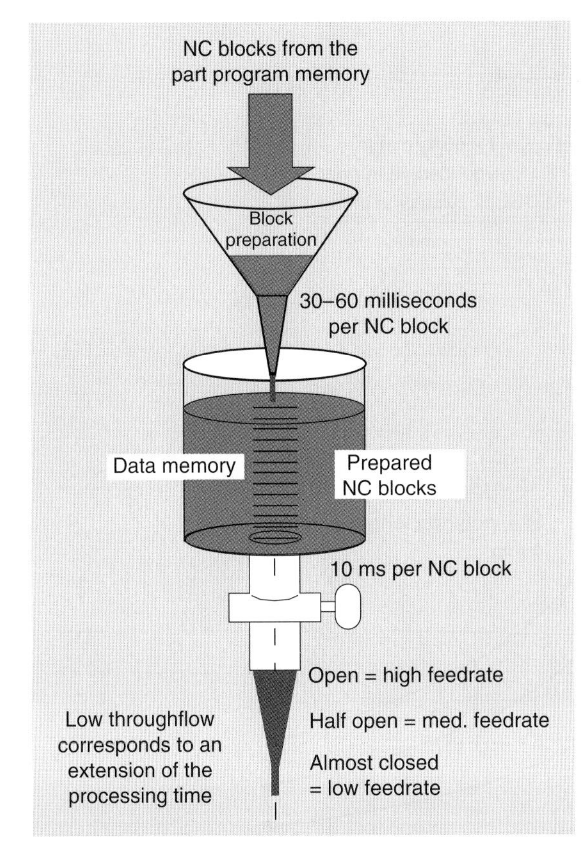

Figure 3.6: Dynamic data-buffer memory with short transmission time in order to ensure short block cycle times.

quickly and without feed fluctuations. If the execution time of a block is shorter than the preparation time for the following block, then the feed rate will be reduced. For this reason, the CNC system has to have a high computing speed and a constant supply of prepared blocks. A dynamic buffer that is constantly refilled will keep an adequate number of prepared blocks on hand, thus preventing so-called stuttering axes.

If this supply is still not adequate, then the feed velocity has to be reduced until the axis moves more slowly but in any case continuously. The interrelationships among the block cycle time, polygon length, and feed velocity are shown in *Figure 3.7.*

Example:
For a polygon length of 0.1 mm and a block cycle time $t = 2$ ms, the maximum feed rate $F_{max} = 4$ m/min.

DNC Interface *(→ Figure 3.3)*

This refers to automatic reading-in and reading-out of part programs, tool compensation, PLC data, status and error messages, and so on. To do this, the CNC system has to have suitable data interfaces.

High-performance DNC interfaces also allow computerized remote control of the machine, for example, when moving to zero points, deleting individual programs, sorting tools in the magazine, and so on (→ *Part 6, Chapter 1: Direct Numerical Control*).

Dynamic Collision Monitoring *(Figure 3.8)*

Complex machine motions in five-axis machining and a high traversing speed make it hard to predict collisions. Therefore, graphic dynamic **collision monitoring in the CNC system** provides a useful function that reduces the load on the machine operator and protects the machine from damage.

Although NC programs from computer-aided design (CAM) do avoid collisions between the tool/tool holder and the workpiece, they generally do not take into account machine components located in the working space—unless an investment is made in external machine simulation software. This does not guarantee, however, that the conditions on the machine (e.g., the clamping position) are exactly the same as in the simulation. In the worst case, collisions will be detected only during machining. In such cases, the controller will interrupt the machining process. This prevents damage to the machine and expensive downtime. Unmanned shifts thus can operate more safely and reliably.

The collision monitoring functions both in **automatic mode** and **manual mode.** When a workpiece is being set up, if the machine is on a "collision course" with a part in the working space, then the axis motion is stopped, a fault message is generated, and the impending collision is indicated.

The software required to describe the **machine components** and the **working space** is provided by the machine

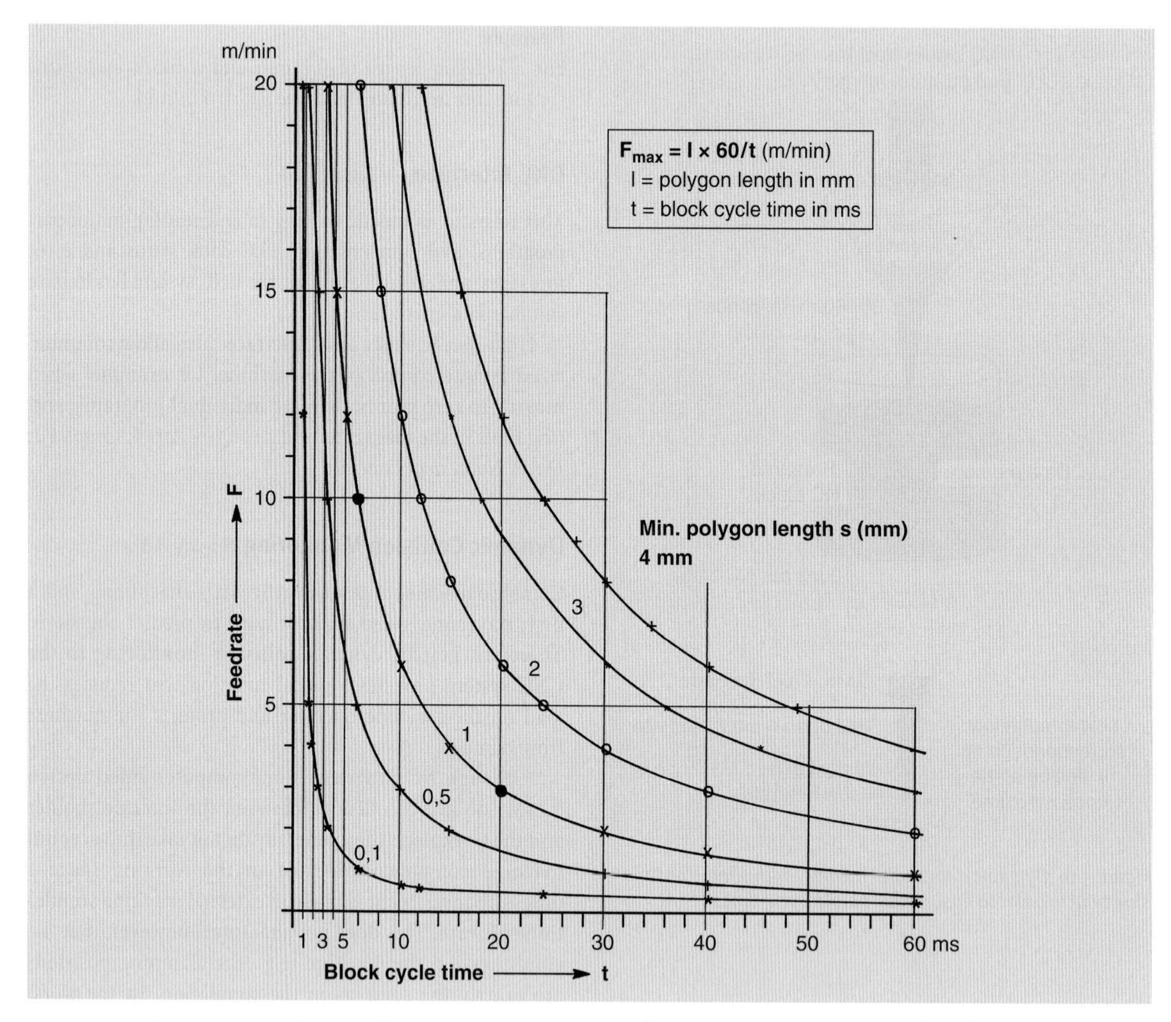

Figure 3.7: Relationship of the maximum achievable feed rate F_{max} to the polygon length S of a curve and the block cycle time t of the controller.

manufacturer. The collision objects are described by means of simplified geometric bodies. The tool is taken into account, for example, as a cylinder with a tool radius (defined in the tool table). For swivel fixtures, the machine manufacturer can use the tables for the machine kinematics to define the collision objects at the same time. In the end, it is determined which machine elements could collide with each other. Because it is impossible for certain bodies to collide with each other owing to the construction of the machine, it is not necessary to monitor all the machine components (→ *Part 5, Chapter 4: Manufacturing Simulation*).

Feed Deceleration at Corners *(Figure 3.9)*

In pocket milling, every time the milling cutter enters a corner, an overload occurs that can damage the tool and the workpiece. Therefore, a so-called corner deceleration is

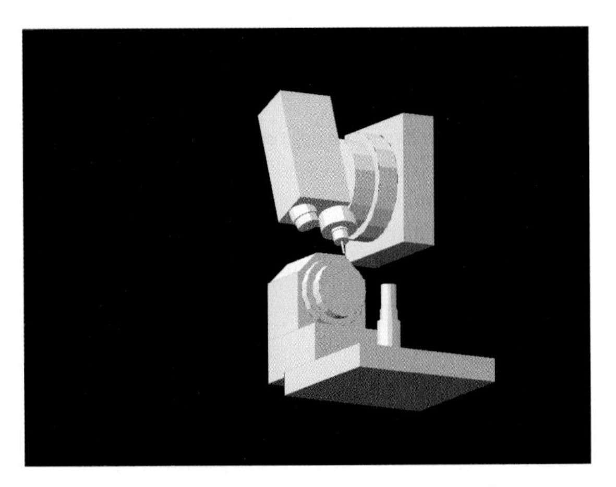

Figure 3.8a: Collision monitoring in a CNC system. (Image courtesy of Heidenhain.)

Figure 3.8b: Collision monitoring in a real CNC system. (Image courtesy of Heidenhain.)

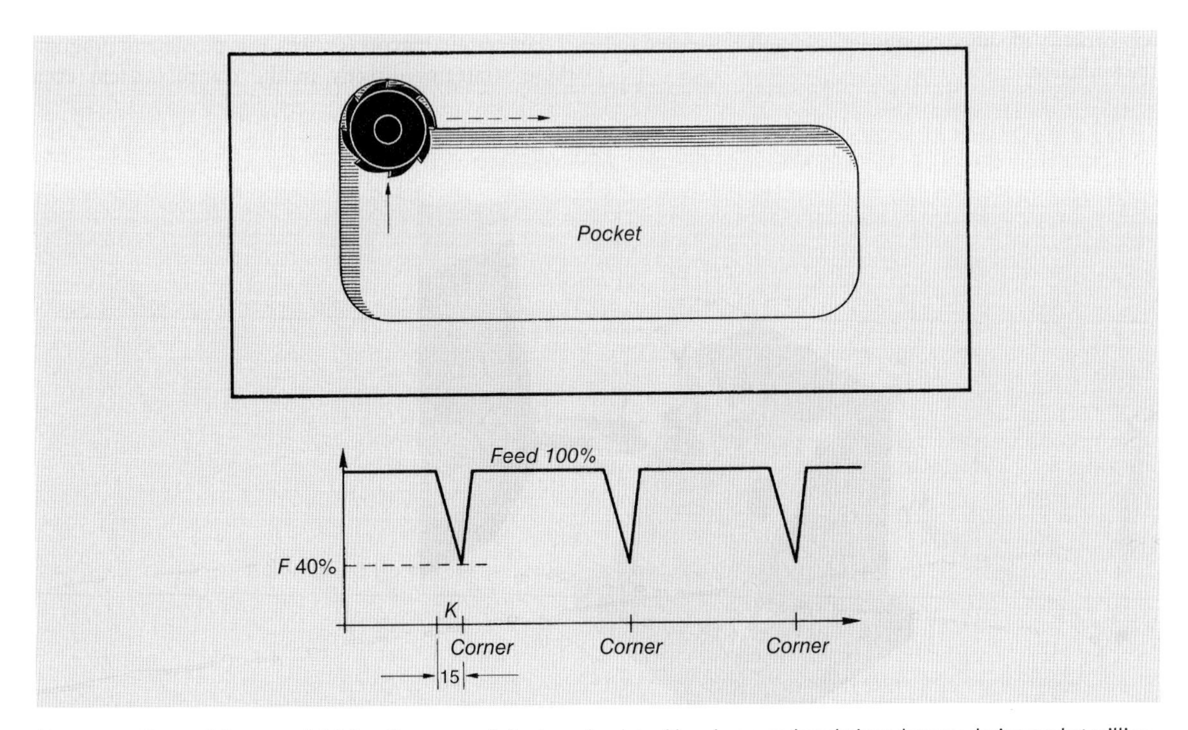

Figure 3.9: Corner delay, a modal G function, prevents tool overload, tool breakage, and workpiece damage during pocket milling.

programmed. This automatically reduces the feed rate to the predefined value at each corner, avoiding overloading the milling cutter.

Example:
N123 G28 K15 F40 $
That is, reduce the feed to 40 percent before each corner point.

These values also can be input manually; G29 switches the corner deceleration off again.

FRAME or Coordinate Offset *(Figure 3.10)*

FRAME is the commonly used term for a coordinate transformation, such as offset and rotation. In order to machine diagonal contours, it is either necessary to use appropriate fixtures to align the workpiece parallel to the machine axes or else, for example, with five-axis machines, to create an appropriately modified coordinate system that is related to the workpiece.

The coordinate system can be shifted or rotated by means of a program using programmable FRAMES. These can be used

- To shift the zero point as desired
- To rotate the coordinate axes and align them parallel to the desired working plane
- To machine diagonal surfaces, drill holes at various angles, and perform multisided machining, all in a single clamping operation

Milling Strategy

This refers to generation of optimized milling paths for special machining tasks. In order to achieve optimal cutting conditions, in CAM systems it is possible for the milling path to be influenced by various milling strategies. The service life of tools can be increased through the use of helical or trochoidal paths, improving the surface and allowing the machining of hard workpieces (up to 65 HRC).

For certain macros (cycles), it is possible to program these strategies directly in the control system even

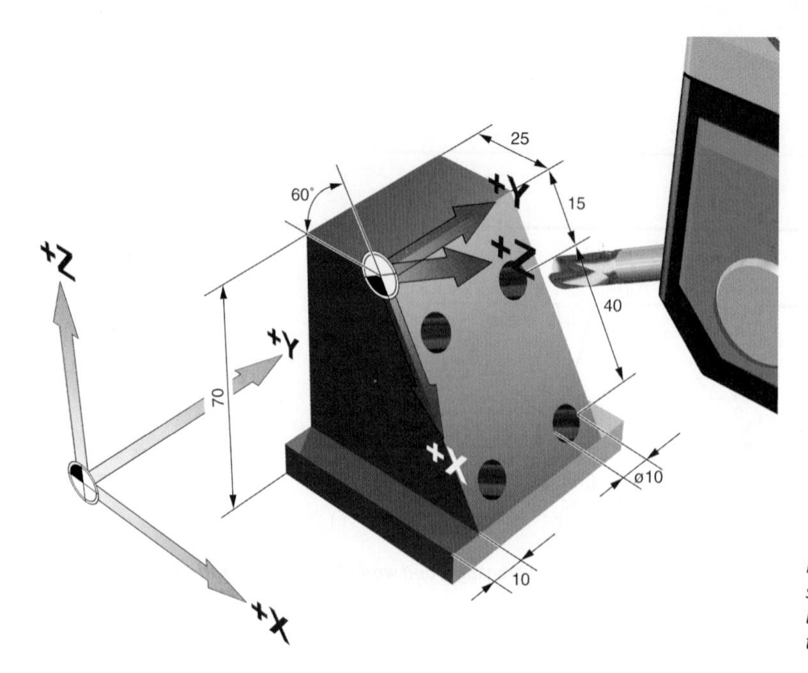

Figure 3.10: Machining of an inclined surface—with high-performance controllers, it is possible to program five axes at the machine.

without a CAD system. Examples of this are macros for pocket milling with helical machining, slot milling with a trochoidal strategy, or plunge milling for large chip removals.

Fuzzy Logic (FL)

This should be understood as an open- or closed-loop control method that takes into account the variable nature and unreliability of human reason and human reactions, such as "somewhat higher," "faster," or "increase rotational frequency and reduce feed rate," in order to keep a process stable despite a large number of interferences. In machining operations, instead of executing mathematically precise responses, **fuzzy logic** decides quickly in favor of one of the specified alternatives and is able to coordinate a number of optimization interventions within a short period of time. This requires constant monitoring of the process and feedback of process parameters.

In the mechanical engineering field, FL was first used in CNC die-sinking EDM machines. There the FL assumed responsibility for automatic control of the electrical discharge machining (EDM) process, thus essentially replacing the analog adaptive control functions that were used previously.

FL also can be used to optimize other machining processes to achieve time savings or greater reliability. The more complex the process or interventions required to optimize it, the more conspicuous are the advantages of fuzzy logic compared with analog control technology.

Tapping Without Compensating Chucks

This function makes it no longer necessary to use compensating chucks that place unnecessary limits on the drilling range (i.e., the drilling depth that can be achieved). By advancing the *Z* axis with interpolation depending on the thread pitch, it is possible to cut threads in blind holes precisely to the final drilling depth, as well as changing the direction of rotation of the spindle and turning the tool out of the thread without exerting any compressive or tensile forces on the screw tap.

Thread Milling *(Figure 3.11)*

When creating internal and external threads with form cutters, helical interpolation is necessary. This consists of two motions: A circular motion in a plane (*X*, *Y*) and a linear motion perpendicular to this plane (*Z*). Here the milling cutter's approach to the workpiece and the feed in the *Z* axis has to correspond to the thread pitch.

Channel Structure

In complex machines with multiple synchronous work units, the CNC system's channel structure is used. Each separate work unit (channel) has a number of (synchronous) axes assigned to it. These axes are interpolated together. Each channel executes its own program and is independent of the other channels.

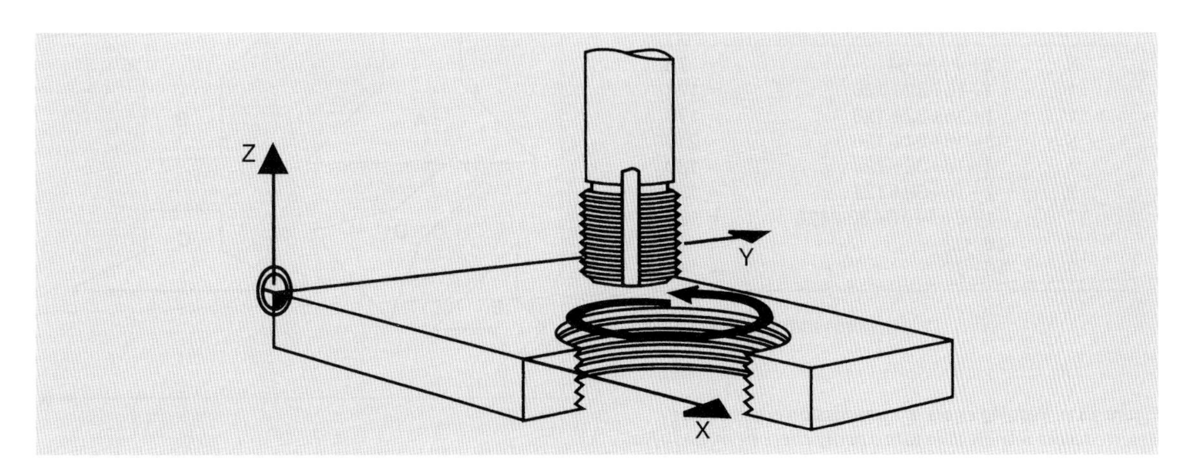

Figure 3.11: Thread milling with form cutter and helical interpolation.

If necessary, individual axes can be taken away from channels and reassigned to other channels. It is thus possible to adapt the logical structure of the machine to the manufacturing task (→ *NC Auxiliary Axes below*).

Detecting Read Errors Using E-Codes *(Figure 3.12)*

In order to detect transmission errors when data are entered into the CNC system, each character undergoes binary evaluation, and the evaluation sum (which is similar to a checksum) is added automatically by the postprocessor at the end of each block after the address E. When the block is read in, the CNC system performs the same calculation, compares the two values, and releases the block for execution only if they correspond to each other. This error-code check is significantly more reliable than the parity check (checking for an even or odd number of bits per character) and is also effective in DNC operation.

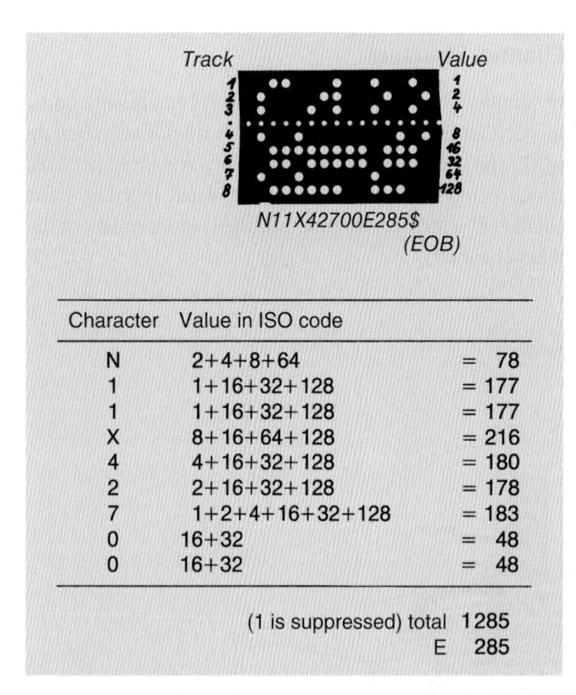

Character	Value in ISO code	
N	2+4+8+64	= 78
1	1+16+32+128	= 177
1	1+16+32+128	= 177
X	8+16+64+128	= 216
4	4+16+32+128	= 180
2	2+16+32+128	= 178
7	1+2+4+16+32+128	= 183
0	16+32	= 48
0	16+32	= 48

(1 is suppressed) total 1285
E 285

Figure 3.12: Reading errors or transmission errors can be detected more reliably than with a parity check by means of an additional E code (E = error). Precondition: Computerized programming with automatic E-code output in each block.

Scaling Factor *(Figure 3.13)*

All the programmed dimensions of a CNC program can be converted using any desired factor—even differently for each axis. In this manner, it is possible to use a single CNC program to manufacture different geometrically similar parts. High-performance CNC systems allow an additional rotational motion by any desired angle.

Scale Error Compensation *(Figure 3.14)*

This is accomplished by measuring the natural error curves $\Delta I = f(I)$ for each CNC axis, and the resulting compensation values are entered in a compensation value memory. The CNC system automatically takes these compensation values into account when the axes are moved. The accuracy that is achieved is thus greater than the accuracy of the measuring system.

Probing Cycles *(Figure 3.15)*

These are process cycles that are stored in the CNC system and used for automatic measurement of boreholes, grooves, or surfaces using a switching sensing probe and to immediately calculate positions, machining accuracies,

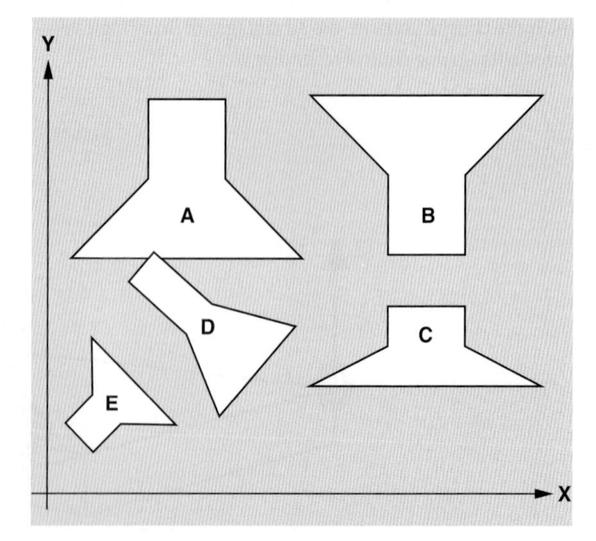

Figure 3.13: Scale factor, for example, enlargement, reduction, or distortion of parts using programmable scale factors.

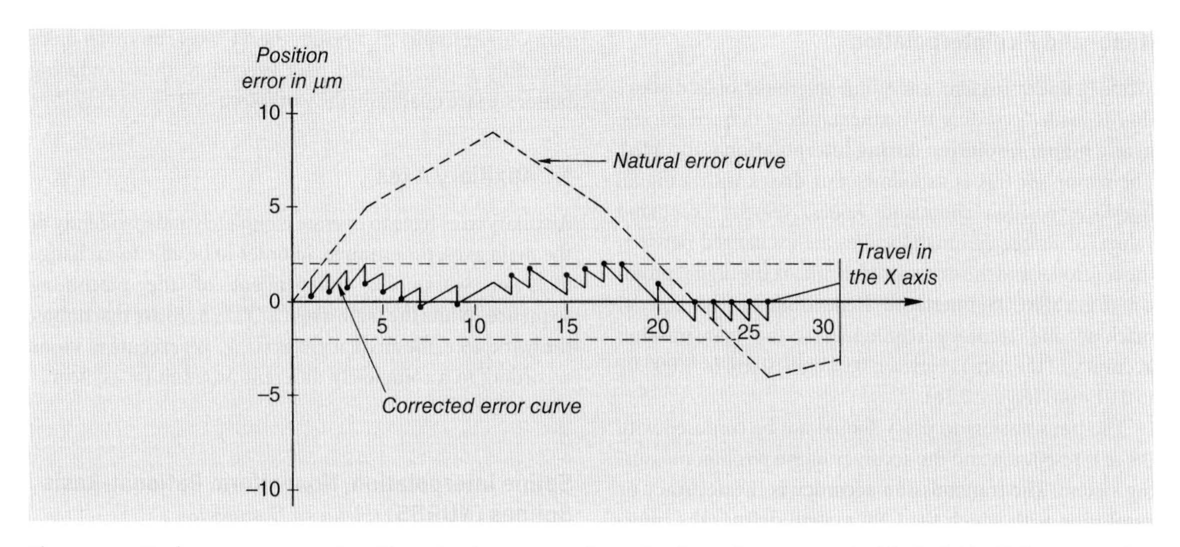

Figure 3.14: Scale error compensation. The natural error curve for each axis can be compensated for individually by means of stored compensation values that are taken into account automatically.

tolerances, centers of circles, center distances, or inclinations of the workpiece. The measurement values thus determined can be output via the data interface. In process measurement, a measurement operation can be used during execution of the CNC program to check compliance of the workpiece with tolerances and, if necessary, to automatically perform correction of the program or tool.

The use of measuring probes requires special calibration and probing cycles in the CNC in order to obtain precise measurement values.

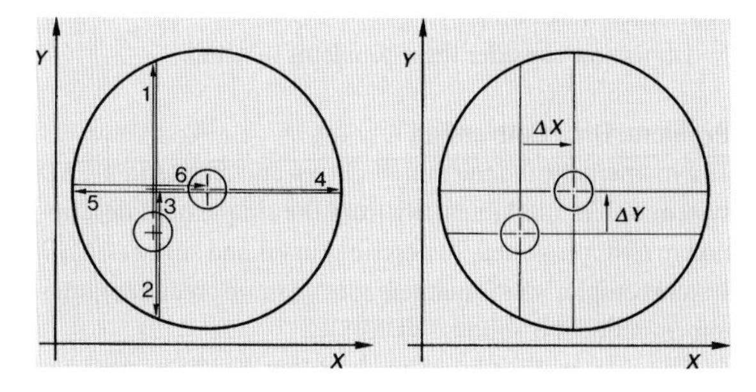

Figure 3.15a: Automatic measurement cycles.

Figure 3.15b: A high-precision measuring probe for machining centers for workpiece monitoring.

Nano- and Picointerpolation

While in linear, circular, and spline interpolation, a distinction is made according to mathematically defined curves; here, a **higher resolution during interpolation** is involved. The reason for this is as follows: For drives with a digital interface, position command values, velocity command values, or torque command values are exchanged between the controller and the drive control unit via the digital protocol (e.g., SERCOS interface). Depending on the type of machine and accuracy requirements, the **transmission accuracy** of the data is between 0.01 and 10 μm (= 0.000.01 to 0.01 mm) (*Figure 3.16*).

The transmission accuracy should not be confused with the axis resolution and the accuracy of the position measuring system. **The transmission accuracy is the accuracy or resolution with which the CNC system defines the interpolated command values** and to which the **drive control unit has to respond.**

To achieve better drive-control characteristics, in modern systems this transmission accuracy is enhanced to the nanometer or picometer level. In **nanointerpolation,** the point specification is 1×10^{-9} m (0.000001 mm), whereas in **picointerpolation** it can be 6×10^{-12} m (0.000000006 mm), for example. As a result, the NC axes run much more smoothly in interpolated path execution, thus producing better surface quality on the workpiece.

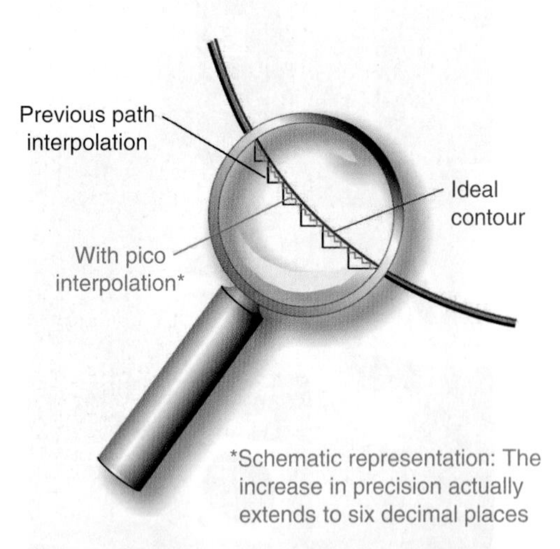

Figure 3.16: Principle of nano- and picointerpolation. The finer resolution of the path in combination with digital drives produces smoother motion of the axes and better workpiece surfaces. (Image courtesy of Andron, Wasserburg, Germany.)

NC Auxiliary Axes

Auxiliary axes have to operate completely independently of the main axes, for example, in order to be able to exchange tools or workpieces independent of the machining sequence. While the main axes (X, Y, Z, A, B) are machining the workpiece, the auxiliary axes (U, V, W) execute motions according to a completely different program (= asynchronous axes).

Spline Interpolation, Nonuniform Rational-Basis Splines (NURBS) *(Figures 3.17 and 3.18)*

This is the concatenation of higher-order mathematical curves with tangential transitions. In this manner, it is possible to depict complex curve forms with fewer NC blocks than is the case with approximation using traverses and linear interpolation. The tangential transitions allow the axes to move more "smoothly." Splines can only be programmed using programming systems that have been configured for that purpose.

Spline interpolation also includes the capabilities of parabolic interpolation.

For geometrically demanding surfaces with small tolerances, such as turbine blades, integral parts for aircraft, or casting molds, the linear machining used primarily today in 3D machining leads to a **number of problems.** These involve primarily the block cycle time of the CNC system, the acceleration jumps of the axes, and the vibrations of the drives.

Appropriate solutions have to be found here, especially for high-speed machining.

Let us now consider these problems in detail.

Problem: Data Conversion

The mathematical depiction of curves and shapes in a CAD system differs fundamentally from the simple path description in CNC programs. To describe curves and surfaces, CAD systems employ spline mathematics or, more precisely, nonuniform rational B-splines (NURBS).

This is a mathematical method that has been used for years now to describe curves and free-form surfaces using

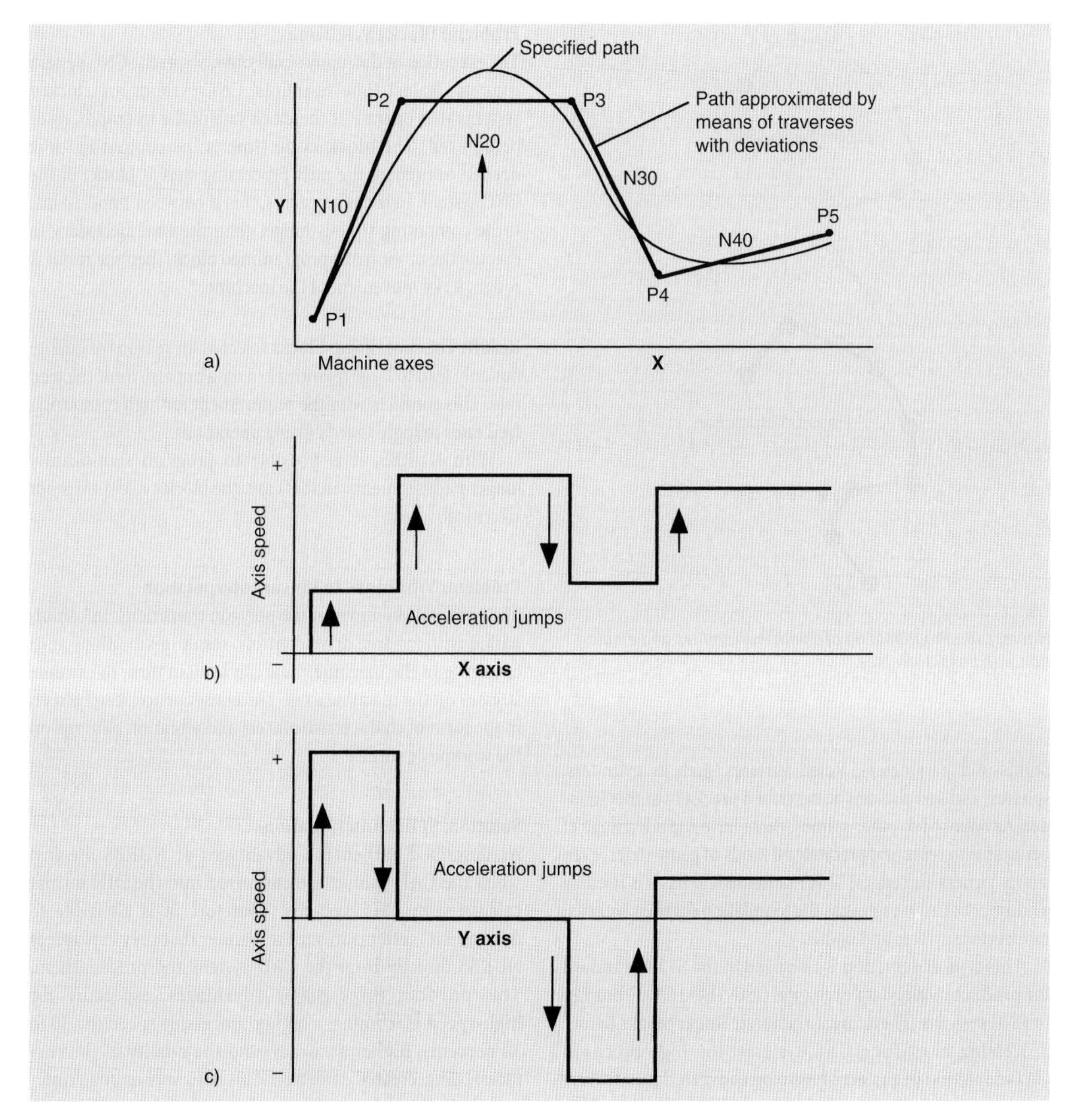

Figure 3.17: Path deviation and acceleration jumps in linear interpolation with constant path velocity.

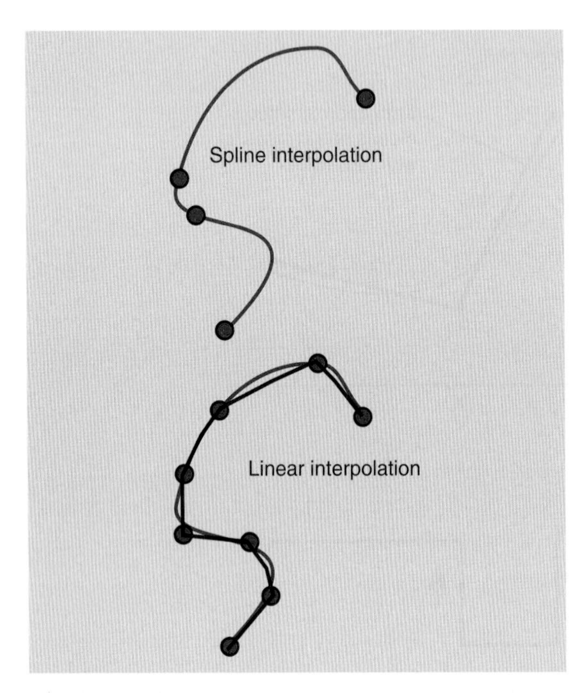

Figure 3.18: Programming of a path with splines and emulation with linear traverses.

points and parameters. Ruled surfaces such as cylinders, spheres, and tori also can be described precisely in this manner. As opposed to other splines, these have the advantage of being able to represent precisely all kinds of geometries, even sharp corners and edges. This mathematical basis is used by a number of CAD systems as the foundation for their internal models for surfaces and bodies.

This type of depiction is also used in the STEP standard for product model data exchange (ISO/IEC 10303) but not in CNC systems. Therefore, to generate linear blocks before processing in existing CNC systems, the high-precision CAD depiction of surfaces has to be converted into traverses and approximated to the shape (→ *Figure 3.17a*). This is one of the tasks of the postprocessor. To achieve high contour accuracy, a small chord error has to be selected for this approximation; this leads to a large number of very small individual steps, thus producing very large CNC programs.

Problem: Block Cycle Time

The execution of the many small traverses in the CNC system runs up against time restraints. CNC systems are clocked systems that operate at the clock rate of the microprocessor being used. The **block cycle time** is the computing time needed to prepare the next processing step (1 block) in the CNC system. In today's systems, this is between 1 and 10 ms. If the computing time is longer than the time necessary to execute the motion of a programmed block, then the result is rough, jerky movement of the machine.

Result: The short linear blocks selected for reasons of quality not only lead to large quantities of data but also limit the feed rate. This conflicts with the requirement for high contouring feed rates in high-speed cutting operations.

With NURBS, it is possible to program significantly longer path segments; in this case, the block cycle time is not so critical.

Problem: Vibrations in Linear Interpolation

The acceleration jumps at the polygon transitions, in combination with lag-free drive controls, result in vibrations and/or jarring in the machine. This can lead, in turn, to extreme stresses on the machine axes. The consequences can be seen in the form of characteristic facets and vibration patterns on the workpiece surface.

Solution: Spline Interpolation

As described earlier, the advantages of NURBS are lost when the CAD splines are converted into the DIN format calculated by CNC systems. Therefore, it is advisable to design CNC systems in such a manner that they can accept NURBS directly from the CAD system and process them. This provides three major advantages, especially for high-speed machining: a higher processing speed (by 30 to 50 percent), higher accuracy, and the quality of the surfaces being created is better. This also means that time-consuming postprocessing of the workpiece is no longer necessary.

Furthermore, the machine's motion sequence is even, without abrupt acceleration peaks. This has a positive effect on the machine load, the workpiece surface, machine service life, and tool life.

This requires, however, a completely different CNC program format, which may look like this:

```
N29 PO[X] = (-3.525, .001) PO[Y]
= (20, -.014, .006)
N30 PO[X] = (-33, -26.371, 26.155) PO[Y]
= 20, 6.947, -3.367)
PO[Z] = (23.977, 25.953, -25.953)
N31 PO[X] = (-33, .265, .095) PO[Y]
= (20, -.034, .012)
PO[Z] = (20.977, -.847, .489)
N32 PO[X] = (-12.155, 36.816, -19.133)
PO[Y] = (20, -7.727, 6.775)
P[OZ] = (20.977, 39.746, -19.808)
```

Here, the coefficients of a third-order polynomial are transferred for each axis, for example, for the X axis:

$$x(t) = at^3 + bt^2 + ct + d$$

Switching Between Languages

All displays and dialogs of the CNC system are stored in two or more languages. This makes work easy for operating personnel, and if it is necessary for service technicians to intervene, then they can simply switch over to their own language.

Languages with different character sets (e.g., Chinese) pose a particular challenge here. In this case, the CNC system has to change not just the contents (text), but also the depiction (characters). In today's CNC systems, there also should be the option for texts in part programs (e.g., comments or notifications) to be in the local language.

Teleservice

This requires a direct data link from the machine manufacturer to the machine at the customer's location. Then it is possible to use teleservice to support personnel via telephone or data link during commissioning, fault diagnosis, and maintenance and repair of machines and systems. This helps the customer-service department to save time and costs by allowing it to quickly diagnose problems and correct errors in CNC machines from a central office located at long distances from the machine, including

- Installation and commissioning
- Troubleshooting
- Transferring new software versions

Teleservice functions can be subdivided as follows:

- Merely display and evaluation-related functions for rapid assessing of the state of the machine and locating errors
- Active, repair-related measures with direct intervention, for example, intervention in the CNC or PLC software

Virtual NC Kernel (VNCK)

Virtual NC kernels of controllers are used as the basis for calculations in simulation systems. For machine room or complete machining simulations, the controller manufacturer makes its CNC kernel available as a virtual environment. This makes it possible to simulate the exact, real control behavior on a PC. The virtual CNC kernel tests the part programs for collisions, surface errors, and so on and makes important characteristics available for production planning. The time required to set up and test the machine is shortened significantly.

Feed Limitation

Feed rates that are too high or too low can damage the tool and workpiece. For this reason, the permissible speed range can be specified in the program. If the CNC program is given values higher or lower than this, then the limit values are automatically observed.

Feed Forward *(Figure 3.19)*

The **following error** or **contouring error** of the CNC axes during path execution results in contour errors on the workpiece. The lag in the system means that the milling cutter has the tendency to leave the desired contour (gray), and the surface thus created (yellow/red) deviates from the desired contour.

The magnitude of the following error depends on the system (e.g., analog position feedback control) and the feed rate. A high $\text{kilo}_{\text{Volt}}$ value and the function "axes feed forward"

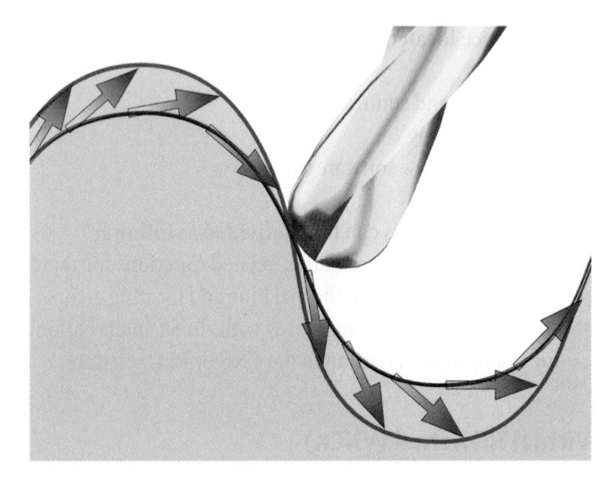

Figure 3.19: The function "feed forward" minimizes the contouring error of the NC axes and improves the contour accuracy during path execution.

Black: Command path,
Red: Form deviation, caused by the contouring error

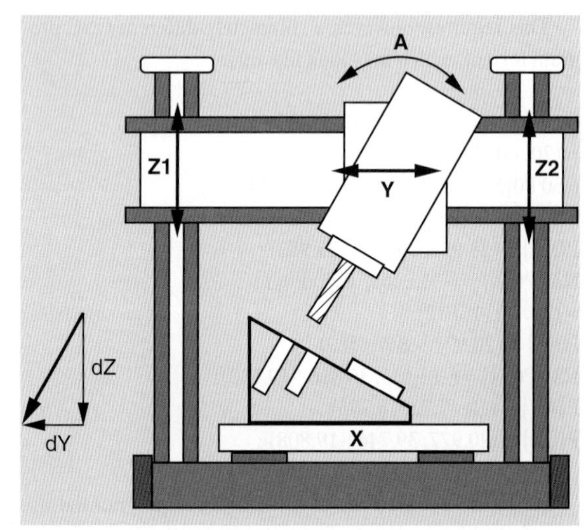

Figure 3.20: 3D tool compensation. Makes it possible to machine inclined surfaces with tool compensations and the drilling of inclined holes with swivel-head machines in which the Z motion is not in the quill (tool axis).

reduce the velocity-dependent following error during path execution toward zero while simultaneously improving the contour accuracy on the workpiece.

3D Tool Compensation *(Figure 3.20)*

This is necessary for CNC machines with four or five CNC axes if one or both swivel motions are performed in the tool axis and the tool length and/or diameter have to be compensated for.

In this case, neither the standard drilling cycles nor the tool compensations will function. Even just moving to a drilling position requires complex calculations, and two or three axes have to be interpolated linearly for the drilling operation. With 3D tool compensation, the operator can enter/compensate for an inclined clamping position of the workpiece on the machine, and the CNC system automatically calculates the resulting positions and motions.

Tool Management

Tool magazines in machines are becoming larger and larger, and their management in CNC systems is becoming more and more complex. The use of interchangeable tool cassettes may involve the following tasks:

- Detecting the cassette size and number of storage locations
- Distinguishing between input and output cassettes
- Moving/resorting tools into different cassettes
- Temporarily disabling specific cassettes
- Adjusting the swivel velocity of the gripper depending on the tool weight
- Checking for collision hazards (length, diameter, profile)
- Monitoring remaining tool life and requesting replacement tools
- Management of alternate tools
- Completing interrupted tool changes properly, even after a power failure
- Automatic exchange of tool data with a tool computer, etc. (→ *Part 4: Tooling Systems for CNC Machines*)

3.4 CNC Displays

Displays are the **user interface.** Therefore, good, informative, clearly structured displays are very important for correct operation of NC machines. Today's CNC systems use **liquid-crystal displays (LCDs) or plasma displays.** They are flat-screen and therefore can be placed in the optimal location. They do not require as high voltages as a picture tube, do not flicker, and are easy to read. The **position display** for each axis is important. This can be used to read the precise current axis position and to determine whether the target position has been reached yet. The standard display in measurement increments of 0.001 mm is superior to all conventional measuring sticks. In grinding machines, measuring systems with measurement increments of 0.1 μm are already in use.

Electronic displays can be set to ZERO or a defined value at any point, thus saving time-consuming calculations on the part of the programmer to convert the dimensions in the drawing to the absolute machine positions. It also must be possible for the operator to use the display to obtain the following information at any time:

- Program number, program name, program memory requirements
- Program contents
- The required and/or remaining memory capacity of the CNC system
- All compensation values, zero-point offsets, and other corrective interventions
- Active feed and speed values
- Active G-functions and M-functions
- Subprograms and cycles
- Workpiece and tool management
- Warnings, status messages, and error messages
- Machine parameter values
- Input and simulation graphics
- Diagnostic programs
- Service and maintenance notes and so on

Depending on the expandability of the CNC system, smaller or larger **graphics screens** can be used as the **display unit.** For some values it is possible to switch the size of the display; graphical displays can be modified by means of "zooming." In some cases, connections for a second or third screen are present. This is very useful in the case of large machines.

3.5 Open-Ended Control Systems

(Figures 3.21 and 3.22)

The definition for *open-ended CNC systems* was debated at length among specialists in the field. It was finally agreed that the open-endedness of a CNC system involves a number of criteria that are of equal importance.

It is necessary to make a fundamental distinction between at least six different **characteristics for the open-endedness of a CNC system:**

- **Open for the operator,** for example, with special graphical support to allow easier programming and optimized operator control
- **Open for the machine manufacturer,** thanks to the ability to create individual user interfaces and displays
- **Open as regards the selection of hardware,** in order to be able to use components from different manufacturers
- **Open with regard to the CNC operating system,** in order to be able to port existing standard software
- **Open when it comes to the input/output (I/O) interfaces,** for example, the data interfaces or drive interfaces
- **Open with regard to the CNC kernel,** to enable machine manufacturers to integrate their process know-how directly into the CNC system

Thus the idea is generally to make CNC systems more flexibly adaptable and more cost-effective, at least when it comes to input interfaces, through the use of series-produced computers and their associated standards (→ *Figure 3.21*).

In contrast to this, today's CNC systems are almost without exception **closed systems.** Specially developed software runs with special applications on special hardware. Few or no standards are applied. Each function has to be developed anew, even if it has been present for a long time in other fields or in previous CNC generations. The consequences of this are high development costs, long development times, fixed specifications, and no scope for customized solutions. Such solutions are too expensive!

If we borrow the term *open* from the field of computer technology, then an open-ended CNC system is a numerical control system in which all the software interfaces are disclosed and described. This is comparable with computers that have an **open system architecture.** This definition is not sufficient for CNC systems, however.

After extensive debates, it appears that an agreement has been reached that in principle a PC should be connected upstream of the CNC kernel controller (→ *Figure 3.22*).

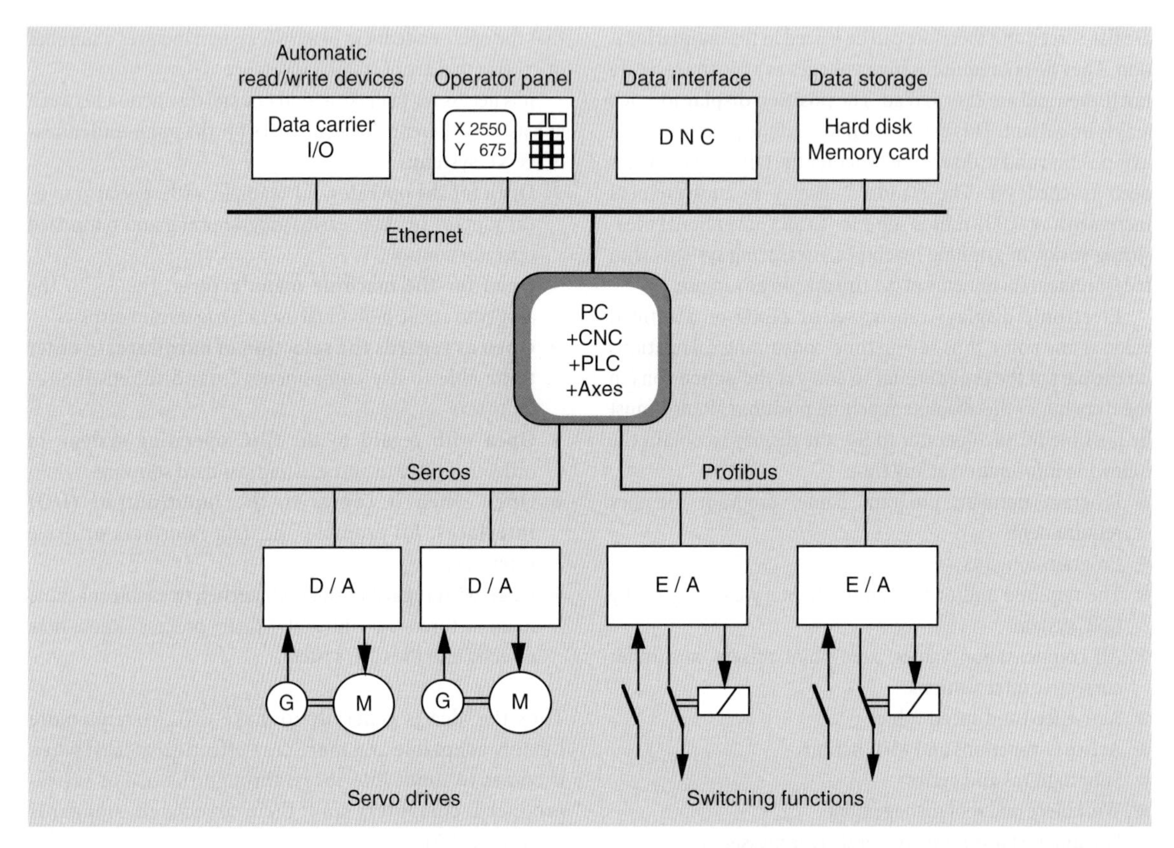

Figure 3.21: Principle of the open compact CNC system "with series-production interfaces" (e.g., Ethernet, SERCOS, and PROFIBUS).

This can provide a number of significant advantages, such as

- A simpler, more economical standard CNC system
- Largely free design of the user interface
- Free scope as regards the SFP system that is implemented
- Simpler linking of computers
- Trouble-free transfer of CAD data
- Use of available standard interfaces for peripheral devices, such as
  - Hard disk drives
  - Floppy disk drives
  - Memory cards
  - Standard screens
  - RS 232C or 242
  - SCSI connections
  - Ethernet

But Open Systems also Have Their Limitations!
It should be clear to everyone that the open part of a CNC system cannot mean more than 20 percent of specially required functions. Therefore, 80 percent of such functions are not subject to individual requests for modifications—this is a major benefit for the systems integrator! After all, systems integrators would suffer greatly if totally free scope were given to modifications: Instead of the desired uniform standard control system for all machines, there would be practically no limits to the diversity of operator control and displays.

3.6 Price Considerations

(Figures 3.23 and 3.24)

Prices for numerical control systems with comparable specifications have fallen by more than 90 percent within

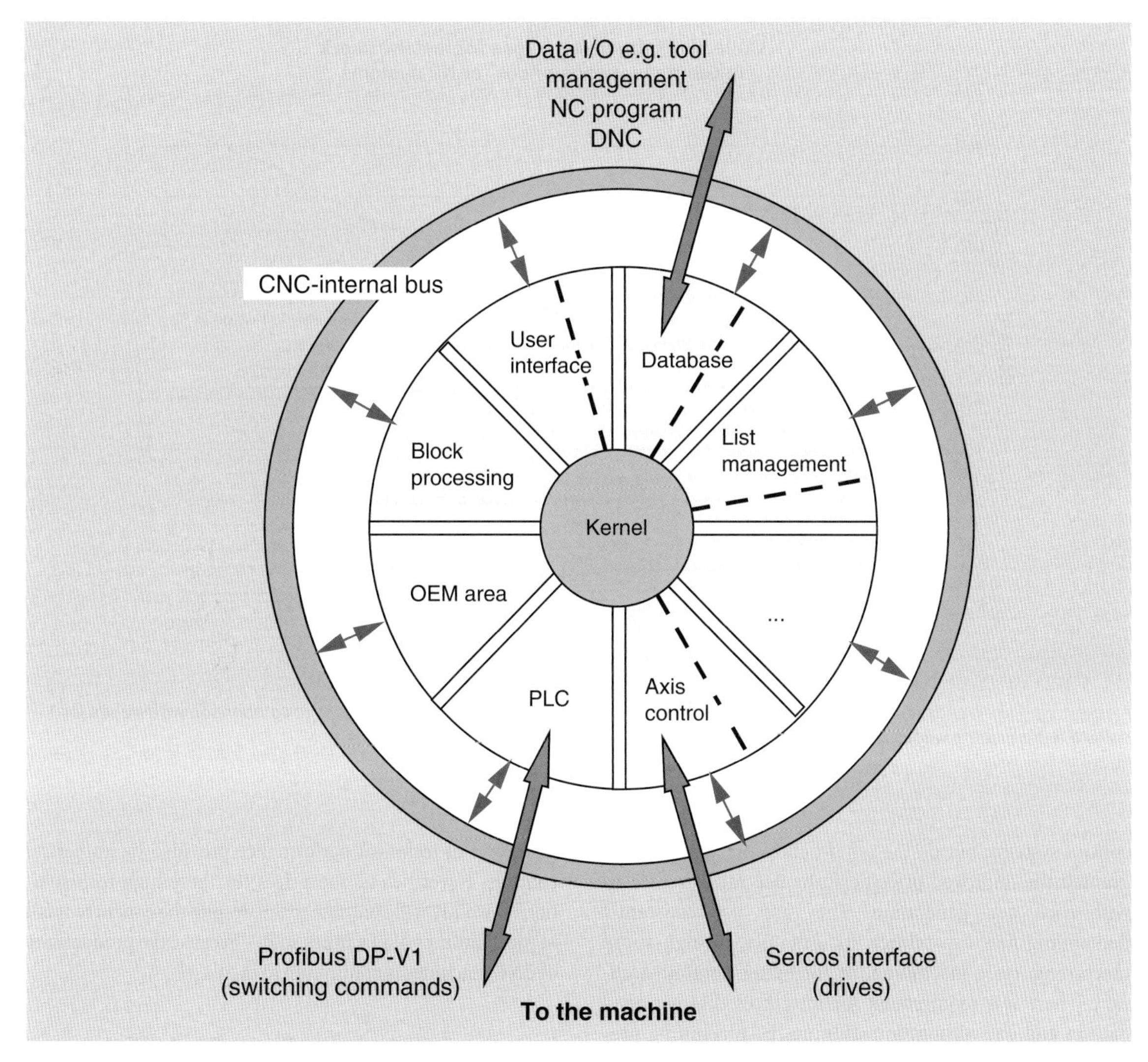

Figure 3.22: Principle of the compact, open CNC system.

20 years. Color graphics monitors, multiple simultaneously interpolatable axes, integrated programming systems, tool and workpiece management systems, supplementary automation, data interfaces, and the practically limitless expandability of data memory have resulted in a large number of new functions that have raised prices again somewhat. Cumbersome interface cabinets have been replaced by small, freely programmable PLCs, and the entire control logic has been implemented as software. This software can be copied quickly, inexpensively, and without errors, thus saving time, effort, and cost. The increased use of surface mounted devices (SMD) technology has helped to drive CNC technology to yet another new level (→ *Figure 3.24*).

To be sure, many users often complain about the excessively short life cycles of the individual generations of

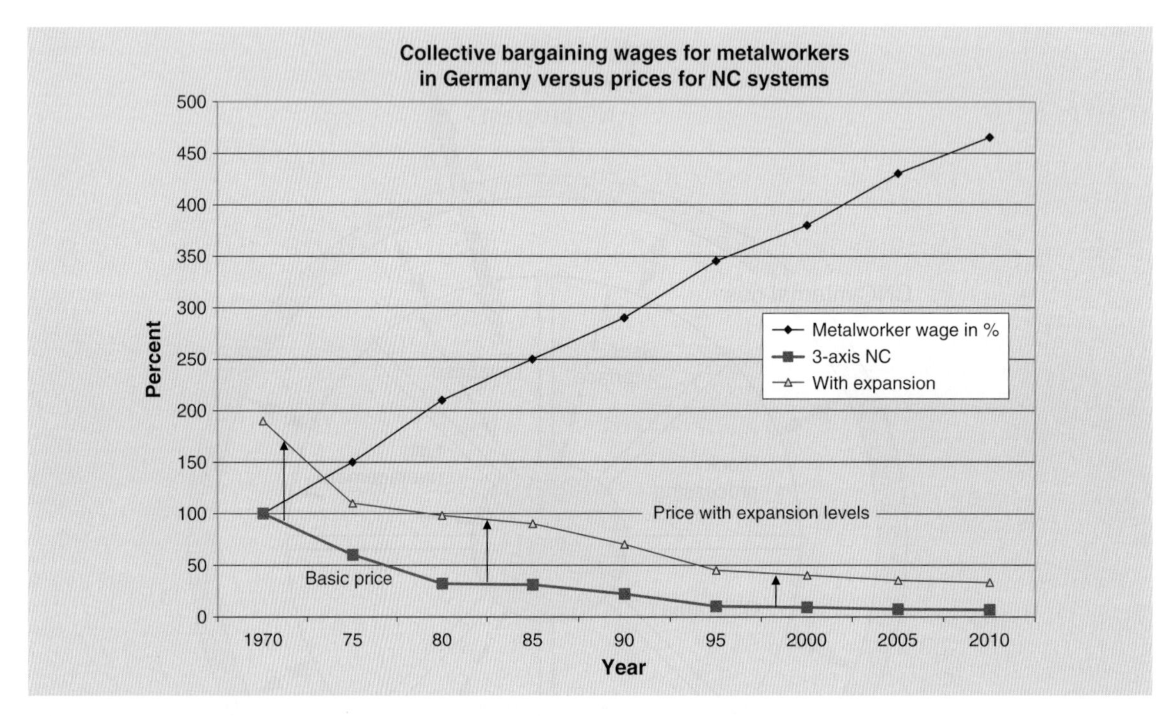

Figure 3.23: Price trends for numerical controllers with a constantly increasing range of functions in comparison with trends for collective-bargaining wages in the German metalworking industry.

control systems, but on the other hand, they also take for granted the improved price/performance ratio associated with each new generation. They are also constantly demanding new capabilities, for example, in high-speed machining, laser machining, or for better contour accuracy. There is a requirement for interfaces that can read data in and out without interrupting the machining process. Turning machines with 7 to 32 CNC axes, driven tools, and two main spindles may not represent the full extent of potential developments. The servo drives have to be controlled digitally because analog technology is too slow and too imprecise. In order to achieve the required dynamic accuracy even at high velocities, the following error of the drives has to be close to zero.

These and many other requirements can only be met using high-performance microprocessor technology.

As a rule, today CNC machines are operated manually in jog mode only for setup purposes. This in itself testifies to the high level of technical performance provided by numerical controls. Nevertheless, the prices for these high-tech solutions certainly will continue to fall despite the concurrent fall in the quantities of machines sold—because the productivity of CNC machines is continuing to increase!

Increases in Processing Power Due to Higher Levels of Integration

Moore's Law

When integrated circuits were still first being developed, Gordon Moore, one of the founders of Intel Corporation, predicted that the number of transistors on a chip would double about every 18 months. This trend will have to end sometime, however – either for economic or physical reasons. The latest this could happen would be the year 2020, at which point the transistors on the chips would be only a few atoms thick.

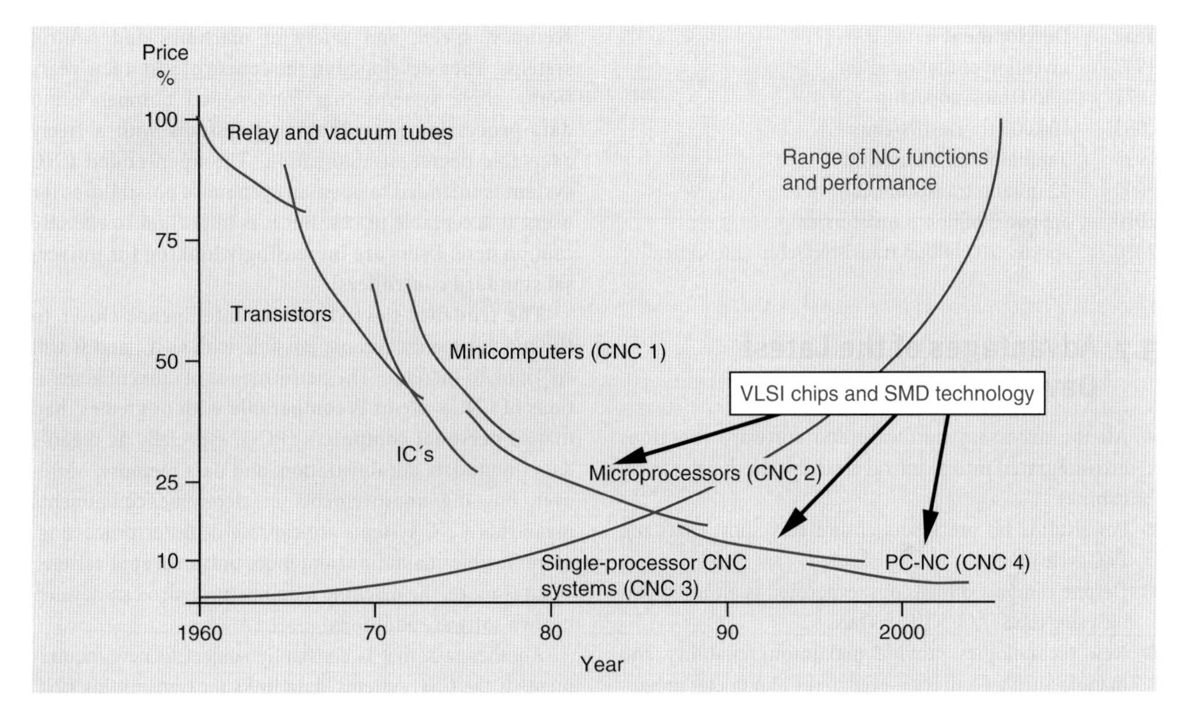

Figure 3.24: Development of prices for numerical controllers with the use of electronic components with ever-increasing levels of integration.

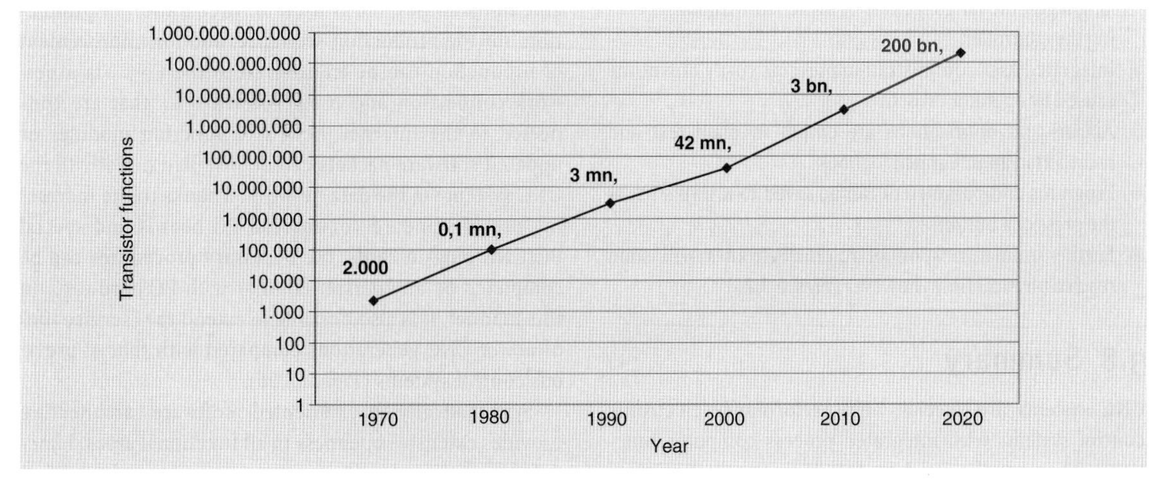

Figure 3.25: Development of transistors/chip within 50 years.

Year	Development
1947	invention of the transistor
1971	2,300 transistors/chip
1982	100,000 transistors/chip
1993	3 million transistors/chip
2000	42 million transistors/chip
2010	approx. 3 billion transistors/chip
2020	approx. 200 billion transistors/chip (theoretically)

3.7 Advantages of the Latest Developments in CNC

- All the necessary hardware and software functions are integrated into a central assembly in the electrical cabinet.
- Powerful 32-bit processors ensure very short CNC and PLC cycle times.
- Despite compact dimensions, expansion is always possible using the SERCOS III axis bus.
- New technologies provide maximum reliability and manufacturing accuracy—even in the nanometer range.
- The HMI software (HMI = Human Machine Interface) provides intuitive display screens for all operator control situations (without any additional operator control PC).
- Convenient editor functions make NC programming and testing easier. The HMI software is available in multiple languages; the language can be switched without rebooting the controller.
- Integrated user administration prevents costly downtime caused by operator errors.
- Alarms and notifications are output in plain text and recorded in an integrated logbook.
- Programs and parameters can be saved to a USB stick at the press of a button.
- Further miniaturization of microelectronics will soon reach a physical limit, however (*Figure 3.25*)

3.8 Summary

CNC systems are special, high-performance electronic control systems with integrated process computer functions. They can be used to fulfill practically any user requirement as regards scope of functions, reliability, accuracy, speed, and safety of machines and control systems. They developed in the course of just a few years from simple systems that "understood numbers" into **data-processing manufacturing systems** with a freely adaptable degree of automation. The **open-ended CNC system** is intended to open up even more possibilities for users at acceptable prices. But it is important to exercise caution here: **Users are increasingly looking for universal standard controllers!**

The tendency toward greater intelligence closer to the machine only became possible with CNC, and it will continue to increase. The performance of current generations of CNC systems is **comparable with** or greater than that of **personal computers (PCs)**, especially as regards data input, data administration, and data memory. Moreover, ongoing improvements in electronic components mean that CNC systems are capable of faster processing, more flexible adaptation, and more universal applications. What is more, the machine manufacturer also can develop its own software additions.

Another task that is constantly subject to new requirements is the CNC systems' **data links** for transmitting CNC programs, drawings in the form of CAD data sets, test plans, quality-assurance data, MDA/PDA data with evaluation, service data, maintenance data, and diagnostic data via teleservice/Internet, etc.

In the widest sense, all the information that is available in databases and is used for preparation of production, for the production sequence, and for improvement of production can be transferred to the CNC via a network connection and used. As with PCs that are connected to the Internet, new CNC function modules or updated software packets can be installed directly to the CNC system via the data network. Thanks to the increasing use of standard commercial PC boards and special PC/CNC plug-in cards, the necessary procedures can be completed by any person familiar with PC hardware. In this manner, it is also possible to extend the "service life" of newer CNC generations compared with that of previous control systems (*Figure 3.26*).

Systematically user-oriented software architecture provides extensive openness in all functional areas. Function libraries and software components based on Windows operating systems allow the user to design individual

Figure 3.26: CNC for up to four axes and a spindle (Siemens 802 Dsl)

user interfaces and programming interfaces custom-tailored for his or her CNC machines. In this manner, it is possible to create manufacturing systems that are optimally coordinated for the specific process by adding or adapting existing functions to suit special machines or areas of application.

Functions of Numerical Control Systems

Important points to remember

1. CNC systems are numerical control systems in which all the control functions are implemented by one or more integrated microcomputers and the corresponding software.
2. Today's CNC systems are characterized in part by the following characteristics:
   - One or more monitors with a full keyboard for data input
   - Multicolored graphics for programming and dynamic simulation
   - Program memory capable of practically unlimited expansion
   - Tool data that can be saved and managed automatically (e.g., service life, wear)
   - A specially adaptable and expandable scope of functions
   - Memory space for customer-specific functions and expansions
   - Probing cycles for measuring probes with evaluation programs and for documentation of safety-related measurement data for workpieces
   - Generally an integrated, machine-specific programming system for programming on the machine
   - Data interfaces for connection to a network (e.g., Ethernet)
   - An integrated or optionally plug-in ASCII keyboard
   - Significantly smaller space requirements and lower heat generation
   - A significantly expanded range of functions
   - Scope for customer-specific functions and expansions
   - Ability to store processing cycles and measurement cycles
   - Many other functions, with more being added every day
3. Modularly structured CNC systems offer a wide range of functions and options. The buyer should check which features are important and useful for his or her application.
4. CNC systems that can be programmed on the shop floor provide powerful, graphically supported programming tools.
5. CNC systems have a number of different, expandable data memory areas for
   - The executive program
   - Part programs with automatic loading
   - Fixed or free cycles
   - Integrated operator prompting
   - Diagnostic software and troubleshooting tools
   - MDA and MDC data
   - Notifications and error displays in plain text
   - Tool and pallet management
   - Zero-point offsets, wear compensation, tool data
   - Machine data and much more
6. The data interfaces that are available for connecting all the possible peripheral devices are very important.
7. The important factors in assessing the speed of a CNC system are the data transmission rate, the processor speed, the block cycle time, the servo scanning rate, and the cycle time of the PLC.
8. In today's CNC systems, the CNC, PLC, and drive functions (closed-loop control outputs) are often housed on a common plug-in card. An industrial PC can be expanded into a CNC system using just such a plug-in card.

4. PLCs: Programmable Logic Controllers

The importance of programmable logic controllers (PLCs) has been increasing steadily. They replace not only the relay controls that were used previously but also perform many additional control functions and diagnostic tasks. Of particular importance are CNC-integrated soft PLCs with data interfaces.

4.1 Definition

Programmable logic controllers (PLCs) are controllers with a computer-like structure for use in industrial environments to implement certain tasks and functions, including sequence control, logical operations, time and counter functions, arithmetical operations, spreadsheet management, and data manipulation. They vary according to their level of performance but are always structured with "neutral" wiring. Like computers, they consist of a central processing unit (microprocessor), program memory (RAM, EPROM, or FEPROM), input/output modules, and interfaces for exchanging signals and data with other systems (e.g., CNC systems, manufacturing host computers, etc.).

The control logic is programmed using computers (PCs) and a system-specific programming software. Input can be in the form of a ladder diagram, instruction list, or function block diagram, by means of graphically supported languages, or using higher-level programming languages, for example, "Structured Text." All these use the graphical support for programming and simulation of the switching functions being generated.

For economical CNC systems, the motion controls for the NC axes and the PLC functions are brought together (creating an integrated software PLC) and controlled by a common processor. The previously preferred method of distributing the tasks to a processor and a coprocessor in the case of high requirements with regard to speed or the scope of functions is no longer practical in today's control systems.

Efforts to create uniformity and international standards for PLC programming languages are documented in International Electrotechnical Commission Standards IEC 1131 (in Europe IEC 61131).

4.2 Origins of the PLC

In 1970, a new type of electronic control system was presented for the first time at the machine tool trade show in Chicago and immediately attracted great interest. While up to that point machines had been controlled using relays with complicated wiring, contactors, or electronic function modules, this new control system had been developed out of computer technology and had some completely new characteristics. The essential feature was that the control logic was no longer defined using "hard wiring" but was rather "freely programmable" like a computer. This was done using a computer-like programming device and a programming language developed especially for that purpose. During the commissioning phase, the control program remained in random-access memory (RAM) units that could be changed quickly. Once the test phase was complete, the program was transferred permanently to erasable-programmable read-only memory (EPROM). Any subsequent changes that became necessary could be made in the same manner without any complex changes to the wiring. The resulting advantages as to the size of the controller, the commissioning time, and ease of modification were so attractive that the only obstacle to immediate success was the high price tag.

Their similarity to numerical controls—which also were quite new at that time—was unmistakable. The continually increasing requirements regarding automation were so complicated and difficult to implement in terms of control functions that these PLCs were put into use very quickly, especially in conjunction with complex NC machines.

Thus, in just a few years, a large number of new PLC products appeared on the market, many of which were designed for special applications.

Unfortunately, the positive experiences that had been gained with the timely standardization of NC technology were not put to use here. As a result, there is no uniformity in the programming of PLCs from different manufacturers. This means that the programs that are created are PLC-specific and will not run on PLCs from other manufacturers. The now-proven principle of object-related "neutral" NC programming and subsequent controller-specific adaptation via postprocessors was never realized with PLCs. The manufacturers put most of their efforts into designing new types of programming to suit the skills and wants of employees in electronics departments; no priority was given to universal PLC programming. In retrospect, the fact that people were "speaking different languages," as well as the high prices for the programming devices, resulted in a delay in market introduction.

The IEC 1131 standard is a first step in the direction of universal PLC programming.

4.3 Structure and Functions of PLCs

Figure 4.1 shows the basic structure of a PLC. As can be seen, PLCs consist of the following functional modules: the power-supply unit, central processing unit, program memory, generally several modules for inputs/outputs, various additional functions (such as markers, timers, counters, or axis modules), and a rack on which the modules are mounted. An interface or coupling module serves to connect the programming device and as a data interface to the periphery. Today, this primarily takes the form of an Ethernet interface. Either direct I/O modules or a suitable field-bus interface are provided for control of the actuators and sensors.

All PLC hardware modules are connected to the power supply and the internal **system bus** when they are plugged into the module rack. The system bus consists of a number of parallel connecting cables and is subdivided into the address bus, the data bus, and the control bus. Data transmission between the individual modules is organized and monitored by the control unit of the central processing unit (CPU).

In CNC-coupled PLCs, the hardware structure has changed greatly as a result of developments in electronic components. The functions of individual plug-in modules for data I/O and data processing, as well as time and counter functions, all have been assumed by the CPU. This created highly integrated **one-board controllers** in which the CNC system, the PLC, and the output of axis command values all were accommodated on the same circuit board (*Figure 4.2*). Such circuit boards generally can function in any PC with a Windows operating system. A field-bus interface is available for connection to the decentralized peripherals; the servo drives are

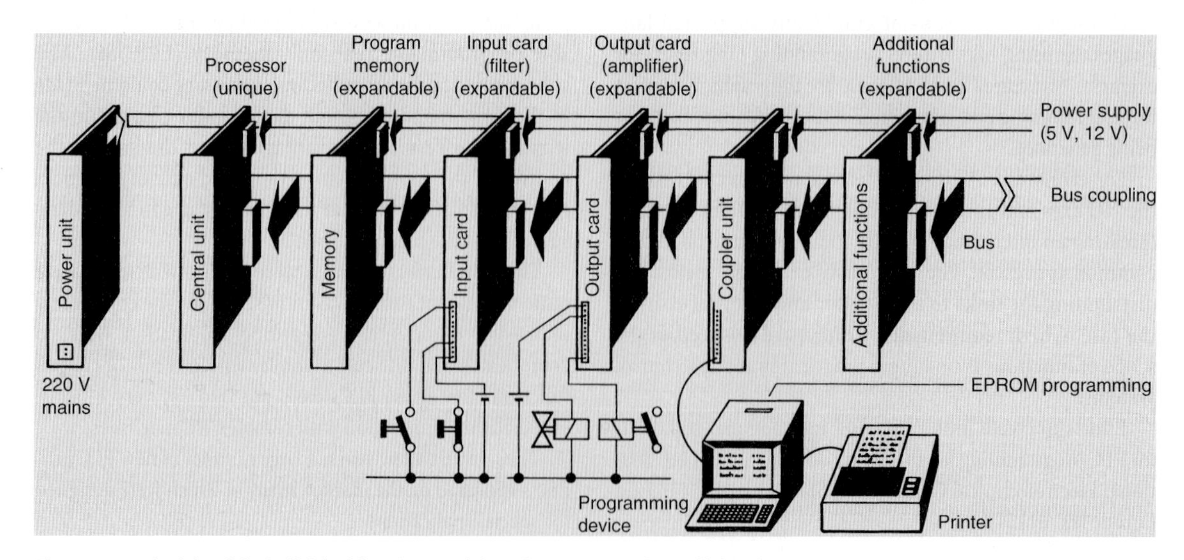

Figure 4.1: Principle of the individual function modules of a PLC. Nowadays, all the electronics are located on a single circuit board.

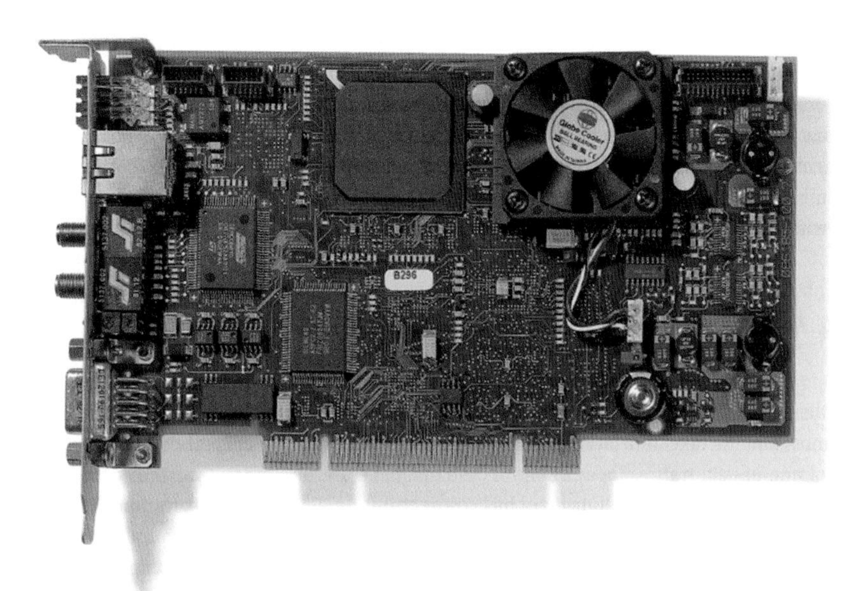

Figure 4.2: Single-board solution: PCI bus plug-in card with MC, CNC, and PLC functions, as well as TCP/IP, PROFIBUS DP, and SERCOS interfaces.

connected via a special interface (e.g., SERCOS interface). The axis control circuits (i.e., position controllers, speed controllers, and current controllers) are located in the decentralized drive amplifier.

The connections for the monitor and keyboard are provided as standard on **industrial PCs (IPCs).** These are PCs that have been constructed for installation on shop floors, with a rugged metal housing and easily accessible connection points.

In principle, PLCs have to fulfill the same tasks as controllers that are constructed based on relays or electronic functional modules:

- Receiving command input and feedback
- Linking, branching, and interlocking these according to a predetermined, programmed matrix
- Using this to generate the corresponding control commands and outputting them to the actuators

Process Monitoring

In state-of-the-art machine tools, PLCs are also used to monitor the process itself. In higher-priority program sections, various CNC data are monitored, for example, in order to detect tool breakage within a fraction of a second and to quickly move it away from the workpiece in the right direction before further damage can occur.

On this basis, it is possible to define two different areas of application for PLCs in machine tools.

- **Program controllers.** These control repetitive sequences in machines according to a fixed, defined, special program, such as the changing of workpieces, speeds, and feed rates on automatic rotary machines.
- **Machine control interfaces.** These are connected between the NC and the machine and have the task of transmitting all the switching functions output by the NC system to the actuators, taking into account the specified links, as long as this can be done without endangering persons, the machine, or the workpiece. This **monitored function control** can include, for example, the automatic processes for tool changing, pallet changing, or handling other process-related equipment. The entire sequence of functions is defined as in a program controller and is only "initiated" by the NC system by means of an output signal.

Similar tasks also apply for setup and manual operating modes, where the commands are issued manually by the operator.

In the case of extensive production plants, the control functions can be distributed among several PLCs. This is sometimes called **multiprocessor PLC** and has the advantage that the individual system components can be tested independent of each other. Depending on the specific configuration, subsequent data communication takes place either directly (within the rack) or via a network (field bus or Ethernet).

4.4 Data Buses and Field Buses

(Figures 4.3 and 4.4)

During the development of PLC systems, it was especially important to take into account the constantly increasing volume of data traffic. Today's control systems make use of the technical and cost advantages of **bus connections.**

These consist of one or more parallel connecting cables and provide bidirectional **data transfer** between multiple "devices" in a system. It was soon realized that it was not possible for a single bus system to cover all the diverse tasks and requirements in the field of manufacturing, and the various tasks were subdivided among a number of different bus systems.

Ethernet became the preferred **data bus** and is now regarded as the industry standard. At present, it uses a data-transfer speed of 10 or 100 Mbaud (**Fast Ethernet**) and can reach 256 or more connected devices directly, depending on the address configuration.

The physical **transmission medium** involves special four-pair twisted-pair cables with eight-pole standard connectors. In today's industrial installations, the single-cable principle is no longer used but rather service-proved network technology, in which each device is connected to a **switch.** Up to 24 devices can be connected to each switch. The advantages of this are that it is possible to transmit large data packets between devices in each switch while using the full bandwidth. Second, the functioning of the overall system is not endangered if there is a fault in one of the devices, and troubleshooting is made easier.

Special **backup procedures** mean that data transmission within a network is absolutely reliable.

Special **field-bus systems**, such as PROFIBUS, InterBus, and CANbus, have been developed for **signal transmission,** that is, for the control of actuators and the feedback of sensor signals

The **SERCOS bus,** which is available especially for controlling servo drives, uses fiberoptic cables as a matter of preference and thus in contrast to copper cables avoids all electrostatic and electromagnetic interference in the signal cables between the NC system and the drive control.

The use of bus connections instead of complicated individual wires minimizes not just the number of cables and contacts but also the number of potential sources of faults. Which bus is best suited for the individual tasks depends first of all on which "level" of the **automation pyramid** (→ *Figure 4.4*) corresponds best to the requirements. The top level involves a small number of non-time-critical data, whereas on the lower level the requirements are reversed: constant transmission of time-critical data packets for open- and closed-loop control of the process.

The typical requirements for a bus are as follows:

- The maximum number of devices (actuators and sensors)
- The required longest response times (short reaction times)
- The scope of the data being transmitted
- The maximum transmission distance

At the control level is **Ethernet,** which is the worldwide standard with the widest distribution; the required interface modules are small, cost-effective, and mass-produced.

The strengths of the CAN, PROFIBUS DP (Decentralized Peripherals), and InterBus-S **field-bus systems** lie primarily in the sensor/actuator area. In contrast, PROFIBUS FMS (Flexible Manufacturing System) is suitable for higher levels, for example, larger data packets, but is increasingly being supplanted by Ethernet.

When establishing networks for large plants, the use of fiberoptic cables is preferred owing to the long distances that have to be covered and in order to avoid interference.

Real-Time Ethernet and SERCOS III

Many companies use a number of networks that either cannot communicate with each other at all or only with great difficulty. It would be useful in any case if a machine could report errors not only to the personnel at the machine but also to production planning systems, materials management, and the manufacturers of the machine and/or controller. A direct link between the developer PC and the controller of a

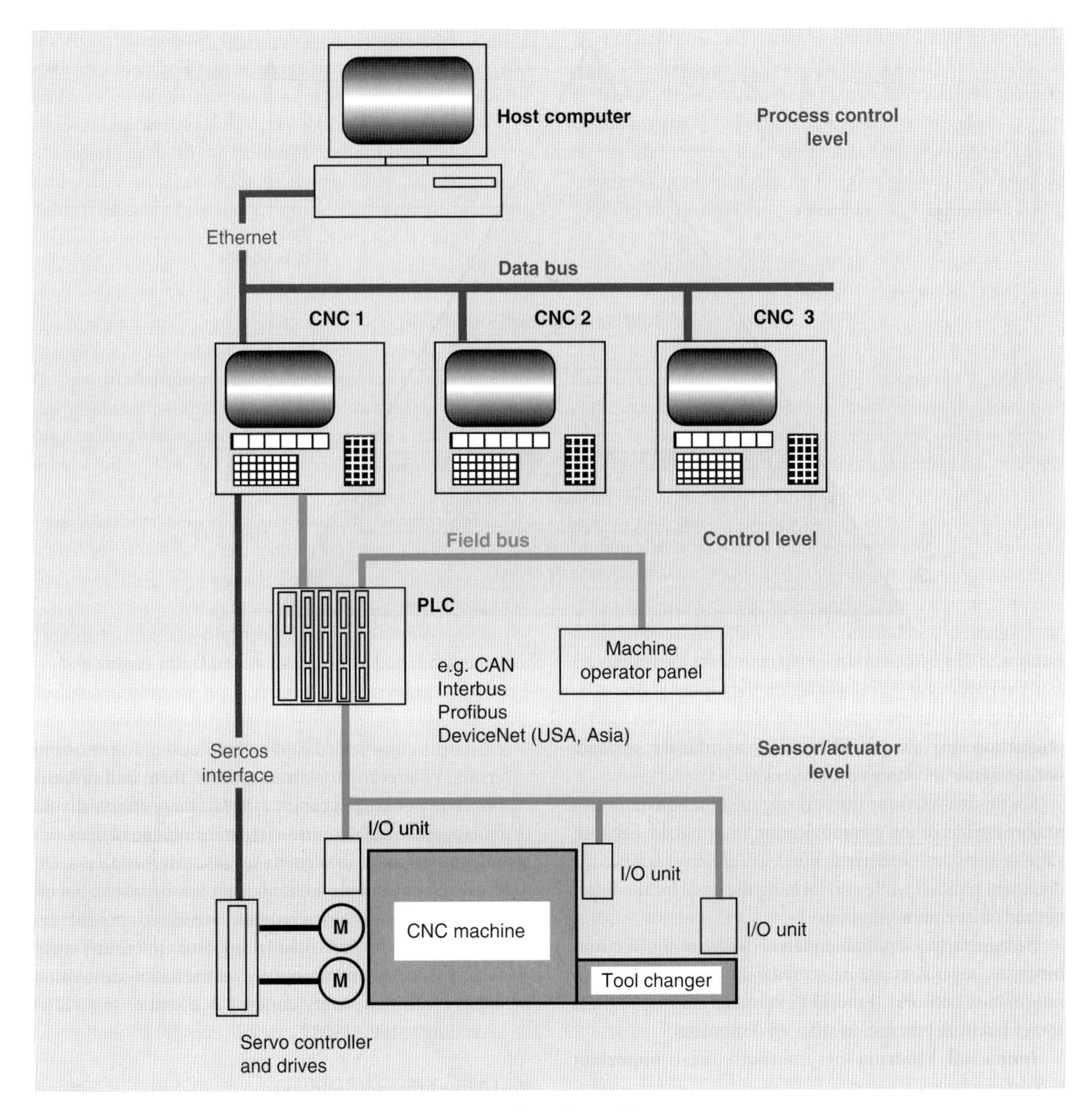

Figure 4.3: Use of various bus systems from the host controller to the machine.

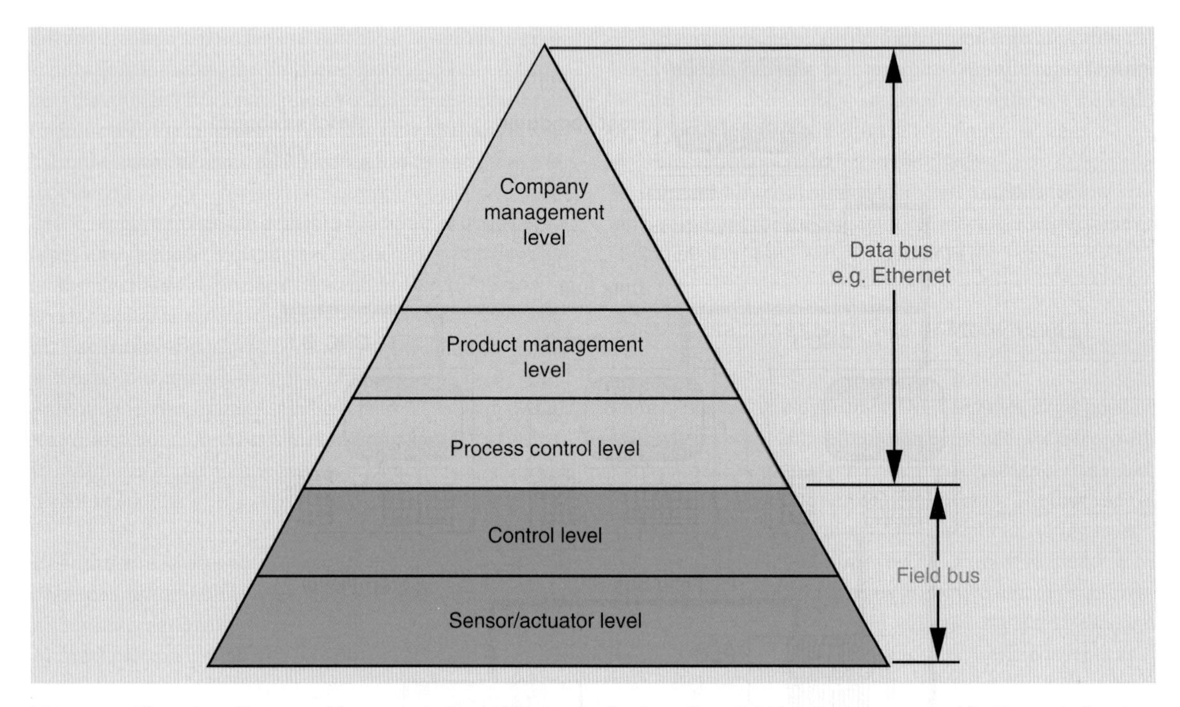

Figure 4.4: The automation pyramid represents the various levels of automation. Field bus systems are used for the control and sensor/actuator levels in combination with PLCs.

production line also could be used to transfer software updates faster and more cost-effectively.

For this reason, companies are relying more and more on a uniform protocol for communicating information between various divisions. **Ethernet** and TCP/IP have achieved a dominant position while also forming the basis for the Internet and office communications.

Because factory automation sets very demanding requirements as regards ruggedness, reliability, and security in information networks, **Ethernet** in production shops differs somewhat from Ethernet in office environments.

Industrial Ethernet has to fulfill three important requirements:

- The most important requirement is **real-time communication,** in order to ensure that important information is transmitted immediately or at the necessary time intervals. Only in this manner is it possible to coordinate complex sequences.
- The second important characteristic is **reliability.** Heat, dust, vibrations, and strong magnetic fields can induce currents in unshielded cables, thus causing transmission errors. Wherever parts are in motion, there is always the possibility of broken cables. None of these effects should endanger the reliable, error-free transmission of data.
- The same applies to **security** against external access. On the one hand, simple communication is desirable, but on the other hand, unauthorized access to production systems must be prevented in any case. To ensure safety for the machine and personnel, secure data transmission using a certified **safety concept** is essential (e.g., SIL3 according to IEC 61508).

Real-Time Ethernet Solutions

The term **industrial Ethernet** in practice can stand for various solutions that differ greatly from each other. Some systems may offer real-time functions but with limited synchronicity; that is, there is no guarantee that a number of connected modules will work at the same system cycle rate. Others are based on a close connection between the control and network functionalities, which limits the user's choices when selecting automation systems. Yet others are in fact open but require rigid network planning with a large number of control units that make it difficult to make changes later and sometimes make even simple, standard communications very slow.

Today, **field buses** and **industrial Ethernet systems** offer basic functions that can be used to perform typical control tasks. Real-time applications that require, for example, precise interaction between a number of servomotors or which have to process sensor data very quickly have to have a network with a very high data throughput and guaranteed synchronicity.

Fast Ethernet with a data throughput of 100 Mb/s guarantees the fast transmission of information. The full duplex property of Fast Ethernet means that it is also possible to communicate directly between the connected devices in order to make response times as short as possible.

Seamlessly Integrated Information

At the **field level,** for example, in the communication between individual drives, sensors, or controllers, **SERCOS III** guarantees the necessary precision. After all, SERCOS III was designed for demanding CNC applications. At the same time, it is extremely simple to link the network to the higher control level.

Via a non-real-time channel, the user can use the full functionality of the TCP/IP data, in addition to the real-time data. Thus it is possible, for example, to operate a webcam for production monitoring via the SERCOS protocol without compromising the real-time function. This makes it possible to perform comprehensive system planning without additional wiring and costs. Various modules can even use the same cables without any compromises as regards security or reliability.

As with previous SERCOS generations, **SERCOS III** relies on tried-and-tested hardware synchronization implemented using logic components (ASIC). This structure means that additional hubs and switches are merely expensive switching points and thus superfluous. **Synchronicity is thus a basic attribute for every SERCOS solution and requires no additional modules or protocols.**

The **cycle time** of 31.25 μs in real-time operation does not mean that any individual module would demand the entire bandwidth for itself alone. Up to eight drives can receive and transmit 8-byte cyclical data in motion-control applications. This is sufficient for even the most demanding tasks that can be imagined today in the field of mechanical engineering. Even fast, high-precision CNC machines are now operating with minimum cycle times of not more than 500 μs. Thanks to the higher efficiency of the SERCOS technology, there are no applications that could appear in the foreseeable future that would require a higher network speed.

Flexibility is a basic requirement in production environments, and Industrial Ethernet is a suitable component that helps to coordinate planning and manufacturing while avoiding long lead times. But, in practice, modern networks have to meet even more demanding requirements. Intelligent controllers make it possible to recombine machines to suit specific requirements, thus grouping individual units to form new cost-effective solutions. This requirement places high demands on the flexibility of the network. Generally, individual components (slaves) are controlled by a control unit (master). The typical linear structure of a production section is derived from this. The individual controllers, in turn, can communicate with each other over a common network segment.

But **direct data exchange** is also possible between a sensor and a drive controlled by different control units. This reduces the load on the central controller and decreases the data traffic in the network. So-called control-to-control (C2C) cross-communication between masters, such as two PLCs, for example, is the basic principle for the decentralized control of complex manufacturing systems. This flexibility in communication, which leads to shorter response times between master and slave devices and thus also shorter response times in the overall processes, ensures constant synchronous axis control even over multiple SERCOS networks.

The technology that enables cross-communication between individual nodes not only contributes to the efficiency and flexibility of SERCOS, but it also increases **safety** because SERCOS III networks can be created with a **ring structure.** This provides a redundant signal path in the event of a broken cable. The SERCOS network coordinates itself and offers a flexible choice of strategies: a classic linear structure to save materials or a redundant ring structure to increase safety. Engineers and planners can choose the appropriate cabling for the specific requirements without having to think about additional elements for the network infrastructure.

The use of a SERCOS interface provides numerous advantages:

- Collision-free transmission through the use of a time-slot method
- A highly-efficient communications protocol
- Closed-loop control of multiple axes with a cycle time of 31.25 μs and a jitter of less than 1 μs
- Interference-free transmission via CAT5e fiberoptic cables
- Digital interface
- Use of an internationally recognized standard
- Standardized interface that is established worldwide
- High transmission rate, up to 100 Mb/s
- High-precision synchronization of the nodes of the SERCOS network (especially for coordinated motion of multiple axes)
- Configurable cycle time (typically 1 or 0.5 ms)
- Ring or linear structure meaning little work required for cabling
- More than 80 suppliers of SERCOS interface products worldwide
- Interoperability of products from different manufacturers thanks to standardized operating data and commands
- Networking of controllers, drives, and I/O station via a common bus
- Networking of safe and nonsafe devices via a common bus

Standardization

Since October 2007, the SERCOS III real-time Ethernet solution has been a component of both the binding IEC standards. This confirms the worldwide importance of the SERCOS interface. SERCOS II had already been standardized worldwide, however. In parallel with the standardization of third-generation SERCOS, the IEC also decided to transfer the existing IEC 61491 standards from SERCOS II to the new IEC 61158/61784-1 series of standards. The SERCOS drive profile also was included in the IEC 61800-7 standard.

The same applies to the **safety of data transmission.** SERCOS inherently provides a certified safety protocol to ensure that information is transmitted reliably. **SERCOS safety** meets the requirements of safety standard IEC 61508 up to safety integrity level 3 (SIL 3). This covers risks that could arise owing to systems failures and which could lead to personal injuries and material damage. Previously, this required separate cables. With SERCOS III, all safety-related information can be transmitted over the existing data cables while guaranteeing that the power supply will be cut off immediately when the "Emergency Stop" button is pressed, for example. This elimination of additional hardware reduces costs with no loss of safety.

SERCOS safety is protected against possible errors such as repetition, loss, insertion, incorrect sequences, corruption, delay, and interchanging of safe data with standard data. The safety protocol is certified according to IEC 61508, and it also has been tested by the TÜV inspection association as regards safety requirements. For safe data transfer, SERCOS safety uses the CIP (common industrial protocol) safety protocol from ODVA (Open DeviceNet Vendors Association). This is used by various communications standards such as DeviceNet, ControlNet, and Ethernet/IP and allows users to employ the same safety mechanisms on various platforms. This allows the seamless connection of multiple CIP-based networks.

Advantages of SERCOS

By combining **high performance, flexible use, and verified safety,** SERCOS III satisfies all the requirements for a modern, integrated automation network. Suitability for everyday use is provided by the service-proven capabilities of the SERCOS protocol, whereas as a real-time Ethernet solution it also offers future compatibility. Thanks to Fast Ethernet and synchronous cycle times of 31.25 μs, SERCOS III features very high-performance data and can handle even very complex automation tasks. For example, up to 330 drives with 4 bytes of I/O data and eight digital I/Os, respectively, can communicate with each other in a cycle of a millisecond. SERCOS III's performance thus already exceeds the requirements of today's advanced production machines. What is more, SERCOS III allows quick processing of process data via decentralized I/O modules in central control systems.

4.5 Advantages of PLCs

At first, PLCs were relatively expensive and complicated to program and had limited performance. Therefore, they were initially used mainly on prototypes, custom-built machines, and special-purpose machines, which, based on experience, would involve large-scale changes in the wiring during the commissioning and test phases. The first experiences with PLCs were not particularly spectacular, but their potential to save time and costs was recognized. Continuous increases in performance combined with reductions in price and space requirements have resulted in PLCs being used in more and more areas. For simpler applications, compact devices with a limited number of inputs and outputs will suffice. Where demanding performance requirements have to be met, PLCs are available in a graduated range of different sizes, with a large number of inputs/outputs and a comprehensive set of instructions.

Today's PLCs have major advantages compared with relay-based control systems or earlier electronic digital control systems. These include the following:

- Compact installation space, smaller electrical cabinets
- Elimination of extensive wiring through the use of data buses and field buses
- Significantly lower power consumption and heat generation
- Greater reliability (e.g., no switching contacts, fewer cable connections/contact points, longer service life of electronic modules, wear-free software)
- Online correction of the PLC program without interrupting operation
- In series-produced machines, the PLC program can be copied with no changes.
- Shorter switching times and response times
- Remote diagnosis and troubleshooting via Internet/Ethernet connection
- High-performance mobile programming devices in laptop format
- Automatic documentation instead of individually created circuit diagrams
- Integrated automatic function test software with error indication
- Overall significantly lower time and cost requirements

Today, PLCs are an indispensible component in almost all industrial machines and plant components. The basics of PLCs and how to work with them is already being taught in vocational and technical schools, with further training being offered by the manufacturers. Trained personnel are absolutely essential for the programming, deployment, and connection of PLCs, as well as subsequent troubleshooting.

4.6 PLC Programming and Documentation *(Figure 4.5)*

As with CNC machines, with PLCs, cost-effective use is heavily dependent on the capabilities of the programming system and the programming language. This requires user-friendly programming, and all the functions have to be programmable. The programs that are generated have to be error-free, and it should be easy to make modifications.

Although attempts were made from the beginning to standardize programming, first with DIN 19 239 and later with IEC 1131, even today, programs are specific to the individual PLC and not interchangeable.

For the simplest low-cost applications, programming devices with **symbol keys and function keys** are adequate. Experience has shown that this does not present any major difficulties to the user as long as he or she has experience in working with relay-based control systems. The ladder diagram shown on the screen during programming is very similar to traditional circuit diagrams. Owing to its limited range of functions, however, this method is not suitable for complex machine control systems.

Wherever the PLC programming for machine tools becomes significantly more complex, laptops with a **Windows interface** are almost always used. The PLC programming software is supplied by the PLC manufacturer. The need in the early days to use a manufacturer-specific programming device for each PLC has now been overcome. This removed what for a long time had been one of the principal obstacles to the quicker expansion of PLC systems.

Laptops as PLC programming devices provide a number of other advantages, such as

- Universal, portable, complete mobility
- Paperless documentation, always up to date and related to the specific system
- Availability of built-in help software immediately on the spot
- Automatic documentation of all changes
- Trouble-free data management, for example, documentation on a central computer via a standard Ethernet interface
- Replaces manuals, drawings, instructions, and handwritten notifications of changes

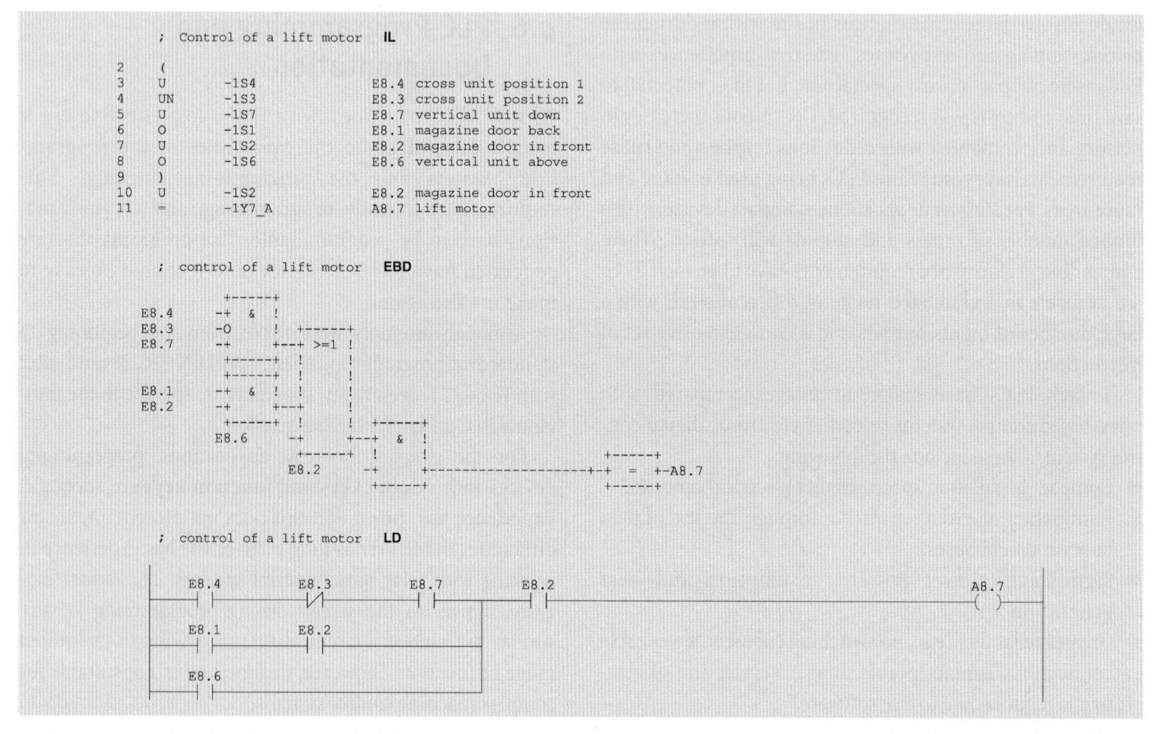
; Control of a lift motor IL
2 (
3 U -1S4 E8.4 cross unit position 1
4 UN -1S3 E8.3 cross unit position 2
5 U -1S7 E8.7 vertical unit down
6 O -1S1 E8.1 magazine door back
7 U -1S2 E8.2 magazine door in front
8 O -1S6 E8.6 vertical unit above
9)
10 U -1S2 E8.2 magazine door in front
11 = -1Y7_A A8.7 lift motor
; control of a lift motor EBD
E8.4
E8.3
E8.7
&
>=1
E8.1
E8.2
&
E8.6
&
E8.2
=
A8.7
; control of a lift motor LD
E8.4
E8.3
E8.7
E8.2
A8.7
E8.1
E8.2
E8.6

Figure 4.5: Programming and documentation of PLCs as an instruction list (IL), function block diagram (FBD), and ladder diagram (LD).

The **Windows operating system** provides a number of options for PLC programming:

- According to IEC 1131 (IEC 61131 in Europe)
- In the high-level language C or using Structured Text (ST), primarily for mathematical tasks
- As with older devices, according to DIN 19 239, as

  IL = instruction list

  SFC = sequential function chart

  LD = ladder diagram

  FBD = function block diagram

Of these, ST, IL, and SFC are classified as textual languages, whereas LD and FBD are graphical languages. All of them are included as standard in all of today's PLC programming systems. It is important for the user to select a PLC that includes the required scope of functions and which also offers personnel the benefit of being able to convert existing circuit documentation into PLC programs without any major difficulties. All systems provide the advantage that when input is made in one form, the other form appears simultaneously. This means, for example, that during programming, as an instruction list, a ladder diagram is also created and that it is possible to switch between each of the input modes.

Once the program has been created, the user can expect further support from the programming system, such as

- Depiction of the **IL** with all comments and device designations
- Depiction of **assignment lists,** which show how the connections are assigned
- Depiction of **cross-reference lists,** in order to determine which input or output is being "talked to" at which address
- Depiction of the **ladder diagram,** which contains the configuration and designations of the contacts, as well as additional written information

- Support when troubleshooting owing to individual steps, interrupt points, display of memory contents, etc.
- Archiving of PLC programs

A significant advantage of so-called **mnemonic languages** is that they do not limit the performance of the PLC in any way. Programming languages based on **Boolean operations** are no longer offered today.

4.7 Programs

The logical connections between the variable input signals and the output signals that are generated in an electric circuit are conventionally depicted in an electrical circuit diagram. Using Boolean algebra and the appropriate mathematical laws, it is possible to convert such ladder diagrams or circuit diagrams into **logic diagrams.** This is the actual task of the programming system.

PLC programs are essentially based on the fundamental functions AND, OR, and NOT. By combining these three logic operations, it is possible to construct further logical functions such as NAND and NOR. On top of these come time and counter functions, shift registers, monostable and bistable clock generators, and so on.

The program generated in this manner is entered into the memory of the PLC's CPU, on which its task is to generate the desired control sequences. Proper execution of the program depends on the sequence of the programmed **commands.** A **command** is defined as one individual program line of the PLC program.

It is a characteristic of PLC systems that the program is executed sequentially, that is, step by step, one command after the other. Although this serial processing takes place at high speed, on average at about 0.1 ms per 1,024 program steps, the execution of the overall program depends on how long it is, that is, on the number of commands. Therefore, for longer programs, the entire program cycle may take several milliseconds. This time is called the **cycle time** and indicates the **response time** of the PLC.

For example, at a cycle time of 20 ms, the program is run through 50 times per second. If the input status of a signal changes immediately after it is scanned, then it will take a maximum of 20 ms until the next time it is scanned. This time can only be reduced by using a "faster PLC" or one with special jump commands.

In addition to the logic operations, the PLC often has other tasks, for instance, those related to "data handling," such as the administration and updating of tables, detection and decoding of barcodes and assigning them to the correct table, correspondence via information network, and so on.

In modern PLC systems, it is possible to define program sections with various cycle times or to make them event-controlled. In this way, it is possible to execute a very time-critical task (e.g., tool-breakage monitoring) with a significantly quicker cycle rate than that for functions that are less time-critical (e.g., compensation values owing to heating).

4.8 Program Memory *(Figure 4.6)*

Today's PLCs use exclusively **semiconductor memory** with various characteristics. For initial testing of new programs, **RAM** with a backup battery is preferred in order to be able to quickly insert and test changes. These memory modules are "volatile" in the event of a power failure and therefore have to have a backup battery power supply. A high level of reliability means that there is no problem if subsequent operation also takes place via the RAM.

Once test operation is complete, the preferred procedure is to transfer the programs to **FEPROMs** for storage and prevention against loss. These data-storage devices, which can be in the form of *memory cards* or *sticks*, can be erased by means of an electrical pulse and then written to again immediately. To do this, it is *not* necessary to take them out of the device or out of their slot; erasure and writing take place on-board. There is no need to manually exchange the storage device.

The **EPROMs** that were used formerly and which could be erased using ultraviolet (UV) light are no longer up to date. They can only be written to again after a waiting period of about 1 hour.

In industrial PC-based PLCs, data also can be saved on the computer's **hard drive.** When the computer is switched on, the PLC program is transferred to the RAM, resulting in a short waiting period.

With today's memory modules, the **storage density,** that is the number of bits or bytes that can be stored on a particular volume, is so large that the program length no longer plays any role. When structuring and optimizing a program, less effort is devoted to minimizing the program

Memory type	Description	Erasing	Programming	When the power is cut off, the memory is:
RAM (SRAM) (DRAM) (SDRAM)	Random access memory Memory with random access Read/write memory	Electrically	Electrically	... Volatile
ROM	Read only memory Read-only storage	Not possible	Via masks during the production process	... Non-volatile
PROM	Programmable ROM Can be programmed once		Electrically	... Non-volatile
EPROM	Erasable PROM Read-only storage, erasable with UV	Via UV light	Electrically	... Non-volatile
FEPROM	Flash EPROM Read-only storage, erasable electrically	Electrically	Electrically	... Non-volatile

Figure 4.6: Read-only storage types and their characteristics.

length but rather to making it clearer and easier to diagnose and to give it a subprogram structure. This is important most of all to the user to allow quick troubleshooting and to minimize downtimes in the production system. It is also possible to provide **diagnostic programs** that record not only the control sequence and cycle times but also the point where the interruption occurred, triggering an exact indication of the fault in plain text. Special self-learning **diagnostic functions** are also available. These first learn the correct sequence for correct functioning and then compare each sequence with the stored one. In the event of a fault, they indicate the program step where the error occurred.

4.9 PLCs, CNC Systems, and PCs in Integrated Operation *(Figure 4.7)*

The most important building block for PLC technology had already been laid by the end of the 1970s through the use of microprocessors. These could be used to process much larger amounts of information in a very small space, and their improved price/performance ratio made PLC technology attractive even for smaller systems. The most important arguments here were the ability to copy existing executive programs quickly and without errors and to insert modifications without difficulty even at a later time.

PLCs of a specific design are even suitable for use in safety-related areas and for systems that have to satisfy high requirements as regards availability. This all meant that PLCs became established as the premiere automation system, especially in manufacturing technology. The ability to save current-switching states in the event of a power failure contributes greatly to fast, error-free, safe restarting of the system once power returns.

Future developments will be directed less toward further miniaturization and increased performance of the PLC and more toward optimization of interactions with

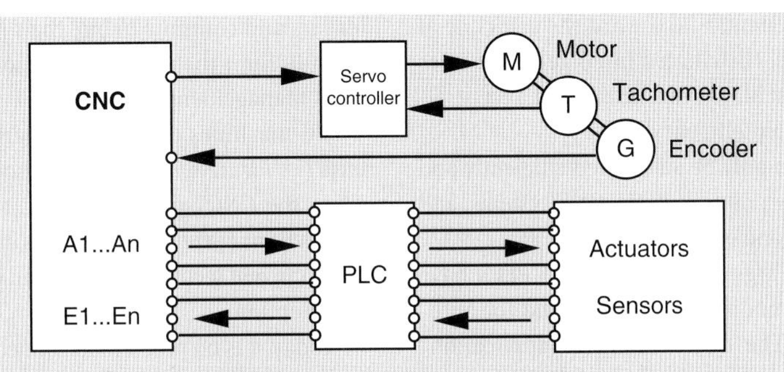

Figure 4.7a: CNC with separate PLC; information exchange takes place via I/O modules.

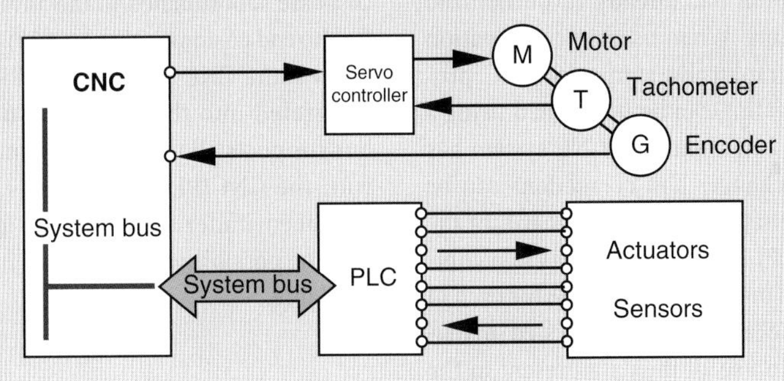

Figure 4.7b: CNC with PLC linked via bus; formation exchange takes place directly via the bus, without I/O modules.

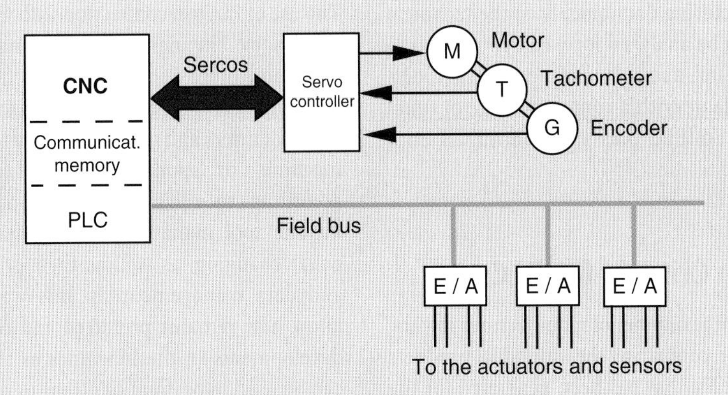

Figure 4.7c: CNC with PLC integrated via software; the information exchange takes place in the CNC/PLC operating software.

Figure 4.7: Options for CNC/PLC integration.

other automation components and toward the principle of **distributed intelligence.** Distributed intelligence involves the relocation of function blocks to a number of decentralized points, for example, intelligent drives.

Today, many tasks can only be automated efficiently and economically if the strong points of PLCs, CNC systems, and computers are combined and used interactively.

The most important arguments in favor of PLCs have always been the ease of modifying programs and the automatic documentation. The development of application-oriented, graphical tools has revolutionized PLC programming in much the same way as graphically supported programming and simulation had already revolutionized the creation of NC programs. When used in combination with CNC systems, the graphically supported PLC programming and function testing can be performed using a separate PC, for example. Later, once the CNC system has been connected, the last changes and corrections can be made using the CNC system's screen and keyboard. In the end, the CNC system and PLC even have access to a common database. Only in this manner is it possible for each of the systems integrated into a computer-integrated manufacturing (CIM) network to use all the data present in the system, to update the data automatically, and to pass the data on without delays. In this way, it is also easy to send machine and manufacturing data quickly to the technical office, where they can be analyzed for weak points and to provide information for management purposes. Today, CNC systems are available with a complete range of functions plus integrated SoftPLC plus axis control, all on *a single* PC plug-in card.

4.10 Selection Criteria for PLCs

PLCs are offered by a large number of manufactures in various countries. However, comparing the individual products with the requirements list that has been generated for the specific customer will cut down the number of choices significantly. It is often possible to avoid this effort by taking a look at large-scale users of PLCs, who test the products that seem most attractive every three to five years and make new selections accordingly. On the other hand, large companies in particular rely on the strengths of market leaders and continue to buy their products. This also helps them to limit their costs for personnel training and spare-parts management.

NC machines often involve very different selection criteria (→ *Figure 4.7*). The most resource-intensive solution is when the CNC system and PLC are linked via individual connections (→*Figure 4.7a*). This does not exploit the potential of current technologies and can no longer be considered a viable option.

Today, it is no problem to connect CNC systems and PLCs from different manufacturers via standardized data interfaces (e.g., Ethernet). In such solutions, the somewhat higher costs are subordinated to the advantages of PLC uniformity for the end customer (*Figure 4.7b*).

In some CNC systems, the PLC functions are already integrated into the CNC as a **software PLC**, and a field bus connects all the actuators and sensors of the machine with the CNC/PLC (→ *Figure 4.7c*). This ideal solution had been under development for a long time but only became possible when standard interface modules became available.

4.11 Summary *(Table 4.1)*

The use of PLCs in connection with CNC is the general standard today. The increasing performance of PLCs has led to more and more of the tasks and functions bcing moved from the CNC system to the PLC. This gives machine manufacturers and users the ability to program and modify machine- or application-specific functions themselves according to their own conceptions. These functions include tool management, tool-changing processes, and pallet changing, as well as their graphical representation and data management. In this way, the manufacturers' know-how remains protected and can be adapted to new developments in the machines as they appear. PLCs also can be used in optimal solutions to requirements arising in connection with the automation of complex manufacturing equipment.

In **series-produced machines,** the fact that tested, error-free PLC programs can be copied means that commissioning time can be much shorter. For these machines, controllers with integrated CNC + PLC + axis command value output are also an ideal solution because few machine-specific program modifications are to be expected. In contrast, for **special-purpose machines and complex systems,** a separate PLC has the advantage that the machine manufacturer can program and test all the process functions of the individual subcomponents even before commissioning of the CNC system.

For further information about bus systems, see

- www.sercos.de
- www.profibus.de
- www.interbusclub.com
- www.ubf.de/ethernet.htm
- www.tecchanel.de

The essential differences between CNC systems and PLCs when used with machine tools

Criteria	Numerical control systems CNC	Programmable logic controllers PLC
1. Designation	(Computer) **N**umerical **C**ontrol	**P**rogrammable **L**ogic **C**ontroller
2. Technical design	Largely standardized hardware and software, but modified on a machine-specific basis for control of a specific type of machine.	Standardized, expandable universally applicable control hardware for all switching function of the machine and periphery, or for groups of machines.
3. Task definition	The main task is the control of machine axes, that is, the relative motion between the tool and workpiece through direct input of dimensions, and additional necessary technological functions (**F**eed, **S**peed, **T**ool, **M**iscellaneous Functions)	The main task is controlling, interlocking, and linking defined, repeating processes in mechanical engineering and systems engineering. The PLC performs coordination of the processes.
4. Functional characteristics	The NC programs for machining the workpieces are created by the machine *user* and can be exchanged or modified as desired.	The PLC program is created by the *machine manufacturer* and is saved to nonvolatile memory. It only has to be modified or exchanged in exceptional cases.
5. Programming	Workpiece-oriented, dimensional programming of the necessary traversing motions of the NC axes and of the workpiece target dimensions in accordance with the workpiece drawing. Optionally, also NC programming at the machine (SFP). Universal NC programming systems generate source programs that can be converted into run-capable NC programs for each manufacturer of machines/CNC systems by means of compilers (postprocessors).	One-time programming and saving of the functions being controlled using instruction lists (ILs), function block diagrams (FBDs), flow diagrams (according to IEC 61131), or ladder diagrams (LDs), as well as structured text (ST) for process data processing. Programming using PCs and company- or device-specific software. No programming during operation.
6. Programs	The NC program contains the geometric machining sequence and the necessary switching functions for feed rate, spindle speed, and tool and auxiliary functions, as well as the existing automation devices (see point 3). The program structure is standardized internationally according to DIN 66025, ISO, and STEP-NC.	The programs have to be created on a PLC-specific basis and cannot be compiled for PLCs from other manufacturers. (The international standard IEC 61131 has been published but is not binding owing to the large number of programs that exist.)
7. Existing programs	Up to several thousand NC part programs per machine are not uncommon. The part programs are created by the machine user; fixed cycles and subprograms are supplied by the manufacturer.	As a rule, only a *single* fixed, system-specific program. The program is created by the machine manufacturer, generally using available function modules.

Table 4.1: Comparison of CNC Systems and PLCs

The essential differences between CNC systems and PLCs when used with machine tools		
Criteria	**Numerical control systems CNC**	**Programmable logic controllers PLC**
8. Areas of application	Flexible control of the machine. The CNC system has to be specially adapted to the type of machine being controlled as regards NC axes, cycles, and subprograms, for example, machines for turning, milling, drilling, nibbling, grinding, cutting, laser machining, etc.	Control, interlocking, and linking of machine functions. The modular hardware does not have to be specially adapted to the machine but only has to satisfy the maximum requirements with regard to the number of inputs, outputs, times, counters, functions, and amplifiers. Machine-specific design is performed via the program.
9. Technical scope	Very extensive! Specification is very substantial, machine-dependent, and requires a great deal of explanation, even when discussing potential sales.	Medium to high, depending on the performance of the device, can be explained with a few standard functions when discussing potential sales.
10. Procurement, purchasing	Customer purchases a complete, functional system.	Customer purchases a hardware module.
11. Project planning	Purchaser and seller have to have extensive knowledge of machine tools in order to be able to adapt the CNC system (3D milling, turning, inputting and processing compensation values, HSC functions, cycle times, servo sampling rates, machining cycles, coordinate transformation, DNC, measuring systems, etc.).	The seller has to have basic knowledge of the construction of specially machines and should be familiar with controller construction and control technology. Advice to the customer primarily involves checking whether the required times, function, processes, etc. can be programmed (selection of the hardware components).
12. Historical development	Replacement for copying systems, control cams, program controls with limit switches, and mechanical automation. The NC programs controls the machine axes and the complete machining process. Fulfills the requirement (a) for quick program changes, short setup times, greater manufacturing flexibility, and greater manufacturing precision, as well as (b) for direct use of CAD design data for programming of machine motions = CAD/CAM, and (c) for full integration of data updating in a closed loop.	Replacement for relay controllers and fixed-programmed (wired) electronic controllers. The term *programmable logic controller* arose owing to the control programs stored in the PLC memory. Fulfills the requirements for (a) smaller space requirement, better reliability, larger scope of functions, lower wiring requirements and fewer sources of errors, more flexible modification options, better documentation, and shorter construction times and (b) adaption of the various I/O bus systems (data bus and field bus).
13. Innovations	From the automated stand-alone machine to the development of complex machines that could not be controlled without a CNC system, such as laser machining, stereo lithography, high-speed milling rapid prototyping methods hexapods, robots, etc.	From small controllers for the simplest functions to computer-aided 32-bit controllers with a very wide range of digital and analog functions. Can be expanded to include computer-integrated systems with complex computing functions. Various I/O bus systems can be used.

Table 4.1: Comparison of CNC Systems and PLCs (Continued)

The essential differences between CNC systems and PLCs when used with machine tools		
Criteria	Numerical control systems CNC	Programmable logic controllers PLC
14. Networking	Increasing networking of multiple systems in one plant area. Process bus: Ethernet for medium distances and small, non-time-critical amounts of data (e.g., NC programs) and TCP/IP, Internet protocol for networking of decentralized systems. Field bus: SERCOS and CAN for drives, RS 485 for sensors and actuators. CAN (ISO/DIS 11 898): Low cost, short response time (developed for the automotive industry).	Networking of PLCs from various manufacturers and of the central controller to the decentralized inputs/outputs. Plant bus: Industrial Ethernet PROFIBUS FMS PROFIBUS (DIN 19 245) PROFIBUS DP (Decentralized Peripherals) InterBus-S (DIN 19 258) CAN-bus for machines and systems ASI = actuator/sensor interface for binary I/Os.
15. Service	Requires hybrid knowledge of machine tools (mechatronics): machine elements, hydraulics, pneumatic systems, electrical systems, electronics, position-measurement technology, logistics, servo drives, control circuits, PLCs, and the associated measuring technology. Being able to work with PCs is essential.	In most cases it is adequate to have knowledge of the control technology and programming. The knowledge required for CNCs is only necessary in exceptional cases. Being able to handle and work with PCs is a basic requirement.
16. Development trends	There is a tendency toward position-measuring systems with a greater measuring accuracy for the NC axes (0.0001 mm or 0.00001°). Interpolation in the nano- and picometer range in order to achieve more precise 3D workpiece surfaces. Use of standard PC hardware in combination with real-time operating systems. Use of digital axis drives and special function modules for maximum contour accuracies in high-speed milling. Use of highly dynamic linear drives.	Strong tendency toward centralized controllers with bus connections to decentralized I/O modules. Use of equipment from various manufacturers for the centralized controller and centralized I/O modules. Internationally standardized, graphically supported programming language according to IEC 61131.

Table 4.1: Comparison of CNC Systems and PLCs (Continued)

PLCs: Programmable Logic Controllers

Important points to remember:

1. PLC stands for **programmable logic controller**.
2. What were formerly NC systems with a separate machine control interface now have become CNC systems with a PLC that is either integrated or linked via a bus.
3. PLCs replace not only relay controls but also perform **additional** control, monitoring, and display functions.
4. **Basic functions** include
   - AND, OR, NOT, SAVE, DELAY
5. **Additional functions** include
   - COUNT, COMPUTE, COMPARE, JUMP COMMANDS, SUBPROGRAM TECHNOLOGY
6. **Higher functions** are table management, analog (A)/digital (D) and D/A conversion, closed-loop control, NC axis modules, and communication via data networks.
7. Important characteristics of PLCs include
   - Range of functions
   - Maximum number of inputs and outputs
   - Cycle time, measured in ms/k commands (ms per 1,024 commands)
   - Number of flags
   - Size of the program memory (for commands)
8. Instead of wiring, PLCs use a **program**, also called **commands**. The program is created by the user with the aid of a PC and system-specific software and specifies how the individual signal inputs are linked and blocked relative to each other so as to generate the required output signals.
9. The **program** is saved in electronic modules: in **RAM** for testing and in **EPROM** or **FEPROM** for subsequent operation.
10. There are five ways to **program** a PLC:
    - As a ladder diagram
    - As an instruction list
    - As a function block diagram
    - As structured text
    - Using a graphically supported language.
11. **Advantages** of PLCs include
    - Hardware can be installed and wired independently of the software.
    - Significantly shorter times are needed for assembly and commissioning.
    - Quick and easy corrections can be made, even during the commissioning phase.
    - Automatic documentation and copying of the software programs is possible.
    - Automatic generation of cross-references, notifications, and other information is possible.
    - There is no wear, thus high reliability.
    - Easy installation and small space requirements.
    - Significantly lower power consumption than relay controls (approximately. 0.1 percent).
    - For series-produced machines, significantly shorter commissioning thanks to identical programs that have been thoroughly tested and confirmed to be error-free.

5. Effects of CNC on Machine Components

CNC has brought about lasting changes to major components of machine tools, leading to new machine configurations and types of automation equipment.

5.1 Machine Configuration

The principal reason for numerical control's influence on machine configurations is the fact that, as with conventional automation, it becomes possible to do without continuous tending and observation of the work sequence by a worker. At the same time, the continued development of cutting tools has led to significantly higher cutting speeds, feed rates, and cutting depths, which, in practice, can only be implemented without a human operator. These improvements offer greatly increased performance but at the same time place new requirements on the machines.

Especially machines used to process smaller workpieces offer many options in this regard. With large workpieces, the weight of the workpiece places tight limits on the feeding and load-bearing characteristics of the machine, as well as on the dimensions, clamping, and large working space.

Thus, for small and medium-sized turning machines, it has become common to place the bed at the rear, whether inclined or vertical, as had already been introduced for copying lathes. In this way, the chip clearance is not obstructed by the bed. This also has the advantage that it is possible to swivel the transverse axis away from the horizontal, significantly improving the access to tools and the workpiece. Space is created under the workpiece for installation of a chip conveyor.

One new machine design was prompted by the development of variable-speed three-phase motors. Here, there is a vertically suspended spindle, with the longitudinal and transverse motions being executed not by the tool but by the spindle. As a result, not only is there good chip clearance even during internal machining, but the motion of the spindle also allows easy tool changing using the pickup principle. Here, too, there is good access to tools and the workpiece and space for a chip conveyor (*Figure 5.1*).

In large turning machines, the configuration of conventional lathes has been retained, that is, a horizontal bed in machines for long workpieces and a vertical design for short workpieces. For longitudinal and shaft turning machines, it may be possible to install a chip conveyor inside the bed. In vertical lathes, the horizontal clamping surface interferes with chip removal. Automatic chip removal is almost impossible, especially during internal machining.

Free program design and the practically unlimited amount of data that can be entered have lead to lathes with two or even three heads that work independently but in coordination with each other. This shortens the production time per piece and increases the number of tools available.

Even lathes developed especially for automated production, in which the tool motions are generated in a conventional manner using cam discs (e.g., multiple-spindle automatic lathes), have in the meantime been converted to numerical control. The fundamental machine configuration was retained, however.

As for drilling machines, the radial drilling machine, with its non-Cartesian motion directions, has been discarded entirely. The standard type is now a design with a vertical spindle and a horizontal workpiece table. The division of the axes among the table and pedestal follows the working space of its conventional predecessors. As a result of automation, however, automatic tool changing becomes almost obligatory, with a corresponding influence on the machine configuration.

In milling machines as well, it is possible to find machine configurations that already existed among conventional machines. However, the freedom of programming in numerical automation set loose a trend toward complete machining and thus a new type of machine: the machining center, a machine that makes possible all types of machining that use recirculating tools. Naturally, this makes it necessary to have

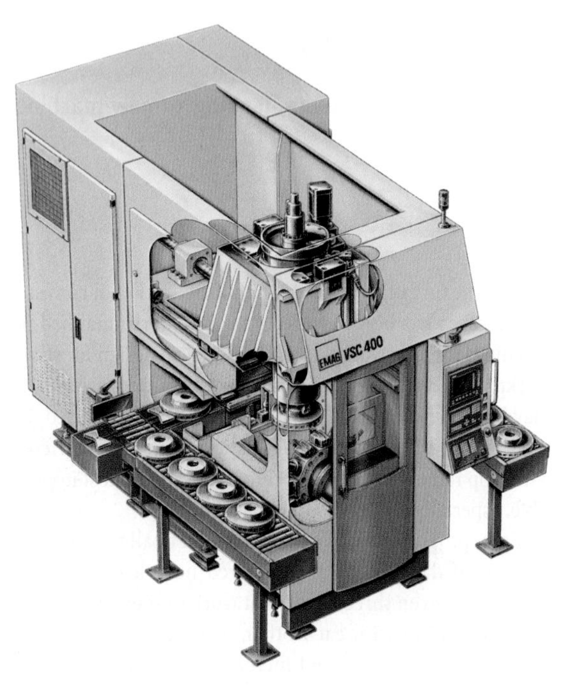

Figure 5.1: Vertical turning machine using the pickup principle for automatic loading and unloading of workpieces.

an appropriately dimensioned tool changer and one or even two rotational axes in addition to the three translational axes. The configurations of these machines also mainly correspond to those of conventional machines, especially to those of boring mills. To expand the automatic sequence, these machines generally are also equipped with an automatic tool-changing system. This additional equipment, of course, also influences the structure of the basic machine.

In all these machines with a horizontal clamping surface, there is a problem with chip removal and installation of a chip conveyor. For this reason, in a few cases in machines for processing smaller workpieces, the workpiece is machined with a vertical clamping surface after being clamped onto the surface in the horizontal position in the clamping station.

The configuration of the original conventional machines generally also was retained in the various types of grinding machines. The reason for this was that priority had to be given to the requirements of the grinding process. Only with smaller external surface grinding machines, in which the longitudinal motion traditionally is assigned to the workpiece, was a different configuration occasionally adopted: that of a lathe with an inclined bed at the rear. Thus the longitudinal motion is executed by the grinding head.

Gear-cutting machines are in their nature single-purpose machines with a fully automatic sequence. Therefore, their configuration was kept completely the same during the transition to numerical control.

No entirely new configuration for machine tools with a completely different structure appeared until after CNC control systems with high-performance computers became available: machines in which the tool was positioned via a parallel kinematic system (→ *Part 3, Section 1.6*). Here the command position values specified in the Cartesian coordinate system had to be converted quickly into the command values for the lengths of the individual jointed arms. Because of the low masses being moved, this kinematic system allows very rapid responses but has serious disadvantages when it comes to the range of motion that is possible, especially for swivel motions.

5.2 Machine Frames

The requirements placed on machine frames are essentially identical to those for conventional machine tools. However, the higher precision required means that optimization of the static and dynamic stiffness is necessary. For trouble-free automatic production, maximum thermal stability and low thermal drift are also important. Temperature changes from the environment or owing to heat sources in the machine should not lead to gradually increasing positional deviations in the machine. The heat sources in the machine can cause particularly noticeable negative effects as a result of high power conversion. These include hot chips that heat up the machine frame locally, or high loads on the main drive motor, or the heat generated by the bearings of a rapidly rotating main spindle. Machine frames made of cast mineral composites are useful here owing to their large mass and the poor thermal conduction of the concrete (*Figure 5.2*).

A certain scope for free design of function-related machine frames comes from the fact that today CNC machines, especially small and medium-sized ones, generally have a machine enclosure on all sides. Thus the visual effect can be ignored in design of the frame.

Figure 5.2: Machine frame made of cast material composites.

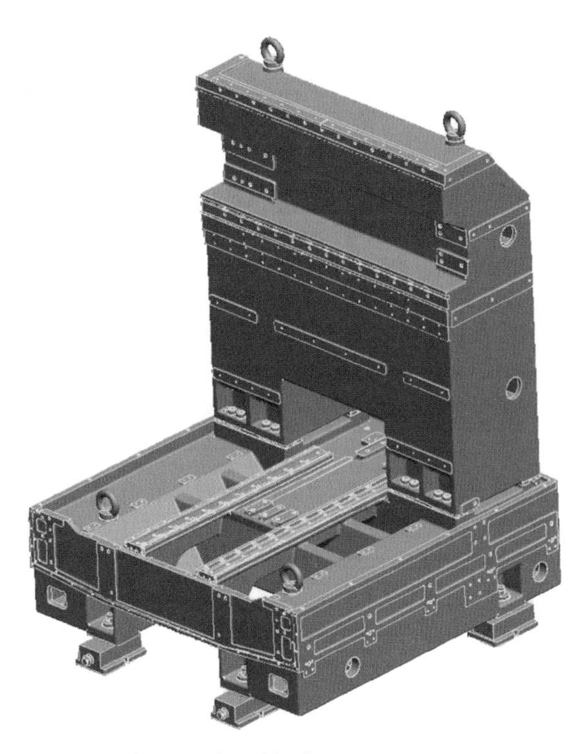

Figure 5.3: Cast-metal machine frame.

Machine parts that are moved in closed-loop position control are subject to particular requirements with regard to their weight, especially those positioned by a linear drive. For machine slides, optimization of stiffness using finite-element method (FEM) analysis and weight reduction by means of topology optimization can produce considerable benefits not only in cast constructions but also most notably in welded designs. Because of their cost, fiber composite materials have not achieved much success so far in series production (*Figure 5.3*).

5.3 Guides *(Figures 5.4 and 5.5)*

As a general rule, guides, especially motion guides that are moved during machine operation, are subject to the following requirements:

- Low friction and no stick-slip effect so as to allow precise positioning
- High stiffness, in order to absorb operating loads without excessive displacement
- High damping to suppress vibrations
- Low wear to ensure precision over a long time period
- Low costs.

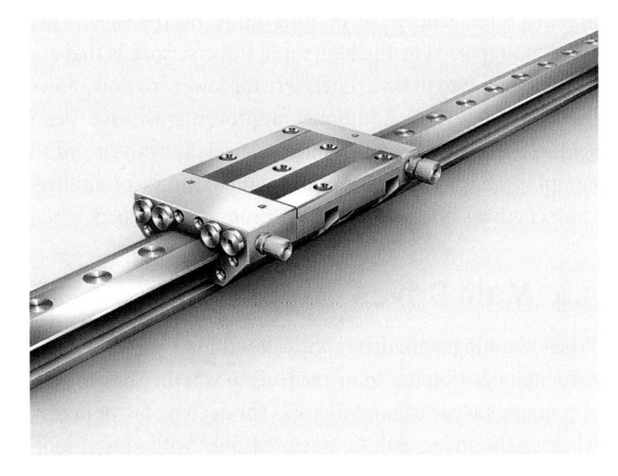

Figure 5.4: Hydrostatic guide with identical dimensions to a linear roller guide in order to ensure interchangeability. (Image courtesy of INA.)

Figure 5.5: Damping test: On the left side, a roller guide; on the right side, a hydrostatic guide. (Image courtesy of INA.)

In conventional machine tools, these requirements were met adequately by sliding guides with a wide variety of designs. They could handle heavy loads, were reliable under operational conditions, and provided good damping. Low friction and the no-stick/slip effect could be achieved by lining them with plastic low-friction liners.

In contrast, closed-loop position control requires particularly low friction and freedom from stick/slip effect in order to achieve high positioning accuracy. Therefore, today, diverse types of rolling guides are used in numerous controlled machines. These guides are supplied by specialist manufacturers and have become quite inexpensive. This trend is supported by the high rapid traverse speeds that are being introduced to save time. Here the lower friction allows lighter feed drives. Additional improvements have been achieved using hydrostatic guides, especially with regard to damping characteristics. Some manufacturers of rolling-contact guides offer these as mass-produced products.

5.4 Main Drives

When planning main drives with closed-loop control, a fundamental decision has to be made as to whether to use synchronous or asynchronous motors. The decisive factor here is whether the motor will be operated only with closed-loop speed control (e.g., as a spindle drive for drilling and milling tools) or with closed-loop position control (e.g., with turning machines with an additional C-axis drive).

Today, all spindle drives in machine tools without exception use electric motors with closed-loop speed control. These have two main tasks:

1. To provide the **torque** and **speed** needed for the work process
2. To allow **interpolation of the speed of the main spindle** with the feed drives, if operation as a C axis is required in turning or machining centers

The automatic work sequence in CNC machines means that the main drives have to satisfy some additional requirements that exceed the requirements placed on drives for conventional machines. These are in particular

- **Automated speed changing.** The automatic work sequence also requires programmable, automatic changing of the speed.
- **Speed changing in very small steps, preferably continuously adjustable.** CNC machines are capital-intensive items of equipment with a high cost per hour. Therefore, it is important to make optimal use of the performance of state-of-the-art tools. For example, to keep the cutting speed constant during face turning or taper turning, continuously adjustable speed changing will be necessary for technological reasons.
- **Large speed adjustment range.** CNC machines are universal machines that are intended to process various types of workpieces using various tools. To do this, the main spindle has to cover a large range of speeds without any intermediate transmission; that is, the entire required adjustment range is provided by the motor alone.
- **Very fast speed changes.** Every change in speed represents a loss of time. This is especially noticeable when there are frequent tool changes on machines with recirculating tools, such as in machining centers. These are equipped with an automatic tool changer, and the spindle has to be stopped for each tool-changing operation.

  As a result, a spindle motor with the shortest possible run-up and braking times is required.
- **High drive output.** The automatic execution of the machining process in CNC machines makes them independent of manual control and reaction speeds. This allows a completely enclosed working space. It is thus possible to achieve working speeds that make full use of modern tools' performance potential. This requires drive

outputs that exceed those of conventional machines by a large factor.

- **Large range with constant output.** The high drive output should be available over as large a range of speeds as possible.
- **High torque at low speeds.** As high torque as possible should be available at lower speeds.
- **Compact dimensions and low weight.** In many CNC machines, the main drive motor is part of a larger mechanical assembly and constantly moves along with it. There is thus a requirement for motors that are as light and compact as possible so as not to compromise the acceleration that can be achieved by the entire assembly.
- **Low heat generation.** The adverse effects that localized heating of the machine has on precision were already been mentioned earlier (→*Part 2, Section 1.3*).

Types of Main Drives *(Figure 5.6)*

In principle, the same types of motors are available as for feed drives. Speed-controlled **asynchronous motors** are preferred for use as standard main drives owing to their positive features, such as low price, their simple, rugged design, and their low maintenance requirements. For speed adjustment, they receive their power from a frequency converter. By now they have supplanted the previously dominant direct-current (dc) motors. They appear in various designs. Depending on the nature of the main drive (→ *Figure 5.6*), they can take the form of housed motors or kit motors with a hollow shaft (*Figure 5.7*).

In order to ensure the full torque at low speeds and even down to a speed of zero, housed motors for main spindle drives are always designed with external ventilation or liquid cooling. Kit motors for direct installation in the spindle are as a rule always liquid-cooled because it is necessary to ensure both high power density in a reasonably small space and a motor with a thermally neutral signature.

Designs for Main Drives

The classic design of the main drive involves the coupling of a housed motor to the tool spindle via a geared or belt transmission, in some cases with multiple stages. This arrangement has the advantage that the motor is thermally decoupled from the machining space and from the spindle. The motor can be installed in a location outside the machining space, meaning that main spindle motors with standardized installation dimensions can be used. However,

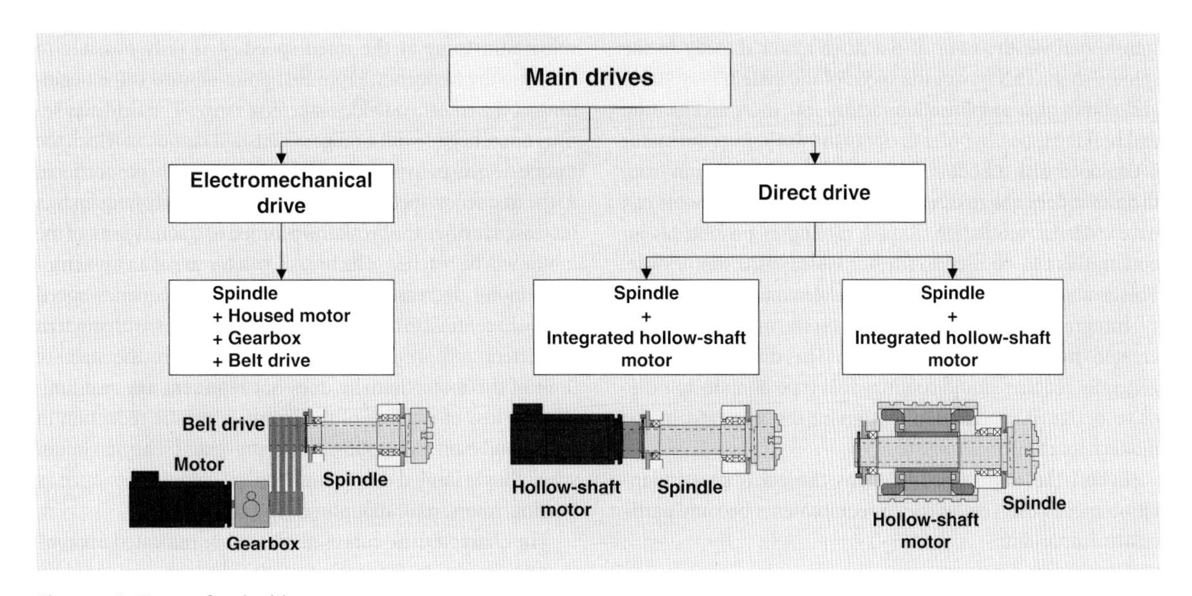

Figure 5.6: Types of main drives.

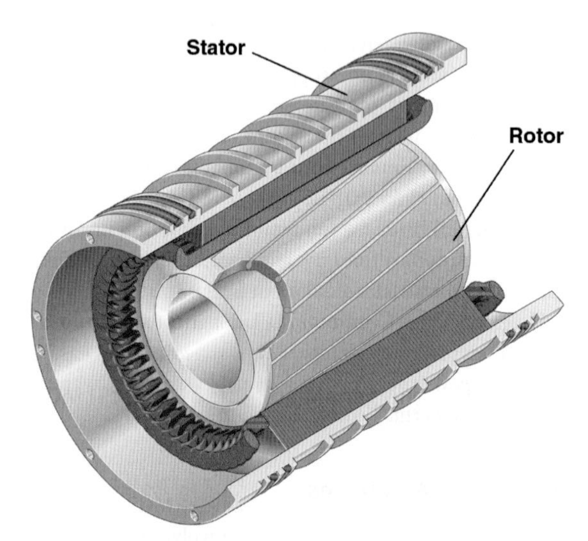

Figure 5.7: Three-phase asynchronous motor as a kit motor with hollow shaft.

the belt drive limits the speed, stiffness, and dynamic characteristics of the drive and thus the productivity of the entire machine tool.

These disadvantages have led to the use of directly driven spindles. The belt or gear drive is eliminated—the torque is transmitted via the rotor of the drive motor directly to the spindle shaft. This makes the speed of the system very stable and allows high amplification factors and short acceleration and braking times. In order to clamp the workpiece, the motor is equipped with a hollow shaft. Because the heat input from the motor does not go directly into the spindle, the motor can have external ventilation. Liquid cooling is possible as an option; this can be used to further increase the motor's use. This arrangement is especially beneficial in machining centers.

Integrating the drive motor directly into the spindle created the so-called **motor spindle**. This direct installation generally requires liquid cooling. This type of main spindle-drive design is increasingly becoming the standard in the modern machine tool industry.

In both these direct-drive designs, the lack of speed adaptation means that the following requirements become particularly important:

- High power density
- Large speed-adjustment range
- Large range with constant output
- High torque at low speeds
- High maximum speed

Three-Phase Asynchronous Motors

Speed-controlled asynchronous motors have developed into standard main drives owing to their positive features, such as low price, simple and rugged design, and low maintenance requirements. The speed of three-phase asynchronous motors can be changed over a broad adjustment range by changing the output frequency and output voltage of the converter that is supplying the power. The speed/torque characteristics of three-phase asynchronous motors operated with converters are provided in the form of characteristic curves (*Figure 5.8*).

Using purely mathematical calculations, the speed-adjustment range then would be "infinite" and up to 1:12 in the field-weakening range, that is, at constant power. The behavior of the characteristic curves is determined by the strength of the intermediate-circuit voltage and by the corresponding motor-specific data, such as inductivity, resistance, motor constants, and breakdown torque. *Figure 5.8* provides a comparison of the two different motor principles.

In the basic speed range, the voltage and frequency increase proportionally up to the rated speed. If externally cooled, the motor generates a constant torque. Once the voltage reaches its maximum value at the rated speed, it is only possible to increase the frequency. From that point onward is the beginning of the so-called field-weakening range. The field-weakening range begins with a range of constant output, in which the torque decreases hyperbolically, that is, inversely proportional to the frequency/speed (rpm). If the speed/supply frequency is increased further, the breakdown torque or stability limit of the motor will be reached. The breakdown torque of an asynchronous motor decreases as the square of the frequency/speed ($1/n^2$). In contrast to operation from power mains, with modern converters and an appropriate control system, the stability limit of the motor initially does not represent any real limit because loss of stability of the motor (a drastic reduction in torque and even motor standstill) is prevented. The maximum speed thus is limited only by mechanical components, such as bearing, rotors, rotor mountings, and so on.

The characteristic curves generally are indicated for continuous operation (operating mode S1) with various duty cycles, frequently 25, 40, or 60 percent.

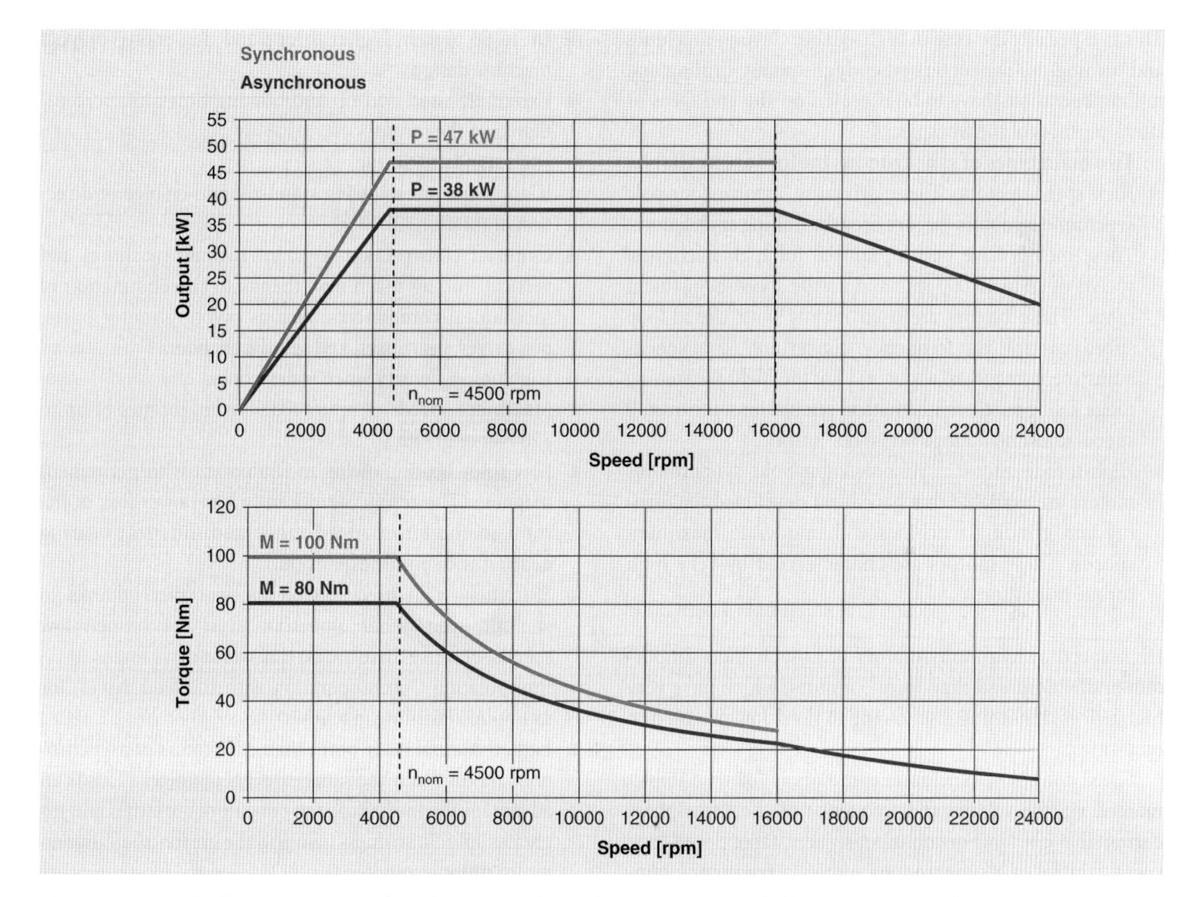

Figure 5.8: Speed-adjustment ranges of synchronous and asynchronous motors with the same output and torque.

The closed-loop control and control structure of main-drive motors are largely the same as those of a modern feed drive. The main-drive controllers encountered today merely have a few supplementary functions, for example, for special field regulation. Thus C-axis operation—the interpolation of the main spindle with the feed drive, as is often required today, especially in turning machines—represents no problem.

Three-Phase Synchronous Motors

In machine tools, synchronous motors are as a rule used primarily as **feed drives** (→ *Part 2, Chapter 1: Implementation of Dimensional Data*).

In special cases, demands for greater power densities and temperature stability have led to the use of permanent-field **three-phase synchronous motors** even in main drives. In the meantime, state-of-the-art practical motor principles and closed-loop control methods have made it possible to implement such drives in series-produced machines. More extensive use is unlikely at present, however, owing to the higher costs. At present, **speed-controlled synchronous motors** are used primarily as main drives **when the following requirements have to be met:**

- Extremely high requirements as to machining quality, precision, and smooth operation
- Extremely short times for run-up or speed changes
- Standstill torque
- Limited installation space

This is especially the case in high-quality turning machines and turning/milling centers when high-quality drilling and milling operations have to be executed on the end faces or lateral surfaces of the workpieces.

Two main types of synchronous motors are available:

- High-speed motors. These primarily involve four-pole synchronous motors that are used for milling applications. These motors have been optimized for high maximum speeds of up to 40,000 rpm and a wide speed-adjustment range. These motors are used mainly with closed-loop speed control via frequency converters. The speed-adjustment range is about 1:3 in the field-weakening range. Using purely mathematical calculations, the overall speed-adjustment range would be "infinite."
- High-torque motors. Six-pole/eight-pole synchronous motors are available that have been developed for turning and grinding machines with moderate maximum speeds. These motors are characterized by very high torque utilization.

High-speed synchronous motors are much more important and widespread than torque motors, which have very high prices, if nothing else, owing to their low production quantities.

Speed control of synchronous motors is likewise implemented via the voltage and frequency of the three-phase current that is fed in. For continuously adjustable closed-loop speed control of a synchronous motor, a **frequency converter** has to be connected upstream of the motor. Both the speed and the rotor position are measured and reported to the converter by a rotary encoder. From this the control electronics determine the necessary "electronic commutation" to advance the rotating field and also the actual speed.

Benefits

Compared with the lower-priced asynchronous motors, synchronous spindle drives offer the following benefits:

- High efficiency
- Low mass moment of inertia and thus good dynamic characteristics
- Low maintenance (in the case of rotors without slip-ring rotors)
- Speed independent of load
- No electrical power necessary for field excitation
- Up to 60 percent higher torque and thus more compact machine designs
- Extremely short run-up and braking times (50 percent) thanks to the torque
- High standstill torque
- High torque even during standstill and change of direction of rotation
- Compact construction (e.g., for turning machines and vertical milling machines) thanks to the elimination of mechanical components such as pivoting motor bases, belt drives, gearboxes, and spindle encoders
- High power density when water-cooled
- Maximum speeds of up to 40,000 rpm, torques of up to 820 Nm or greater
- Low rotor heating owing to equipment with permanent magnets (The result: Significantly less power loss in the rotor in the lower speed range and thus less bearing heating and spindle expansion.)
- Extremely high precision on the workpiece thanks to smooth, even spindle operation even at extremely low speeds, because there are no transverse drive forces
- Interpolating C-axis operation with the feed drives, for example, in turning machines
- Larger internal rotor bores than cage rotors of asynchronous motors with the same external diameter (This is an advantage for the bar capacity of automatic lathes and for greater spindle stiffness owing to the greater shaft diameter for milling spindles.)
- Increased stiffness of the spindle drive thanks to mounting of the motor components between the main spindle bearings
- Less cooling output necessary for the same power output compared with asynchronous motors, that is, greater efficiency
- Only one encoder (hollow-shaft measuring system) for detecting the motor speed and spindle position
- Easy servicing by exchanging complete motor spindles

Further potential for improved efficiency, such as faster part machining times and smaller footprints, can be achieved through the optimized combination of synchronous spindle motors, drive controls, and CNC controllers.

On the other hand, there are a few disadvantages:

- Expensive magnet materials, therefore high procurement costs for permanent magnet motors

- Elaborate closed-loop control requirements (frequency converters)
- Possibly an annoying whistling noise from the motor

5.5 Machine Enclosures

During our discussion of main drives, we have already noted that automatic work sequences are necessary for full utilization of the high-performance characteristics of state-of-the-art cutting tools. One result of this is that chips are thrown outward at high speeds, presenting a risk of injury to persons in the vicinity. This fact must be addressed through the use of appropriate enclosures and by securing the working space.

On small and medium-sized machines, such enclosures generally are a sheet-metal construction attached to the body of the machine, often combined with the electrical cabinet and even enclosed on the top. Thus they form a single transport unit together with the machine, a so-called machine suitable for single-point lifting. In such cases, the machine enclosure not only serves as protection against chips but also retains the vaporized coolant and provides good protection against process noise. These are a key means of preventing accidents and thus are subject to a large number of regulations. On the other hand, they are also intended to make operation, servicing, and maintenance of the machine easier. All this means that they have to satisfy a number of extensive requirements that are often mutually conflicting.

For example, machine enclosures have to make the working space easily accessible for setup work and tool changing. For this purpose, they generally have large doors to allow access to the working space. These doors, however, must be locked when the machine is operating, or at least the work sequence has to be interrupted immediately when they are opened. The doors and often other parts of the fixed enclosure are provided with windows to allow safe observation of the work sequence. These windows have to withstand "bombardment" by chips without becoming obscured. This is only possible with silicate glass. Then again, in the event of collisions that cannot be avoided completely, they have to be able to survive impacts by large parts, such as flying workpieces, clamping equipment, or tools. This works better with elastic plastic windows. Therefore, composite windows are often used in such cases. It is also important for the window to be firmly anchored in its frame to prevent it from being pressed out too easily.

Coverings that enclose the entire machine often obstruct access to areas that have to be accessible for cleaning and maintenance work. They thus have to be easy to remove or open.

Frequently, the operator panel with the keyboard, equipment for the CNC, and additional control elements for manual actuation of the machine motions for setup and maintenance work are also integrated into the machine enclosure. After all, the coverings are decisive in determining the external appearance of the machine and therefore display an important starting point for design. Like all components of the machine, they are naturally also subject to heavy pressures to reduce costs, all the more so because it is easy to view them more as a necessary evil and not as an important functional component.

5.6 Coolant Supply

Because in CNC machines the tool can move freely in the working space, the coolant has to be coupled with the tool. In turning machines, this is done by feeding the coolant over the turret to the tool holder that is located in the working position, which, in turn, feeds it to the cutting edge via a preadjusted tube. For recirculating tools, the coolant is fed to the tool via the main spindle. Owing to heavy misting, however, there is at present a trend toward dry machining.

5.7 Chip Removal

As a consequence of the high productivity of numerically controlled machine tools, large quantities of chips are produced in any given time interval. These chips have to be removed from the machine without adversely affecting the work sequence. The issues associated with free chip clearance and installation of the associated chip conveyors in the machine have already been discussed in connection with the machine configuration.

Various types of conveyors are used depending on the shape of the chips. The most common and widely distributed ones are hinged belt conveyors. Scraper conveyors are more suitable for very small and friable chips. Magnetic conveyors can only be used for steel chips.

5.8 Summary

The automatic work sequences made possible by numerical control have extensive effects on the design of machines because there is no need for continuous tending and observation by workers. This makes it possible to focus the design of the machine on optimal execution of machining, making full use of the increased performance of modern tools. This requires a main drive with correspondingly high output and a completely enclosed working space. The high productivity of such machines results in a high level of chip generation, and the chips have to be removed automatically.

Automatic work sequences require a higher level of precision, which has to be taken into account, especially in the design of machine bodies and guides. CNC closed-loop position-control systems also place requirements on machine designs. Moving parts have to be as light as possible, especially when driven by linear motors. The guides should have low friction and be free from stick-slip effects. Machine designs also must take into account automatic tool changing and workpiece changing, which are often associated with considerable space requirements.

Effects of CNC on Machine Components

Important points to remember:

1. CNC machines are machines that operate automatically. They do not have to be tended by a worker. For this reason, they often have a different configuration than conventional machines. In particular, their working space should be completely enclosed on all sides. In part, this is due to their very high cutting speeds.
2. Machine frames should have high static, dynamic, and thermal stability in order to achieve trouble-free automatic production.
3. The demanding requirements on guides have resulted in increased use of rolling guides.
4. Main drives are primarily frequency-controlled three-phase asynchronous motors.
5. Increasing use is being made of directly driven motor spindles, especially when positioning tasks or synchronization with feed motions is demanded at the same time.
6. Machine enclosures have a very wide range of functions:
   - Protection against flying chips
   - Containing the coolant vapor
   - To allow observation of the work sequence
   - To allow machine setup and clamping of the workpiece
   - To contain flying parts resulting from collisions
7. The coolant in-feed has to be coupled with the tool; with recirculating tools, it must be fed through the spindle.
8. The chips have to be removed in such a way that they do not interfere with the work sequence and so that they do not cause local heating in the machine frame.

Part 3

Types of Numerically Controlled Machines

1. Computer Numerical Control Machine Tools

Over the years, the influence of numerical controllers on the construction of machine tools has led in some cases to completely new machines and mechanical automation equipment. Today, computer numerical control (CNC) machines are the basic cornerstones of modern manufacturing equipment.

The various types of machines will be discussed below in descending order of their importance on the market. This is based on sales figures from the German Machine Tool Builders' Association (Verein Deutscher Werkzeugmaschinenfabriken).

1.1 Machining Centers, Milling Machines

Machining centers are machine tools that arose only as a result of the development of numerical control (NC) systems. They were developed from machine tools with rotating drives, that is, drilling machines, milling machines, or boring mills. The goal was to be able to perform as wide a range of machining operations as possible automatically in a single setup. The definition therefore is as follows:

A machining center is a machine tool that is numerically controlled in at least three axes and is equipped with an automatic tool-changing system and a tool magazine.

If the machine is also capable of machining with a rotating workpiece, then it also can be called a **turning center** or **turning/milling center**.

Nowadays, milling machines without automatic tool changing are more the exception than the rule. Today, simpler, smaller machines of these types are also produced for training purposes, thanks to their lower cost and compact construction compared with conventional machines.

There are many possible variants for machining centers. First of all, one can make a distinction between **horizontal and vertical machines** based on the position of the main spindle. Vertical machines are preferred for the machining of flat sheets or very long workpieces, whereas horizontal machines are used more for machining of box-shaped workpieces. There is an important difference here: In vertical machines the *Y* axis is oriented horizontally, whereas in horizontal machines it is vertical. Thus, for horizontal machines, a column-type construction with a roughly cubical working space predominates, as in *Figure 1.4.* For vertical machines, *XY* tables are used most often or, for large working spaces, a gantry or portal construction (→ *Figure 1.6*).

A distinction is made based on the number of feed axes:

- *Three-axis machines:* Three linear axes, the most basic configuration for a machine with a rotating tool.
- *Four-axis machines:* Three linear axes and one rotary axis; the purpose of the rotary axis is to allow machining on all sides. For horizontal machines, this is in the form of a rotary table, and for vertical machines, it is in the form of a reversible clamping device for the machining of cylinder barrel surfaces (→ *Figure 1.3*, center) or for the machining of small workpieces on three sides (→ *Figure 1.3*, above).
- *Five-axis machines:* Three linear axes and two rotary axes; this allows the tool to be moved in any desired direction relative to the workpiece. This means that milling is possible in any plane regardless of its orientation in space, and holes can be drilled in any diagonal orientation. The two rotary axes can be assigned as desired to the workpiece holder and the tool spindle. As a result, a large number of different machine configurations are possible.

Today, machining centers are without exception equipped with **continuous-path-control** systems in at least three to five axes, including spatial (simultaneous) interpolation in all axes. For this reason, programming is also performed primarily using computerized **programming systems** and machine-specific postprocessors. Thanks to its greater flexibility on the shop floor, for relatively simple machining tasks, experienced users prefer **shop-floor programming (SFP).** Parameterizable milling and drilling cycles, graphical support when inputting contours, graphical simulation of the machining sequence, and technological programming tools have, by now, become practically a part of the standard equipment.

Essential CNC functions include tool length compensation, cutter diameter compensation, automatic or programmable tool monitoring, and in many cases temperature error compensation. The **user friendliness** of the CNC system is also important so that restarting after interruptions, data input, and handling as a whole do not develop into time-consuming, expensive problems.

Three-Axis Machines

Three-axis (milling) machines with three linear axes are the most basic configuration for machining centers. The implemented constructions as a vertical machine are shown in *Figure 1.1*. Such machines are very seldom delivered to customers with only three axes because a fourth or fifth axis can be added for relatively little expense compared with the overall cost of the machine. If a tool changer and/or workpiece changer is also installed, the result is a machining center with a high degree of automation.

Four-Axis Machining Centers

These generally consist of **three linear CNC axes and a rotary table,** making it possible to machine cubical workpieces on four sides in a single setup. If a horizontally/vertically swiveling tool head is used, it is also possible to machine the fifth side. *Figure 1.2* shows various types of four-axis machines in a horizontal configuration.

All types of chip removal can be performed, such as face milling, drilling, boring, finish rolling, and tapping. With the addition of further equipment contour milling, diagonal boring or thread turning is also possible. Speeds and feed rates have to be programmable for each tool (*Figure 1.3*).

The tools are kept in a **tool magazine** attached to the machine. The tools are found automatically by the program and exchanged into the main spindle automatically. Tool magazines have a wide range of designs and storage capacities. The designs used most often are chain, disc, and cassette-type magazines.

The use of additional **workpiece-changing equipment**, generally implemented in the form of pallet changers, shortens machine idle times during tool changing. The workpieces are clamped and released during the productive time outside the machine's working space.

More complex machining centers are equipped with additional equipment, such as a second rotary table, a swivel fixture for the workpiece, or a tool head that can be adjusted horizontally/vertically to any desired angle.

Today, users can choose among a wide variety of designs and sizes. The primary distinction to be made is between machines with a **horizontal or vertical main spindle.** While vertical main spindles are generally more suitable for plate-shaped parts, machines with horizontal spindles are used primarily for the machining of cubic workpieces on four or five sides.

In the case of machines with a horizontal spindle, the X (longitudinal) motion and the rotary motion are performed mainly by the workpiece, whereas the Y and Z motions are made by the tool. The resulting axis designation is $X'YZB'$.

In the case of machines with a vertical spindle, the predominant design is $X'Y'ZA'$; that is, only the vertical spindle motion in the Z axis is performed by the tool, and all other motions are made by the workpiece.

The potentially universal versatility of machining centers can only be fully exploited with continuous path control. As workpieces become more complex and new machine concepts are developed, today's machining centers require **three-dimensional (3D) control systems.** These systems must, as a minimum, be capable of linear interpolation in all the axes simultaneously. When a swiveling tool head is used, three axes have to perform linear interpolation in order to drill diagonally. One or two additional axes are necessary when using surfacing heads. To allow unlimited use, **multiple-compensation-value tables** for the tool length, cutter diameter, remaining tool life, and cutting parameters for all tools are often required. On newer machines, the CNC system also has to save the tool weight, tool identifier, tool contour, and many other characteristics to allow for proper **tool management.**

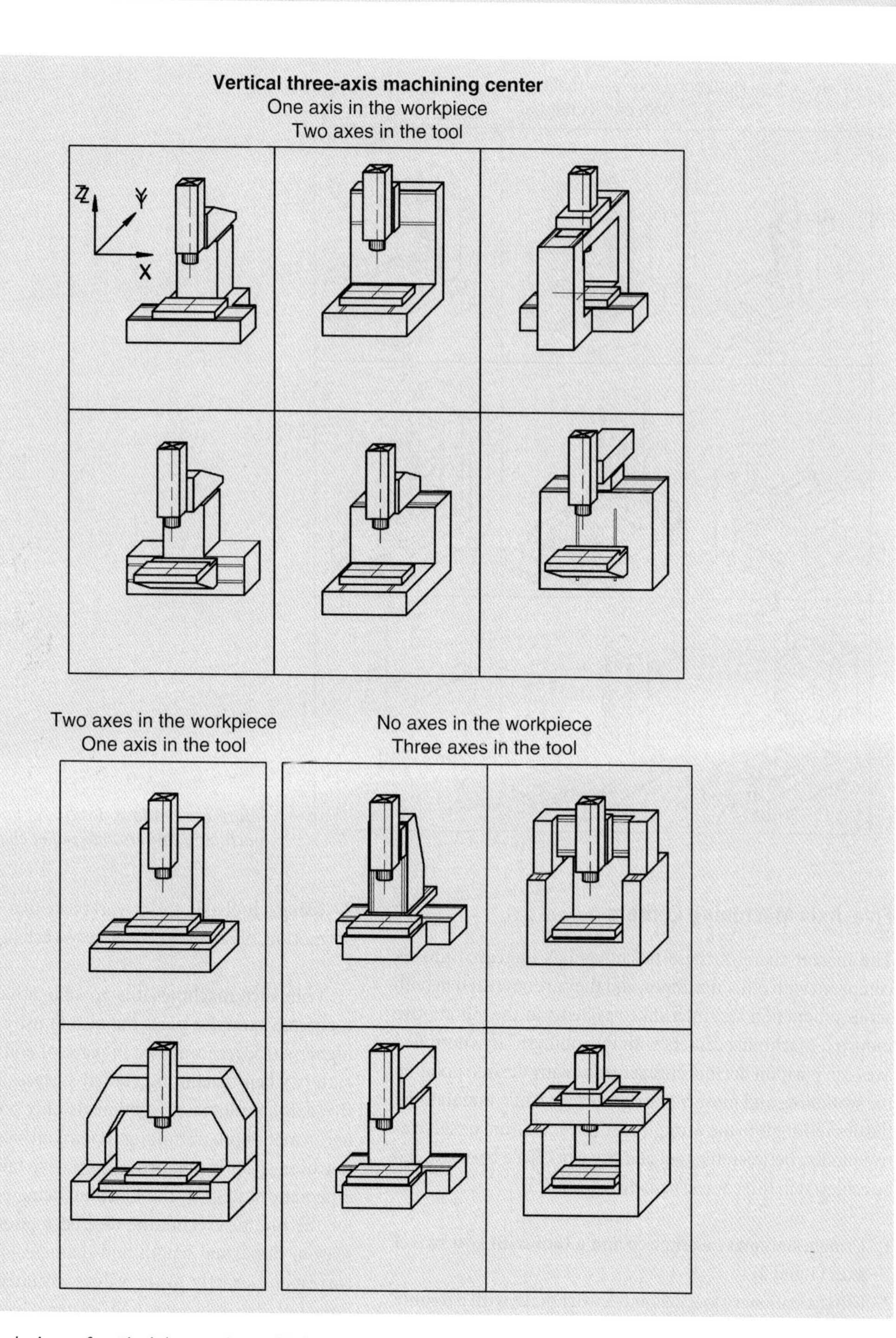

Figure 1.1: Various designs of vertical three-axis machining centers.

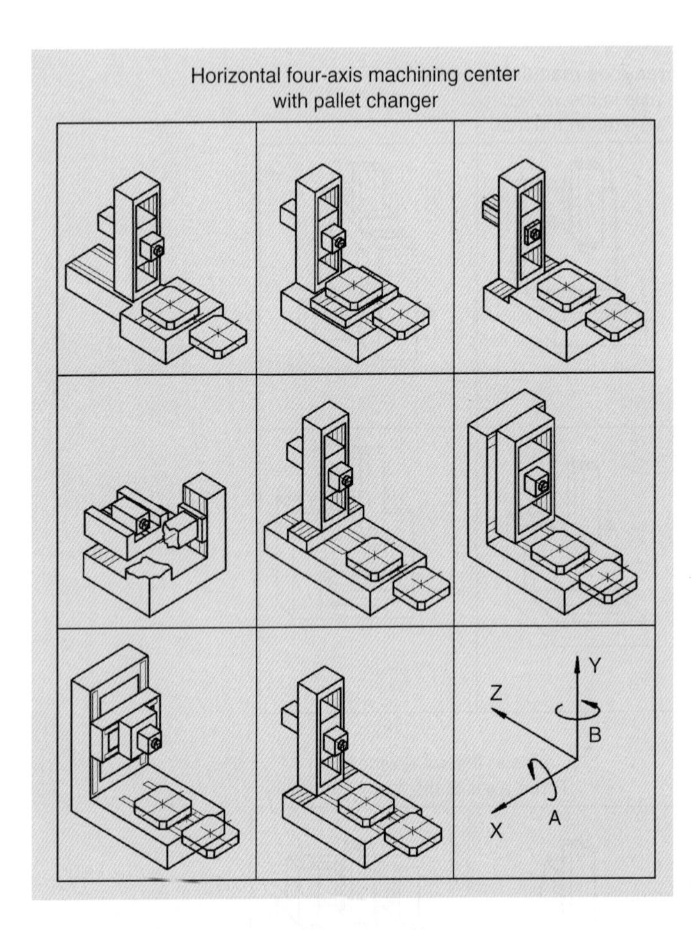

Figure 1.2: Various designs of horizontal four-axis machining centers with pallet changers.

Five-Axis Machining Centers *(Figure 1.4)*

The market share of these machines has increased greatly compared with other machines, and they are now used in both series production and interlinked systems in the automotive industry. Machining centers with five numerically controlled axes can position the tool engagement to any desired point on the workpiece and move along the surface while maintaining the desired angle to the workpiece surface. This universal relative motion between the tool and the workpiece basically can be achieved in three ways, namely (*Figure 1.5*)

1. Using a stationary workpiece and **a tool with two swivel axes** (1 and 2).
2. Using a stationary tool axis and **a workpiece with a double swivel motion**, for example, via a swiveling rotary table (3)
3. Using a tool axis and a workpiece that each have a swivel motion, offset by 90° relative to each other (4)

With such machines it is possible not only to create geometrically complex parts but also to use cutter heads with a higher stock-removal rate in place of end mills or ball-nose cutters when machining curved surfaces. **Programming** of five-axis simultaneous motions is only possible when high-performance programming systems are used. In this case, the machine-specific **postprocessor** has to take into account the kinematic system of the machine being controlled in order for the tool to execute the desired motion exactly. For this reason, the actual length and diameter of the tool have to correspond exactly to the values assumed in the programming because only a few CNC systems have the capability to

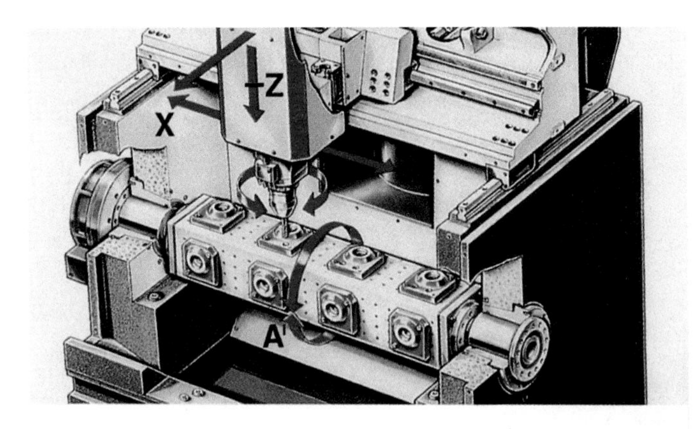

Multiple clamping bridge for three-sided machining of smaller workpieces

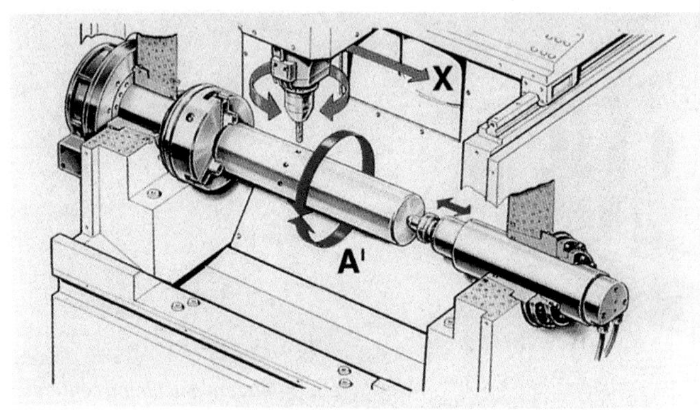

Reversible clamping device for machining of cylinder barrel surfaces (A' axis)

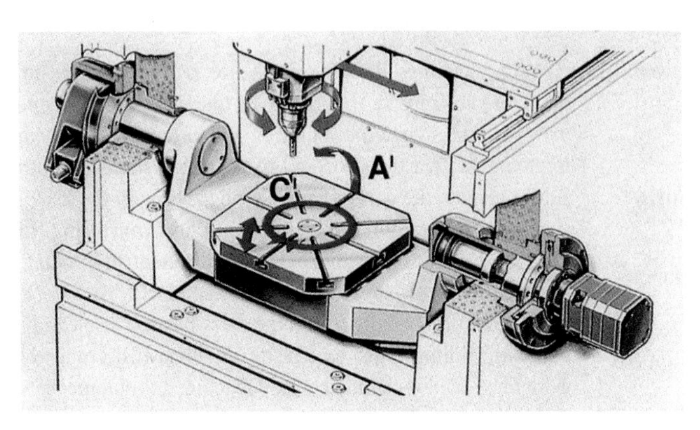

Swiveling rotary table for five-sided machining (A' and C')

Figure 1.3: Several expansion levels of a three-axis machine in order to implement the fourth and fifth axes.

Figure 1.4: Modern machining centers with linear motors in the X, Y, and Z axes.

perform the spatial tool compensation that otherwise would be necessary.

Gantry Milling Machines with a Movable Gantry

(Figure 1.6)

The use of this type of machine is preferred when the following conditions have to be taken into account:

Workpiece: Flat components or components with similar lengths

Shop floor: Limited floor space

Operation: User-friendly operation of the machine from a position close to the milling spindle because the operator station and CNC move along with the operation

Planning: The option to lengthen the machine subsequently

With machines above a certain size, two **feed drives** are necessary to traverse the gantry in the *X* axis, that is, one drive on each side of the gantry. A **cant-monitoring system** implemented via the CNC and measuring systems on both sides prevents the gantry from canting. Through the use of synchronized or counterdirectional motions (mirroring) in the *Y* and *A* axes, it is possible to **simultaneously** produce two identical or two mirror-image components.

In other designs, the rear of the cross-beam is also equipped with milling units, as a result of which two groups of workpieces can be clamped on and machined simultaneously during the same *X* motion.

For machines of this size, a **mobile control device is essential for setting up the machine.** These machines make use of the largest and most expensive **3D continuous-path-control systems with** all the available features, such as parallel axes, cant monitoring, and temperature compensation.

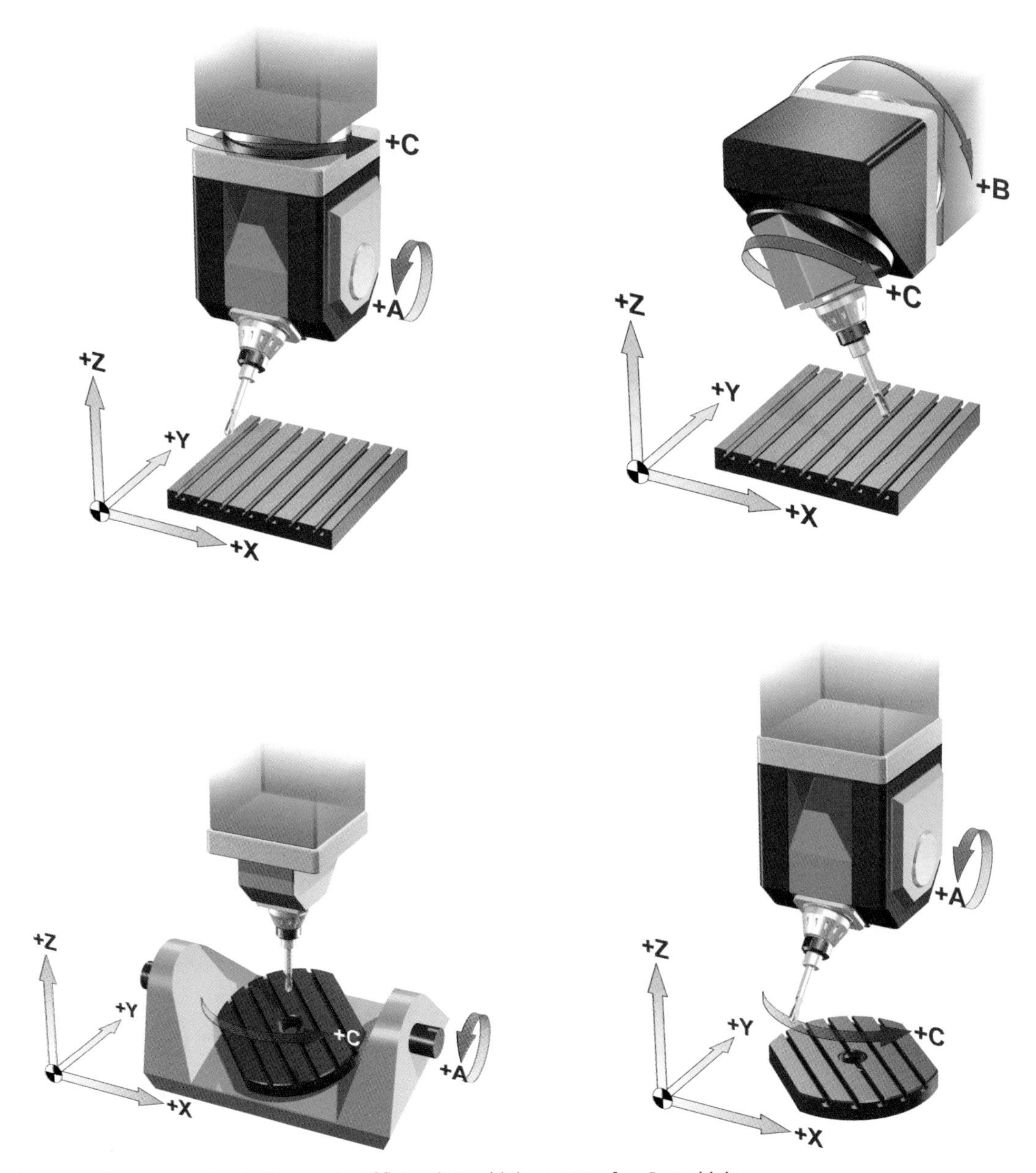

Figure 1.5: Four options for the kinematics of five-axis machining centers for 3D machining.

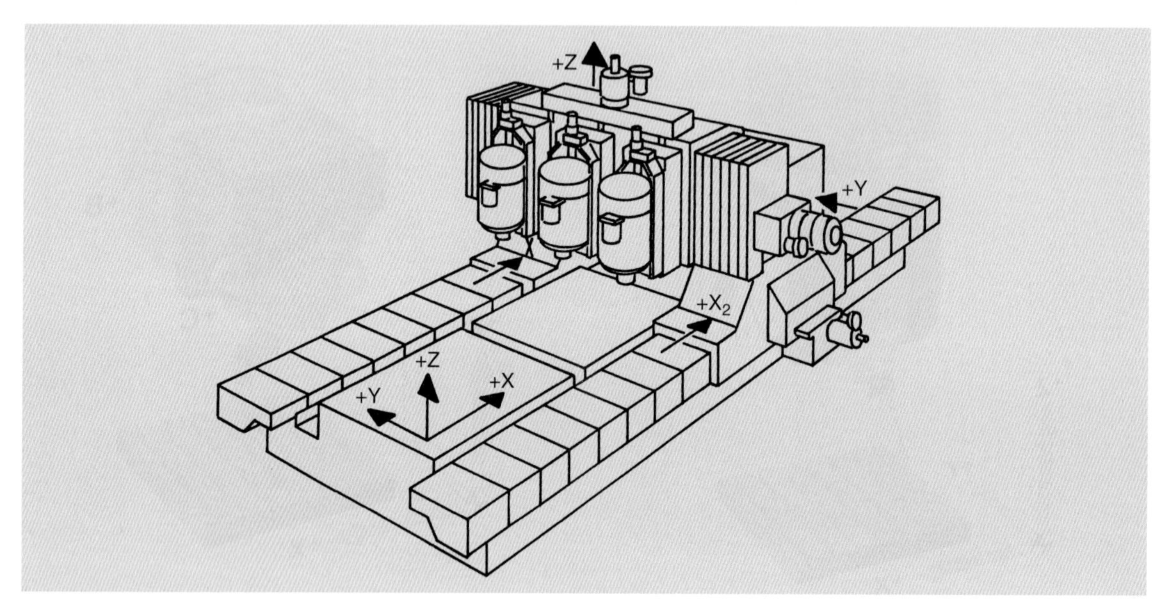

Figure 1.6: Three-axis gantry milling machine with three milling units.

Figure 1.7: Modern milling/turning centers are multifunction machines. (Image courtesy of STAMA.)

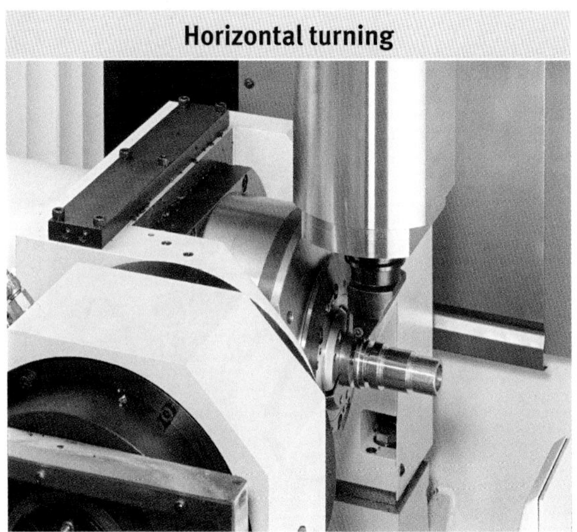

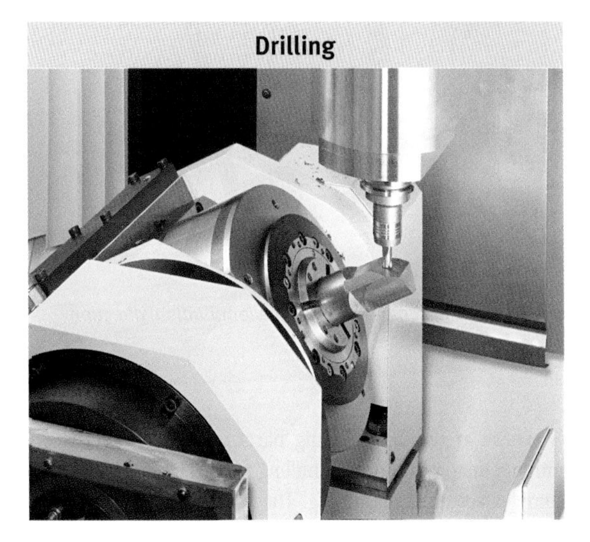

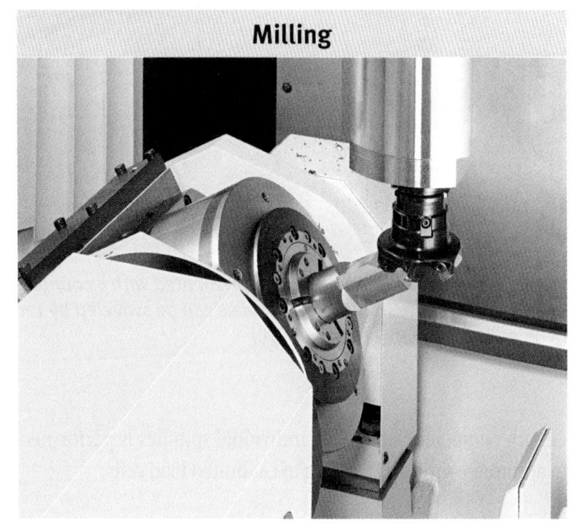

Figure 1.8: With modern milling/turning centers, all machining technologies can be implemented with rotating or fixed cutters. (Image courtesy of STAMA.)

Multiple-Spindle Machining Centers

All the machines mentioned earlier can be designed as two-, three-, or four-spindle machines in order to process a number of identical workpieces at the same time. Two-, three-, and four-spindle machines are used most of all in large-scale series production; this also requires the use of multiple clamping fixtures.

With **multiple-spindle machines,** all the tools have to have uniform dimensions. An identical length is achieved either by means of preadjusted tools or individually calibratable spindles.

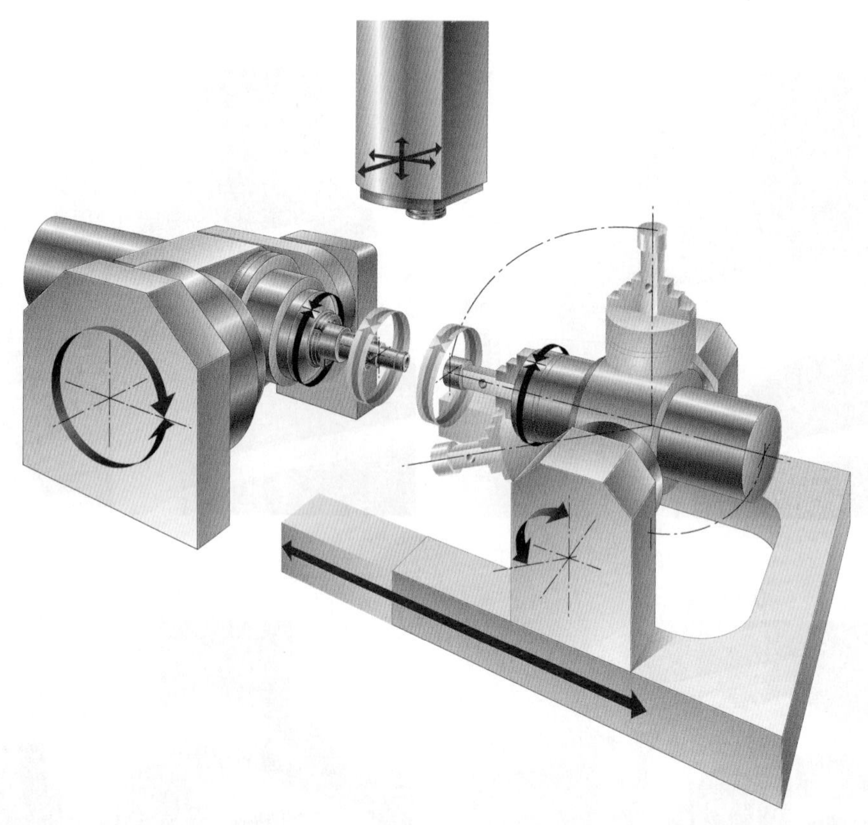

Figure 1.9: Milling/turning center implemented with a counterspindle for machining of the sixth side. The counterspindle (rotational) likewise can be swiveled by 120° (–30° to +90°) and can be moved in the longitudinal direction (X axis). (Image courtesy of STAMA.)

Length compensation for the individual spindles is performed in automatic mode by moving to calibrated load cells.

Milling/Turning Machining Centers *(Figures 1.7 to 1.10)*

As a contrast to turning/milling centers (→ *Part 1, Section 2.2*), an interesting family of machines has emerged that had its origin in machining centers for milling. The point of departure for the development of such machines was the analysis of the ranges and families of parts that on the one hand involve large production quantities but on the other hand have to be machined in different ways. An excellent example of this is the production of tool systems for CNC machines (→ *Part 4, Figure 1.1*).

These are milling/turning parts with an emphasis on milling, that is, not turning/milling parts. These parts require complex six-sided machining. They are often repeat parts or small batches that require a CNC program with high flexibility. The workpieces generally move within a diameter of 60 mm and a length of 100 mm. These centers are also often called *bar-machining centers* because a wide variety of highly complex workpieces can be produced directly from bar stock. The machining focuses on milling processes, but turning operations can be performed with equal efficiency because the rotary/swivel units are equipped with integrated turning spindles.

The bar stock is fed from the bar magazine into the main rotary/swivel unit. At this time, the first five sides of the workpiece are machined simultaneously in five axes. To

Figure 1.10: Workpieces created on a turning/milling center.

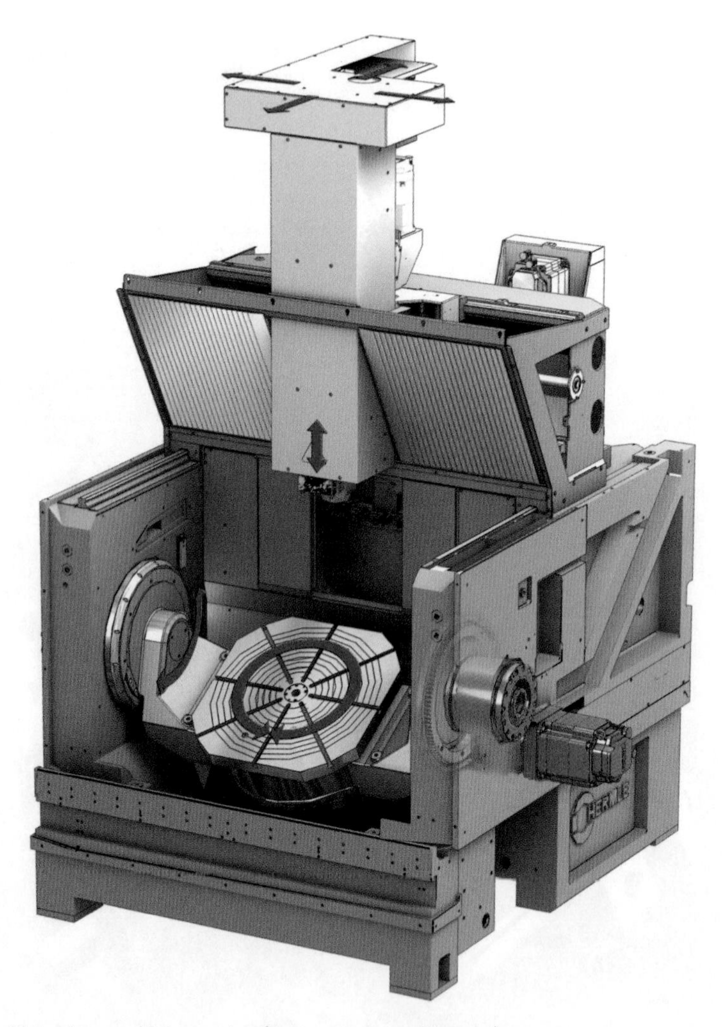

Figure 1.11: High-speed machining in a machining center. (Image courtesy of Hermle.)

machine the sixth side, the workpiece is transferred to the second rotary/swivel unit. Then the sixth side is machined.

To ensure efficient use of the machine, a tool magazine with about 100 tools is recommended.

Trends in State-of-the-Art Machining Technology

High-Speed Machining (High-Speed Cutting [HSC])

High speed is generally defined in terms of the cutting speed. The cutting speeds are greater than the conventional speed range by a factor of 5 to 10. This depends on the material, however. In high-speed milling, a relatively small effective tool diameter is used, resulting in significantly higher spindle speeds than in conventional machining. Thus the definition of HSC in milling is extended to include the factors of high tool speed in conjunction with high feed rates (*Figure 1.11*).

Dry Machining

Dry machining is becoming more and more important as a way to protect the environment. Research projects are underway to investigate the potential for using more environmentally friendly auxiliary substances and reducing the amounts employed in manufacturing (e.g., the use of minimal lubrication). At the same time, tools for dry machining have been developed. The importance of dry machining is especially increasing in the automotive industry and its component suppliers.

Hard Machining

Hard machining is likewise increasing in importance. Hard machining makes it possible to perform high-precision machining by cutting with a geometrically defined cutting edge. Owing to the development of suitable tools and technologies, it has become possible to machine workpieces with a hardness of up to 62 HRC via turning and milling, with quality equal to grinding.

High-Performance Machining/High-Performance Cutting (HPC)

This has the goal of reducing machining time in order to achieve considerable cost savings, in some cases more than 50 percent over previous technologies. The increase in the stock-removal rate based on new tool types and optimized machine components increases the availability of the machine tools being operated.

Requirements for Machine Tools

- High stiffness and the maximum possible damping in order to avoid vibrations and resonances (This requires short projections and slide units with high stiffness.)
- Vibration-free spindle drives owing to extremely high rotational frequencies
- Acceleration of relatively small masses to achieve acceleration values of up to 3*g* and K_V factors of 2 to 4
- Vacuum extraction units, especially for HSC machines that are used to machine aluminum

Requirements for Controllers

- Short block cycle times in the 1-ms range, for example, a processing rate of about 100 CNC blocks per second. At high feed rates, the CNC blocks have to be processed in rapid succession, and thus the time to read in and present them is very short.
- A look-ahead function to detect edges and corners in a timely manner. In order to avoid contour violations, the feed automatically should be reduced temporarily while at the same time adapting the spindle speed.
- High stiffness of the feed drives with a high K_V value to achieve the necessary accelerations and precision.
- Zero-lag axis motion without contouring error to achieve good contour accuracy despite the high feed values.

The increased use of computer-aided design (CAD) systems recently has given rise to an additional requirement, namely, the direct **processing of the geometric data generated by CAD systems.** These essentially consist of DXF data, nonuniform rational-basis splines (NURBSs), or Bezier formulas. These mathematical data then no longer have to be converted into linear vector elements by a postprocessor because they are transferred directly to the CNC and processed there. In this manner it is possible, despite the high accelerations and speeds, to achieve much smoother machine motion. This also has a positive effect on the surface quality of the workpieces.

1.2 Turning Machines

Scientists originally assigned little importance to numerically controlled turning machines. This was so because this type of machine had already been automated to a large extent through mechanical means, and it did not seem practical to try to further increase this high degree of automation. But the experts were mistaken. Within a short time, the CNC turning machine had already become the most popular type of CNC machine. For many years, CNC turning machines made up more than 50 percent of all the CNC machines manufactured in Germany. In time, CNC systems were used to equip not only drilling and milling machines but also grinding, nibbling, gear-cutting, and erosion machines. Only then did the proportion of turning machines sold fall back to just under the 50 percent mark. Another reason for this is the fact that modern CNC machine designs have become more and more efficient and productive, so fewer and fewer machines are needed to produce a given quantity of parts.

Although turning machines have always been quite versatile and highly automated, they have become even more universal and flexible through NC and especially CNC. These high production quantities also explain the great interest among controller manufacturers in developing CNC systems with special functions for turning machines. For this reason, CNC turning machines now feature the most advanced level of automation, including the programming and control technology that this automation requires. When CNC manufacturers cannot meet the requirements of machine manufacturers, then the latter develop their own software and hardware modules.

Turning machines are available in a number of **different types and designs.** A distinction can be made based on the following criteria:

- Horizontal or vertical design
- Flat bed or slant bed
- For producing bars, bushings, or shafts
- With one, two, or more spindles
- With one or more cross-slides and tool turrets
- With or without an **auxiliary spindle** for machining the cutoff end

Turning machines also can differ greatly as regards their **degree of automation** (*Figure 1.12*). For example, the following **automation components** are available:

- **Workpiece magazines** with automatic workpiece changing
- **Tool magazines** with automatic tool exchange between the turret and the magazine
- **Driven tools**, generally in conjunction with an additional CNC axis (*Y* axis) and a controlled **spindle representing the *C* axis**
- Automatic tool monitoring
- Automatic **jaw changing** in the chuck
- **Numerically controlled steadyrest and tailstock**
- **Equipment for linking** multiple machines of the same or different types

Turning Machines with Two or More Heads

(Figure 1.13)

Long before the development of CNC machines, it was possible to use heavy-duty lathes to machine a single workpiece with two or three tools simultaneously. The advantage of this is that production time can be reduced considerably. The downside of this is that not all tools will work at their optimal cutting speeds.

Turning machines with two CNC axes were found to be unsuitable for the simultaneous use of two tools. For this reason, they were soon equipped with two separate heads and thus four axes so that the two tools could work independent of each other. In modern CNC turning machines, both turrets are designed and arranged to allow simultaneous operation

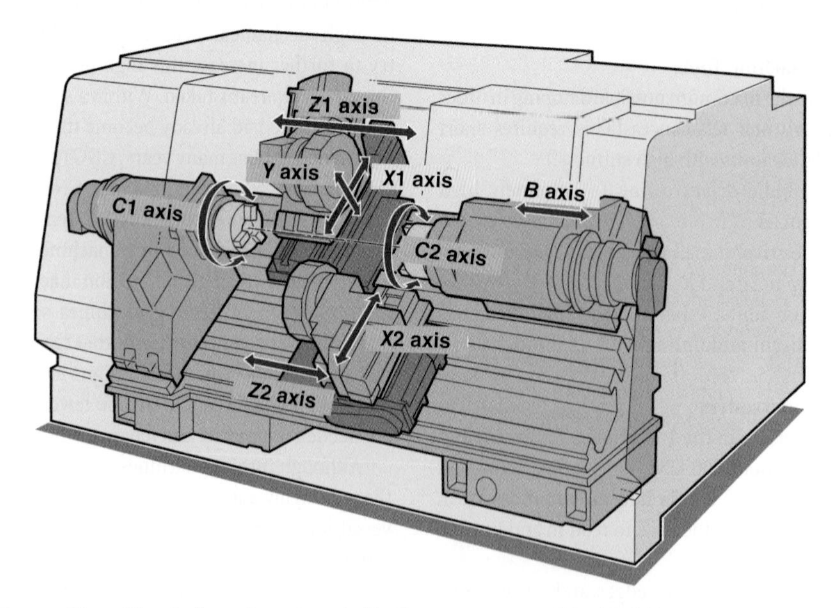

Figure 1.12: Turning machine with spindle and counterspindle. (Image courtesy of Mori Seiki.)

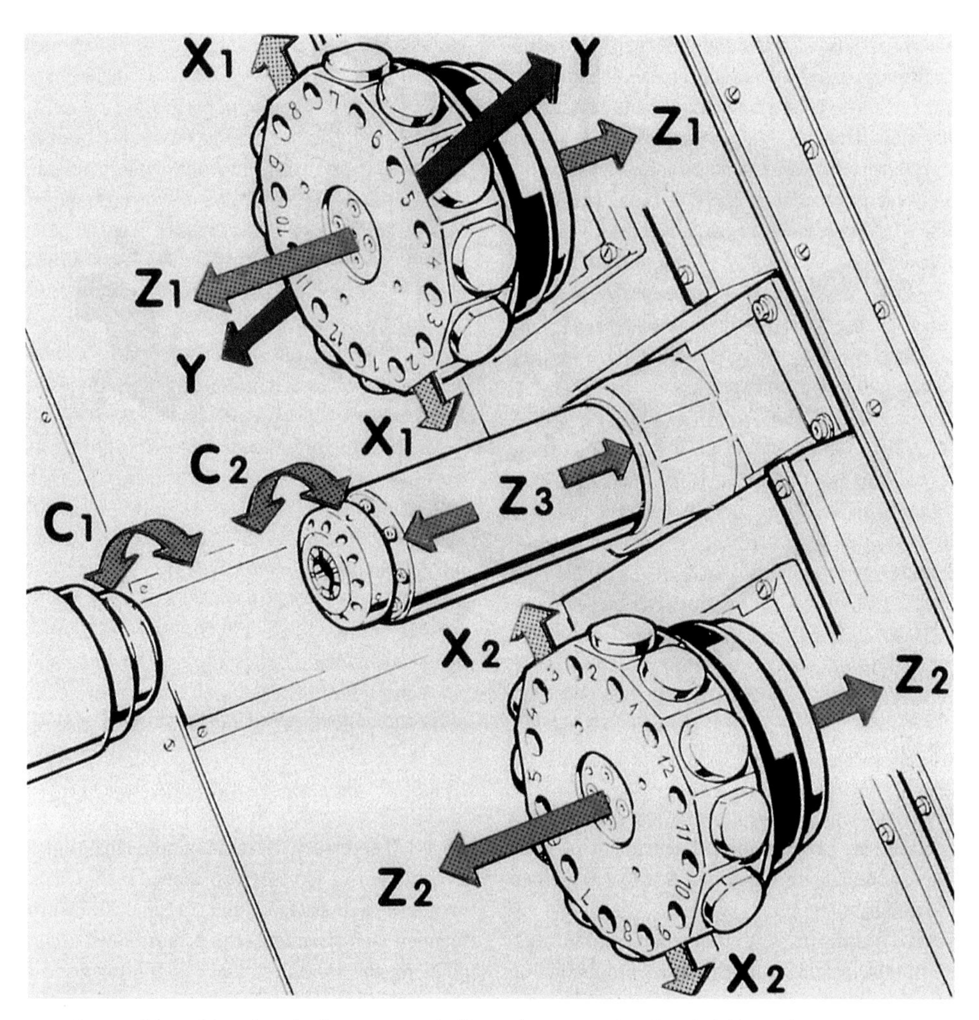

Figure 1.13: Turning machine with main spindle, counterspindle, and two turret heads, of which one has a Y axis for machining of cylinder barrel surfaces (total of eight CNC axes).

with a minimal risk of collision. This allows shafts and bushings to be machined by two tools simultaneously. Collisions between the two turrets are prevented by taking appropriate safety precautions, including software-based monitoring features and additional proximity switches on the heads.

The control of these machines requires special CNC systems that can interpolate 2×2 axes independent of each other. The initial experience with 2- $\times$ 2-axis turning machines was very negative because the programmer always had to think in terms of four axes in order to coordinate and intermesh the timing of the two tools. The reason for this was that there was only one punched tape and one tape reader, and the entire program had to be read in block by block. This problem has been solved by using CNC systems that assign the two head programs to separate storage areas. At points where one head has to wait for the other, this can be accomplished by means of a special *G* command in the program. This makes programming significantly easier because the

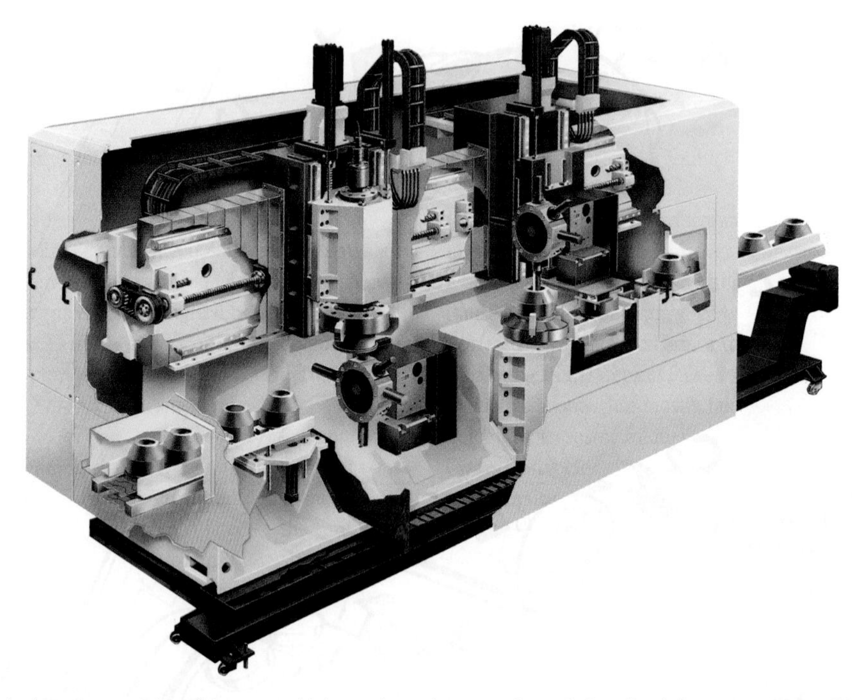

Figure 1.14: Vertical turning machine with two machining units and automatic workpiece feed. (Image courtesy of MAG-Hessapp.)

two tools' motions are programmed independent of each other, and these motions are coordinated with each other only at critical points.

In summary, it can be said that multiaxis turning machines are suitable primarily for the series production of medium to large quantities. Whether shop-floor programming is a practical approach depends on the performance of the programming system (*Figure 1.14*).

Turning/Milling Centers

In addition to turning operations, the high performance of modern CNC systems has made it possible to use suitably equipped turning machines to perform milling and drilling operations on the workpiece while it is held in the lathe chuck. **This principle also has been applied to the use of grinding spindles** for finishing operations on turned parts.

This is done by equipping the turret with **driven tool spindles** that hold the necessary cutters, drills, taps, or grinding wheels. If necessary, the main spindle can be linked automatically to a sensor accessory, positioned as the *C* axis in a similar manner to a rotary table, and continuously controlled. In this manner, the driven tools can be positioned accurately to any point on the workpiece and can mill or grind any desired shape. This requires a CNC system that can perform **coordinate transformation.** This function allows milling and drilling operations to be programmed in Cartesian (linear) coordinates. The CNC system transforms the motions into polar coordinates (rotation of the *C* axis) (*Figure 1.16*).

Turning/milling centers are especially suitable for the production of **small, complex parts** from solid stock. They are offered by all the major manufacturers of turning machines.

Various configurations of turning/milling centers are also available for large parts. These applications also take advantage of the fact that all the necessary turning and milling operations can be performed on the workpiece without rechecking and with a higher stock-removal rate (*Figures 1.15, 1.16, and 1.17*).

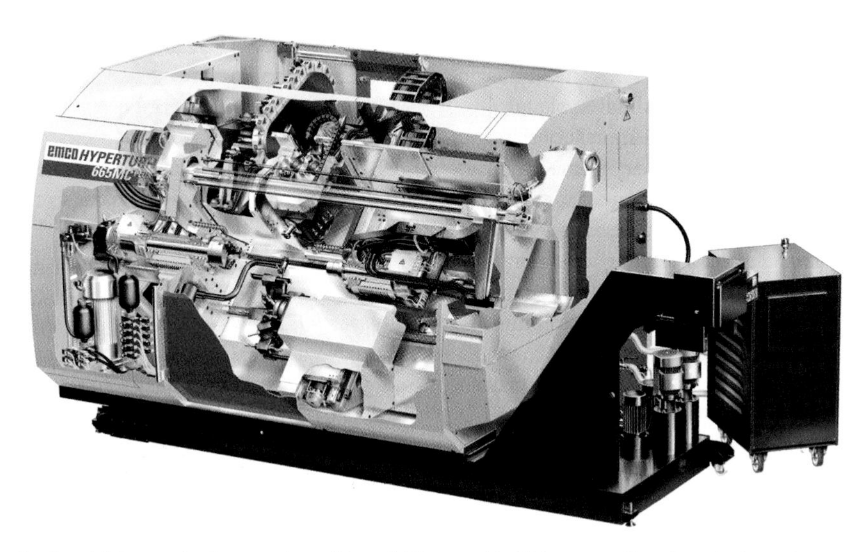

Figure 1.15: Turning/machining center for complex milling, drilling, and hobbing operations in a single work process. (Image courtesy of EMCO.)

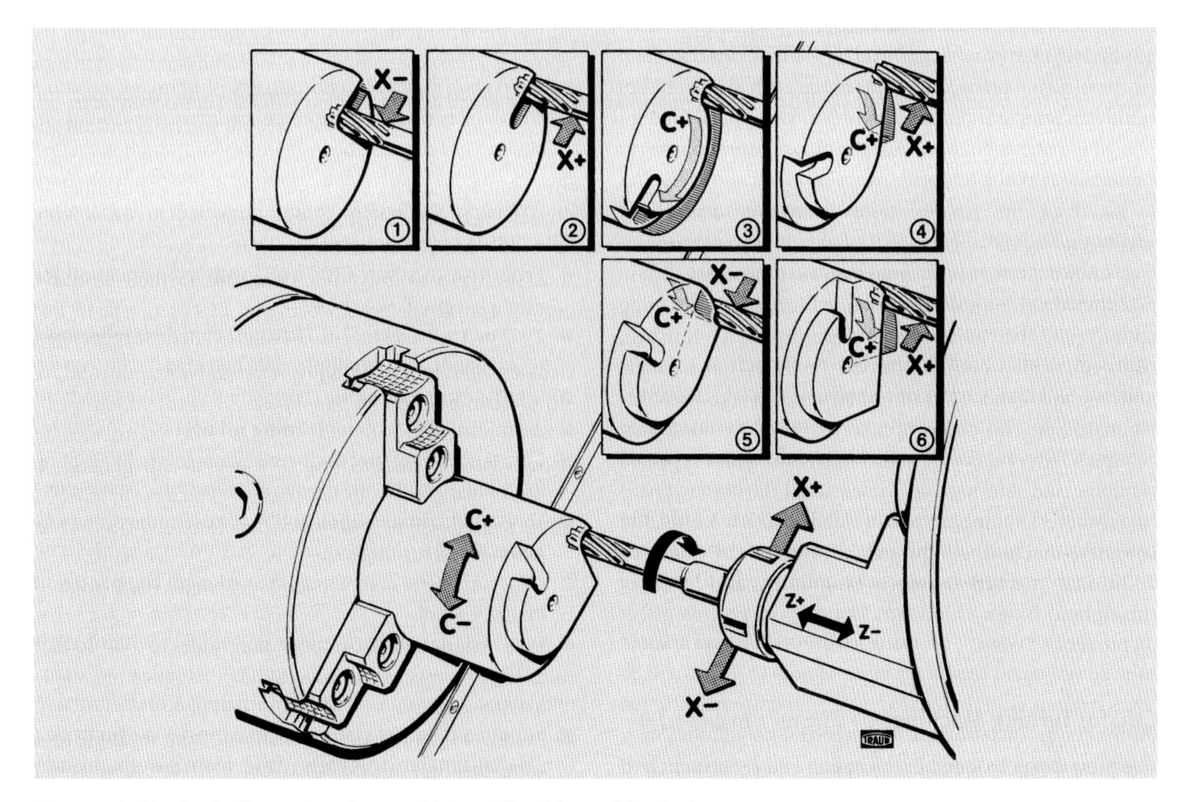

Figure 1.16: Turning/milling center: face machining with a driven milling tool.

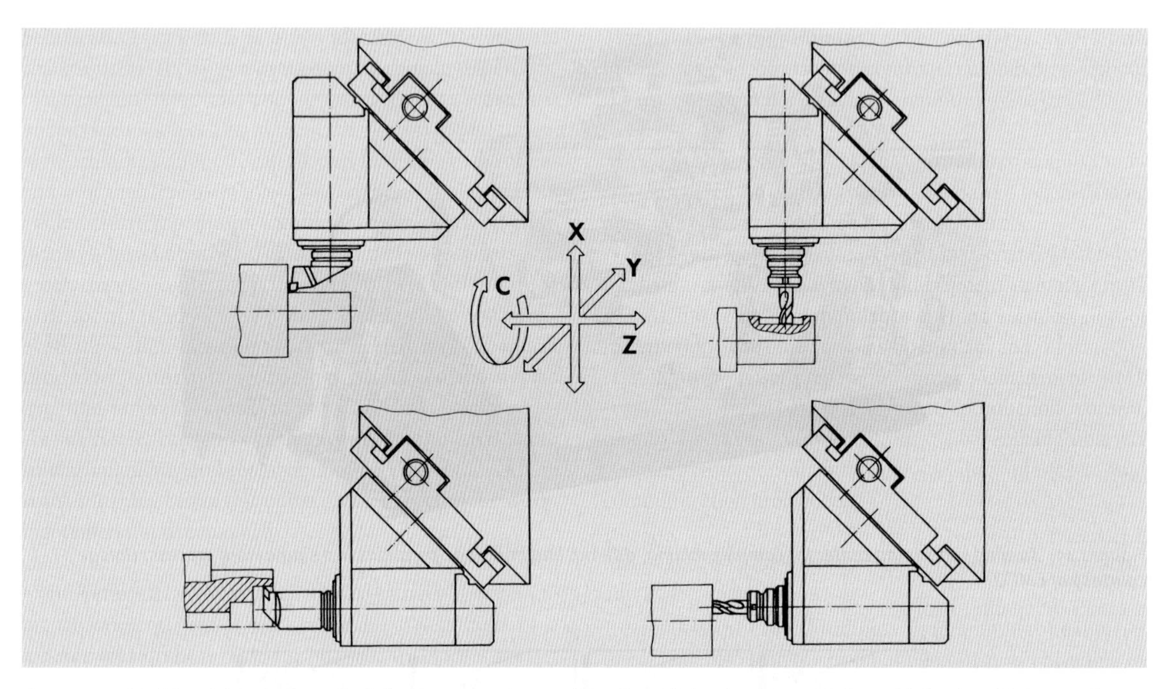

Figure 1.17: Tool head that can be swiveled 90° on a turning machine for radial and axial machining with fixed and driven tools.

Based on the familiar terms *machining center* and *manufacturing cell*, highly automated turning machines are also known as **turning centers** or **turning cells.** These machines provide an almost unlimited variety of machining options. The great advantage for the user is that a finished, ready-to-install workpiece can be produced on a single machine and in much less time than it would take on multiple machines. This also eliminates any need for additional clamping fixtures, as well as the time required for repeated clamping and unclamping. Dimensional deviations resulting from reclamping are also avoided. In other words, **the complete-machining approach eliminates the need for additional machines, improves quality, and reduces throughput times.** This naturally also has a positive effect on production costs.

CNC for Turning Machines (→ *Figures 1.16 and 1.17*)

The great diversity found in the design and construction of turning machines has carried over into NC systems. Even the most basic CNC system is now expected to fulfill **very demanding requirements:**

- **From two to seven CNC axes,** with as many as 30 for multiple- spindle machines
- **2 × 2 or 3 × 2** axes that can be interpolated independently of one another (for multiple-slide machines)
- Spindle controlled as the *C* axis
- Additional CNC axes for **loading robots**
- **Constant cutting speed** achieved by automatically adapting the spindle speed to the turning diameter
- Tool-offset compensation and tool-nose-radius compensation for all turning tools
- Cutter-diameter compensation and length compensation for driven tools
- **Free assignment** of compensation values to tools so that different compensation values can be assigned to a single tool if necessary
- Ability to simultaneously take into account **multiple tool compensation values** such as tool offset, tool nose radius, and tool wear

- Monitoring of tool cutting edges and tool breakage
- **Tool life monitoring** and automatic call-up of a replacement tool when the service life of the original tool has expired
- **Feedback of measured values** to the compensation-value memory and automatic readjustment of those values

Another very valuable function of turning machines is NC **thread cutting**. For this purpose, the main spindle has to be equipped with a measuring system, usually in the form of an incremental pulse generator. This provides the CNC system with feedback as to the speed and exact angular position of the spindle. An additional **reference pulse** is generated once per rotation. In thread cutting, this reference pulse ensures that the feed always begins at a predefined position of the main spindle and that the individual cuts are made at exactly the desired thread pitches (which are turned in advance). NC also can be used to produce **tapered threads, multiple start threads, and progressive or regressive pitches.** This eliminates the changeover time and the expense of mechanical thread-cutting devices. The CNC system controls the absolute synchronization of the spindle rotation and the feed motion by processing the pulses emitted by the pulse generator (or spindle encoder).

Programming of Turning Machines

Although this summary has been far from exhaustive, it should be clear now that within a very short time, simple lathes have developed into highly complex **CNC machines.** Therefore, another primary objective has been to develop a simple, clear, easy-to-learn approach to programming. These efforts were greatly aided by progress in the development of powerful desktop computers. **Color graphics monitors** helped these computers to quickly become accepted by users. It is no longer necessary for the programmer to learn an "artificial language" to program a CNC machine. Instead, the programmer works **interactively,** answers questions posed by the system without any mathematics or *G/M/F/S/X/Z* functions, and immediately sees the results of his or her input displayed graphically on the screen. The desired part geometry is entered first, followed by the dimensions of the unmachined part and the machining operation to be performed. Any desired point in the machining sequence also can be simulated **graphically and dynamically** on the screen. Errors can be corrected quickly, and the results of this correction can be checked. This can be repeated as many times as necessary until a complete, error-free machining operation has been entered. The system then generates the CNC program and can output it to any desired data storage medium for any suitable machine. All calculations, positioning motions, retrieval of compensation values, and other special functions are generated automatically by the computer. Errors are largely excluded. With **shop-floor programming (SFP)** it is now possible to program the CNC system directly at the machine.

Multiple-Spindle and Multiaxis CNC Turning Machines

Today, the advantages of CNC technology are also exploited in a type of turning machine that was considered a typical non-CNC machine until recently: the indexing drum machine. These rotary index machines are especially well suited for the mass production of small, complex, high-precision workpieces. The indexing drum advances the workpieces (blanks or material supplied in ring or bar form) through a sequence of 14 machining stations, where they are machined in simultaneous, parallel operations.

By machining at all the stations simultaneously, the total machining time is reduced to the length of a single cycle. Special swiveling chucks allow final machining of the workpieces on five or more sides. In addition to cutting operations, this type of machine also can be used to assemble parts, drive screws, create flanges on discs, and insert pins.

Until relatively recently, all these operations were controlled primarily by means of cams. This is still an economical solution for single-purpose machines or sufficiently large series. The problem becomes evident with smaller lot sizes, that is, when frequent changeovers are required. Long downtimes quickly reduce the profitability of this highly productive type of machine. Conversion to NC of the machining units makes the operation independent of cams and cam changes. Feed-path lengths, feed rates, and spindle speeds can be programmed freely. Other advantages of CNC include the ability to machine different diameters at a single station, as well as the creation of radii, edges, and tapers.

The time savings during setup and changeover can amount to 85 percent, which can equate to four hours or more. In a matter of minutes, up to 20 CNC axes can be converted to produce an entirely different workpiece simply by entering a different program. Machining units that can be exchanged quickly at the individual stations are also conducive to short changeover times.

1.3 Grinding Machines

(Dr.-Ing. Heinrich Mushardt)

Designs and Requirements

Grinding is generally considered to be the classical process for fine machining and hard machining and is used for a diverse range of applications. For example, grinding machines are able to meet demanding requirements as regards precision and surface quality in the tool and die industry. They can machine even the hardest materials to create and resharpen tools. In the manufacturing of gears, they can create complex, high-precision tooth geometries. In automatic series production, they can achieve short cycle times.

The machine tools produced today generally are numerically controlled. In contrast to conventional grinding machines, in which the operator is able to intervene to make corrections, CNC machines are expected to ensure the required quality requirements in automatic mode while also being largely immune to external interference and variable process input parameters. For example, in view of high requirements in terms of precision, it is necessary to limit all deformation resulting from, among other things, variable grinding forces or heating of machine components, coolants, and the surrounding environment.

A distinction is made between various grinding processes based on the shape of the surfaces being created. These correspond to different types of machines. The machines used most commonly are the flat and profile grinding machines depicted in *Figure 1.18*, as well as cylindrical grinding machines and tool-grinding machines. Special designs also exist for particular applications, for

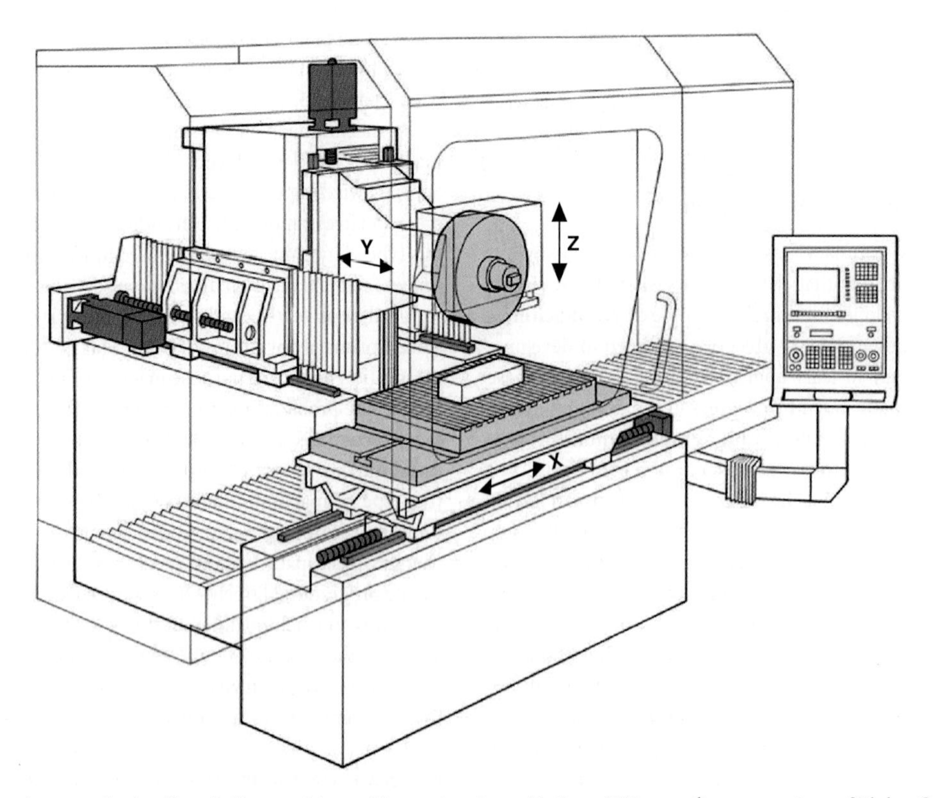

Figure 1.18: Planomat flat/profile grinding machine, table construction with three CNC axes. (Image courtesy of Blohm-Schleifring.)

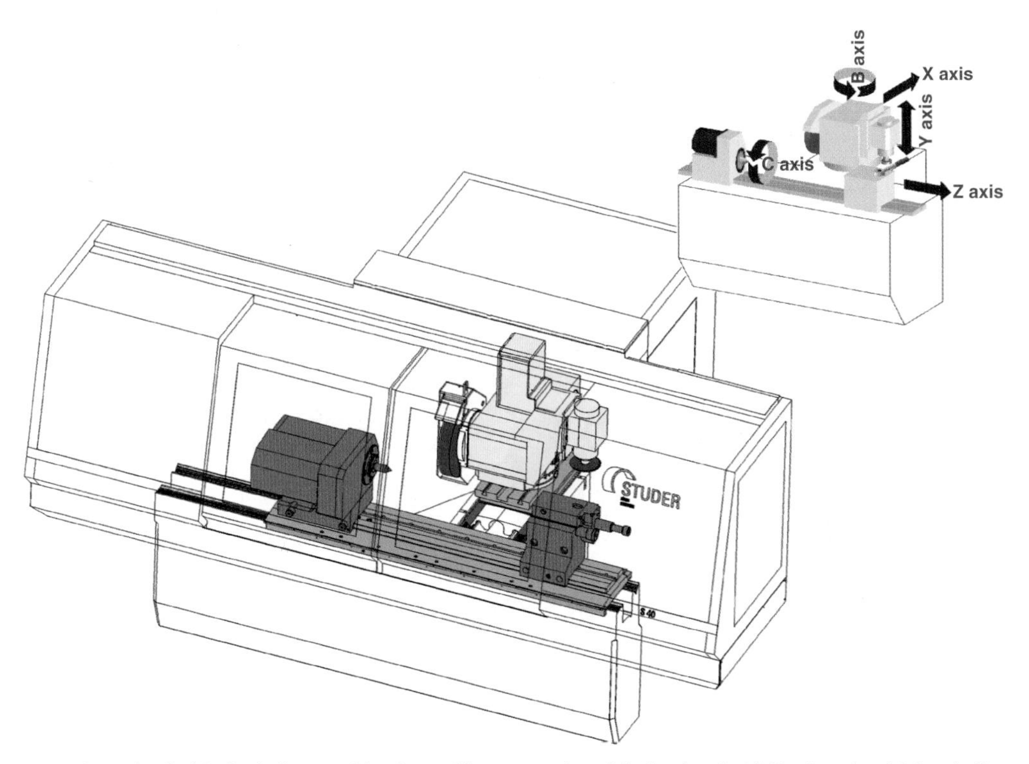

Figure 1.19: Universal cylindrical grinding machine S40, table construction. Grinding head with B axis and multiple grinding spindles. (Image courtesy of Studer-Schleifring.)

example, for grinding of threads and screws, toothing, cams, and eccentrics. *Figures 1.18 through 1.20* present schematic diagrams of a surface grinding machine, a universal cylindrical grinding machine with an axis *B*, and a tool-grinding machine, respectively.

The high-precision machining and special operating conditions of these grinding tools place particular demands on the stiffness of the frame and slide components. What is more, grinding machines are characterized by very precise main spindles, guides, and measuring systems that allow precise positioning and exact feed paths. Various guidance principles have to be adopted depending on the specific areas of application for the machine and the required axis speeds. The main emphases are on the operational use of play-free, low-friction rolling guides but also on plastic-coated sliding guides with low stick/slip and good damping properties or frictionless hydrostatic guides. For feed drives, servomotors and ball screws are preferred. Increasing use is also being made of linear and torque motors as drives for rotational axes. These motors can operate with transmission elements that otherwise would introduce play. This allows for very quick and precise positioning and compliance with paths. Linear scales are practically obligatory, at least in the axis directions that are decisive in determining the dimensions, that is, the *Y* axis for flat grinding and the *X* axis for cylindrical grinding.

To ensure unblemished surfaces, it is very important for the machine structure to have good damping characteristics and that all components move without vibration. Because most grinding operations involve coolants, it is necessary for spindles, guides, and measuring systems to be protected against penetration by water and oil, as well as against fouling by the grinding chips that are produced. It is also necessary for the entire workspace to be screened off.

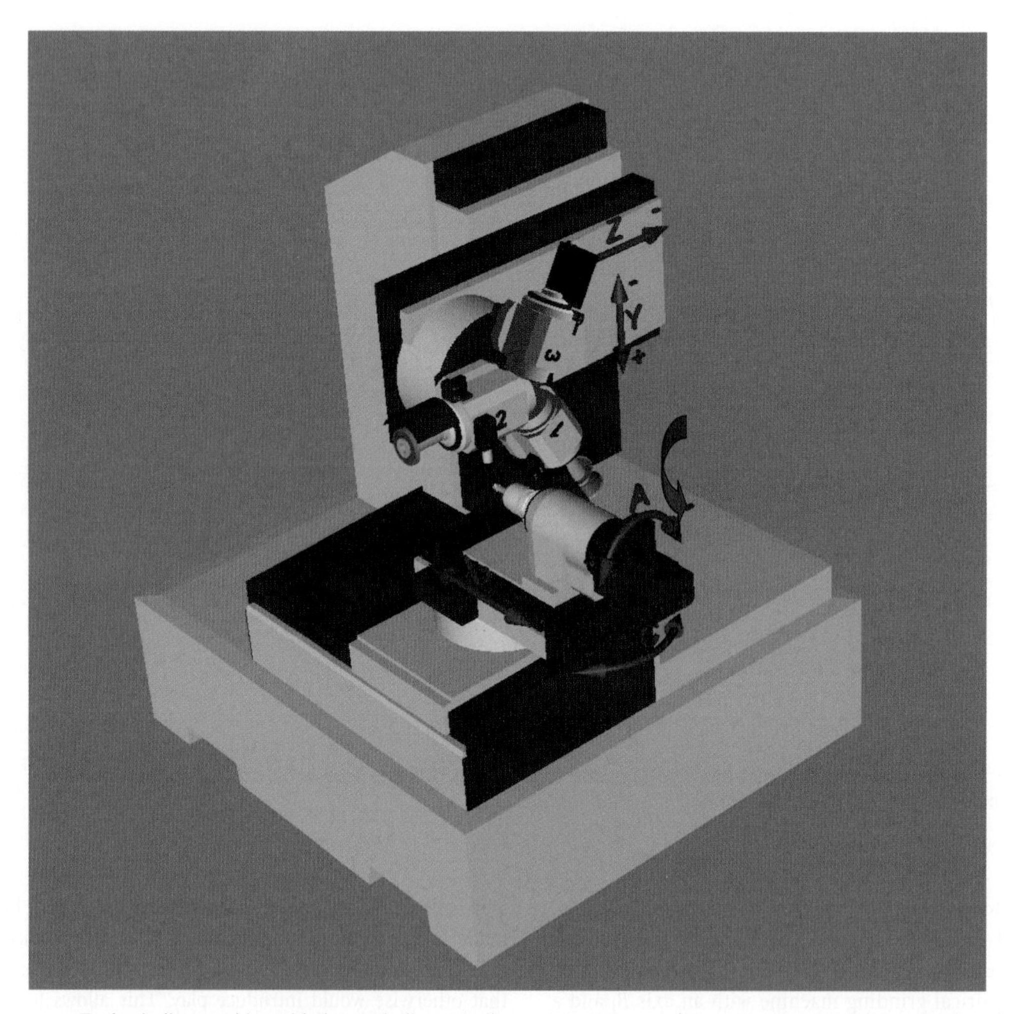

Figure 1.20: Tool-grinding machine with three grinding spindles and six CNC axes. (Image courtesy of Walter-Schleifring.)

Control Functions for the Use of Grinding Wheels

The particular features of grinding tools and grinding processes place special requirements on control systems. The grinding tools most often used are grinding wheels, followed by grinding belts. The peripheral speed is an important parameter setting and therefore has to be taken into account in process optimization. For this reason, a large proportion of grinding machines is equipped with variable-speed spindle drives. Because speeds generally are very high, it is necessary to balance the grinding tools to prevent vibrations during the grinding process. This can be done using automatic devices integrated into the grinding spindles or grinding-wheel mounts.

Grinding wheels, with the exception of wheels with a single-layer coating, are dressed after they have been clamped in the machine and thus prepared for the grinding process. This gives them the necessary precise concentricity and axial run-out, the profile appropriate for the grinding task, and the required abrasiveness. Sharp-edged abrasive particles can be used to machine even the hardest technical materials. In doing so, they become worn and blunt. Once the profile and abrasiveness have decreased owing to wear to such a point that the workpieces can no longer be produced within the required tolerances, the profile and abrasiveness are restored by means of dressing.

Dressing and grinding mean that two different machining processes are being executed in the grinding machine. The control system has to be able to handle both these processes while also taking into account the changing abrasiveness and diameter of the tools. With variable-speed spindle drives, the control system can correct the speed to keep the peripheral speed constant when the wheel diameter is changed owing to dressing.

To protect the operator, the peripheral speed of the grinding wheel has to be limited in a failsafe manner. To do this, the control system has to know the current diameter. This is input and then confirmed by the operator during setup or when a wheel is changed. After dressing, it is calculated automatically based on the position of the dressing tools.

Dressing is mostly performed using diamond-coated tools. Stationary or rotating tools may be used. A profile can be created with continuous path control by moving the dressing tool over the grinding wheel under two-axis control. To allow maximum flexibility and in order to reach the opposite edges, the dressing tools also may be given a swivel function. *Figure 1.21* shows a dressing device with this capability. Here the dressing tool has a radius profile and is aligned precisely on a swivel axis. The inevitable residual deviations in alignment and deviations from the radius profile can, if necessary, be measured and compensated for using controller functions to improve the precision of the grinding-wheel profile being created.

As an alternative, it is also possible to dress grinding wheels using profiled diamond rollers and a plunge-cut process. Diamond profile rollers are expensive and have to be tailored for a specific workpiece, but the time savings involved means that they pay for themselves very quickly in series production. With diamond profile rollers it is possible to sharpen the grinding wheel continuously during an ongoing grinding process. When machining high-strength materials that cause a high degree of wear while also requiring particularly sharp cutting edges, continuous dressing of the grinding wheel makes it possible to achieve higher stock-removal rates. This, in turn, can lead to shorter cycle times. In this case it is the CNC system's task to compensate for the changes

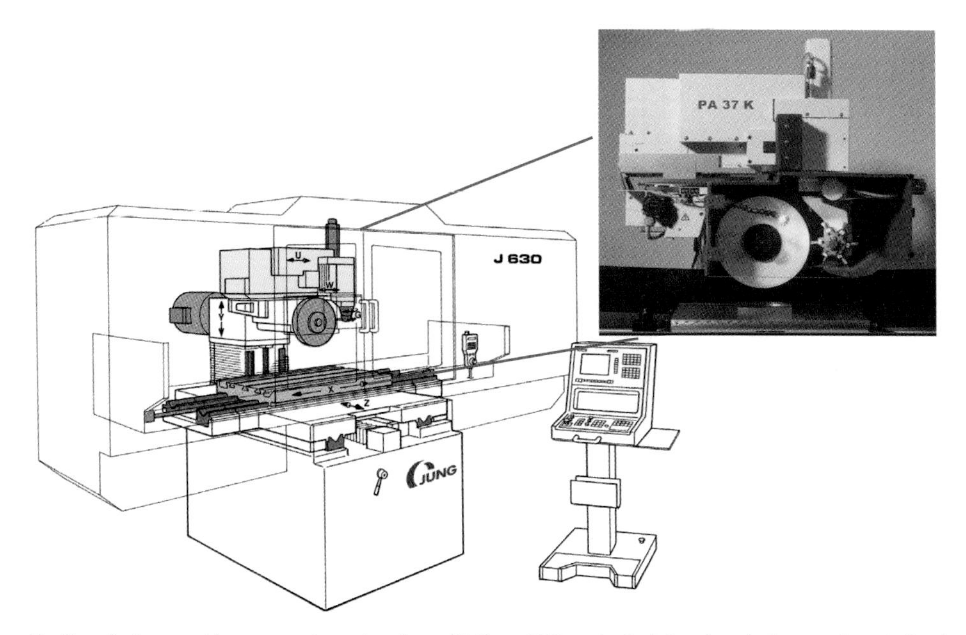

Figure 1.21: Profile grinding machine, support construction with three CNC-controlled dressing devices on the grinding head. (Image courtesy of Jung-Schleifring.)

in diameter caused by dressing and to adjust the speed of the grinding wheel to keep the cutting speed constant. In high-performance processes, it is also necessary to adjust the positions of the coolant nozzles (*Figure 1.22*).

Control Functions for Setup, Programming, and Optimization of Grinding Processes

User-friendly control systems make it easier to set up machines, to operate them, and to optimize their processes. Other requirements include manual intervention into automatic processes, for example, to shift the reversal positions in the event of oscillation movement, to superimpose in-feed rates in the case of variable allowances, or to insert dressing cycles when the grinding wheel is becoming dull.

During setup, the positions in the workspace of the workpieces, dressing tools, and grinding wheels are determined precisely. The CNC system can support this measurement process by managing the signals from measuring probes and other sensors, for example, acoustic emission sensors. These signals can be used to detect contact between the grinding wheel and the dressing tool or workpiece. The subsequent automatic conversion into machine coordinates is especially helpful for machines with swivel axes. This reduces the risk of errors (*Figure 1.23*).

Dressing and grinding cycles can be supported in many ways by means of software. To create profiles on grinding wheels, geometric data should be transferred from the drawing files to calculate dressing paths. Programmers can use the menu-driven functions of programming systems to manage the grinding wheels, to select suitable dressing tools, and to optimize the setting parameters for dressing. They also can perform calculations for collision monitoring and to compensate for the effects of errors.

When grinding noncircular shapes, variable engagement conditions arise that trigger force variations, which, in turn, can lead to form errors on the workpiece. These effects and other systematic error components can be minimized by

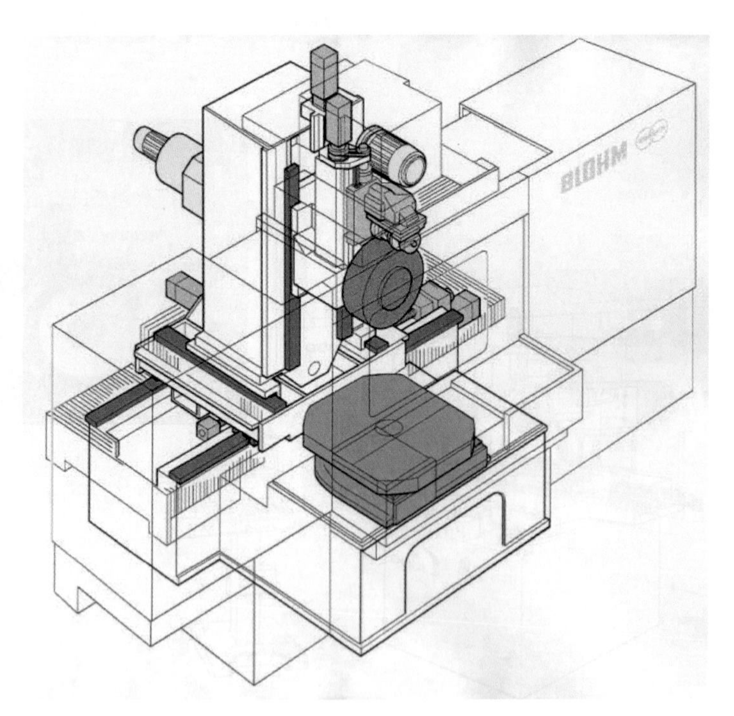

Figure 1.22a: Profile-grinding machine with diamond roller dressing device on the grinding head and tactically for loading and unloading during machining. (Image courtesy of Blohm-Schleifring.)

Figure 1.22b: Dressing device with diamond profile rollers. The coolant nozzles are adjustable.

means of tailored speed profiles and by allowing for path correction. When high precision is required, remaining form deviations on the workpiece are measured and compensated for in further optimization steps (*Figure 1.24*).

The wide variety of drilling, countersinking, and milling tools and their high geometric complexity place extremely high demands on the programming of tool-grinding machines. The complex calculations required for

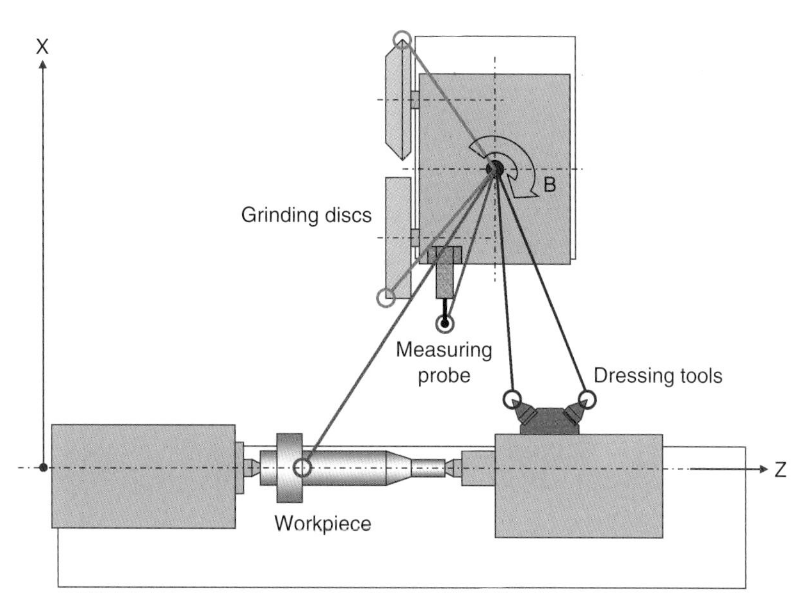

Figure 1.23: Support for the CNC system via determination of the reference points for the workpiece, grinding disc, and dressing tools by means of software. (Image courtesy of Studer-Schleifring.)

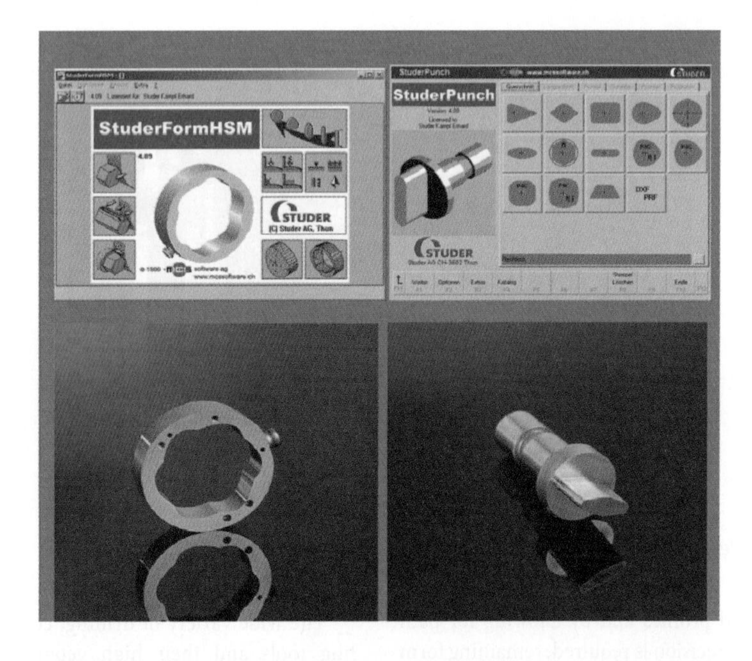

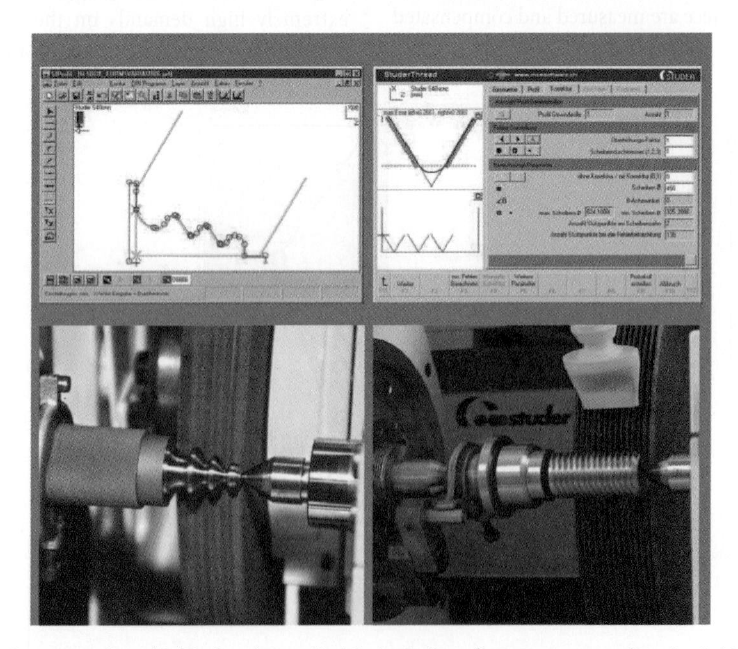

Figure 1.24: Programming system for cylindrical and noncylindrical grinding. (Image courtesy of Studer-Schleifring.)

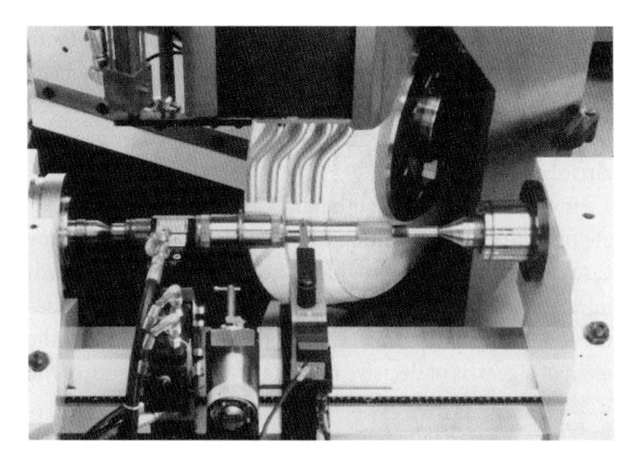

Figure 1.25: Measurement-controlled grinding of a shaft. (Image courtesy of Schaudt-Schleifring.)

feed paths can be performed by up-to-date programming systems. These programming systems also check the engagement geometries and analyze the workpiece profile that is generated. By checking the programs on the screen, it is possible to detect any errors and opportunities for optimization. This means that collisions in the workspace can be detected even before the program is put into operation on the machine, thus reducing risks (*Figure 1.25*).

The possible causes of dimensional and form errors include different allowances, grinding-wheel wear, and thermal deformation. In cylindrical grinding, their effects can be eliminated through the use of measurement-control devices that switch over from roughing to finishing feed rates when preset allowances are reached and which terminate the process when the specified dimensions are achieved. In contrast to turning and milling, with grinding, it is possible to use an in-feed that approaches zero. This largely eliminates the grinding forces and deformation. At the same time, workpiece roughness is improved owing to greater overlap of the cutting-edge engagements (*Figure 1.26*).

A typical plunge-cut grinding process comprises several steps: Until contact with the workpiece, the grinding wheel is advanced at a higher speed to minimize the amount of unproductive time. Contact can be determined by means of either force measurement, power measurement, or acoustic-emission measurement. Once this occurs, the grinding wheel is switched back to the roughing feed rate within a fraction of a second. During roughing, the objective is to make full use of the potential of the machine and the grinding tool and to grind off most of the allowance as quickly as possible. After that, finishing is done at a lower feed rate to improve workpiece quality and ensure that the tolerance limits are met at the end of the machining process.

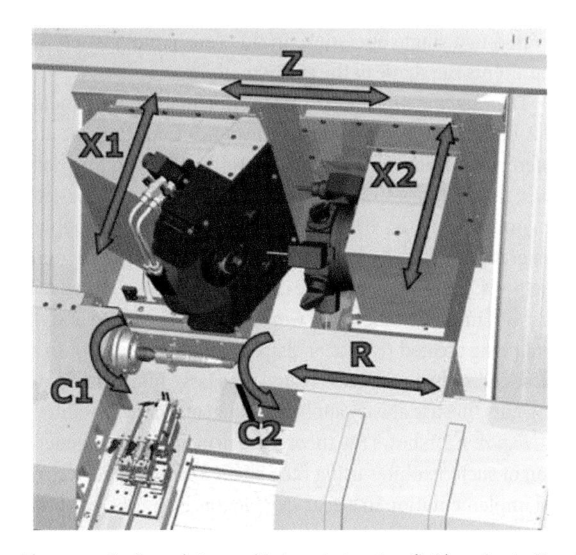

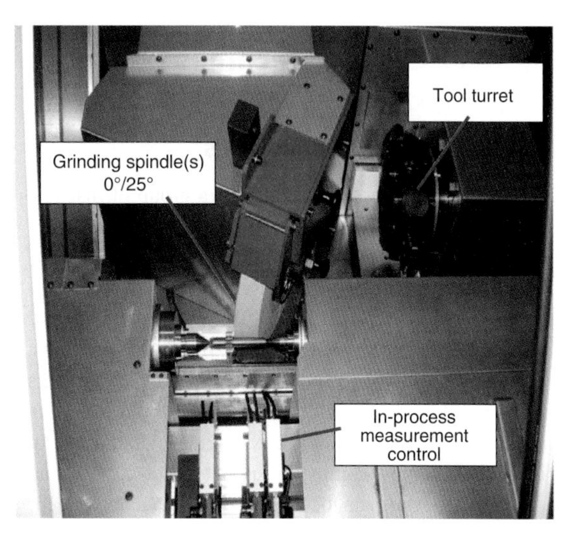

Figure 1.26: Complete machining via turning (left) and grinding (right). (Images courtesy of Schaudt-Schleifring.)

Automatic Workpiece Changing and Tool Changing

In series production, workpieces are mostly changed automatically. For variable and prismatic workpieces, automatic changing is often possible only with the use of clamping pallets. In contrast, cylindrical workpieces generally can be gripped and clamped directly. Shafts are often exchanged using gantry loaders or robots. Chucking components also can be picked up by the workpiece spindle directly from presentation positions.

In cylindrical grinding, clamping is performed either between centers, in chucks, or centerless using bearing rails and support rollers. For slim workpieces that require additional support, steadyrests are used. If they have to be supported at points that are also going to be ground, a controllable design that compensates for the stock removal must be used.

Workpiece magazines and automatic loading units make it possible for the Helitronic Power tool-grinding machine to operate in unmanned operation. A number of profiled wheels are required for complete machining of a workpiece. These can be clamped as a set on the spindle next to each other. An alternative option is automatic changing of the grinding wheels. In the case presented here, the grinding spindle takes arbors with hollow-shank taper holders directly from a disc magazine. At the same time, the associated coolant nozzles are picked up and connected.

1.4 Gear-Cutting Machines

(Dr.-Ing. Klaus Felten)

Basic Principles and Applications

Gear-cutting machines are defined as a group of machine tools whose purpose is to create very precise tooth flanks. These machines appear in a very wide variety of forms depending on the type of part (i.e., bevel gear, spur gear, or straight or helical gearing) and the technological process (i.e., cutting/noncutting or soft or hardened material). For these reasons, this discussion will first go into the fundamental requirements and the resulting designs and characteristics of gear-cutting machines. Gear-hobbing machines for the manufacture of spur gears, the most commonly used type of gear-cutting machine, will be the main focus of the discussion. Subsequent sections will go into the special features of hard-finishing machines and machines used to produce bevel gears.

Tooth flanks are curved surfaces. In spur gears, the tooth profile is an *involute*; the tooth direction is straight or helical. The involutes are also called *evolvents*. They are formed based on the tip of a taut string that is developed from a base circle.

In the end state, the tooth flanks are only allowed to deviate from the ideal condition by a few micrometers. In many cases it is necessary to provide corrections for both the involutes and the tooth direction (tapers and crowning). Besides the tooth flanks, the *pitch*, that is, the distance between the teeth of a gear, is of decisive importance for the quality of the workpiece being produced.

There are basically two process variants that can be used to produce tooth flanks:

- Shaping processes (e.g., noncutting processes, broaching, form milling, form grinding)
- Hobbing and hollow-cut processes (e.g., hobbing, gear shaping, hob peeling, hob grinding, generation of gears by planing)

The simpler variant based on the kinematic system of the machine represents the shaping process. Almost all noncutting methods fall into this category. In cutting processes, forming tools are used that have the same shape as the tooth space. These tools then are used to create tooth spaces one after another using the dividing method. Another shaping process is *broaching*, in which a complete internal gear can be created in a single operation. In this case, a tool is used that represents the shape of the entire toothing.

The *hobbing* process has higher kinematic requirements. It is based on the generation of involutes from the so-called reference profile. This reference profile is in the form of a rack; that is, it has straight flanks, and in some processes, it is an immediate part of the gear-cutting tool (e.g., hobbing, generation of gears by planing, hob grinding). In other methods, the reference profile can be viewed as a toothed rack that can mesh with both the workpiece being created and the tool being created (e.g., gear shaping and hob peeling). In all these cases, the involute is created via a large number of profiling cuts that are always applied as tangents to the involute.

Figure 1.27 shows the theoretical principle for the generation of such involutes using rack-shaped tools and its technical implementation in a gear-cutting machine. The left-hand part of the illustration shows three cross sections of a straight-flanked tool that are always positioned tangentially

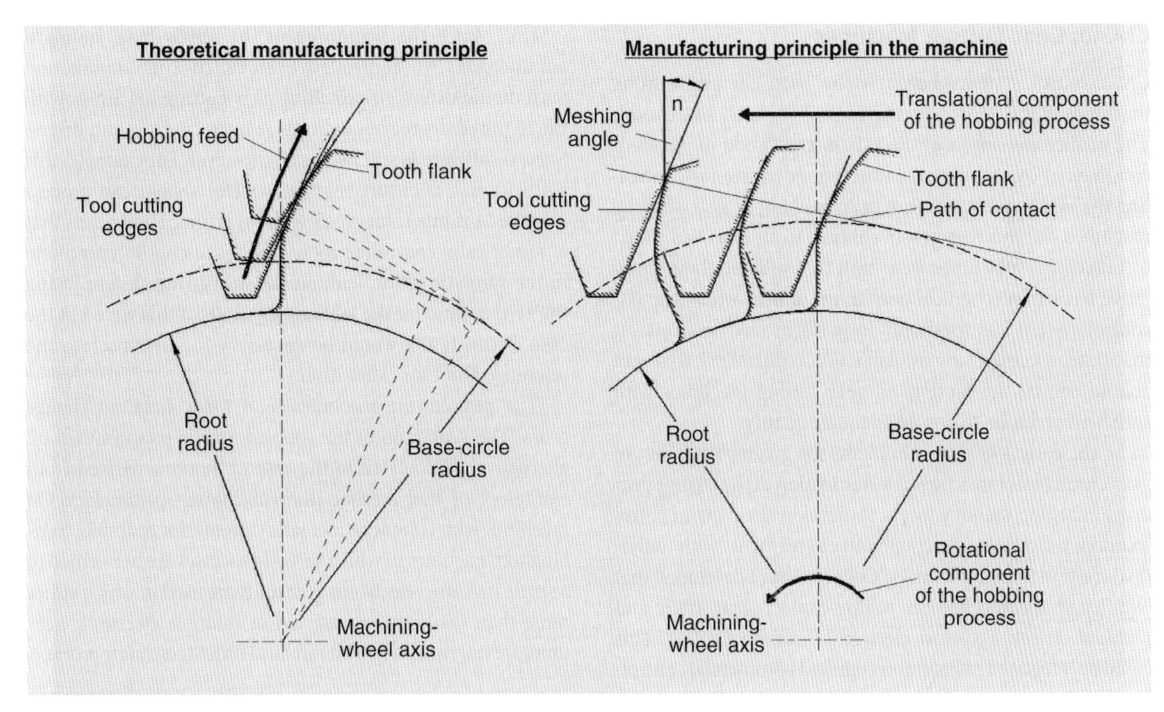

Figure 1.27: Manufacturing principle for an involute flank. (Source: Beck.)

to the involutes, thus creating them in a hollow-cut process. This relative motion between the tool and workpiece is implemented in a gear-cutting machine, as shown in the right-hand part of the illustration. Here, two individual motions are combined so that together they produce exactly the same relative motion between the workpiece and the tool. This consists of

- The rotational component of the hobbing as a rotation of the workpiece
- The translational component of the hobbing as a linear advance of a rack-shaped tool

Using this motion characteristic, it is possible to create involutes in a continuous hobbing process. Creating gear teeth, however, does require some additional feed motions that take place at the same time. Thus the teeth are cut from the tip to the base by advancing the tool radially. What is more, the axial feed means that the tool is guided over the entire width of the gear workpiece, which means that the entire wheel is toothed. If a helical gear is being produced, the axial motion must be coupled with rotation of the workpiece in such a way that the workpiece is always cutting in the direction of the tooth; that is, the workpiece table is subjected to an additional rotation on top of the hobbing rotation that depends on the angle of inclination and the axial position of the tool.

The demanding requirements as regards workpiece quality when four or more axis motions are being executed continuously have forced design engineers to find even more special solutions for gear-cutting machines. Thus, from the beginning, synchronous motions of multiple axes were implemented by means of mechanical gear couplings, as a result of which a single motor could drive a number of motions by means of gear branching. Changes in the gear ratio, for example, to produce a gear with a different number of teeth, were accomplished using plug-in change gears. In this regard, the mechanical-axis couplings meant that gear-cutting machines were always automated machines.

CNC for Gear-Cutting Machines

CNC control systems began to be used on gear-cutting machines in the 1970s with the control of purely linear axes. The quality and speed of the axis interpolation of standard commercial control systems were not adequate for controlling the instantaneous axes of gear-hobbing or gear-shaping machines. For this reason—in contrast to all other industrial technologies—machines were built that still had their qualitative basis in mechanical gear trains and in which the NC control system was used only to perform feed motions. At that time, commercially available CNC controllers were not able to control the motions of gear-cutting machines with sufficient precision to ensure adequate quality.

In the early 1980s, requirements for greater flexibility of gear-cutting machines through elimination of the change gear drove machine manufacturers to develop their own control solutions or to enter into cooperative agreements with individual controller manufacturers. Concepts were developed that eliminated the mechanical rolling-contact gear train and replaced it with a special electronic *hobbing module*. This could be integrated either into a standard commercial controller or into a special control system. It was not until this point that a genuine CNC philosophy could be implemented, in which each motion is driven by its own axis module. Besides the term *hobbing module*, terms such as *electronic gear unit* and *digitally constrained motion of two axes* also were used.

Only since the beginning of the 1990s have standard commercial CNC controllers been capable, in combination with digital drives, of providing gear-cutting machines with the required degree of quality as regards controls and drives. Special software is still necessary, however. For example, this software has to ensure that the control system can process so-called infinite rotary axes without errors. This means that a numerically controlled rotary table can continue to rotate in the same direction without the measuring system losing the zero point after the table rotates 360°; if this were not the case, it constantly would be necessary to move back to the reference point for calibration.

The gear-cutting machines built today have no change gears. The gear ratio of the coupled axes is programmed or, alternatively, calculated in the case of motions derived from workpiece or tool data or from machining parameters. On machines with complex kinematic systems, for example, combination machines in which several processes are performed in a single machine, electronic gear units are used in which either more than two axes are coupled or multiple electronic gear units are interlinked in a hierarchical order (*cascading gears*).

Machining of Spur Gears/Hobbing before Heat Treatment

Figure 1.28 shows the most commonly used hobbing methods in machining processes before heat treatment using a rack as

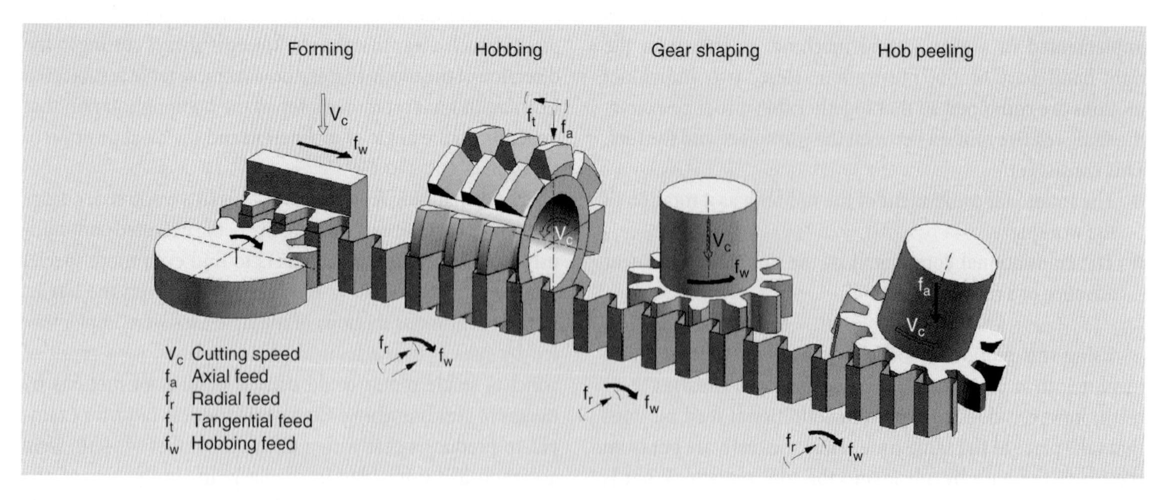

Figure 1.28: Principles of the four hobbing methods for soft machining

the common reference profile. To understand this illustration, please view the workpiece as first engaging at the left-hand end with the rack-shaped cutter, second with the gear hobber, third with the cutting wheel, and fourth with the peeling wheel. In gear hobbing, gear shaping, and hob peeling, the fact that the tool length is practically unlimited in the hobbing direction means that continuous hobbing processes are possible, in which workpieces can be hobbed completely with no interruptions. In planning, the limited tool length within a workpiece means that it is likewise always necessary to perform a partial operation.

If one also takes into account the formation of the tooth gap in the direction of the teeth, then there are significant differences between, on the one hand, the gear-hobbing and hob-peeling processes, and, on the other hand, the gear-shaping and gear-planing processes. Both shaping and planing motions involve a straight-line cutting motion. This results in chip removal that essentially proceeds in the direction of the tooth. In gear hobbing and hob peeling, the axial slide moves the rotating tool in the longitudinal tooth direction. In gear hobbing, owing to the relatively low rotational speed of the workpiece, markings may be created in the direction of the teeth depending on the axial feed. Together with the above-mentioned profiling cuts of the involute profile, these hobbed tooth flanks produce a faceted appearance. Hobbed and hob-peeled chips are always short, whereas chips from shaping and planing machines are removed along the entire breadth of the gear. As a basic rule, hobbing tools can be used with the same module to produce different numbers of teeth.

Gear-hobbing machines are gear-cutting machines that operate continuously. From a geometric point of view, the tool used here is an involute worm whose helices are interrupted by chip flutes. The flanks and head of the cutting teeth are relieved to create the clearance angle needed for machining. The tool and the gear workpiece mesh with each other as in a worm drive. The rotation of the milling cutter generates not only the cutting motion but also the translational component of the hobbing process via a tangential screw motion on the tooth flanks.

The production of straight-toothed gears is governed by the following parameters:

- The gear hobber and workpiece engage at crossed axes in a similar manner to a worm and a worm gear.
- The axis of the gear hobber is inclined by the helix angle of the milling cutter relative to the transverse plane of the workpiece.
- The gear hobber and the workpiece are positioned opposite each other according to the tooth depth.
- The gear hobber and the workpiece are rotated according to the ratio between the number of teeth of the workpiece and the number of cutter threads.
- Either the gear hobber or the workpiece is moved at a feed rate parallel to the workpiece axis, and chips are removed.
- After enough passes have been made around the workpiece, all the tooth gaps will have been cut out along the entire length of the workpiece.

With helical gears, the milling slide also moves parallel to the axis of the workpiece. Therefore, in this case, the axis of the gear hobber has to be inclined by the milling helix angle and by the spiral angle of the toothing that is being milled. In order to ensure that the gear hobber always cuts in the direction of the teeth during the axial feed motion, the workpiece is given, in addition to the rolling motion, an additional rotational motion by means of a differential gear unit. In modern machines this is always implemented using an electronic system.

The axial feed of the tool during gear hobbing creates feed marks. The distance between two marks shows the distance traveled by the gear hobber during a single rotation of the workpiece. There is an interrelationship here between quality and machining time insofar as higher axial feed rates reduce the machining time but, on the other hand, lead to more pronounced feed marks, with the end result being lower quality.

During gear hobbing, only a few teeth of the gear hobber are involved in the creation of a gear because the teeth are guided parallel to the workpiece axis over the complete width of the gear. In order to ensure even loading of the milling cutters and to reduce wear, it is possible to shift the tool either continuously or at certain time intervals in the longitudinal direction of the milling cutter, that is, tangentially to the workpiece. This motion is called *shifting*. The module and meshing angle are defined for each individual gear hobber. By varying the machine settings, the same milling cutter can be used to create all the required workpieces with different numbers of teeth, profile offsets, and spiral angles.

Figure 1.29 shows a state-of-the-art CNC gear-hobbing machine for use in mass production. Such machines require at a minimum five numerically controlled axes, which in some cases have to operate simultaneously. This involves the following axis motions:

- *A* axis—tangential motion of the tool (shifting)
- *B* axis—rolling motion of the tool (cutting motion)
- *C* axis—rolling motion of the table or the workpiece
- *X* axis—radial motion of the tool
- *Z* axis—axial motion of the tool

As a rule, both rolling motions and axial motions are intercoupled by means of electronic gearing. If greater flexibility is required, for example, to mill different workpieces or to correct the angle of inclination, then swiveling of the milling head can be used as the sixth CNC axis.

Play-free table drives are very important in producing high-quality toothing. The manufacturers of gear-hobbing machines have taken various approaches when it comes to minimizing play. As a result, duplex worm drives, double worm drives, tensed cylindrical gear drives, and drives with hypoid gears are all in use. However, the future definitely belongs to direct drives, which provide, among other things, high speeds, which means that they are also suitable for other technologies, such as gear-flank grinding. With the exception of very large machines, these days all new machines employ direct drives for all their instantaneous axes.

The trend toward dry machining means that new machine concepts with changed axis configurations have been created. The common goal of these conceptions is to allow free removal of chips. For this reason, the workpiece spindles are arranged either horizontally or vertically.

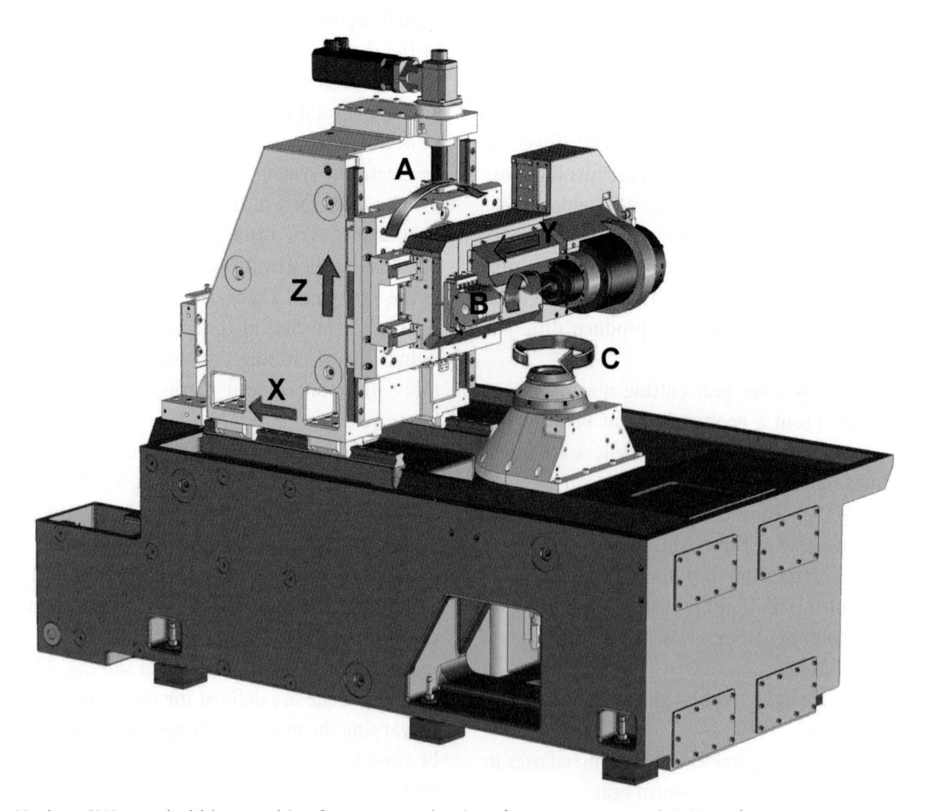

Figure 1.29: Modern CNC gear-hobbing machine for mass production. (Image courtesy of Liebherr.)

The advantages of this design are counterbalanced by the disadvantage that these axis positions are not suitable for universal platform designs that could be used on the same basis to implement various types of workpieces or even different technologies, for example, gear hobbing and gear shaping. As such, these products have to be regarded as single-purpose machines whose use is limited in various ways.

Figure 1.30 shows a modular product system for the gear-hobbing, gear-shaping, and hob-grinding processes. It is also very possible to perform the associated shaping process on the machines. All the hobbing drives are implemented as digital drives. The uniform machine basis means that the workpiece feeding, the automation, and other auxiliary functions such as deburring and wheelslips are performed independent of the toothing process.

Hard-Fine Machining of Spur Gears

The machining of hardened tooth flanks basically follows the same principles as soft machining. The goal of hard-fine machining is to eliminate the deformation created during heat treatment and to generate any necessary edge corrections. The processes created for this include hob grinding, profile grinding, honing, skiving, peeling of hardened materials, and broaching of hardened materials. Hard machining requires additional machine and control functions. These are of decisive importance for the quality of the tooth flanks.

The unmachined parts for a hard-finishing machine have already been toothed, and the tool has to be integrated into the existing tooth gap with great precision in order to be able to machine all the flanks. The position of the workpiece is detected with the aid of a sensor that has been impressed with the set-point position based on the first workpiece and which brings all subsequent workpieces into the same position by correcting the angular position of the workpiece table. Various geometries of the individual tooth gaps are averaged out. For this reason, these are called *threading devices* or *centering devices*.

For working with adjustable tools such as worm-grinding wheels made of corundum, dressing devices are integrated into gear-cutting machines that can in themselves be multiaxis devices and which can return a tool to its original form according to certain strategies (e.g., after the production of a specified number of workpieces or when a certain level of wear is reached). Dressing tools are diamond-coated.

Creation of Bevel Gears

Whereas the generation of involute spur gears can be described as engagement of the straight-flanked reference profile

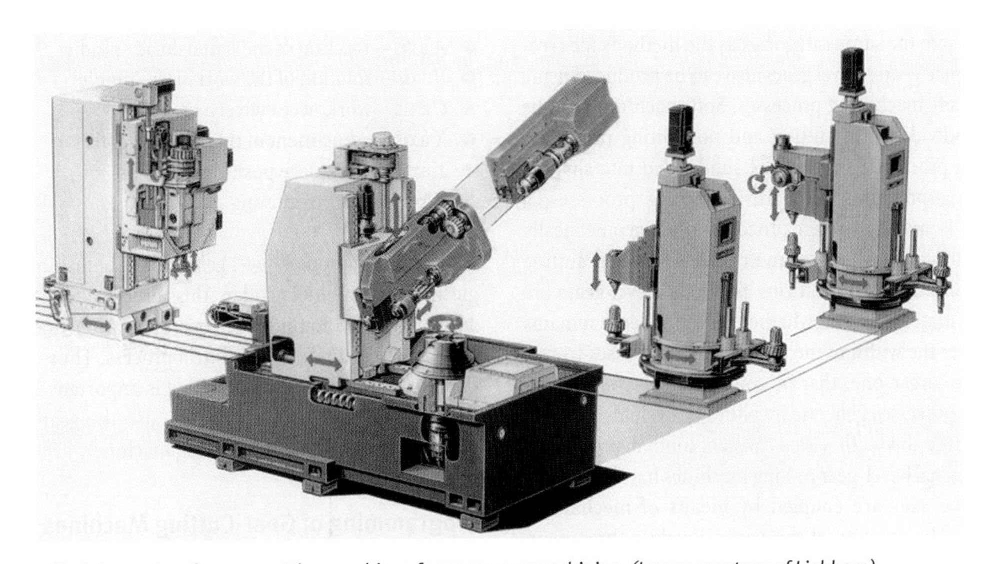

Figure 1.30: Modular system for gear-cutting machines for spur-gear machining. (Image courtesy of Liebherr.)

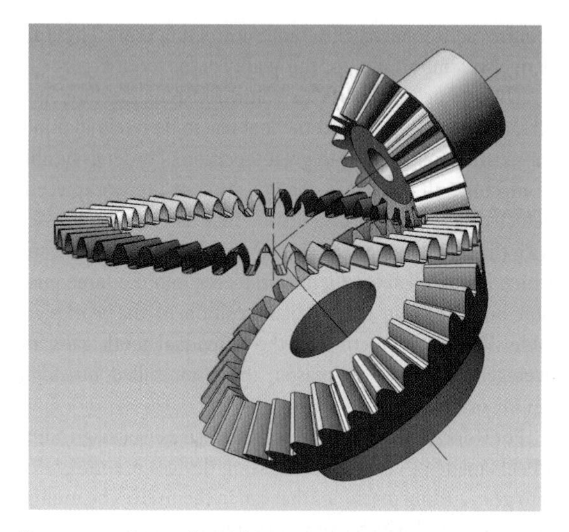

Figure 1.31: Gear pairs with intersecting axes.

with the workpiece, the production of spiral-toothed bevel gears involves the meshing of the workpiece with the so-called generating crown gear. Both wheels of a bevel gear pair are based on the same crown gear. The center point of the crown gear is identical to the common tip of the two pitch cones (*Figure 1.31*). In the production process, the cutting tool generally represents only one tooth or one gap in the crown gear.

The methods for creating bevel gears basically can be subdivided into the same categories as the methods for creating cylindrical gears. Bevel gears also can be produced using hard- and soft-machining processes. Soft machining can be further subdivided into cutting and noncutting processes, and cutting processes, in turn, are subdivided into shaping and hobbing processes. In hard-machining processes, a distinction is made between processes with geometrically defined cutting edges and geometrically undefined cutting edges. Besides the manufacturing methods, bevel gears are also differentiated based on whether the tooth depth remains constant over the width of the tooth face or increases from a smaller to a larger one, that is, conically. The continuous-production processes generate involutes or cycloids as a longitudinal curve and teeth with a constant tooth depth.

Conventional bevel-gear milling machines have from 10 to 12 axes. The axes are coupled by means of mechanical gear trains. The creation of the flanks involves three main elements with mutually associated motions: the roller drum, the cutter head, and the gear workpiece.

The kinematic basis for the bevel-gear production process is engagement of the generating gear with the workpiece. The axis of the roller drum is identical to that of the generating gear. The rotation of the roller drum brings about the rotary motion of the imaginary generating gear.

The actual cutting motion is performed by the cutter head. The path of the cutter in the area of engagement between the workpiece and the cutter head describes a tooth in the generating gear. The cutter head's rotary axis is positioned eccentrically and is not always parallel to the rotary axis of the roller drum. Both the milling cutter axis and the generating axis are rotated during the cutting process. The gear workpiece rotation is composed of the gear ratio between the gear workpiece and the generating gear and a relative motion that takes into account rotation of the roller drum.

The increased use of CNC technology has led to machine concepts with simpler mechanical designs and more flexible applications. These machines no longer have a roller drum but still use cutter heads as their tool. They only have six CNC axes. The complex relative motions between the tool and workpiece are implemented solely through the use of complex control systems and drive technology. Such state-of-the-art CNC machines can be used for all the usual processes (e.g., the dividing method and the continuous method) and tooth shapes (e.g., arcs and cycloids), provided that suitable tools are used. *Figure 1.32* shows an example of this type of machine. The following axes are involved in the production process:

- *A* axis—rotation of the cutter-head spindle
- *B* axis—rotation of the workpiece spindle
- *C* axis—workpiece swivel axis
- *X* axis—adjustment of the milling depth
- *Y* axis—workpiece positioning
- *Z* axis—tool positioning

The production of bevel gears does not just involve a single machine or tool. Today, this almost always involves a special type of manufacturing organization that functions similarly to a closed-loop control process. Thus it is not just the gear-cutting machine itself that is important in achieving the necessary level of quality but also the gear-measuring machine and the tool-sharpening machine.

Programming of Gear-Cutting Machines

The programming of modern gear-cutting machines is heavily automated and often has a graphical user interface. The

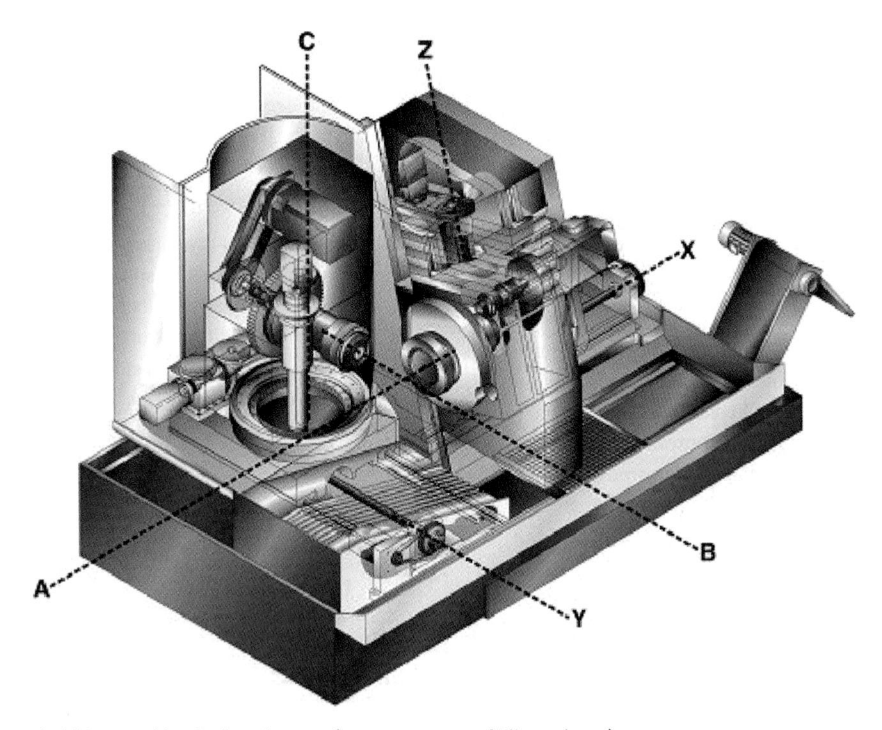

Figure 1.32: Gear-hobbing machine for bevel gears. (Image courtesy of Klingenberg.)

programmer or operator generally has to enter the set-point parameters for the workpiece and the actual dimensions of the tool in screen masks. For standard applications, the control system retrieves the cutting parameters from a database. Once created, the program also indicates the expected machining time. Programs are also used to simulate chip formation and thus provide an indication as to tool wear. Manual intervention by the programmer is possible at any point to optimize the process or to take special circumstances into account.

1.5 Drilling Machines *(Figure 1.33)*

All drilling machines have two common design features:

1. A spindle head with a vertical drill spindle that holds the tool and executes the feed motion (vertical *Z* axis)
2. A machine table to which the workpiece is clamped and positioned under the drill spindle in the *X* and *Y* axes

The drill pressure therefore is directed vertically onto the machine table. The advantage of this is that the workpiece is pressed firmly onto the table. Thus there is no horizontal shifting of the workpiece, fixture, or axis drives.

Drilling machines are especially well suited for the machining of plate-shaped workpieces with single- or multiple-spindle tools. These machines are also expected to perform simple milling tasks.

In their simplest configuration, CNC drilling machines feature only one numerically controlled coordinate table. The drilling depth is preset by means of mechanical stops or cam rails. On machines of this type, the drilling cycle is initiated by a signal that indicates "table in position" when positioning has been completed. This cycle is executed by mechanical or electrical means, without CNC. When the drilling operation has been completed, another signal is sent to the CNC system indicating "spindle up." Then the next positioning operation can begin. Program components include chip-removal cycles with programming of the *Z* axis (e.g., tool length, rapid traverse positioning, etc.).

Those who don't mind spending additional money on their machine can opt for another advanced feature known as

automatic tool changing. This feature has finally made it possible to limit the duties of the operator to pure monitoring functions so that machining becomes automatic in the truest sense.

However, this requires the presence of two additional features in the machine and controller. The first of these is the programmable selection of the **spindle speed** and **feed rate.** Only then is it possible to adapt to the cutting conditions for the workpiece and the selected tool in the best possible way without requiring manual intervention.

If extensive milling operations have to be performed (which may include milling along contours), the point-to-point control and straight-cut control discussed earlier must be replaced by **continuous path control.** Usually 2½D continuous path control is adequate for this purpose.

Mirror imaging is an advanced feature that simplifies programming of axially symmetric drilling patterns. A pattern that can be "reflected" in mirror symmetry in an adjacent quadrant can be subjected to mirror imaging about one or two axes.

This function either is activated manually with one switch per axis (known as a **mirror switch**) or is programmed with a G-function. This makes it possible to machine the entire drilling pattern by programming only one-half or even one-quarter of it. When machining has been completed in one quadrant, the symmetry function is switched on (for "mirror imaging about the *Y* axis," for example), and machining is repeated with the same program but in the second quadrant. Similar mirror imaging then is performed to execute the machining operation in the third and fourth quadrants.

Powerful CNC systems also allow **rotation and tilting** by a programmable angle, as well as to-scale **enlargement and reduction** of drilling patterns. A drilling machine also should provide repetitive **drilling cycles.** These represent a kind of fixed subroutine. Variations of these cycles are called by **functions G81 through G89** and then are repeated automatically at each new *X*/*Y* position. The programmer terminates these cycles with function G80. Example with a fine-boring cycle: Stop the spindle at the lowest point, move the workpiece slightly, and only then move the tool out to avoid scoring during the withdrawal operation.

Tool-length compensation is also absolutely essential. This allows the use of tools whose actual lengths differ from those specified for them in the program. The difference has to be entered into the **memory area** in the controller reserved for **compensation data** in order to achieve the drill depth specified in the program. These compensation values can be assigned to the respective tools in a fixed manner or retrieved by means of the H address (*Figure 1.33*).

Drilling Centers *(Figure 1.34)*

Customers often want to achieve greater degrees of automation without giving up the special advantages of drilling machines. This demand has led to the development of the drilling center. An additional rotary table or index table with a horizontal axis allows **four-sided machining** of cubical workpieces, thus turning the machine into a drilling center. To reduce idle time, especially for large lot sizes, these machines are often augmented with two **multiple fixtures** and a tilting table. This allows workpieces to be loaded and unloaded manually or automatically outside the machine's work area during productive time. Output can be increased relative to single-spindle machines by equipping the drilling center with **two or three main spindles.**

Drilling Machines for Printed Circuit Boards

A special variation of the drilling machine has been developed for the machining of printed circuit boards (PCBs) in the electronics industry. On these machines, multiple circuit

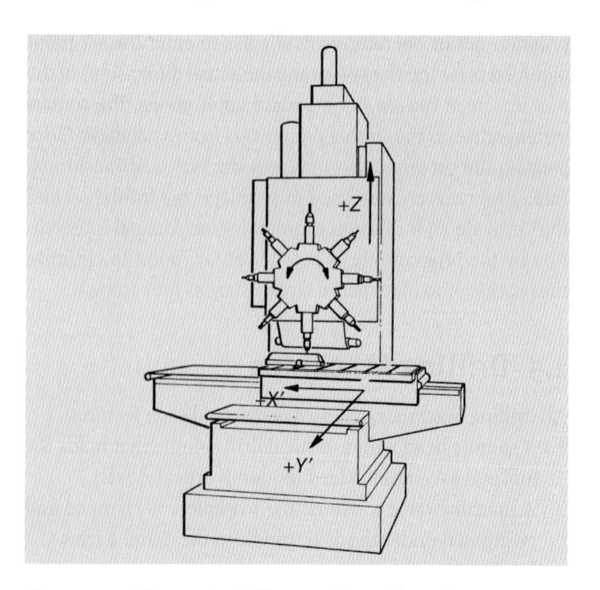

Figure 1.33: Three-axis drilling machine with tool turret.

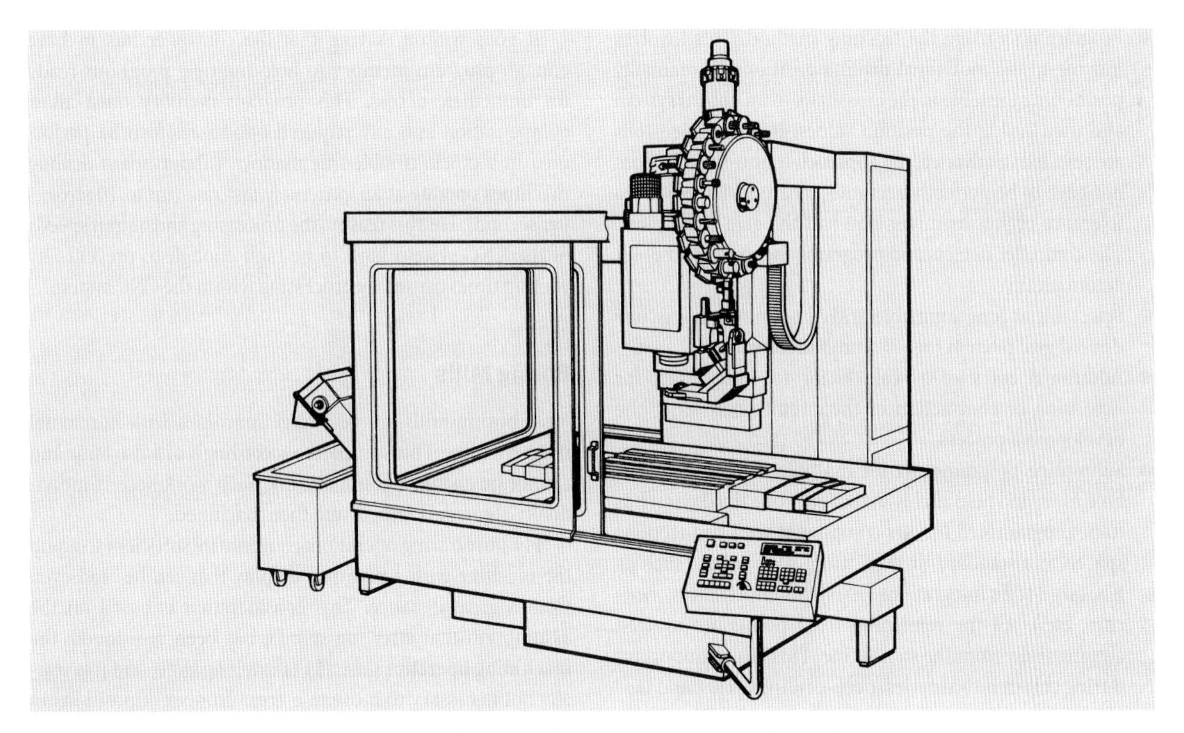

Figure 1.34: Vertical drilling machine in basic configuration with rotary magazine and X, Y, and Z axes

boards of a similar type are clamped on top of and next to each other. They are then drilled and milled simultaneously using multiple drill spindles. The drill spindles are mounted on a cross-beam, have a speed range of 15,000 to 60,000 rpm, and in some cases feature tools that can be changed automatically. Tools are picked up and deposited by moving to the magazine positions located at the front and rear ends of the machine table. All the spindles pick up their tools or deposit them in the magazine at the same time in these locations. It is thus possible to drill the circuit boards and then mill any desired cutouts or borders in a single setup.

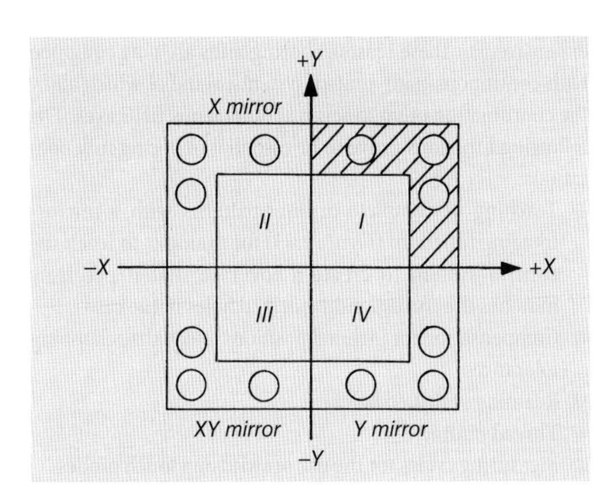

Figure 1.35: Mirroring of the dimensional data.

Numerical control of this special type of drilling machine therefore requires certain special functions, including the following:

- Control of two axes (*X* and *Y*) in rapid traverse and when milling along a path at a controlled feed rate. The desired spindle is retrieved once the desired position has been reached. The upper and lower limits of the drill spindle motion (*Z* axis) are set using external devices and do not require programming.
- Programmable retrieval of individual spindles or spindle groups so that various selected spindles remain in their retracted positions.

- Programming using the teaching method. This involves moving to the individual drilling positions by manually positioning the table in the crosshairs of an optical system and then saving the position values. If grid dimensions are selected in advance, the approach to the position does not have to be perfectly accurate. It can deviate within a tolerance of plus or minus one-half the grid pitch value. The controller detects and programs the exact grid point automatically.
- Selectable or programmable grid dimensions, that is, any desired grid pitch in inches or millimeters.
- Additional entry of a scale factor to compensate for expansion or contraction of the circuit board after the development process.
- Subroutine programming for multiple drilling patterns, hole rows, and hole circles so that only one fixed point has to be programmed for any component with connections spaced at a standard distance from each other (such as ICs, large-scale integrations [LSIs], potentiometers, switches, etc.). All the remaining fixed points are derived automatically from the subroutine. This is also important during chip-removal cycles, deep-drilling cycles, and tapping cycles.
- In some special cases, rotation of hole patterns by freely programmable angles.
- Programming and storage of customer-specific cycles, drilling patterns, or milling patterns.
- Character milling, for example, labeling the circuit board with a model designation, serial number, date, or similar information. This requires a special character generator so that any desired text can be entered directly, without any complicated programming. It is desirable for this function to provide variable character sizes, if possible.

Finally, the CNC system also has to control the following machine functions:

- Spindle selection and speed
- Feed-rate selection for drilling (*Z*) and milling (*X*, *Y*)
- Picking up tools from the magazine and replacing them there
- Drilling on/off
- Milling on/off

These functions are controlled by means of freely assignable M-functions.

It goes without saying that the controller has to have enough program memory to hold multiple programs ready for immediate access. This program memory must allow compensation and any required optimization to be performed in a quick, trouble-free manner. Circuit-board drilling machines operate at an extremely fast rate (up to 10 strokes per second). For this reason, the processing and output speeds of the CNC system have to be high enough to prevent any waiting times caused by processing within the CNC system.

Boring Mills

Most boring mills are very large machines with horizontal main spindles. The workpieces accordingly are also large and cannot be changed automatically. Each workpiece is individually aligned, clamped, and then machined.

For pure boring operations, it would be sufficient to equip the machine with a **position display** that can be read even from a great distance. This consideration is based on the assumption that after the spindle has been positioned, the machining operations usually take a long time, and it is usually not necessary to move to a great number of positions in rapid succession. The machine operator would have time to preselect the next *X*/*Y* position while the machining is taking place and then perform positioning. This would be even easier using an automatic **positioning-control system.**

Over the course of time, however, large workpieces also have become more complicated, machining operations have become more diverse, and control systems have become less expensive. For these reasons, boring mills are now equipped with continuous-path-control systems, some of which allow the coordination of up to seven simultaneous CNC axes. The following functions prove to be essential in boring mill operations:

- Teaching of complete drilling patterns with a mirror-imaging function, (*Figure 1.35*) for example, to teach the control system to create a cover so that it precisely matches the corresponding drilled holes in the case
- Compensation for inclined positions to equalize clamping tolerances
- Reaming of drilled holes
- Thread milling
- Measuring cycles for the use of switching touch probes
- Programming supported by graphics and machining simulations

Special **cycles** are provided for special machining operations. These cycles can be called up on the screen, provided with the required parameter values, and also modified, if necessary. Typical examples include hole circles, hole rows, mold milling, and pocket milling. Easy operation of the controller is also very important. This allows a running program to be interrupted and then resumed at a later time, for example, in order to replace used tools, taking into account the corresponding compensation values. The CNC system also should allow the user to create his or her own specific **subroutines**, including graphical support and parameterization.

The most commonly used **measuring systems** are linear scales. These are associated with stringent accuracy requirements and scale error compensation in the CNC system.

The following enhancements provide **increased flexibility** in large horizontal boring-milling centers:

- **Tool magazines with automatic tool changing**
- Interchangeable **accessory heads** for drilling, milling, planetary thread milling (thread whirling), and facing
- Pickup stations for special tool heads
- A movable control station
- Rotating and sliding **worktables**
- **Combination with other large machines,** such as a second boring-mill column, a vertical boring mill, or a moving-column machine at a large clamping plate

1.6 Parallel Kinematic Machines

In contrast to conventional machine tools, in which the individual axes are arranged serially, that is, one after the other, in parallel kinematic machines, all the motion axes have a direct effect on the tool-carrier unit that is being moved. In conventional machines, the motion axes display the same arrangement as the workpiece coordinate system; that is, they are arranged in an orthogonal relationship. In contrast, parallel kinematic machines are characterized by a **nonorthogonal arrangement of the axes.** The orientation of the motion axes does not correspond to the rectangular coordinate system. As a result, a positional deviation along one motion axis affects the accuracy that can be achieved in all directions. However, most of the machines that have been implemented so far in practice involve hybrid solutions. For this reason, only a few of the feed axes are parallel axes—the remaining axes have a serial structure. *Figure 1.36* provides an overview of the various classifications of parallel kinematic machines and potential combinations of assignments to hybrid motion axes.

Tripod designs are shown in *Figures 1.37 and 1.38*. Three coupling arms guide the kinematic struts of the cardan-jointed spindle platform. The CNC control system uses the three motor-driven coupling arms to move the milling spindle and the tool as necessary both longitudinally in the *Z* axis and in a swivel motion about the *A* and/or *B* axis. Combined with the *X* and *Y* axes of the machining center, this produces five-axis simultaneous machining with maximum dynamics. This type of tool head has been implemented in a number of applications for the production of large aviation components in integral construction. The feedstock materials include aluminum plates with thicknesses between 60 and 300 mm and lengths of up to 30 m. Removal rates of up to 98 percent can be achieved.

Parallel			Hybrid (serial – parallel)		
Machine coordinate system (degrees of freedom)					
3	5	6	3	5	
Workpiece coordinate system (axes)					
X/Y/Z	X/Y/Z A/B	X/Y/Z A/B/C	X/Y/Z	X/Y/Z C/A	X/Y/Z A/B
↓	↓	↓	↓	↓	↓
TRIPODS	**PENTAPODS**	**HEXAPODS**	**Types**		
			I	II	III

Figure 1.36: Parallel kinematic machines: definition and categorization.

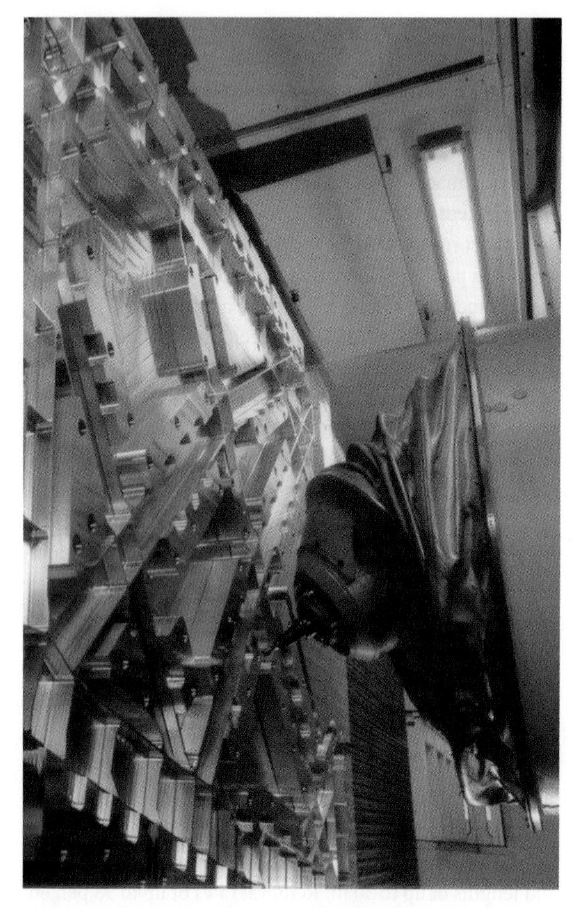

Figure 1.37: Tripod in use in high-speed milling on an aluminum-alloy aircraft component. (Image courtesy of DST.)

Figure 1.38: Tripod: graphical depiction. (Image courtesy of DST.)

If one considers only the tool head, these machines do not exhibit any conventional, perpendicular *X/Y/Z* axes or axes of rotation. Therefore, the CNC system has to take into account the kinematics of the given machine and perform interpolation along all the CNC axes, even for manual or programmed operation along linear paths. This requires a very high processing speed and also makes it difficult to automatically incorporate compensation values for systematic mechanical errors. Therefore, the CNC system has to be capable of controlling all the axes simultaneously to produce a relative motion between the moving spindle and the fixed workpiece. This is accomplished by lengthening and shortening the struts.

The following advantages are noteworthy:

- High stiffness
- High speed
- Low mass that has to be moved

These have to be balanced against the following disadvantages:

- Only a small part of the machine work area can be used.
- The spindle can be inclined only within a narrow range of angles.
- Five-sided machining is problematic.

1.7 Sawing Machines

(Dipl.-Ing. Armin Stolzer)

Machine tools with sawing functions are used to cut materials such as bar steel, plates, sheets, and the like. As a process with geometrically determined cutting edges with multiple-edge tools (DIN 8580/8589), sawing also can be considered as a kind of prefabrication in the context of manufacturing organization.

In the view of users, for a long time now cutting generally has played a lesser role in comparison with the other production processes because it was mostly not considered to be a part of production but rather of warehousing. This view has undergone a fundamental transformation in view of the comprehensive rationalization of production processes and development of tools and machines in the field of circular saws and band saws.

Circular Saws

Circular saw machines are offered for sale in various kinematic designs for the use of HSS and carbide-coated saw blades. What is more, some circular-saw models can perform miter cuts. Because, in general, only a third of the diameter of the saw blade can be used effectively, large cross sections require proportionally large tools. These days, circular-saw machines generally are used for material diameters that are less than 150 mm. They are used as universal saws with the ability to perform miter cuts, as a fast saw for repetitive cuts, or as a production-to-order saw with a high degree of automation.

Band Saws

In heavy-duty production applications, a parallel arrangement of sawing units and broad saw blades are preferred. Horizontal band-saw machines with a stiff two-column design and modern linear guide elements are used most often at present. The use of cast-mineral composites in saw slides also provides a significant performance boost for band-saw machines.

For production band-saw machines with high stiffness, high-performance band-saw tools made of bimetal or with a carbide coating are now available. If a material is suitable for a carbide band-saw tool, then the sawing performance can be increased by a factor of two or three through the use of carbide bands, as long as the machines are configured for their use.

Designs and Variants

The core of every sawing machine is the **sawing unit** that is used to control and drive the sawing tool. The important point here is a design with maximal stiffness. To achieve this, in **circular-sawing machines,** a compact gear unit with hardened and ground gears is used. The sawing unit has to be guided precisely and with low vibration either by a swivel bearing or by a linear guide. Depending on the specific operation, the cut can be performed from below, from the side, diagonally from above, or vertically from above.

In the case of **high-precision production belt-sawing machines,** the usual practice involves a linear feed motion. When it comes to smaller work areas and longitudinal cutting machines, vertically operating saw blades are used. In other areas, horizontal saw blades are used more commonly.

Sawing machines can be subdivided into three groups based on their **degree of automation**:

- *Manual sawing.* The operator maintains control of the cut.
- *Semiautomatic sawing.* A single cut is performed automatically; the machine switches off automatically at the end of the cut.
- *Automatic sawing.* A specified number of cuts is performed with no operator input.

Depending on the application, the degree of automation can be increased further by means of a suitable in-feed and removal periphery. The machine operating times in unmanned operation can be extended by feeding materials via a universal magazine, fully automatic loading from the sawing center, or else a manipulator drawing from a long goods-storage facility, thus achieving greater cost-effectiveness.

Control and Technology Settings

Significant progress has been made not only with regard to mechanical systems but also when it comes to controllers, drives, and technology in general. Classical contactor controls now have been replaced with modern, high-performance CNC control systems. For simple automated machines, series-produced controllers are available with a liquid-crystal display (LCD), function keys, manual control functions, and plaintext diagnostic functions. These controllers moreover can be programmed for compatibility with Siemens programmable logic controllers (PLCs). These controllers can be used to define and call up a large number of order data sets (length/quantity combinations) via a keyboard. At the same time, the saw-feed technology can be set directly via the central operator display (*Figure 1.39*).

When feed magazines and sorting equipment for cut-off sections are used, they are combined with high-quality display screens with touch-screen control and a wide range of functions. These are usually PC based. Sawing-machine controllers for highly automated applications are also characterized by Windows operating systems, an interface to higher-level computer systems, and extensive options for visualization and diagnostics. An important subset of this is the technology control, which makes it easy for the operator to set the desired cutting and feed parameters.

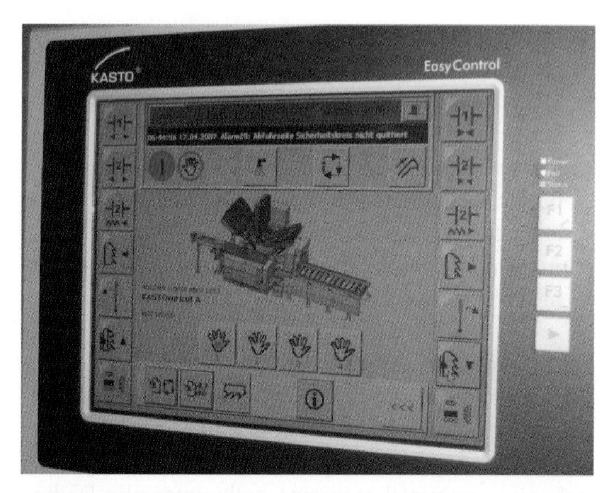

Figure 1.39: Modern sawing-machine controller. (Image courtesy of Kasto.)

Application-Oriented Configuration of Sawing Machines

Production-to-Order Sawing in Industry and at the Point of Sale

For high-performance saws in industrial applications, production-to-order cuts in small to medium-sized quantities require quick handling, good accessibility, simple operation, and high cutting performance. For longer production runs in unmanned mode, double-roller tables and magazines are available onto which materials with various cross sections can be placed and processed (*Figure 1.40*).

Fully automatic, unmanned sawing of warehouse materials is possible by means of sawing centers. These involve combining shelf storage and retrieval machine to handle the individual bars, a CNC sawing machine, and a quick-changing station to couple the two machines.

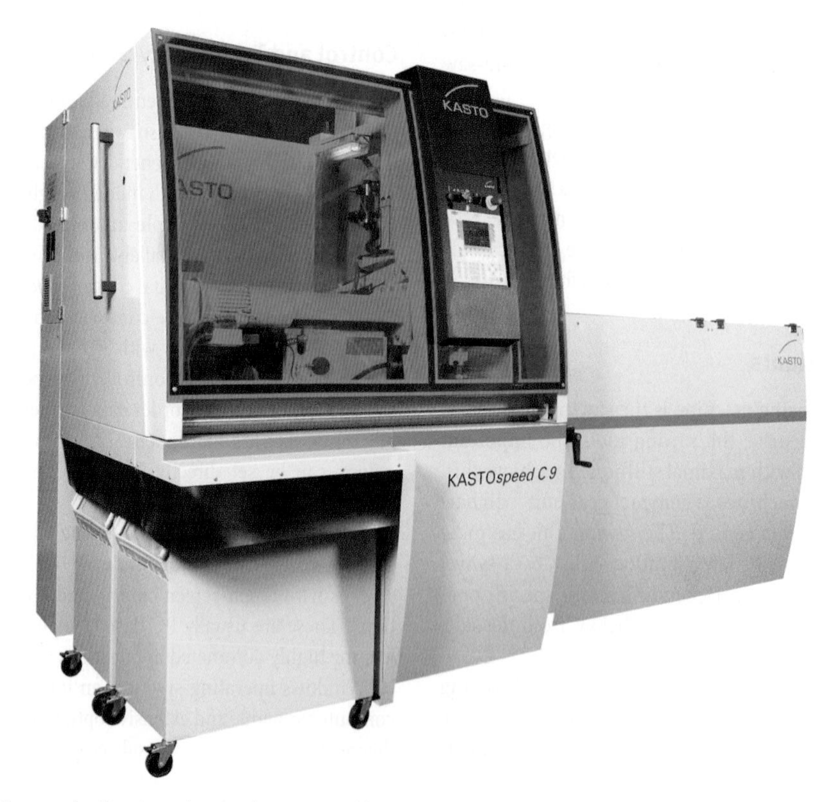

Figure 1.40: Modern production-to-order circular-saw machine.

Automatic warehousing systems and high-performance sawing machines are linked via fully automatic computer-controlled bar-stock manipulators. This means that the saws are provided with a continuous, flexible supply from the bar-stock warehouse. No matter what or how much has to be sawed for a given order, the manipulator brings the right material in the right quantity to the roller table of the saw in question. Whether from a cartridge or a cantilever arm compartment, all types and shapes are brought to the saw reliably, and residual materials are removed again.

Today's state-of-the-art systems can mark the cutoff sections, sort them into containers, turn slices around and stack them, and set down shafts in a space-saving manner. Sorting with robots provides a cost-effective way to stack cutoff sections in transport containers very compactly without increasing the cycle time of the sawing machine. Modern software concepts make it possible to palletize an entire range of goods without additional work by the operator. The system responds intelligently to every new change in dimensions. If a gripper change is necessary, the system detects this automatically and carries it out. Even the stacking pattern of the cutoff sections in the containers is generated automatically for each set of dimensions in order to optimize it in relation to the fill level in the containers. Robots also can be used to implement additional tasks such as container management or marking or deburring of the cutoff sections (*Figure 1.41*).

1.8 Laser Machining Systems

Definition and Physical Principles

The word *laser* is an acronym for "light amplification by stimulated emission of radiation." The function of the laser is based on the property of electrons of a laser-active medium to emit a photon when they go from a higher energy level to a lower one. The practical implementation of this is to place a laser-active material in a resonator between two mirrors, and then excite it using an energy source (*Figure 1.42*).

The laser beam is a contact-based tool that can be used to machine almost any material. The major functional components of a laser machining center are as follows:

- Laser beam source
- Beam guidance system, including the machining head
- Motion axes for relative motion of the laser beam and workpiece
- Workpiece support surface

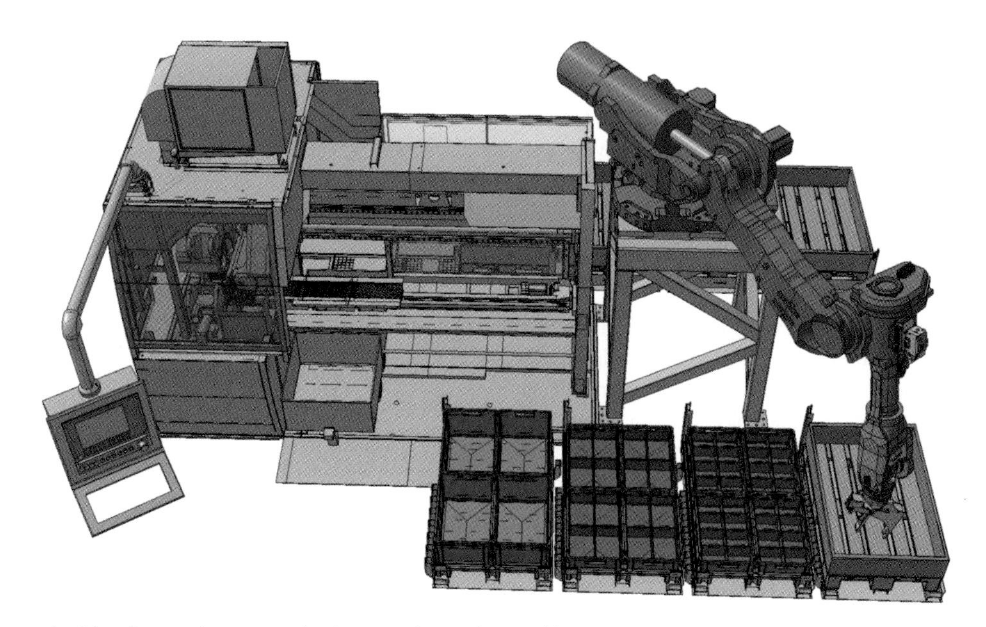

Figure 1.41: Flexible robot sorting at a production-to-order sawing machine.

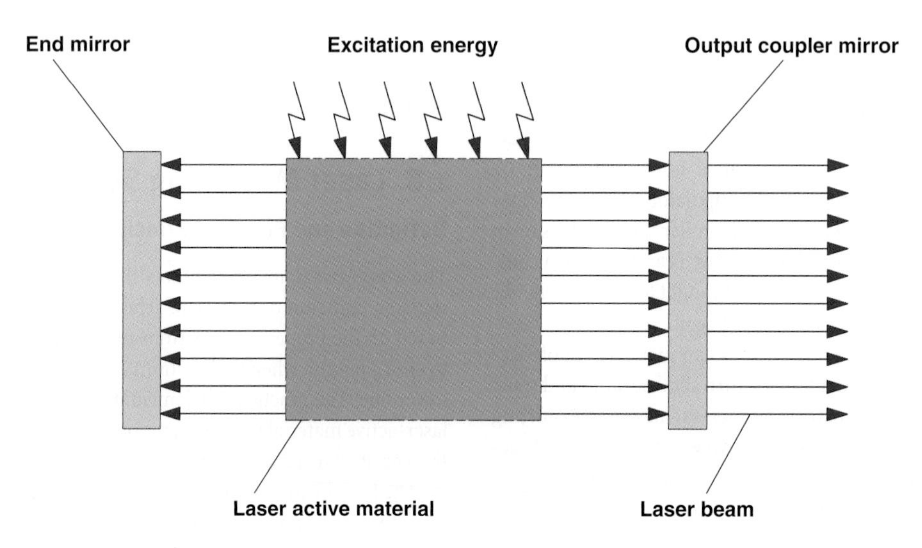

Figure 1.42: Generation of laser beams.

- Extraction and filtering system
- Protective booth

Beam Sources

Given their high laser output power, CO_2 and Nd:YAG lasers are used primarily for the machining of materials. For surface finishing, for example, hardening, coating, alloying, welding, and brazing applications, diode lasers have been seeing increasing use.

CO_2 Lasers

Carbon dioxide (CO_2) lasers are gas lasers. The gas for this laser consists of carbon dioxide, nitrogen, and helium. The nitrogen molecules are excited by electrodes and pass their energy on to the carbon dioxide molecules. The carbon dioxide is the actual laser-active material that emits laser light when excited. The residual energy of the carbon dioxide is released in the form of heat. Therefore, the gas has to be cooled continuously during operation. The power range can be anywhere from a few hundred watts to 20 kW of laser power. CO_2 lasers generally are used in the welding and cutting of metals (*Figure 1.43*).

Nd:YAG Lasers

Neodymium–yttrium aluminum garner (Nd:YAG) lasers are solid-state lasers. The laser-active body is an artificially created monocrystal made of yttrium aluminum garnet (YAG) in which a portion of the yttrium atoms is replaced with atoms of neodymium (Nd). The crystal rod is excited via flashtubes or diodes (*Figure 1.44*).

Disk Lasers

Disk lasers are a type of solid-state laser that is playing an increasingly important role in the metal-working industry. Their advantages include high efficiency and good beam quality. The laser-active material (Nd:YAG) has the form of a thin disk. The advantage of this disk geometry is that efficient cooling is possible (*Figure 1.45*). Disk lasers and fiber lasers are also Nd:YAG lasers. In disk lasers the crystal rod takes the form of a thin disk, whereas in fiber lasers the crystal rod is elongated.

Fiber Lasers

The principle of fiber lasers is as follows: Instead of a rod, thin fibers are used. The advantage here is that fibers, unlike

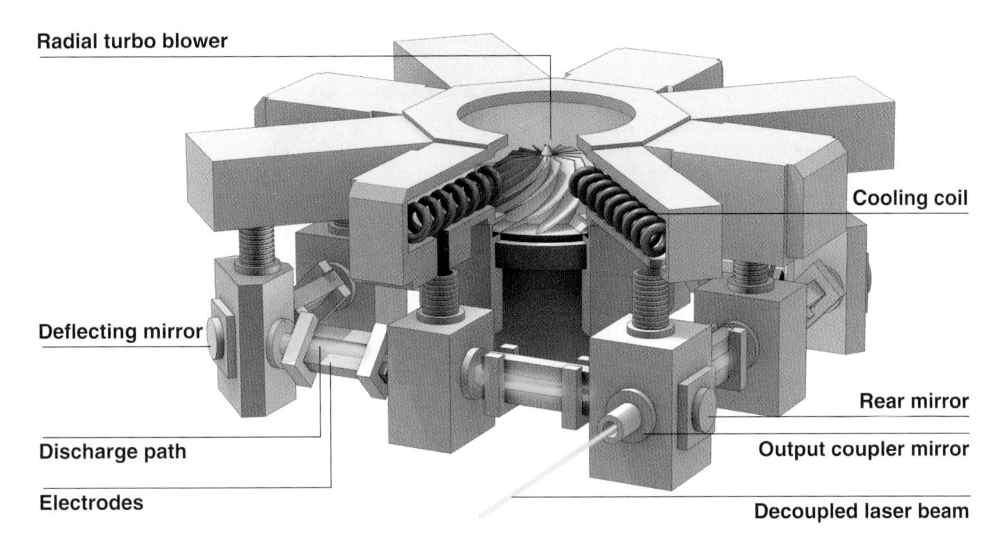

Figure 1.43: View of the interior of a diffusion cooled CO_2 laser. (Image courtesy of Trumpf.)

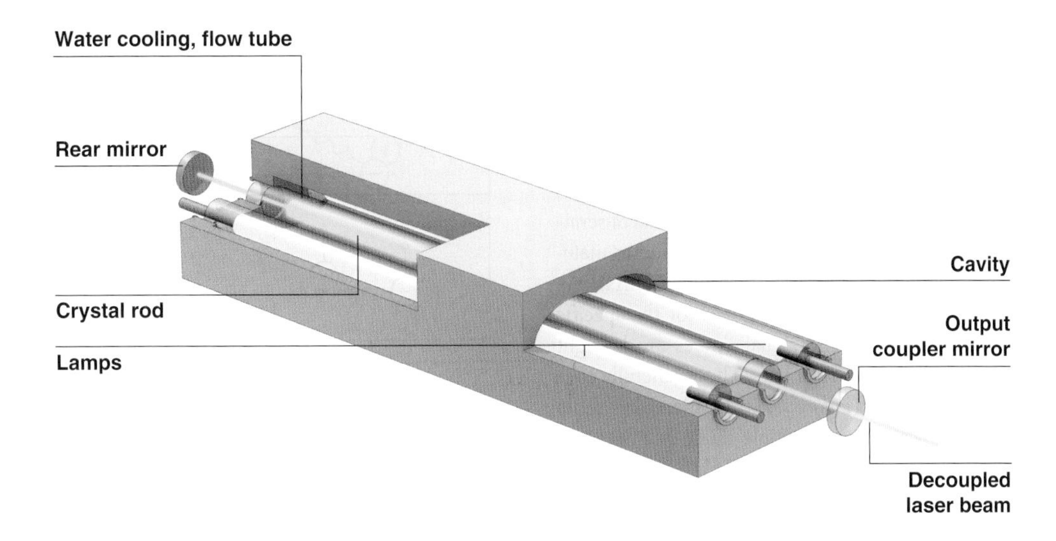

Figure 1.44: Nd: YAG laser as a rod laser. (Image courtesy of Trumpf.)

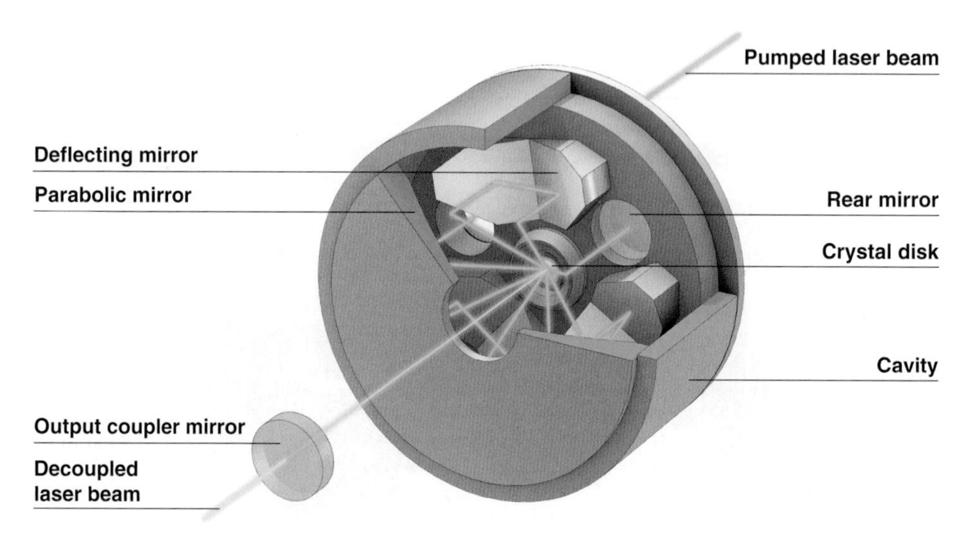

Figure 1.45: View of the interior of a disk laser. (Image courtesy of Trumpf.)

rods, do not require elaborate cooling systems. Because their surface is very large compared with their volume, heat emission to the surrounding air provides sufficient cooling. Ideally, the resonator of a fiber laser consists of only a single long, thin quartz glass fiber. The beam source can be attached directly to a transport fiber, for example, to the glass fiber of a laser-light cable. Beam sources for metal machining can achieve powers of up to several kilowatts by coupling the power of many individual fibers in parallel.

Diode Lasers

Here the laser beam is generated by laser bars made of semiconductor material (*Figure 1.46*). The semiconductor material consists of a gallium-aluminum-arsenide (GaAlAs) crystal. In order to obtain the power required for the specific production technology, the individual diodes are brought together into so-called diode laser bars. These generally consist of 10 to 20 diode lasers (*Figure 1.47*). In order to boost the power into the kilowatt range, multiple-diode laser bars are connected in parallel, thus forming diode laser stacks. The laser beam is transmitted via an optical fiber. The laser beam is available after it passes through the output coupling optics.

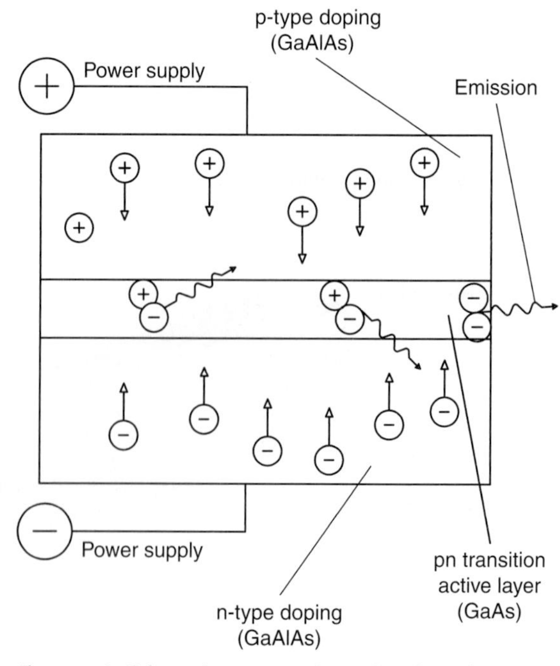

Figure 1.46: Schematic structure of a semiconductor laser.

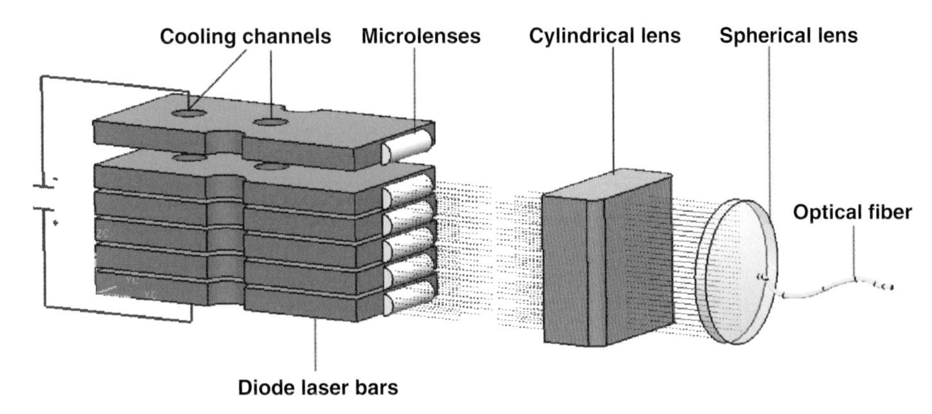

Figure 1.47: Structure of a diode laser.

The advantages of diode lasers are compact dimensions, low weight, and the fact that they are maintenance-free.

Beam Guidance Systems

Beam Guidance Using Deflecting Mirrors

In CO_2 lasers, the beam guidance system consists of a number of deflecting mirrors. The beam path is enclosed by a gas-filled system of pathways made up of tubes and bellows.

Beam Guidance Using Optical Fibers

For Nd:YAG lasers and diode lasers, beam guidance can be implemented by means of an optical fiber. Input coupler optics ensure that the laser light is fed into a very thin fiber (e.g., 100 μm). The output coupling optics on the end of the optical fiber form the beam as appropriate for the machining process.

Laser Machining Heads

Laser machining heads are used at the end of the beam guidance system to focus the laser beam on the workpiece. This can involve spherical or parabolic mirrors or lenses. In addition, process gas and, if necessary, shielding gas, are fed to the machining head. Sophisticated sensor systems (e.g., for distance control) also make the machining head an extremely important component when it comes to process control.

Motion Axes

The following kinematic arrangements are possible for positioning the machining head relative to the workpiece (*Figure 1.48*):

- Moving optics with a stationary workpiece
- Stationary optics with a moving workpiece
- Moving optics with a moving workpiece

Laser Cutting Systems

Thermal cutting with laser beams can be subdivided into the following three processes:

- In **sublimation cutting,** the material in the area of the kerf is vaporized. Since no appreciable melt is created, the result is smoothly cut edges, and the workpiece can be processed further with practically no posttreatment.
- In **fusion cutting,** the material in the area of the kerf is transformed into a molten state and ejected by a jet of gas (e.g., nitrogen) under high pressure (up to 30 bar). Much greater cutting speeds can be achieved than with sublimation cutting.
- In **flame cutting,** oxygen is used as a process gas. The oxygen assists in the cutting process, thus enabling high cutting speeds. The cut edges become oxidized, however. This may mean that postprocessing is necessary, or downstream processes such as painting may become more difficult.

	2D Laser machines			3D Laser machines			
Type	Flying optics: Moving cross member	Fixed cutting head in C frame	Cantilever machine with rotary axis	Cantilever machine	Cantilever machine with rotary axis	Gantry-type machine	Jointed-arm robot, can move freely in space
Application	Typical flat-bed machine for the machining of metal plates. Also suitable for very heavy workpieces	Laser machines or punching/laser combination machines. Workpiece is moved; weight and sheet thickness are limited as a result	Laser machine for 2D machining of tubes. The laser beam cuts only perpendicularly to the workpiece	Laser machine for the machining of three-dimensional workpieces, for example deep-drawn components	Laser machine for 3D machining of tubes. The laser beam can also cut diagonally to the workpiece	Laser machine for the machining of very large 3D workpieces	Cutting of 3D contours in automated production lines
Motion	Optics: 3 axis	Workpiece: 2 axis Optics: 1 axis	Workpiece: 2 axis Optics: 2 axis	Optics: 5 axis	Workpiece: 1 axis Optics: 5 axis	Workpiece: 1 axis Optics: 4 axis	Robot arm: 6 axis Optics: 1 autonomous axis
Graphics	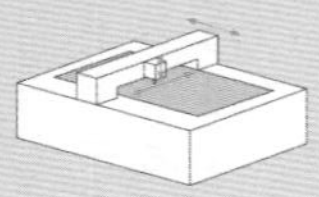	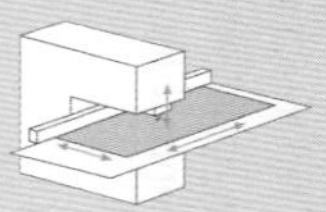	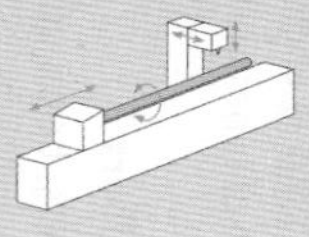	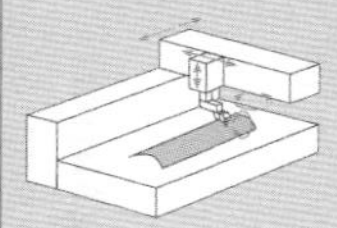	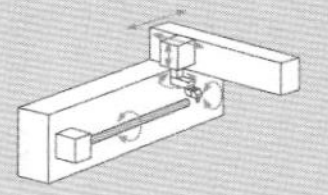	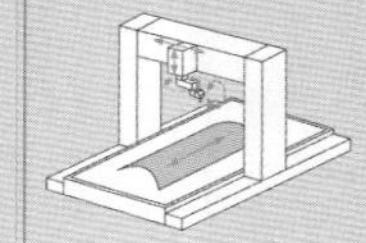	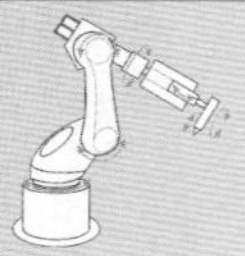

Figure 1.48: Motion axis and machine concepts for 2D and 3D machining. (Image courtesy of Trumpf.)

An Overview of Laser Cutting Systems

Given the various types of machines and laser units, laser cutting systems come in an extremely wide variety of forms. They do, however, all have the same basic components.

In sheet-metal manufacturing, the machine concept most often encountered is a flat-bed laser cutting system operating with a CO_2 laser (*Figure 1.49*).

- Basic machine with drives that carry and move all the components and the workpiece
- Laser unit that supplies a laser beam with sufficient power and in the correct wavelength
- Beam guidance system that directs and shields the laser beam
- Cutting head that focuses the laser beam (Cutting gas is supplied to the cutting head.)
- Workpiece support or holder that carries the workpiece
- Extraction and filtering system that removes cutting fumes and particles of slag
- Protective booth that protects the operator against reflected radiation and metal spatter

Application Examples

Flat-Bed Laser Cutting Systems

Among laser cutting systems, the most frequently used system concept is the flat-bed laser cutting system. These are used to machine flat sheets of metal. With these two-dimensional (2D) parts it is sufficient to have motions in a plane and vertical motions in order to reach all the required points. Furthermore, it is possible to move the workpiece while the cutting head remains mounted in a fixed position. Systems with flying optics are still more common, however. Here, the cutting head is moved over the workpiece.

3D Systems with Laser Beams

If the laser beam is being used to cut contours in 3D workpieces, the optics have to be very flexible. 3D laser systems have

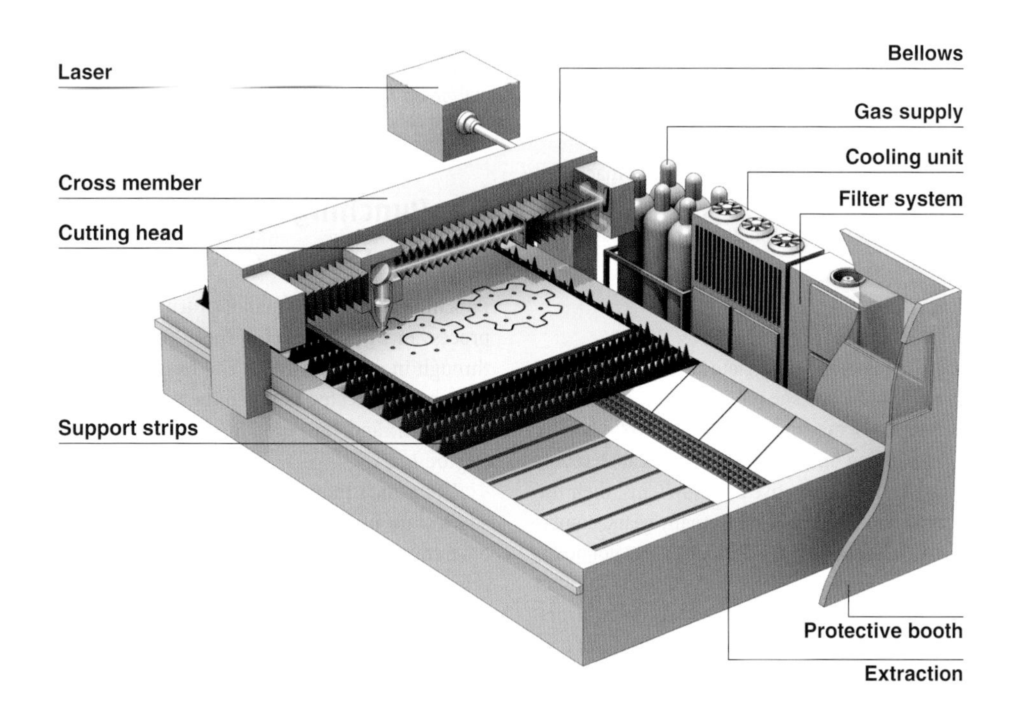

Figure 1.49: Overview of the main components of a flat-bed laser cutting system. (Image courtesy of Trumpf.)

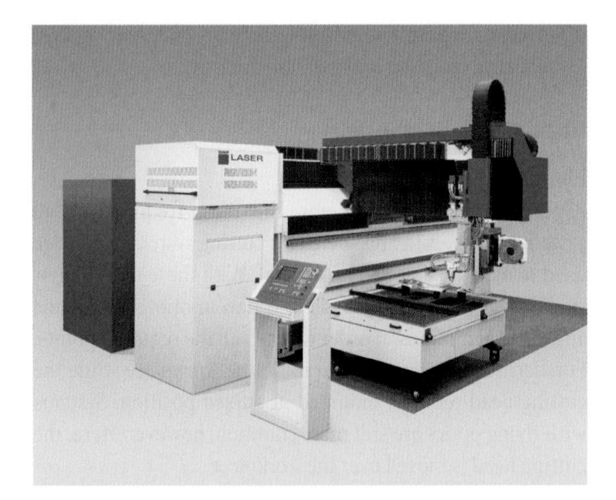

Figure 1.50: 3D Laser cutting system with 5-axis machining optics. (Image courtesy of Trumpf.)

optics with at least five motion axes. In addition to the three spatial axes, there is a rotary axis and a swivel axis. In special cases, it is possible to move not only the optics but also the workpiece (*Figure 1.50*).

Tubes and Profiles

Tubes and profiles up to 6 m (9 m) long can be machined using special tube-cutting systems or 3D systems. The machining optics have two to five motion axes. In these systems, the workpiece is always also moved.

Robots

The combination of robots and solid-state lasers is best known from the automotive industry. The robots operate in automatic mode in transfer lines, machining 3D body parts. They are suitable both for welding and cutting tasks. Because robots constitute a cost-effective alternative to 3D systems, their use is becoming more and more popular. For many years, combining robots with CO_2 lasers was very difficult. It was not possible to guide the laser beam via an optical fiber, as a result of which it was carried to the cutting head via tubes and mirrors. Compact, diffusion-cooled beam sources have solved this problem. These are so compact and lightweight that they can be mounted directly on the robot arm. (*Figure 1.51* shows a robot that operates with a solid-state laser.)

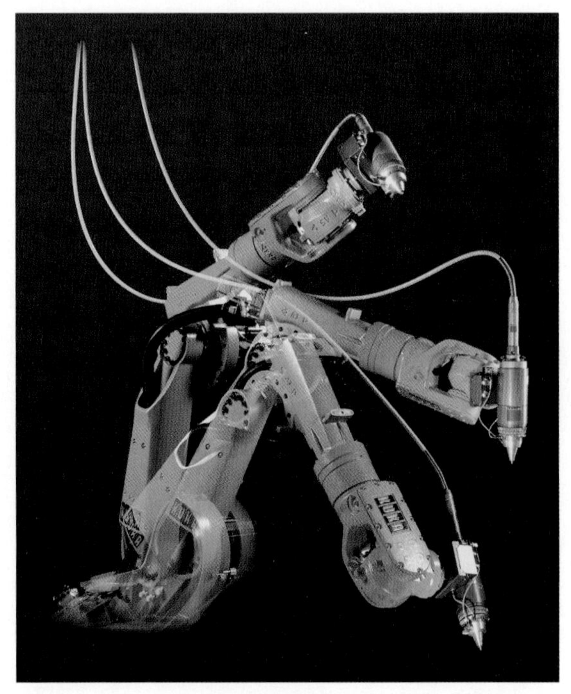

Figure 1.51: Alternative for 3D machining: this robot carries cutting optics with an integrated distance control system. (Image courtesy of Trumpf/Kuka.)

1.9 Punching and Nibbling Machines

Principle of Punching

Punching is a chipless manufacturing process. It is a cutting process in which a sheet of metal (or other material) is cut through in a single stroke. The forming tool has two parts. The sheet is located between the upper tool, also known as the *punch*, and the lower tool, also known as the *matrix* (*Figure 1.52*).

The punch moves downward and into the matrix. The edges of the punch and matrix move parallel to and past each other, thus cutting the sheet. Punching is therefore one of the group of processes called *shearing*.

Principle of Nibbling

In nibbling, punched holes are made next to each other in such a way that they overlap. This makes it possible to create

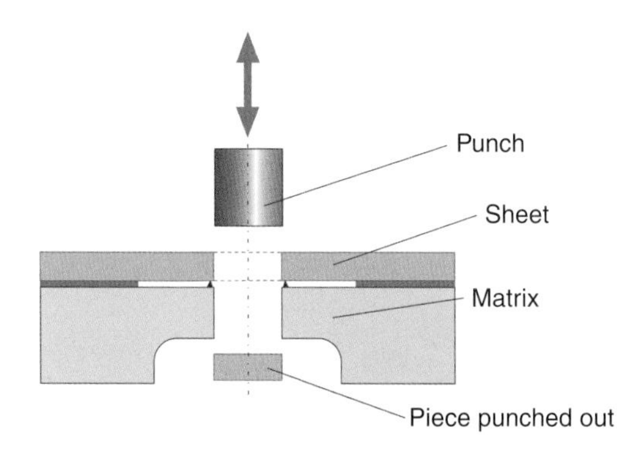

Figure 1.52: Schematic structure of a punching tool.

apertures and contours in any desired shape. The sheet is gradually shifted by a fraction of the size of the punch as soon as the punch rises up out of the sheet. In this manner, a strip is nibbled out of the sheet at a rate of up to 1,200 strokes per minute.

The tool that is used can be a small round punch, for example. Whether a round, slot-shaped, or square tool is used depends on which punch is best for creating a particular geometry.

Punching and Nibbling Machines

These machines can be used to machine plates up to 12-mm thick, with a size of up to 1.5 × 3 m. This is done by fastening a plate in a coordinate guide using a number of clamps and positioning it under the machining head with the tool using two axes.

In punching, the shape of the punch leaves holes with a diameter of up to about 100 mm. Form-punching tools are used when it is necessary to create a large number of small apertures. Such operations are more efficient with punching than with nibbling.

Structure of Punching Machines

If you compare punching machines made by various manufacturers, you will encounter a number of different machine concepts. Each concept can be characterized based on the design of the tool frame, the punching head, and the tool holder. *Figure 1.53* shows the basic structure of a punching machine.

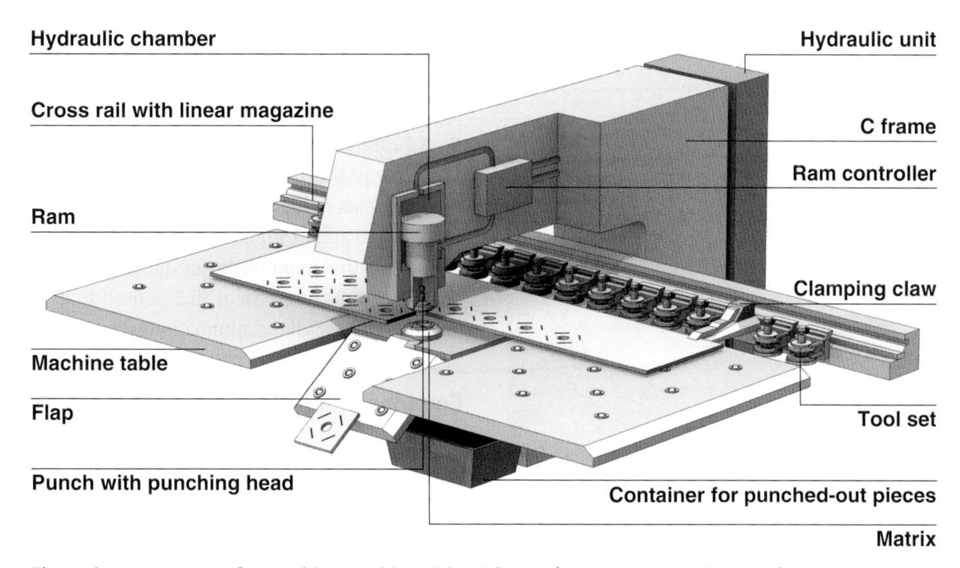

Figure 1.53: The main components of a punching machine with a C frame. (Image courtesy of Trumpf.)

The Machine Frame

The machine frame generally is constructed from centimeter-thick steel plates because it has to be able to transmit dynamic forces (i.e., acceleration forces and vibration) of several hundred kilonewtons. The frame is C- or O-shaped to allow easy access by the operator.

The Punching Head

The punching head is the central element of the machine. The punching head includes the ram and the drive that moves the ram. For high-end machines, a rate of 1,200 strokes per minute is possible today. The ram is driven hydraulically or electromechanically. As with other machine tools, however, there is a trend toward electrical or fully electrical machines.

Tools

A punching tool consists of a punch, a matrix, and a stripper. Sophisticated tools have been developed to allow machining centers to execute complicated machining operations fully automatically and in shorter and shorter times. These tools include

- Rotatable tool holders so that the tools can be moved to any desired angular position at high speed.
- Multitools. Up to 10 punching tools in a single holder, which can be rotated to any desired angular position. This reduces the machining time for parts in which small holes of different sizes have to be punched (*Figure 1.54*).

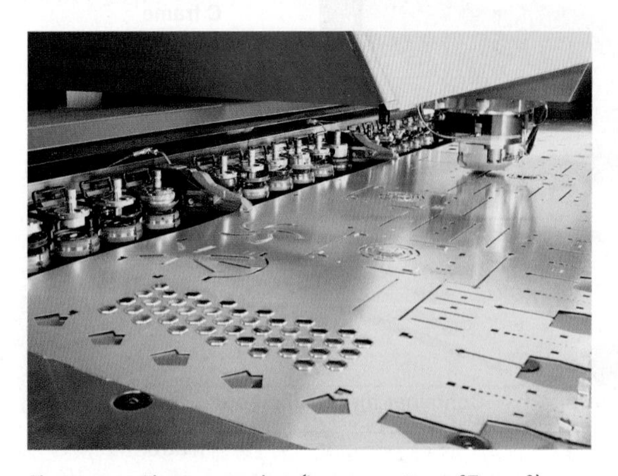

Figure 1.54: Linear magazine. (Image courtesy of Trumpf.)

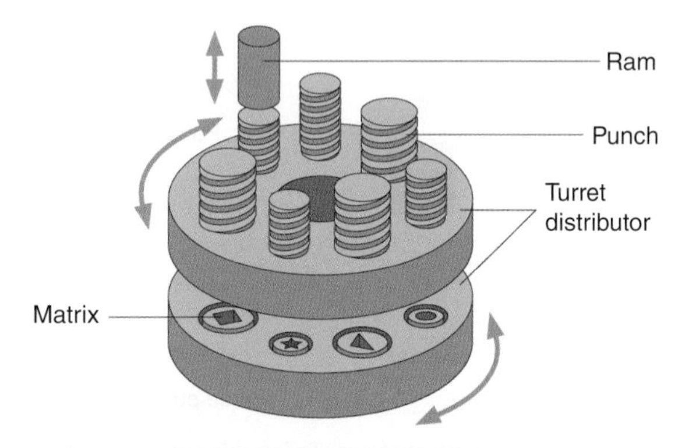

Figure 1.55: Schematic depiction of a multitool system.

- Tool magazines with tools that can be exchanged into the machine with no conversion time (*Figure 1.55*).

Flexible Machining Cells

Automation with individual components is possible. Machine manufacturers offer additional components that can be used to turn series-produced machines into flexible machining cells (*Figure 1.56*):

- Loading and unloading unit with sorting function that inserts the sheets and then removes the parts separately and sets them down sorted into the appropriate places
- A number of parts containers into which small parts or waste can be sorted via a slide
- A gripper that removes the waste lattice
- An external tool magazine with a tool changer that can exchange the tool sets in the linear magazine
- A compact or high-bay warehouse from which material is removed and in which the finished parts are stored
- Sensors that monitor the production process
- A programming system that is used to generate the NC programs for all the automation components

Combined Punching and Laser Machines

Punching/laser machines are constructed according to the same principles as punching machines. However, the C- or O-shaped frame is broadened in such a way that it can hold two machining stations: the punching station and the laser machining station (*Figure 1.57*).

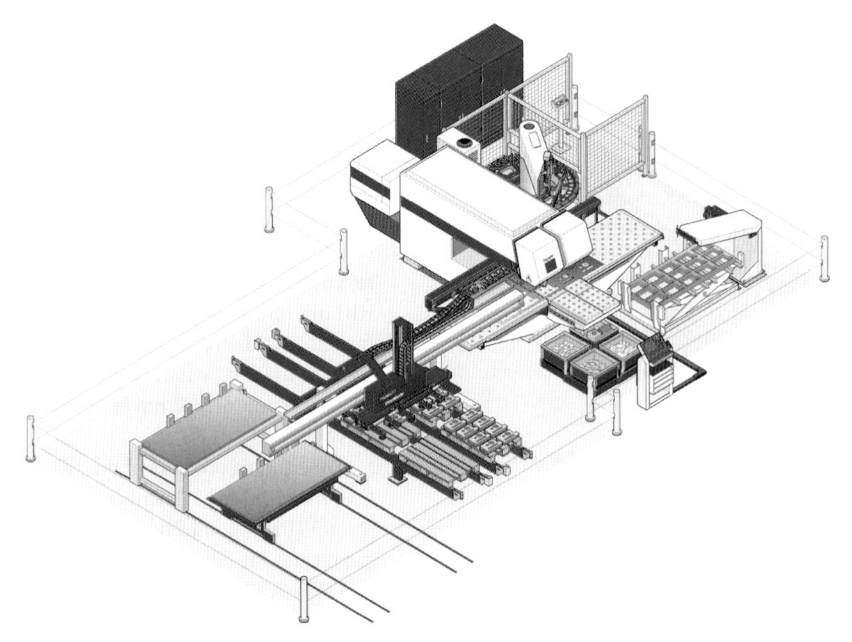

Figure 1.56: Flexible machining cell, consisting of a punching machine with loading and unloading unit, parts container, waste lattice removal device and external tool magazine. (Image courtesy of Trumpf.)

Unlike laser flat-bed machines, though, here it is not the laser beam that is moved but rather the metal sheet. There is an opening directly under the laser station. The extraction unit can use this to vacuum out residual slag and cutting fumes.

The advantages of combination machines become evident in operation: Complex inside and outside contours are cut by the laser. Punching is performed when standard contours have to be machined quickly, for example, punchable round holes.

The following table shows that the laser power has to be adapted depending on the capabilities of the punching processes and the user's requirements.

1.10 Tube-Bending Machines *(Figure 1. 58)*

Bent tubular parts are used as structural elements or for conducting fluids in a wide range of industries, for example:

- In the aircraft industry for airfoils, engines, control surfaces, and brakes
- In the machine-tool industry for hydraulic systems, compressed-air systems, and heat exchangers
- In shipbuilding for conduits of all kinds (freshwater, raw water, seawater, fuel, hydraulic fluids and lubricants, and sprinkler systems)

Material	Punching, nibbling	Laser cutting
Structural steel	Up to about 8 mm	Up to about 30 mm (depending on the laser power)
Stainless steel	Up to about 8 mm	Up to about 25 mm (depending on the laser power)
Aluminium	Up to about 8 mm	Up to about 15 mm (depending on the laser power)
Plastics	Conditionally, if not too brittle or weak	In principle yes, but problematic due to the generation of toxic gases

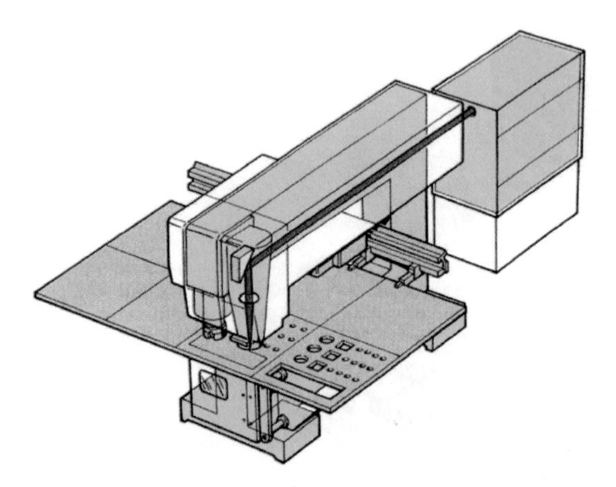

Figure 1.57: Combined punching and laser machine with a C frame. (Image courtesy of Trumpf.)

- In the automotive industry for exhaust systems, fuel-tank fill pipes, seat frames, stabilizers, front-end guards, and handlebars for bicycles and motorcycles
- Cooling systems and heating coils in appliance construction
- Sporting goods, toys, and lawn furniture

These parts require an accuracy of ±0.1° for the bending angle and torsion angle and ±0.1 mm for the distances between the individual bends. In order to satisfy these requirements despite the relatively large tube tolerances and the great differences in spring-back behavior (as measured by the spring constant), additional compensation values have to be entered for the longitudinal position, the bending angle, and the torsional angle at each step of the program. These motions are functionally independent and therefore require only relatively simple NC systems for three to six axes without interpolation. State-of-the-art CNC tube-bending machines with multiple tool levels have 12 to 15 controlled axes. Hydraulic cylinders or hydraulic swivel motors can be used as axis drives, but the preferred solution is electric servo drives.

The bending program of this type can be entered into the CNC memory via keyboard or directly via a direct numerical control (DNC) connection. A bending program of this type can include the following individual steps:

First block. The tube is inserted (longitudinally positioned along the *Y* axis) up to the beginning of the first bend that will be created.

Second block. The tube is bent to the preset or programmed bending angle (in the *C* axis).

Third block. The tube is advanced by the length of the straight piece between the individual bends (*Y* axis).

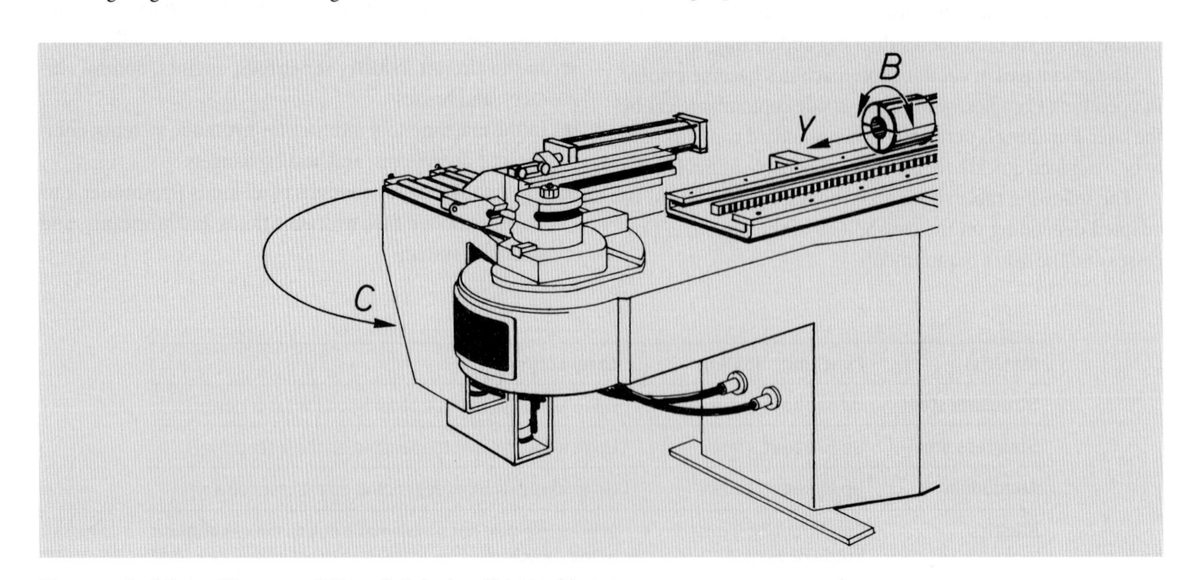

Figure 1.58: Schematic representation of a tube bending machine.

Fourth block. The tube is rotated into a different bending plane (*B* axis). The processing of block 4 could begin simultaneously with the processing of block 3 if permitted by the tube being bent because the bent tube normally has to be removed from the groove of the bending tool before being twisted into another plane.

Fifth block. The second bend is created and so forth to the end of the program.

A bending program can be created using the following methods:

- On the basis of the workpiece drawing, which already contains the bending data
- On the basis of a bent model or a wire model on a special pipe-measuring machine that calculates and outputs the complete pipe-bending program
- By creating a drawing and dimensioning in a tube isometric program, followed by automatic calculation of the bending data
- By measuring a bent model or a wire model on a special pipe-measuring machine that calculates and outputs the complete pipe-bending program

A three-axis pipe-measuring machine mainly consists of

- A computer with monitor and printer
- A measuring probe mounted on a movable arm with various joints or in state-of-the-art designs with automatic laser scanning
- A support table on which the pipe or wire model can be secured (*Figure 1.59*)

To create a program on the basis of a model, the geometry of the model is traced manually using a measuring probe. The computer automatically calculates the required bending data, which then appear in plain text or also can be saved as a file. If there is a direct connection between the bending machine and the pipe-measuring machine, then the resulting program also can be transferred directly into the control system for the bending machine.

After the first pipe has been bent, it is measured at the pipe-measuring machine. The computer compares the actual measured values with the target values from the model, and the resulting deviations can be fed into the NC system of the pipe-bending machine as compensation values.

Numerically controlled pipe-bending machines are capable of meeting demands for accuracy, quick compensation, and quick changeovers. For the purposes of fully automatic manufacturing, they also can be provided with features such as magazine feeding, weld-seam positioning, and ejection devices for the finished pipe, thus allowing the operation of multiple machines in such cases.

The Fraunhofer Institute has, in cooperation with Tracto-Technik GmbH, developed a measuring system for integration into tube-bending machines to determine the bending angle. The bending angle is measured within the bending machine immediately after the bending process. The measurement and evaluation process takes only a few fractions of

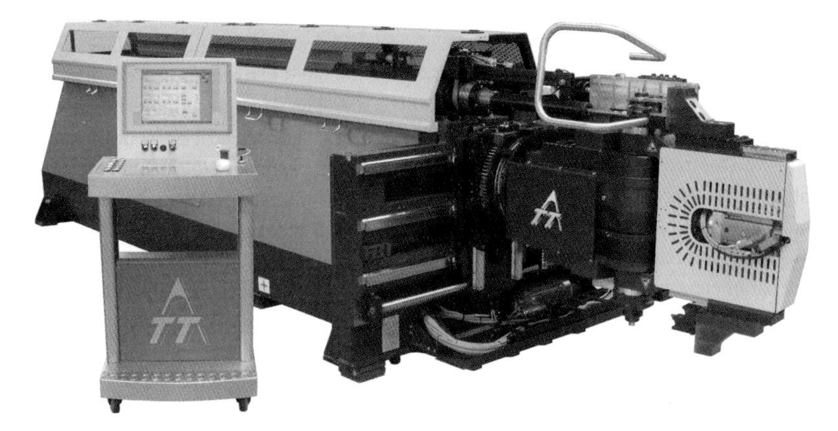

Figure 1.59: 3D Tube bending machine. (Image courtesy of Trakto-Technik GmbH & Co. KG.)

a second. The value of the spring-back at the tube bend depends on a number of parameters including the material, the machine, and the process. This means that it can be corrected only by specifying the corresponding compensation values. At the same time, the user is able to capture and log the bending results during the process and to integrate the bending results into the subsequent bending process.

1.11 Electrical-Discharge Machines

Electrical-discharge machining (EDM) is one of the most important methods used today to manufacture punching tools, injection molds, die-casting tools, compression molds, blow molds, and dies used in extrusion and other processes. EDM methods can be subdivided into **wire and die-sinking EDM.** These methods have especially proven useful in areas where parts exhibit the following characteristics:

- Complex forms.
- High-strength materials.
- Manual machining is problematic.
- No other automatic machining method is possible.
- High accuracy is required.

Electrothermal discharges between an anode and a cathode cause surface particles to evaporate. The EDM process takes advantage of this physical effect. Admittedly, this process takes much more time than machining with a cutting tool. Nevertheless, its cost-effectiveness resides in the fact that machines can run around the clock in a fully automatic, unsupervised mode, even when machining complex parts. It is important to prevent the development of an arc during the process because this could destroy the electrode and the workpiece.

Highly qualified operating personnel are required to set up and operate an EDM. This means that metal workers will need several weeks of specialized training to master the peculiarities of EDM. One reason for this is that the EDM process does not allow the operator to view the area in which the machining actually takes place. In comparison with cutting methods, the operator is required to set more unfamiliar, abstract electrical parameters and has to monitor the process by means of instruments.

Wire EDM

In the wire EDM process, material is removed from the workpiece by a wire electrode with a diameter of 0.1 to 0.3 mm. This removal involves neither contact between the workpiece and electrode nor application of mechanical force. Thousands of discharges take place every second, melting and vaporizing tiny particles of material, which are then condensed in a dielectric fluid and washed away. Deinonized water is used as a **dielectric,** which serves several purposes at the same time. For one thing, it creates the breakdown resistance that is necessary for the discharges to occur. The dielectric also flushes the removed particles from the cutting zone, cools the stressed machine parts, and improves the antifriction properties in the areas of the wire guides and current-supply wires.

The electrical pulses that are required for the discharges are produced by a **generator.** A feed regulator ensures that the required spark gap is maintained between the wire electrode and the workpiece during relative motion. When a short-circuit occurs, that is, when the cutting wire comes into contact with the workpiece, the wire has to be retracted while maintaining the correct path until the short-circuit is eliminated. Only then can the cutting process be resumed. The shape accuracy and surface quality of the cuts depend on the feed rate and feed consistency (*Figure 1.60*).

The **wire electrode** is worn away by the discharges. Therefore, new wire is continuously passed through the cutting zone at a constant rate. A spool of wire weighing 6 kg is sufficient for more than 100 hours of cutting time. The drive and guidance devices for the wire electrode are very important in obtaining precise results.

Numerical control of the *X/Y* motion ensures close adherence to the desired cutting path. When cutting on a slant or in three dimensions at variable angles of inclination, a second motion in the upper *U/V* plane is superimposed on the lower *X/Y* guidance plane. Conical cuts and continuous changes in inclination are also possible.

The **workpiece material** processed in wire EDM can be any type of electrically conductive or semiconductor material. The advantages of this method often consist of the fact that machining is performed **after hardening** and that **extremely high accuracy** and high **surface quality** can be achieved.

The simplified sketch (→ *Figure 1.60*) shows the arrangement of the five axes and the wire feed.

Die-Sinking EDM

In this method, a shaped electrode is moved down toward the workpiece from above, and the electrical discharges produce

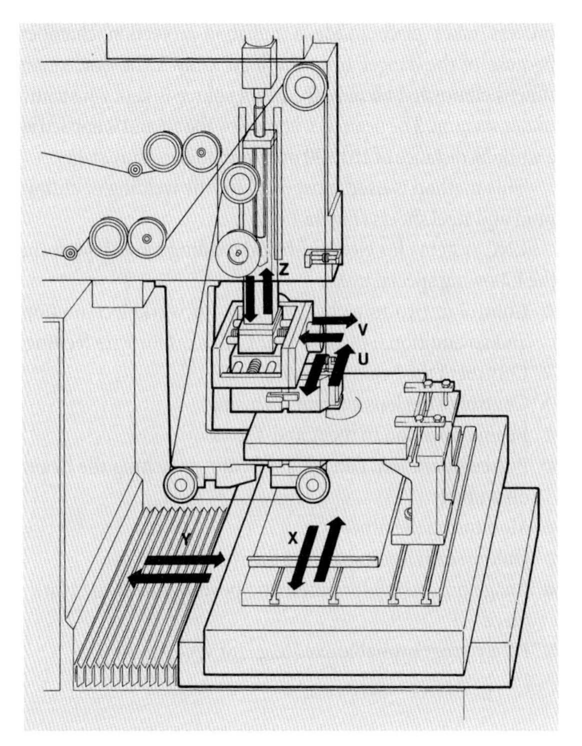

Figure 1.60: Wire electrical-discharge machine (EDM) with five CNC axes.

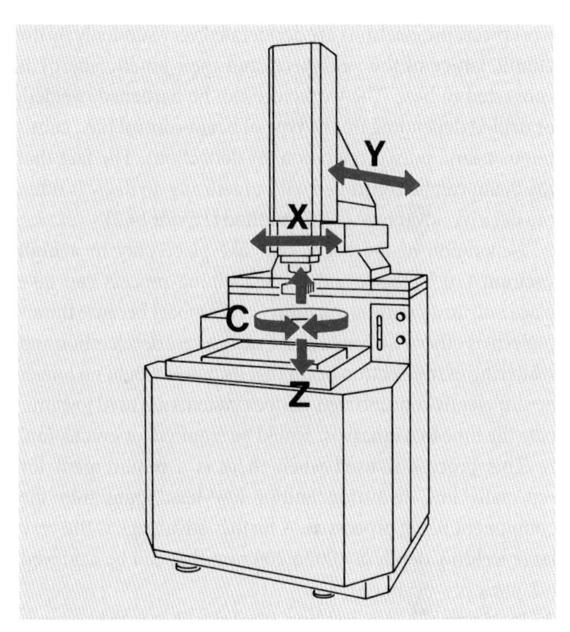

Figure 1.61: Die-sinking EDM machine with four CNC axes.

a "negative" impression on the workpiece. Like cutting operations, EDM also involves the removal of large volumes of material in the roughing operation and smaller volumes in the finishing operation. To produce superfine surface quality, an additional polishing operation can be performed. The polishing step also works on the principle of electrical discharge but uses very low pulse energies. The CNC system guides the electrode in an orbital (circular) motion in the plane while also providing a very slow in-feed in the *Z* direction (*Figure 1.61*).

In contrast to wire EDM, die-sinking electrical-discharge machines are manufactured not only as **knee-and-column machines** but also in larger designs as **gantry machines.** This type of construction is necessary for extremely heavy workpieces and electrodes, such as those found in the construction of automobile bodies and large dies, for example.

To create shapes and contours, the CNC system guides the electrode along straight lines and circular paths (in the *X/Y* plane) and monitors the in-feed in the *Z* direction. When a short-circuit occurs, the CNC system retracts the electrode and then immediately extends it again. This process runs constantly and requires special control-loop signals to be exchanged with the generator. Small machines also can be equipped with automatic tool changers (i.e., electrode changers) and pallet changers for the workpieces. The CNC system is often programmed directly at the machine. In this programming, the operator has access to stored cycles and can check his or her input on graphical displays.

1.12 Electron-Beam Machines

These machines have been used for 50 years now for welding and drilling. In isolated cases, they also are used to harden materials or improve them by surface remelting. The "tool" is a narrow, sharply focused beam of electrons moving at a high velocity and with high energy. In principle, the beam is generated in a manner similar to that which occurs in the neck of a television picture tube, but the beam power is higher by several powers of 10, ranging between approximately 1 and 100 kW. This high power is concentrated as a small focal spot with a diameter of 0.1 to 2 mm, yielding a power density of 10^6 to 10^9 W/cm^2. When this electron beam strikes the surface of a workpiece, the electrons are decelerated very suddenly by the atomic lattice of the workpiece, and their kinetic energy is converted to heat. The workpiece can be hardened, welded, or drilled depending on the type of beam control (i.e., continuous beam, pulsed operation, or deflection). The fact that the beam can be controlled with practically no delay is often the decisive advantage of this method (*Figure 1.62*).

Generation of the beam can take place only in a **high vacuum** (10^{-5} mbar), and the machining process can take place only in a **medium vacuum** ($<10^{-2}$ mbar) because the air molecules otherwise would cause such severe deceleration and scattering of the electron beam that its power would no longer be sufficiently concentrated. Airlock systems are used to eliminate the time that otherwise would be required for evacuation.

This process is used when there is a requirement for especially low distortion and/or low heat input into the component being processed. A further advantage is the very large welding depth of 100 to 200 mm that can be achieved (→ www.pro-beam.de).

In **electron-beam welding in atmosphere,** the vacuum chamber is replaced by a pressure-stage system, and the welding process takes place inside a radiation-protection chamber because of the danger from the hard x-rays. The evacuation time is eliminated because the beam source is kept constantly under vacuum. The beam is scattered by the air and is applied at a working distance of 10 to 20 mm from the pressure stage.

This method is used most of all for the welding of aluminum and steel sheets *(Figure 1.63).*

CNC systems for electron-beam welding have to perform the following special tasks:

- Control of two to eight CNC axes for workpiece motion and an additional three to four CNC axes for wire feeding
- Control of the beam current
- Control of the lens current (focusing)
- Beam deflection in the *X* and *Y* axes
- Pulsed operation, continuous beam, switching the beam on and off
- Monitoring the process
- Online seam finding during the welding process
- Image processing for automatic positioning of the workpiece
- Freely programmable axes (e.g., for vectorization)

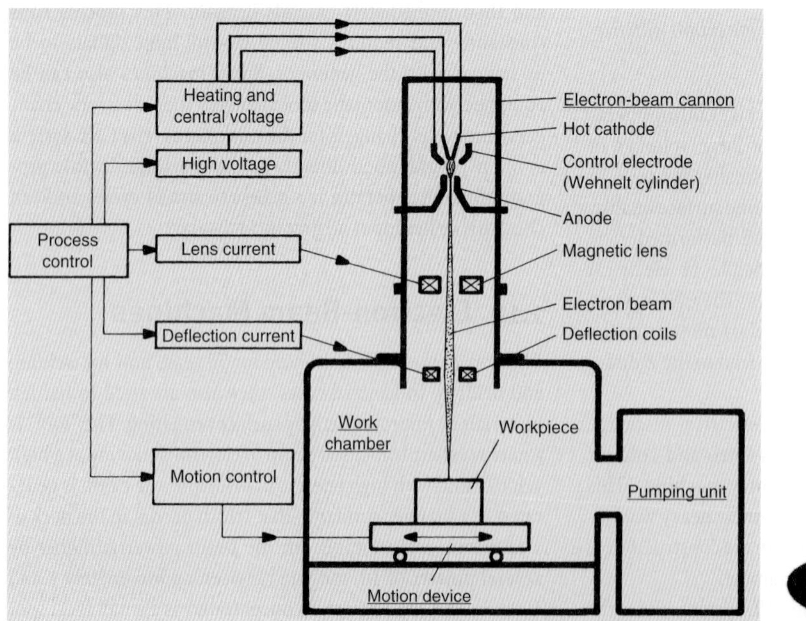

Figure 1.62: Diagram of an electron-beam machine, to the right an illustration of an electron-beam cannon.

Figure 1.63: Electron welding system.

1.13 Water-Jet Cutting Machines

Conventional cutting tools can hardly be used to cut soft or weak materials such as rubber, leather, paper, foamed plastics, or Styrofoam. They are also poorly suited for cutting polyvinyl chloride (PVC) or plastics that are reinforced with carbon or glass fibers. Water-jet cutting represents a suitable modern alternative in these cases. The principle is simple: Water is forced through special nozzles under a pressure of 4,000 to 9,000 bar. The diameter of the nozzles ranges from 0.1 and 0.3 mm. The exit velocity of the water jet is 800 to 900 m/s, which is more than twice the speed of sound. When directed against the workpiece material, the jet functions as a thin, invisible blade. It bores into the material and can cut in all directions with equal effectiveness. The kerf measures only 0.1 to 0.3 mm. The cutting speed can range between 1 and 500 m/min depending on the type and thickness of the material. Water consumption is about 1.5 liter/min. This can be purified using microfilters and then fed back into the cutting process.

If the water jet alone is not sufficient, an extremely fine grade of abrasive powder can be added to the water. This is called **abrasive cutting.** This method can be used to cut steel up to approximately 80 mm, as well as titanium, marble, and glass (*Figures 1.64 and 1.65*).

The advantages of water-jet cutting include

- The ability to machine flat and 3D workpieces
- High-quality cuts (better than flame cutting when used to cut steel)
- Clean-cut edges without burrs
- Low amounts of material loss at the cut edge
- No chips, no accumulation of dust, and no deposition of dust on the workpiece material (although the abrasive does leave behind a fine, dustlike deposit)

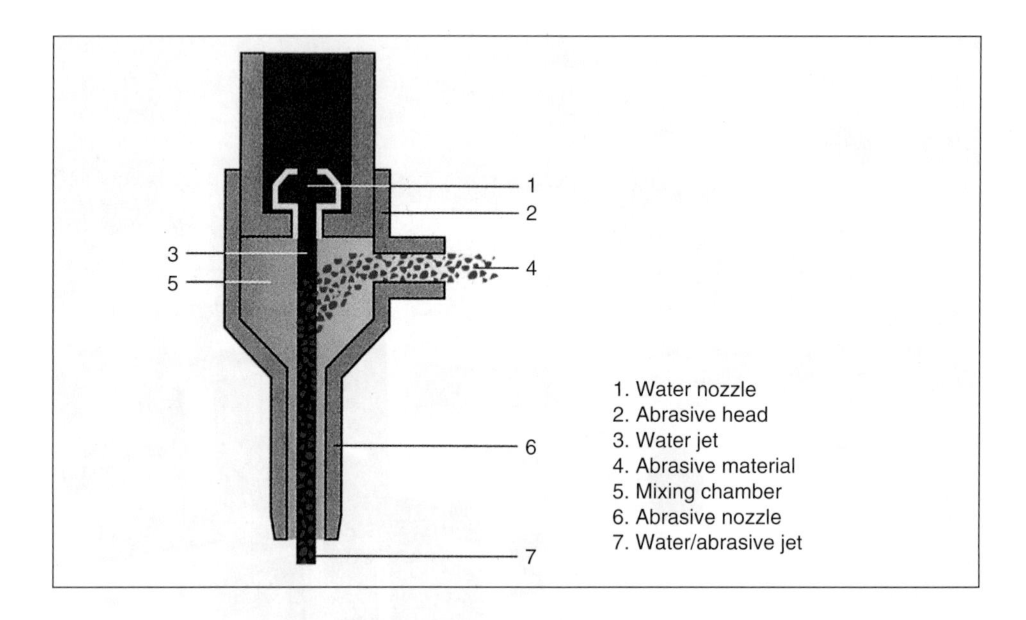

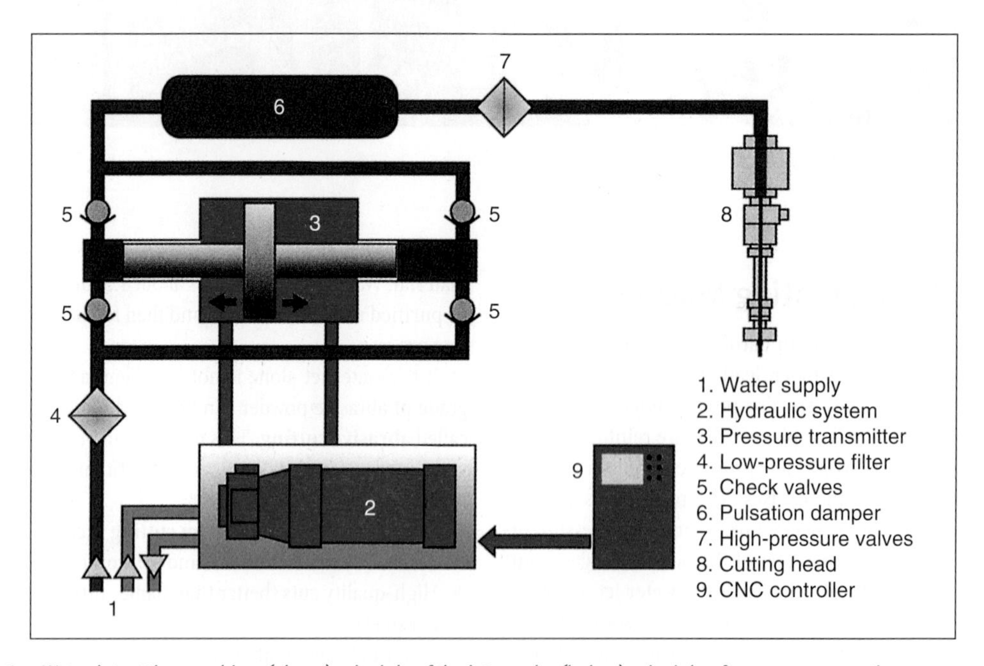

Figure 1.64: Water-jet cutting machine: (above) principle of the jet nozzle; (below) principle of pressure generation.

- High feed rates
- No high cutting temperatures and thus no thermal distortion of the workpiece
- No feed forces or cutting forces and thus no deformation of soft materials during cutting
- No electrical charge imparted to the workpiece [Therefore, even PCBs can be cut using this method without damaging sensitive components already mounted on the board (*Figure 1.65*)].

Most water-jet cutting machines have the basic structure of a three-axis gantry machine with two additional swivel axes for the jet nozzle. However, it is also easy to combine the cutting nozzle with a robot to achieve optimal flexibility.

1.14 Development Trends in Numerically Controlled Machine Tools

Reconfigurable Machine Tools

One of the current research projects toward the development of new machine tools is the development of reconfigurable machine tools and their components. The objective is to make it possible to adapt these machines extremely quickly to changing requirements. This can involve adding or exchanging complete technology modules.

Hybrid Machine Concepts *(Figure 1.66)*

Hybrid machining means the use of different physical effects within a given machine. The laser machining units are kept ready for use in a magazine between the two workspaces of the machine. The laser units can be used either via the main spindle of the machine or by means of an integrated jointed-arm robot. Another robot, which is installed in a fixed position on the machine bed, can tend both the workspaces. The workspaces, in turn, are equipped with rotary swivel tables. Besides machining with lasers, the robot also can be used for workpiece handling. This results in a high degree of utilization for all the components. What is important is that the hybrid machine structure has to be taken into account during the CAD/computer-aided manufacturing (CAM) planning of a component. Within the machine itself, the control systems of the robot and machine are coupled. Via this coupling, the CNC control system ensures that the components work together quickly and reliably (→ Machine tool colloquium in Aachen, Germany, 2008).

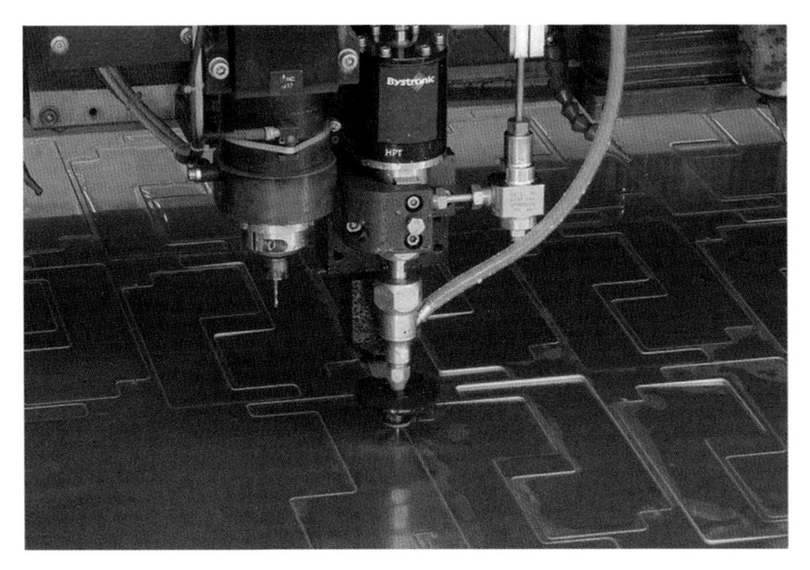

Figure 1.65: Combined, cost-effective production with water-jet abrasive cutting and milling. The parts are premanufactured in a nested manner, and only precise fits are reworked. (Image courtesy of Bystronic.)

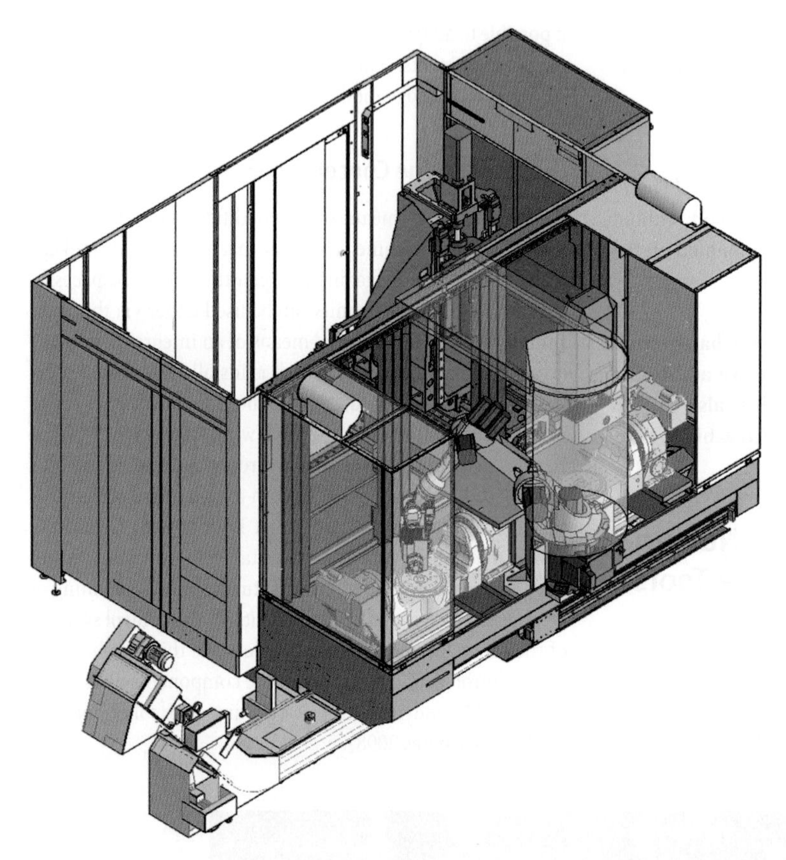

Figure 1.66: Hybrid machining center with laser unit and robot for laser head changing and workpiece changing. (Image courtesy of Chiron.)

Energy Efficiency

Because energy is constantly becoming more expensive, machine builders have to develop machines that are as energy efficient as possible. Unlike in previous years, the requirement is no longer for machine tools to be "as fast as possible" but rather "fast, but energy-efficient and cost-efficient." Machine manufacturers and users can make use of simulations and diagnostic software to help meet these requirements.

Explanation of Figure 1.67

For reasons of clarity, *Figure 1.67* shows only a subsection of the machining process for a standardized part. The machining process involves the milling of a square block of aluminum. Outer and inner circles with various diameters are milled in various positions, as well as rectangles that are slightly rounded. This can be seen in the foreground.

The diagonal lines are the motions to and away from the tool changer in the rear part of the machine. The four-dimensional (4D) visualization shows the program sequence in the form of a 4D graph. Axes *X*, *Y*, and *Z* show the motions of the assigned NC axes. The gross output of the machine is shown in color for every path. Red indicates a very high output demand; from orange to yellow, a low output demand; green a very low output demand; and blue a negative one.

As can be seen based on the graphic, the energy consumption at the top left is very low. This is where the tool changing takes place, that is, transfer of the tool between the spindle and the magazine. During motion toward the tool-changing point, that is, precisely when the spindle is being braked, energy is fed back into the plant-power grid (blue color). Once the tool change has been executed and the tool has been moved to the workpiece, in other words, on acceleration of

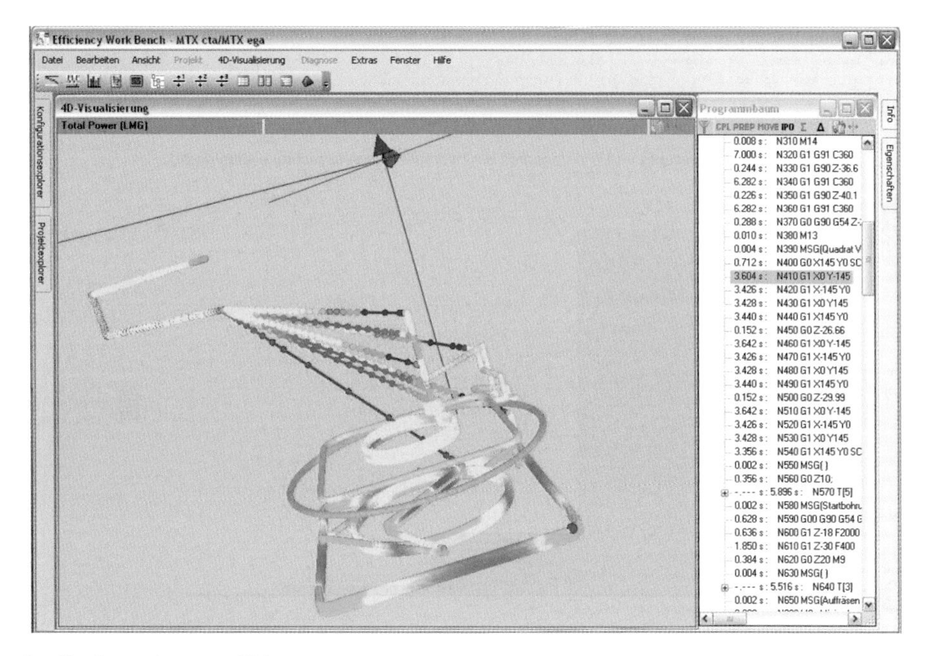

Figure 1.67: Visualization and energy efficiency.
The illustration shows analysis software for reducing the cycle time and energy consumption of the machine. There are two windows open: the 4D visualization of the machine motions to the left and the associated program tree to the right.
(Image courtesy of Bosch Rexroth Electric Drives and Controls, IndraMotion MTX.)

the feed axes and especially the spindle, an increased output demand can be seen (red color).

The program tree shows the associated NC program. On the left is the program tree, which shows the block execution time in seconds for each NC block. On the right is the programmed NC block. Both these displays can be synchronized if desired; that is, clicking on the 4D path calls up a red cursor (see the lower right). At the same time, the associated NC block is highlighted in blue in the NC program (see the program tree). In turn, clicking on the NC block in the program tree will display the associated start point of the block in the 4D graphic.

Explanation of Figure 1.68

For this analysis, the electrical power fed into the hydraulic system (*P_Hyd_zu*) and the volume flow and pressure of the hydraulic system during the production of a typical workpiece have been recorded for this machine. Subsequently, by measuring the volume flow and pressure, the power drawn (*P_Hyd_ab*) and the efficiency are calculated and displayed in a diagram.

These analyses on this turning machine showed that the hydraulic pump had low efficiency, as a result of which it was replaced with a variable-speed pump.

Explanation of the Principle

With appropriately equipped CNC systems, all the data (signals) within the CNC system, the PLC, and the SERCOS drives can be recorded in each interpolation cycle (or a multiple thereof). It is also possible to read in data from external measuring systems, such as output-measuring devices, temperature sensors, and measuring devices for pressure and volume flow. These data can be read in via the drive field bus or the input/output (I/O) field bus. In each measurement, which can, if necessary, go on for several hours, up to 100 signals can be recorded. This also can include the set point and actual positions of the *X*, *Y*, and *Z* drives and the gross output of the machine.

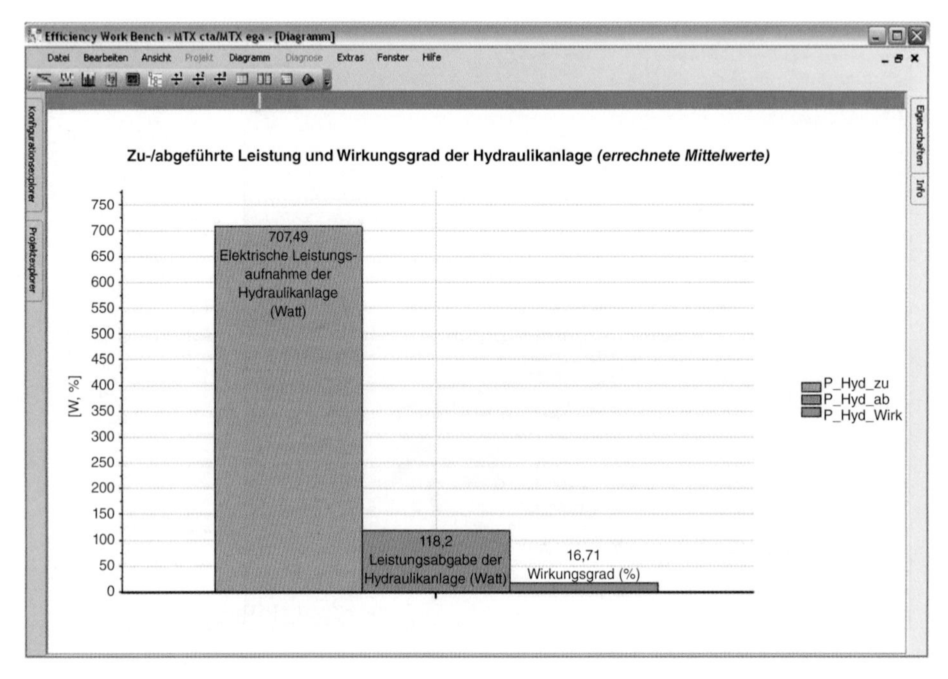

Figure 1.68: Energy consumption and efficiency.
The illustration shows the results of an output measurement on the hydraulic system of a double turning machine.
(Image courtesy of Bosch Rexroth Electric Drives and Controls, IndraMotion MTX.)

The signals to be recorded are selected using a wizard that guides the operator through the menus. During this process, the operator enters, among other things, the desired start and end triggers and uses drag and drop to select the desired signals from a tree representing all the data available on the machine.

1.15 Measuring and Testing

Measuring Machines and Cycles *(Figures 1.69 to 1.71)*

The introduction of numerically controlled machine tools in a plant usually also requires far-reaching changes in the area of quality control. Simple and complex workpieces have to be tested for compliance with stringent quality specifications. The focus of testing constantly alternates among these different parts. Sometimes a measuring machine is implemented for the sole purpose of determining whether a part should be accepted or rejected. In other cases, the machine is intended to provide data that will be used to correct various parameters in the manufacturing process.

The goal in these applications is for the measuring time to equal approximately 30 to 100 percent of the machining time (depending on part geometry), thus allowing any necessary corrections to be carried out quickly and preventing unnecessary stoppages of expensive machines.

CNC measuring machines—equipped with the necessary extra features related to this special measuring task—were the first to fulfill the 11 **most important requirements** of any automatic measurement process:

1. Universal measuring probes must be provided for multiple axis directions.
2. Operation must continue without interruption when measuring programs are being loaded.
3. Measuring must be executed at high speeds. This may require a positioning speed as high as 3 m/min. Fine

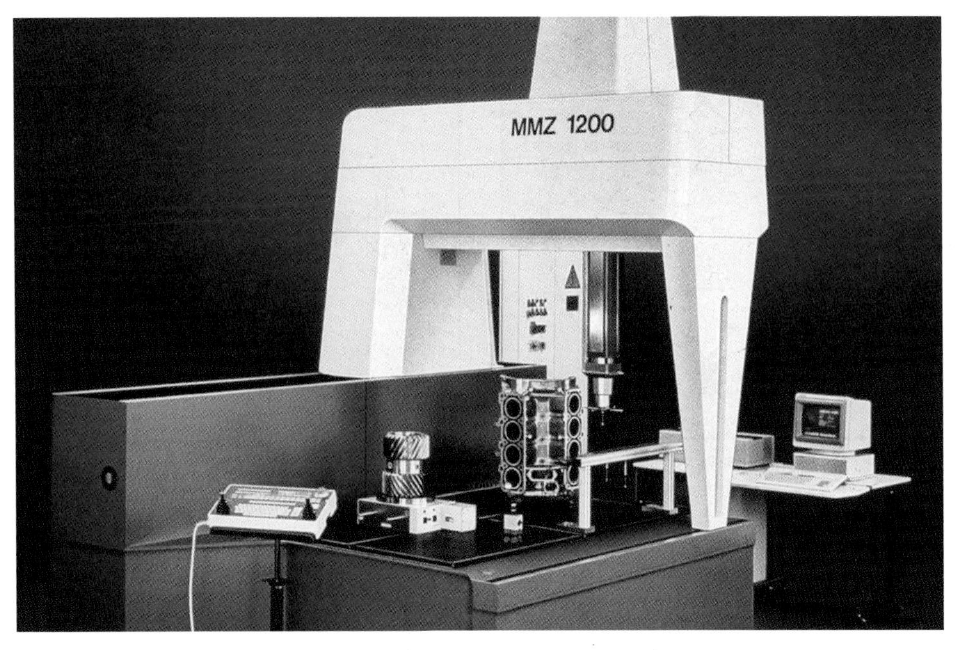

Figure 1.69: CNC-controlled gantry measuring machine. (Image courtesy of Karl Zeiss.)

programming must achieve a level of precision that corresponds to the resolution of the measuring system. Coordinates must be captured quickly after probing.

4. Measurements must involve a low degree of uncertainty and a high degree of accuracy.
5. There must not be an excessive number of measuring points or repetitions required to increase the accuracy of measurements.
6. It must be possible to change quickly from one type of workpiece to another.
7. The measurement data that are output must be usable without requiring conversion to a different form. This output must be useful for purposes of fast, reliable evaluation and correction of production quality.
8. Measuring programs for new types of workpieces must be created quickly.
9. Systematic measuring errors must be avoided.
10. The machine must be designed with a view to the future, including such features as high-precision data types, a universal probe system, expandability for special measuring tasks, software for control and data analysis, easy adaptability to all potential requirements, and last but not least, the capability of creating special customer-specific computer programs.
11. Measurement records must be created and printed out automatically. These are more reliable than manually created records and eliminate the need for subsequent analysis.

The **programming** of the sequential measurement program is usually performed by using a sample workpiece on the measuring machine. After loading an appropriate executive program (processor), the measuring machine temporarily becomes a programming workstation. The workpiece-specific sequential program can be created by manually scanning the first part. This scanning does not require any special programming knowledge, but it does require an understanding of geometric relationships and metrology. All positioning and measuring operations are stored in sequence and saved on a magnetic-tape cassette, if necessary, so that the program can be repeated later.

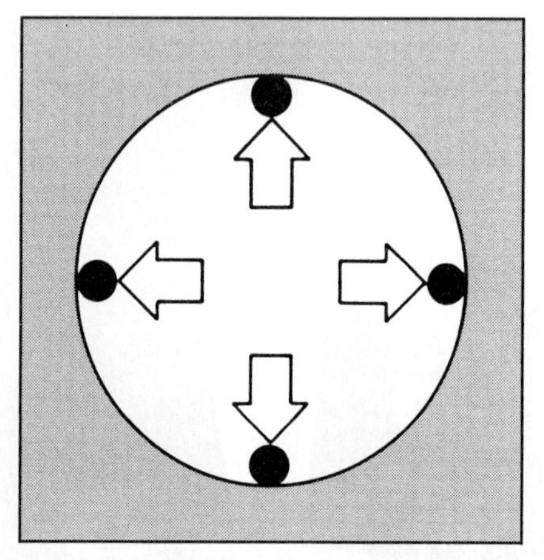
Measuring cycle for centering a borehole.

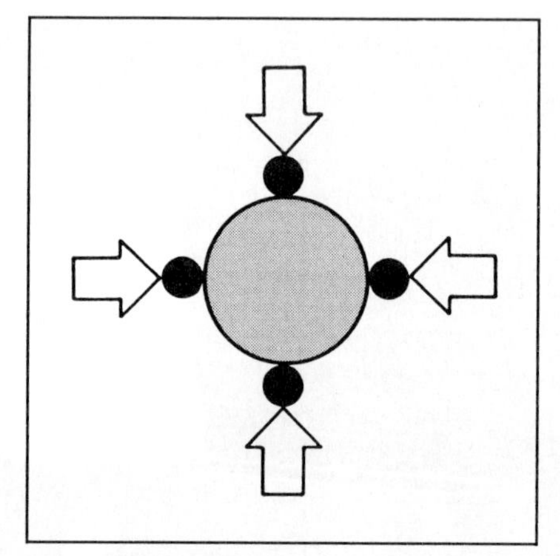
Measuring cycle for centering a shaft.

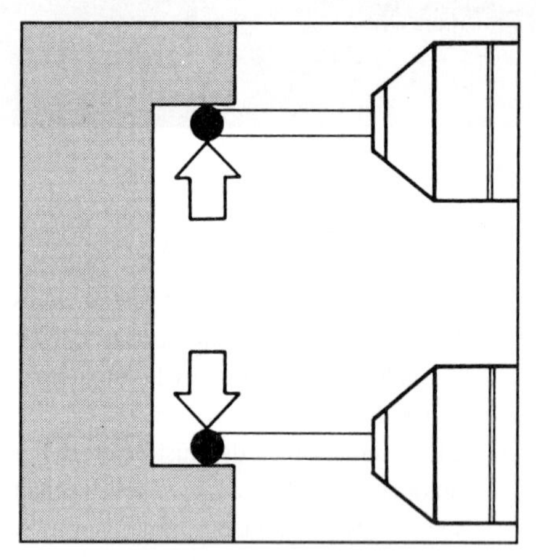
Measuring cycle for centering a groove.

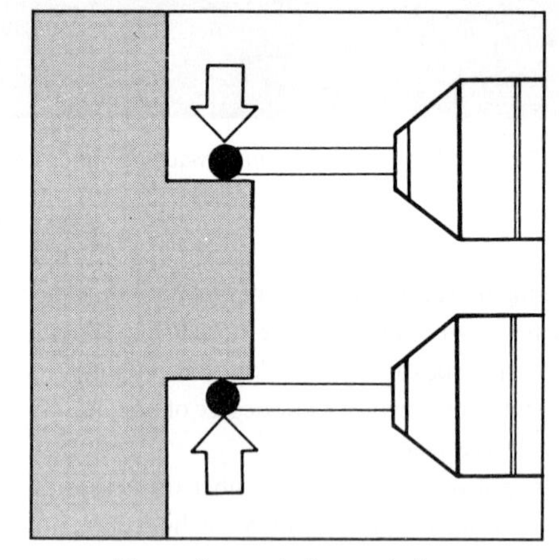
Measuring cycle for centering a web.

Figure 1.70: Examples of various measuring cycles.

The probe is the most critical element of any 3D measuring machine. It determines the machine's accuracy and range of applicability. When the probe contacts the workpiece, the probe either acts as a control element, providing position feedback control for the axes in reference to its own zero point, or it provides measurement values for the *X*, *Y*, and *Z* axes to augment the positional values of the measuring machine.

In connection with special software, the computer fulfills numerous tasks in addition to processing of the machine coordinates and stylus coordinates. These tasks include

1. Recognition of the measuring plane or measuring axis with spatial transformation of coordinates (3D alignment)
2. Distinguishing between the inside and outside contours of circles and cylinders

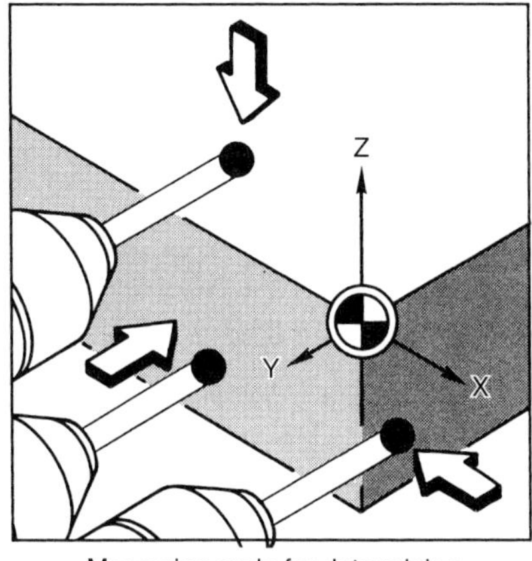

Measuring cycle for determining individual positions.

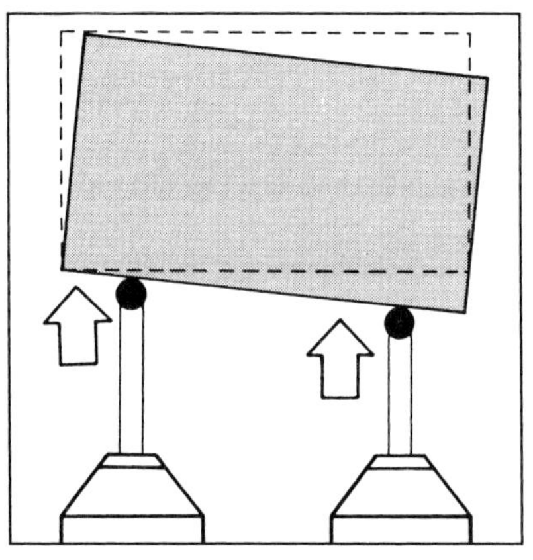

Measuring cycle for determining an angle.

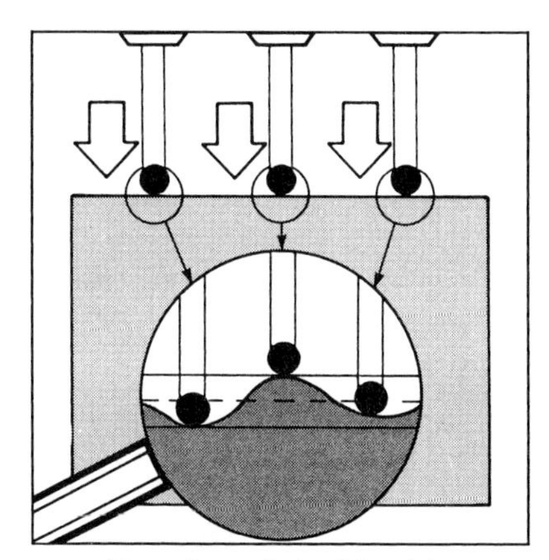

Measuring cycle for determining multiple points parallel to an axis.

Figure 1.71: Examples of various measuring cycles.

3. Detection of problem conditions such as undesirable probe collisions, absence of drilled holes, or failure to reach a certain point on the workpiece, as well as overtravel protection and standby mode when measurement data are not being collected, etc.
4. Storage of measurement data
5. Processing and analysis of measurement data and output of these data in the desired format
6. Subroutines for measurement of
   - Spatial elements such as cones, spheres, cylinders, and flat surfaces
   - Plane sections of spatial elements such as ellipses, circles, straight lines, and points of intersection
   - Coordinate points and relationships between them such as distance angle, symmetry, etc.

The CNC system of a measuring machine differs from the CNC system of a machine tool in a few small but important details. The machining program for a machine tool is based on the assumption that the tools and machine geometry conform to the assumed values. By contrast, measuring machines have to determine

- The magnitude of the deviations between the target values and actual values on the workpiece
- Whether the drilled holes and surfaces are perpendicular to one another
- Whether a particular drilled hole, bevel, or surface is present at all
- Whether a compensation value should be factored into the process for purposes of conforming to tolerances and, if so, when and how this should occur
- Whether the production process should be stopped when relatively large deviations indicate a problem in the machine tool

In order to switch between the measuring and programming modes of the measuring machine, the executive program has to be interchangeable, and the control system must have adequate data-storage capacity. The processing speed must be sufficient to allow all the various processor operations to be executed in the shortest possible time.

Scanning

This term refers to the **continuous probing of a surface.** To do this, the CNC system guides the probe over the workpiece surface in a continuous, line-by-line fashion. At the same time, the computer stores all the measured values at predefined time intervals or at certain distances along the measuring path. The CNC system constantly adjusts the probing axis by means of the servo drives so that the probe always remains within its measuring range. For this reason, a **scanning probe** must be used in place of the touch probe. The displacement of the scanning probe is constantly measured and used to correct the measured values. Given a resolution of 0.1 μm in the measuring probe, a measurement accuracy of ±1 μm can be achieved (*Figures 1.72 and 1.73*).

The measurement values can be analyzed in **measurement records** or used directly by suitably equipped CNC systems for milling of a similar surface. Further development of measuring machines and measuring probes will lead to increased use of **optical probes** that function without contacting the workpiece. These optical probes are especially likely to make use of laser technology. The measuring machine itself will be located closer and closer to the production machine so that correction values can be fed back by the shortest, fastest route.

Figure 1.72: Scanning head for measurement of dimensions, shape, and position. Optionally, single-point and scanning multiple-point measurement.

Figure 1.73: The switching central scanning head for rapid single scans.

1.16 Summary

Modern CNC machines exhibit a wide range of impressive design characteristics. These include not only automatic tool and workpiece changing but also the machines' static and dynamic precision, the acceleration and speed of the CNC axes, and the reliability and safety of these complex machines as a whole. The development of all these components has been very cost intensive. The goal has been to satisfy continuously increasing requirements and to allow more economical production. Workpieces are placing ever-increasing demands on manufacturing technology. Current demands also include the use of previously unknown tools, such as water, ultraviolet (UV) and laser light, graphite, and wire, not to mention improved conventional tools made of coated carbide and ceramics.

The high level of development of CNC machines has only been possible through close coordination and integration into the entire industrial context. It was not enough to convert conventional machine tools for use with CNC. Machine-tool manufacturers and CNC manufacturers worked together to create new requirements and new solutions. But even successful CNC machines quickly became unattractive to purchasers if their manufacturers stuck to existing concepts for too long. Enormous sums had to be invested continually in ongoing development in order to implement new ideas. This tough competition has led to ever-improving solutions.

Thanks to the high performance of the machine tools currently on the market, design engineers are creating workpieces of greater and greater complexity. This challenge, in turn, provides a motivation to further improve the performance of the machines. This field certainly can look forward to an exciting future!

CNC Machine Tools

Important points to remember:

1. **Machining centers**
   a. Are a type of machine tool that arose only as a result of the development of numerical control systems.
   b. A machining center is a machine tool that is numerically controlled and is equipped with an automatic tool-changing system and a tool magazine.
   c. Today, machining centers are without exception equipped with continuous-path-control systems in at least three to five axes.
   d. The essential distinction is between horizontal and vertical machines based on the position of the main spindle.
   e. A distinction is made based on the number of feed axes:
      (1) Three-axis machines: Three linear axes are the most basic configuration for a machining center with a rotating tool.
      (2) Four-axis machines: Three linear axes and one rotary axis to allow machining on all sides.
      (3) Five-axis machines: Three linear axes and two rotary axes. This makes it possible to perform milling in any plane, regardless of its orientation in space, and holes can be drilled in any diagonal orientation.
   f. Milling/turning machining centers: A family of machines that had its source in machining centers for milling, conceived for milling/turning parts with an emphasis on milling, for families of parts that require complex six-sided machining.
   g. According to the German Machine Tool Builders' Association (VDW), based on turnover figures, machining centers are the most produced machine tools in Germany.
2. **Turning machines**
   a. The most produced machine tools based on the number of machines.
   b. Turning machines have the following main designs:
      (1) Horizontal or vertical design
      (2) Flat or slant bed
      (3) For producing bars, bushings, or shafts
      (4) With one, two, or more spindles
      (5) With one or more cross-slides and tool turrets
   c. Turning/milling centers: In addition to turning operations, the high performance of modern CNC systems has made it possible to use suitably equipped turning machines to perform milling and drilling operations on the workpiece while it is held in the lathe chuck.
3. **Grinding machines**
   a. Grinding is generally considered to be the classical process for fine machining and hard machining.
   b. A distinction is made between various grinding machines according to the shape of the surfaces being created. The most common ones are
      (1) Flat and profile grinding machines
      (2) Cylindrical grinding machines
      (3) Tool-grinding machines.
   c. The particular features of grinding tools and grinding processes place special requirements on control systems.
   d. Because of the wear on the grinding wheel, a large number of grinding machines are equipped with variable-speed spindle drives.
   e. Because speeds are high, it is necessary to balance the grinding tools to prevent vibrations during the grinding process.
   f. Dressing and grinding mean that two different machining processes are being executed in the grinding machine. The control system has to be able to handle both these processes while also taking into account the changing abrasiveness and diameter of the tools.
4. **Gear-cutting machines**
   a. The type of machine with the most complicated kinematic systems, especially when producing bevel gears.
   b. Only since the beginning of the 1990s have standard commercial CNC controllers been capable, in combination with digital drives, of providing the required

degree of quality as regards controls and drives. This requires special software that has to ensure, for example, that the control system can process so-called infinite rotary axes without errors.

c. The generation of involute spur gears can be described as the engagement of the straight-flanked reference profile with the workpiece.

d. The kinematic basis for the bevel-gear production process is the engagement of the generating gear with the workpiece.

e. The programming of modern gear-cutting machines often has a graphical user interface. The programmer generally has to enter the set-point parameters for the workpiece and the actual dimensions of the tool in screen masks. For standard applications, the control system retrieves the cutting parameters from a database. Manual intervention by the programmer is generally possible at any point in order to optimize the process.

5. **Drilling machines**

a. All drilling machines have two common design features:

b. A spindle head with a vertical drill spindle that holds the tool and executes the feed motion

c. A machine table to which the workpiece is clamped

d. Drilling centers: A drilling machine with a higher degree of automation with an index table for four-sided machining or a tool changer

e. Drilling machines for PCBs: A special variation of the drilling machine for the machining of PCBs in the electronics industry.

f. Boring mills: Generally very large machines with horizontal main spindles.

g. Special cycles are provided for special machining operations. These cycles can be called up on the screen, provided with the required parameters, and also modified, if necessary.

6. **Parallel kinematic machines**

a. In contrast to conventional machine tools, in which the individual axes are arranged serially, that is, one after the other, in parallel kinematic machines, all the motion axes have a direct effect on the tool carrier unit that is being moved.

b. Most of the machines that have been implemented so far in practice involve hybrid solutions. For this reason, only a few of the feed axes are parallel axes—the remaining axes have a serial structure.

(1) *Advantages:* High stiffness, high speed, low mass that has to be moved.

(2) *Disadvantages:* Only a small part of the machine work area can be used, the spindle can only be inclined within a narrow range of angles, five-sided machining and axis compensation are problematic.

7. **Sawing machines**

a. Circular sawing machines are available in various kinematic designs for use with HSS and carbide-coated saw blades. Today, circular sawing machines are used primarily for material diameters of under 150 mm.

b. For heavy-duty production work, band saws are preferred. These have a parallel arrangement of the sawing units and broad saw blades.

c. Classical contactor controls have now been replaced by modern CNC control systems and can be programmed for compatibility with PLCs. With these control systems it is possible to define and call up a large number of order data sets via a keyboard.

8. **Laser machines**

a. *Laser* is an acronym for "light amplification by stimulated emission of radiation."

b. Given their high laser output power, CO_2 and Nd:YAG lasers are used primarily for the machining of materials. CO_2 lasers are gas lasers. The gas for this laser consists of carbon dioxide, nitrogen, and helium. Nd:YAG lasers are solid-state lasers. The laser-active body is an artificially created monocrystal made of yttrium-aluminum-garnet (YAG).

9. **Punching and nibbling machines**

a. Punching is a chipless manufacturing process. It is a cutting process in which a sheet of metal is cut through in a single stroke. The forming tool has two parts. The sheet is located between the upper tool, also known as the *punch*, and the lower tool, also known as the *matrix* (→ *Figure 1.52*).

b. In nibbling, punched holes are made next to each other in such a way that they overlap. This makes it possible to create apertures and contours in any desired shape. In this manner, a strip is nibbled out of the sheet at a rate of up to 1,200 strokes per minute.
c. Punching and nibbling machines can be used to machine plates up to 12 mm thick, with a size of up to 1.5 × 3 m. In punching, the shape of the punch leaves holes with a diameter of about to 100 mm. Form-punching tools are used when it is necessary to create a large number of small apertures. Such operations are more efficient with punching than with nibbling.
d. Combined punching and laser machines were developed to increase productivity. They are constructed according to the same principles as the punching machine. However, the frame is broadened in such a way that it can hold both the punching station and the laser machining station.

10. **CNC tube-bending machines**
    a. These machines are used for bent tubular parts in the aircraft industry, in the machine-tool industry, in the automotive industry, and in appliance construction.
    b. State-of-the-art CNC tube-bending machines with multiple tool levels can have up to 15 controlled axes.
    c. In modern machines, the preferred axis drives are electric servo drives.
11. **Electrical-discharge machines**
    a. Electrothermal discharges between an anode and a cathode cause surface particles to evaporate. The EDM process takes advantage of this physical effect.
    b. In the wire EDM process, material is removed from the workpiece by a wire electrode with a diameter of 0.1 to 0.3 mm. This removal involves neither contact between the workpiece and the electrode nor the application of mechanical force.
    c. In the wire EDM process, a shaped electrode is moved down toward the workpiece from above, and the electrical discharges produce a "negative" impression on the workpiece.
12. **Electron-beam machines**
    a. The "tool" is a narrow, sharply focused beam of electrons moving at a high velocity and with high energy.
    b. The beam power is between 1 and 100 kW and is concentrated on a focal spot with a diameter of 0.1 to 2 mm.
    c. The workpiece can be hardened, welded, or drilled depending on the type of beam control.
13. **Water-jet cutting machines**
    a. *Principles:* Water is forced through nozzles under a pressure of 4,000 to 9,000 bar. The diameter of the nozzles ranges from 0.1 to 0.3 mm. The exit velocity of the water jet is 800 to 900 m/s.
    b. Suitable for cutting rubber, leather, paper, foamed plastics, or Styrofoam.
    c. *Abrasive cutting:* By adding an extremely fine grade of abrasive powder, it is possible to cut steel up to 80 mm, titanium, marble, and glass.
14. **Measuring machines**
    a. Used for 3D measurement of any desired workpiece contours and for cost-effective process monitoring for large components with high requirements as to precision.
    b. The programming of the sequential measurement program is usually performed by using a sample workpiece on the measuring machine.
    c. In order to switch between the measuring and programming modes of the measuring machine, the executive program has to be interchangeable, and the control system must have adequate data-storage capacity. The processing speed has to be high enough to allow a large number of processor operations to be executed in the shortest possible time.
    d. *Scanning:* Continuous probing of a surface. To do this, the CNC system guides the probe over the workpiece surface in a continuous, line-by-line fashion.

2. Additive Manufacturing Processes

*Prof. Dr.-Ing. Michael F. Zäh, Christian Eschey, Imke Nora Kellner, Harald Kraus,
Toni Adam Krol, Michael Ott, Johannes Schilp, Stefan Teufelhart, and Sebastian Westhäuser*

Additive manufacturing processes are based on the fundamental concept of building up parts layer on layer; that is, the part is built up additively by generating individual layers. Fabrication of geometries is performed from amorphous materials (e.g., fluids or powders) or materials with a neutral form (e.g., strips, wires, paper, or film) using chemical and/or physical processes directly from the digitally generated data models via a CAD/CAM link.

2.1 Introduction

Additive manufacturing processes were first presented to the public in 1987 in the United States; in 1989–1990, the first machines were delivered to Europe. At that time, this generally involved stereolithography (SLA) machines. Additional process variants were developed over the next few years, including selective laser sintering (SLS), electron- and laser-beam melting, and laminated-object manufacturing (LOM). Based on known processes (→ *Section 2.4*), new or modified processes will be developed further in the next few years because additive manufacturing has not reached its full potential, for example, when it comes to processing of multiple materials.

The process sequence also can be summarized as follows (→ *Section 2.3*): Parts are built up layer on layer through the consolidation of an amorphous or neutrally shaped feedstock material to which energy is applied layer by layer. All layer manufacturing processes must meet the following requirements with regard to data processing (CAD/CAM link):

- The starting point for the additive manufacturing of a part is a **three-dimensional (3D) computer-aided design (CAD) model** that depicts all the workpiece data in digital form.
- For the construction process, the 3D solids have to be **broken down into individual layers**, thus **reducing them to two dimensions.** These layer data define a **process-specific computer numerical control (CNC) program.**
- The subsequent manufacturing process is performed on a **numerically controlled machine** that executes the information layer by layer, thus generating a part.

Comparison with conventional manufacturing processes shows the economic and technical potential of additive manufacturing: While simple solids can be produced cost-effectively in large quantities using familiar processes such as turning, milling, or casting, for smaller quantities and more complex components, the use of *layer manufacturing* becomes more economical. Moreover, a few highly complex components, for example, those with internal geometric features, can only be produced using *additive manufacturing* processes.

Workpieces generated by means of additive manufacturing can be used for a very wide range of functions in various fields of application:

- Models:
  - **Concept models** are employed to visualize the dimensions and general appearance of the new product being developed at the earliest possible stage.
  - **Design models** are employed to depict the CAD model accurately as regards shape and dimensions. The surface quality and the position of individual elements are important.
- Prototypes:
  - **Functional prototypes,** which are largely the same as the series-production sample, are used to verify one or more of the planned functions of the later-series part.
  - **Technical samples,** which only differ from the later-series parts in the manufacturing processes that are employed, are used to verify that the requirements are met.

- Components:
  - **Tools and molds** can be used to create end products in a subsequent manufacturing process (e.g., injection molding).
  - Fully functional customer-specific **single parts and series parts** with close to the final contours can be employed.

As this overview of the areas of application shows, additive manufacturing processes can be used in all phases of product development.

2.2 Definition

Additive manufacturing processes are defined as all processes that are used to create 3D models, prototypes, and components additively, that is, through the joining or layering of multiple solid elements. Many other terms can be encountered in practice and in the literature; these have been brought together and standardized in VDI Standard 3404, which was introduced in 2010.

For historical reasons, it is a common practice for additive manufacturing processes to have the word *rapid* added to them in order to convey that additive processes are faster (for small and medium quantities) than the conventional alternatives. Conventional processes generally require tool making. By avoiding this, additive processes are not only faster but also in most cases provide large potential cost savings. In short, this is why additive manufacturing processes are also called **rapid technologies.**

As described earlier, additive processes can be used economically anywhere in the product-creation process (*Figure 2.1*). Rapid technologies can be subdivided into rapid prototyping, rapid tooling, and rapid manufacturing.

Rapid prototyping is an application of rapid technologies for cost-effective, quick production of test parts and prototypes. These components generally have limited or special functions. Their design can be but does not have to be optimized for series production. With rapid prototyping, it is also not necessary to use the generally more expensive material that is used for series production. The term **rapid prototyping** thus covers only a small subset of additive applications and therefore should not be used as a synonym for the term *additive processes.*

In **rapid tooling,** additive processes are used to create tools and molds (e.g., casting molds, injection molds, and deep-drawing dies) for the production of prototypes, preseries parts, and series parts. For the most part, selective laser melting is employed here, allowing not only rapid manufacture but also efficient use of the shaping options provided by

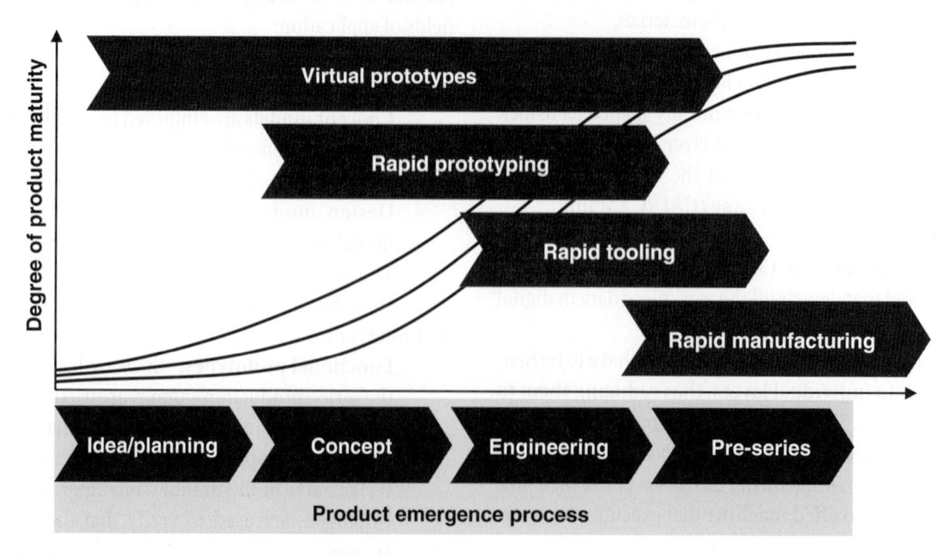

Figure 2.1: Rapid technologies in the product emergence process.

rapid technologies. In order to achieve the necessary precision and/or the required surface characteristics, conventional processes such as high-speed cutting (HSC) milling are often implemented to rework the tools and dies created by means of the additive process. This is called *direct* rapid tooling. *Indirect* rapid tooling is the production of tools by molding master patterns created using additive processes. Processes in which computer numerical control (CNC) programming and subsequent HSC milling are used to create tools from solid stock within a short time are also considered to be rapid technologies, along with additive manufacturing processes. Because they involve chip removal, however, they should not be confused with additive processes.

Rapid manufacturing is the additive manufacturing of end products for single-part or series production. The parts are manufactured in the genuine material based on the design data and have all the characteristics of the final product. Besides the ability to produce parts rapidly, building up parts using additive processes allows the creation of certain product design characteristics (e.g., cooling channels that are close to the surface or curved boreholes) that are difficult or impossible to create using conventional production methods.

This means that with rapid technologies it is possible to expand and master the various options for the manufacture of new design elements and to move on directly to the manufacture of end products for single-part or series production. Besides shortening the time required for product development and product emergence through the employment of rapid technologies in the creation of prototypes and tools and for direct manufacture of the final parts, the use of additive manufacturing processes also facilitates the logical interlinking of order data processing. Not only the manufacturing time can be described as rapid compared with conventional processes—the direct CAD/CAM link also simplifies and speeds up production planning, for example, for generating and converting the manufacturing data.

For metals, the rapid technologies with the greatest potential for the future are beam melting (also known as *laser forming*), selective laser melting (SLM), LaserCusing, electron-beam melting (EBM) or direct metal laser sintering (DMLS) *(Figure 2.2)*, and direct metal deposition (DMD). These are suitable both for prototype production and for repairing or modifying tools and molds. DMD makes it possible to process spatial surfaces by fusing powdered metal layer on layer in a laser beam. The amount of heat input into the workpiece

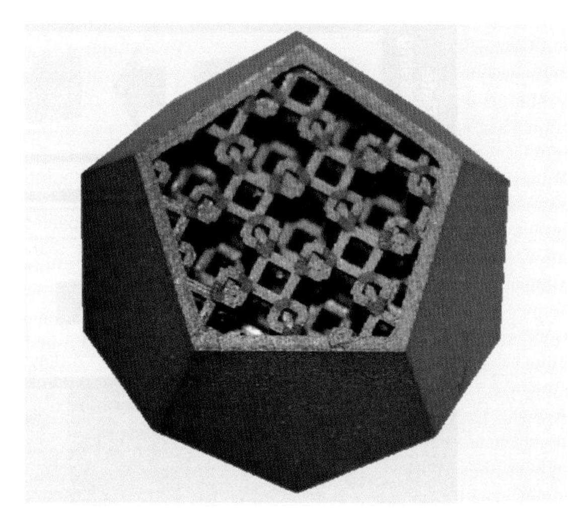

Figure 2.2: Dodecahedron produced by means of beam melting with a geometrically complex internal structure made of metal, without mechanical finishing work.

is minimal. The two process variants described earlier both have the following characteristics:

- Powdered metal as the feedstock material
- Complete melting of the powdered metal by the laser
- Mixing and application of various powdered metals to base materials of a different type
- Fully automatic production, with no manual work
- Generation of parts directly from the 3D CAD data
- Manufacturing close to the final contours with minimal rework on the functional surfaces

2.3 Process Chain

The generation of models and the process chain follow similar principles for all process principles in additive manufacturing. This is illustrated in the process sequence shown in *Figure 2.3*. The process for generating additive components can be subdivided into the following areas.

Preparations for the Construction Process

A complete, dimensionally accurate 3D CAD solid is the basis and prerequisite for all additive manufacturing processes.

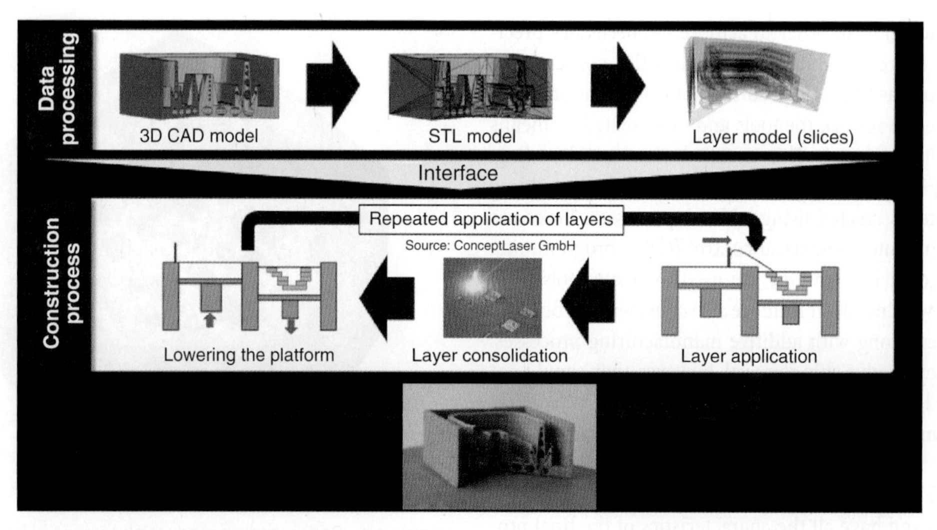

Figure 2.3: Sequence of data processing and the construction process from the 3D solid to the completed part.

The CAD program is selected based on the available options for data export. Formats used with rapid technologies include STL, Initial Graphics Exchange Specification (IGES), and Standard for the Exchange of Product Model Data (STEP). The format most often used is STL, which is based on approximation of the geometry using triangles (triangulation). This format is also supported by all the commonly used CAD programs. The triangulation process used here involves approximation of the outer geometric surface by means of triangles. These triangles are defined via the position of their three vertices and the associated normal vector pointing away from the solid of the part. The aggregate of the triangles and normal vectors embodies all the surface information. Nevertheless, the geometry of the parts always should be verified after triangulation. As a rule, all the software programs for data preparation provide repair functions for use in the event of triangulation errors.

The STL data set for the part is the input information for the slice process. This breaks the part down into individual layers. The thickness of the layers depends on the process being used and the required surface quality. This means that with slicing, the geometric information for the construction process is generated for each layer. Where curvatures, sculptured surfaces, and blunt angles appear, the layer-by-layer construction means that a "stepped" effect is created. This leads to lower surface quality. The greater the layer thickness, the greater is the stepped effect. On the other hand, increasing the layer thickness reduces the construction time and thus also the cost of the part. It is necessary to find a compromise for each individual construction process.

The last step in the preparation for the construction process is to place the part virtually in the machine space using the system software. At the same time, the information for the individual layers is transferred to the control data of the system. The systems integrator then defines system-specific parameters, such as the travel speed or the temperature of the working space.

The Construction Process

As a general principle, all additive processes are performed in three process steps: (1) The material is applied, (2) the layer is consolidated, and (3) the platform is lowered to allow application of the next layer. On the other hand, application of the material and the joining process vary depending on the specific system and manufacturing technology. Moreover, the basic states of the materials may vary and can include powdered, liquid, and solid materials. The feedstock material is consolidated through the use of an energy source or by applying a chemical activator. After the first layer from the preparation for the construction process has been consolidated, the platform is lowered by the thickness of one layer, and a new layer of feedstock material is applied. In order to create as even a base layer as possible, this is generally performed using a broad application mechanism, such as a roller or a wiper. This is followed by consolidation of the new layer in accordance with the data specified in the preparation for the construction process. At the same time, the layer is joined to the underlying layer. In additive processes, this method leads to anisotropic properties in the material because the joining of the feedstock material in the *X/Y* plane generally is stronger than in the *Z* direction.

Postprocessing

Even when the stepped effect is not a problem, at present, many additive processes can provide only relatively low surface qualities. In most cases, the parts have to be reworked after the construction process. The reason for this is the stepped effect inherent to the process and the limited dimensional accuracy of additive manufacturing technologies. For example, fixed points should be provided during design work as reference points for subsequent rework. This makes it possible to set up an auxiliary coordinate system as a basis for finishing operations using a CNC machining center. The design engineer should select these fixed points so that that they can be produced accurately using the additive process. Furthermore, the anisotropic properties in the material can be reduced or eliminated by means of thermal posttreatment processes.

2.4 Classification of Additive Manufacturing Processes

The additive manufacturing processes in use today can be classified based on two aspects: the feedstock material and the shaping method.

Classification Based on the Feedstock Material

As shown in *Figure 2.4*, three types of feedstock materials are used today:

- Powdered granulates
- Liquid synthetic resins
- Solid feedstock materials

In processes where **powdered or granulated** feedstock material is used, sintering or bonding processes are employed. Here, targeted laser beams are used to melt and consolidate materials that are "applied" or fed onto each other in thin layers. In 3D printing, the consolidation is performed via the targeted introduction of binding agents, for example, water in plaster.

In the case of **liquid** materials, for the most part, targeted laser beams or heat (ultraviolet [UV] beams) are used to consolidate (polymerize) synthetic resins and join them to the underlying existing layers. However, this process also includes the use of solid feedstock materials (plastics) to build up the existing model layer on layer by means of melting followed by rapid cooling. The semiliquid plastic is sprayed on top of itself layer by layer.

When **solid** feedstock materials with a neutral form are used, this generally involves layers of film or paper that are bonded on top of each other layer on layer and then cut out to the exact contours using a laser or cutting blade. Both conventional bonding processes and partial polymerization (bonding via heating) are used here.

Classification Based on the Shaping Method

A distinction is made here between processes that can be used to create 3D forms directly and processes that generate the final shape layer by layer. *Figure 2.5* provides an overview of the various processes.

All the processes used today function in two dimensions, that is, with individual layers. Models are built up layer on layer, thus creating a third dimension. This also applies to processes that generally also would be able to work in three dimensions directly (e.g., fused deposition manufacturing). The reason for this is that the 3D software required for this is significantly more complex and therefore not available at this time.

2.5 Introduction to the Principal-Layer Manufacturing Processes

Beam Melting

Process Description

In additive manufacturing processes that use beam melting, the feedstock material is in the form of a powder.

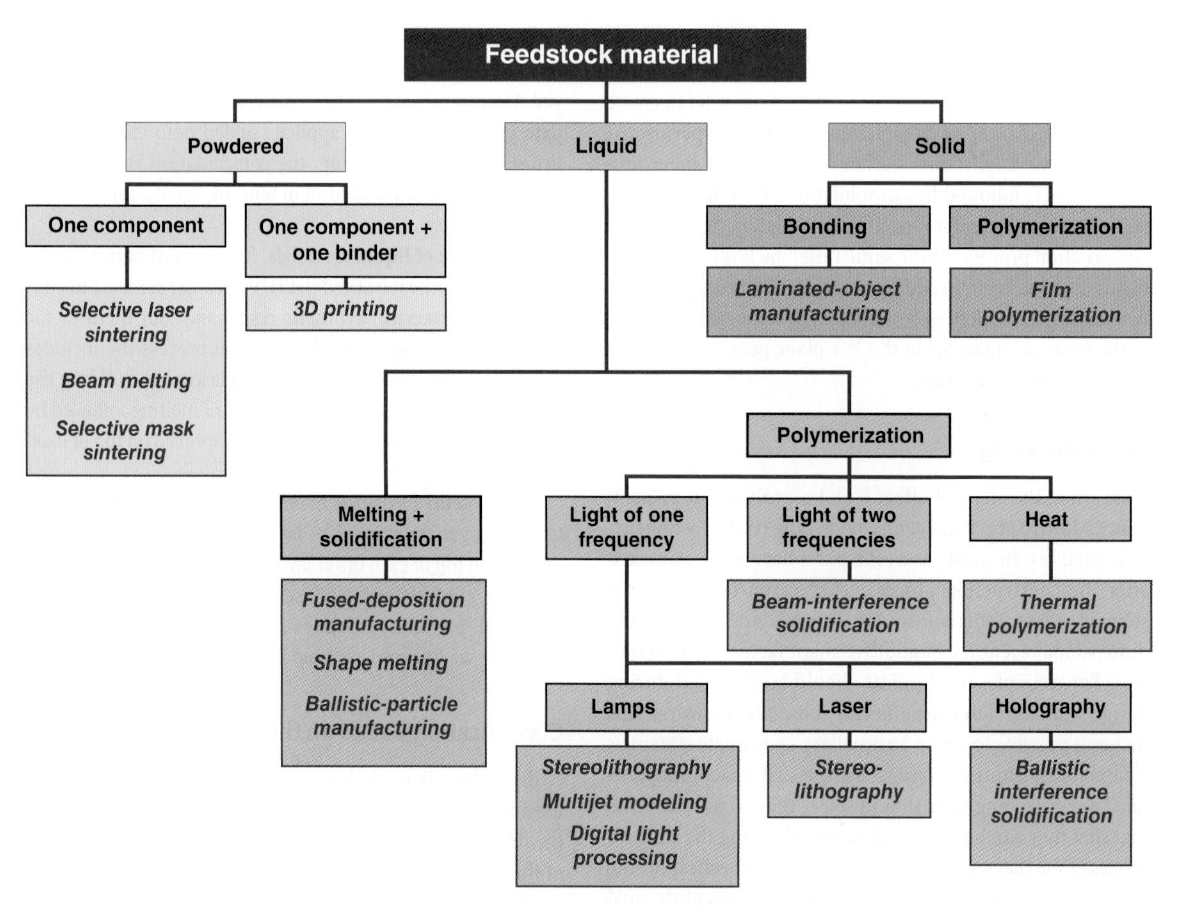

Figure 2.4: Classification of currently known additive manufacturing processes according to the feedstock material.

At the beginning of the process, a layer of powder is applied to the platform using an application mechanism (e.g., a doctor blade). At the points where the component is subsequently to be created, this layer is fused and thus consolidated on the substrate using an electron beam or laser beam depending on the process method. The part is created layer on layer by lowering the platform repeatedly, each time applying a new layer and melting the component volume (*Figure 2.6*).

The effects that appear when the powdered feedstock material is fused locally are characterized by complete transformation of the feedstock material into the molten state, which is what differentiates this process from sintering processes. Instead of a two-stage sintering process (→ *"Laser Sintering" next*), a single-stage beam melting process now has become the established practice in industrial applications. Various terms are used to describe this process. Whereas EOS GmbH uses the term *direct metal laser sintering* (DMLS), other companies prefer the terms *LaserCusing* (Concept Laser) or *selective laser melting* (SLM) (MTT Technologies). However, the process sequence is the same for all manufacturers: The feedstock material is always a single-component powdered metal that is completely fused during the construction process. In this way, it is possible to create a nearly pore-free component whose material properties are similar to those of a component created from the same material using conventional processes (e.g., casting). Unlike with the IMLS process (→ *"Laser Sintering" later in the chapter*),

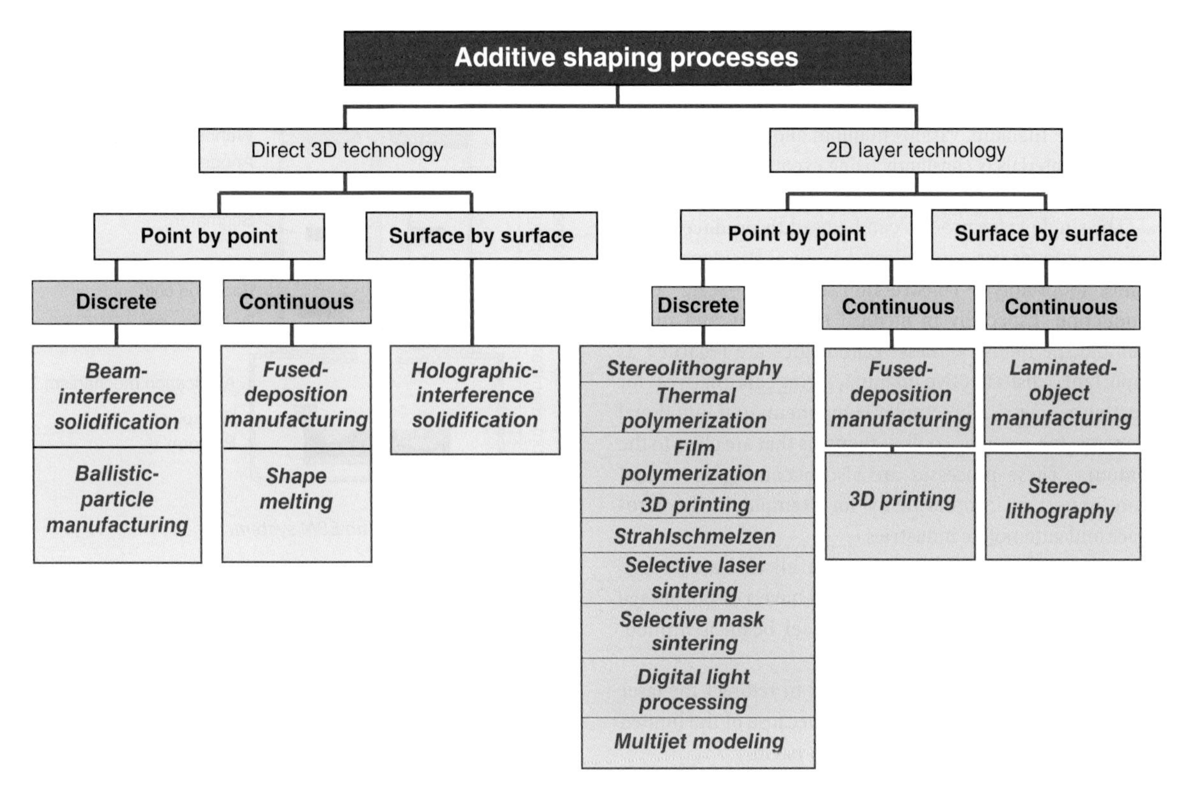

Figure 2.5: Classification of currently known additive manufacturing processes according to the shaping method.

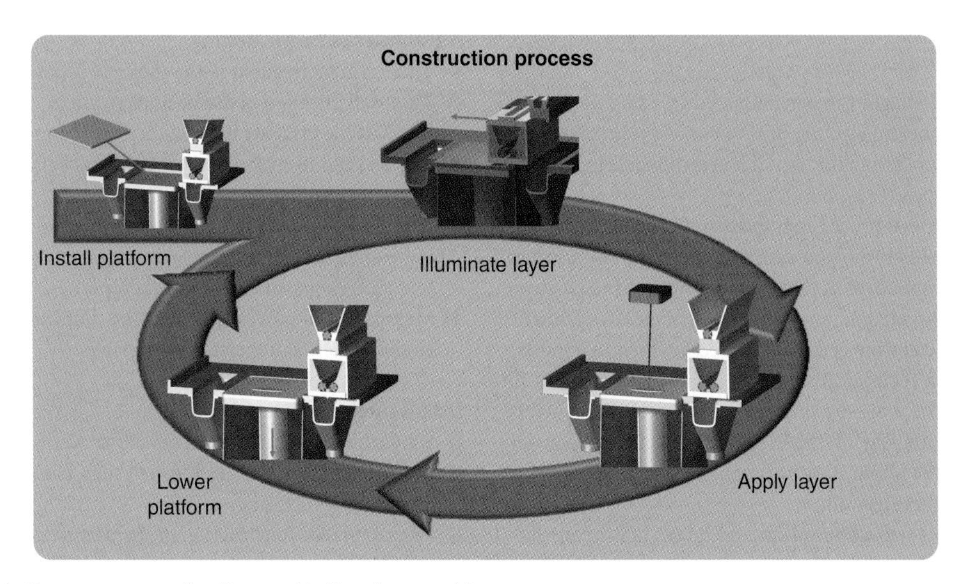

Figure 2.6: Process sequence for direct and indirect beam melting processes.

no additional processing is necessary. At present, the powdered feedstock materials that are available and usable include various tool steels, stainless steels, aluminum and nickel alloys, pure titanium, various titanium alloys, and gold. The range of materials is constantly being expanded as a result of numerous research and development (R&D) projects.

The single-stage process can be used to produce usable components for various applications, in particular when it comes to prototype construction and small-scale series production. Especially in medical technology and in the tool and die industry, these technologies are regarded as important, cost-effective manufacturing alternatives for creating geometrically complex components and functional elements, for example, cooling channels that are close to the contours. These processes are also becoming more and more important in other fields, for example, in the aerospace and automotive industries.

It should be noted, however, that all the laser-based beam melting processes just described have a disadvantage in that the traversing speed of the laser beam is limited. There are two main reasons for this:

- The mechanical mirror optics used to redirect the laser beam limit the power of the beam because of the limited thermal load capacity of the mirror system.
- The mass moments of inertia in the mirror optics limit the traversing speed of the laser beam because increased deflection speeds have an adverse effect on the accuracy of travel.

These disadvantages can be eliminated by using not a laser beam but rather an electron beam in an *electron-beam melting* (EBM) process. In this way, it is even possible to increase the process speed (*Figure 2.7*).

In this type of processing, the beam is generated using a so-called electron-beam cannon, which can be used to apply the beam current, that is, the power of the electron beam, in a targeted manner. By means of electromagnetic lenses, the electron beam is shaped to form a circular cross section focused at the focal point and deflected in the plane. The workspace is in a vacuum chamber—this prevents scattering of the electron beam. The powder reservoirs, the application mechanism, and the platform are located there as well.

The higher deflection speeds and power density of electron beams allow higher process speeds. The high deflection speed also offers a number of options for improved process control, such as quasi-parallel illumination and freely configurable beam shaping to control and optimize heat input to the part. Thanks to these advantages, the EBM process is currently being subjected to further investigation and development with the goal of establishing it as an industrial application on a larger scale.

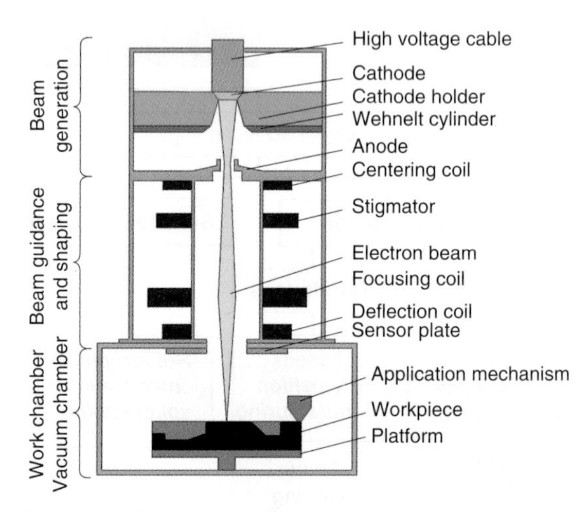

Figure 2.7: Structure of an EBM system.

Advantages of Beam Melting

- There is great freedom in the choice of geometric shapes.
- Thin wall thicknesses can be implemented.
- It is possible to create usable functional components.
- It is feasible to make internal cooling channels that are close to the contours.
- It is possible to process materials that are difficult or impossible to process using conventional methods owing to their thermal and mechanical properties.
- Multimaterial processing and the implementation of graded material properties are possible.

Disadvantages of Beam Melting

- Supporting structures are necessary at overhanging areas.
- A platform has to be used, which then has to be cut off during postprocessing.
- Layer-on-layer construction of the part results in a stepped effect.

- There are high production costs for each part with long process times.
- Internal stresses occur in the part owing to high temperature gradients during cooling of the fused powder.
- Rough surfaces are produced in some cases, with the associated need for rework on functional surfaces.
- There is a limited work area and thus limited part sizes (at present 300 × 350 × 300 mm maximum).
- A shielding-gas atmosphere or vacuum is necessary (for electron-beam melting).

Laser Sintering

Process Description

Sintering is defined as a consolidation process for crystalline, granular, or powdered materials through coalescence of the crystallites by means of appropriate heating. This is done by heating the powder bed, in some cases to several hundred degrees celsius. This process can be used for both metals and plastics.

For metals, this involves a two-step process. Indirect metal laser sintering (IMLS) is used to fuse a plastic binder contained in the powdered metal surrounding the metal particles. This initially produces a so-called green compact with low strength. In order to generate an adequate metal part from this green compact, a subsequent heat-treatment process is required. Here, the plastic binder is expelled and so-called sinter necks are formed between the metal particles. At the same time, bronze is infiltrated into the part, thus creating a stable structure consisting of about 60 percent steel and 40 percent bronze.

Laser sintering (LS) is also known as *selective laser sintering* (SLS). This process is based on the powder-bed principle and can be used to produce prototypes and functional components from plastics in just a few hours *(Figure 2.8)*. The materials used most often in this case are polyamide and polystyrene. In contrast to beam melting processes, in which the feedstock material is fused solely by means of the beam, with LS the feedstock material is first heated to a temperature just under the melting point via a large-area heat radiator. The feedstock material then is fused locally using a low-power laser (up to ~30 W). The laser beam is deflected by the scanner optics. After the iterative production of the individual layers of the part, the support structure is slowly cooled down to room temperature. The time required for this cooling

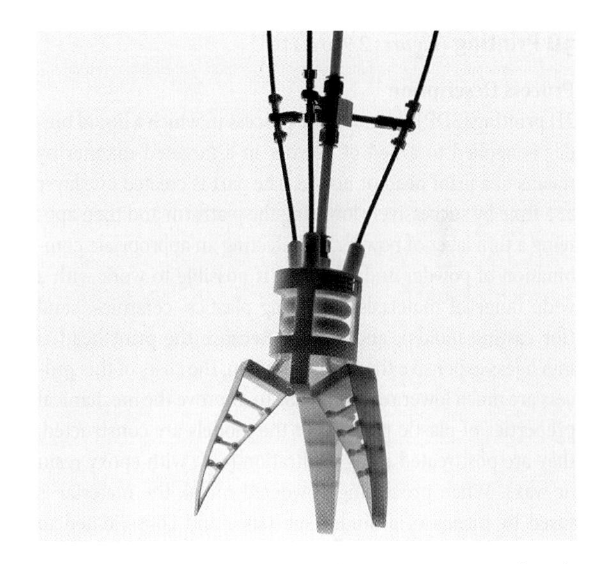

Figure 2.8: Grippers produced via additive manufacturing (Festo).

process is approximately the same as the time required to build up the part. If the cooling is too rapid, the temperature gradients will be too steep, thus resulting in excessive distortion in the part. After the cooling process come the process steps for unpacking, cleaning, and postprocessing of the part. The sintered part sits in loose powder, from which it can be removed. The part then can be cleaned using compressed air to remove powder residues. Any feedstock material that has not been fused can be reused. For optimal results, approximately a 1:1 mixture of old and new powder should be used.

Advantages of Laser Sintering

- Short throughtime compared with beam melting processes
- Possible to create complex, functionally integrated parts
- Usable, functional components with complex geometries can be manufactured.
- Wide range of materials
- No need for support

Disadvantages of Laser Sintering

- Shrinkage and distortion of large components owing to the thermal construction process
- Porous surface
- Aging owing to the effects of UV light

3D Printing *(Figures 2.9 to 2.11)*

Process Description

3D printing (3DP) is an additive process in which a liquid binder is applied to a bed of powder in a targeted manner by means of a print head or nozzle. The part is created one layer at a time by successively lowering the platform and then applying a thin layer of powder. By selecting an appropriate combination of powder and binder, it is possible to work with a wide range of materials, including plastics, ceramics, sand (for casting molds), and metals. Because the print head is much less expensive than a laser system, the costs of this process are much lower relative to LS. To improve the mechanical properties of plastic parts, after the models are constructed, they are posttreated using infiltration (e.g., with epoxy resin or wax). When processing powdered metal, the material is fused by means of a binder substance and consolidated to form a green compact. This is then heat treated in a similar manner to IMLS and infiltrated with bronze.

Advantages of 3D Printing

- High construction speed
- Many different materials can be used.
- Large work areas are possible.
- A considerable number of system manufacturers (*Figure 2.9*)

Figure 2.9: 3D printer made by voxeljet (voxeljet).

- Colored parts can be created.
- Economical process

Disadvantages of 3D Printing

- Low surface quality
- Mediocre mechanical properties owing to low density

Fused Deposition Modeling *(Figures 2.12 and 2.13)*

Process Description

In extrusion processes, a liquid or softened material is applied to a platform via one or more nozzles. The subsequent cooling process gives the part its strength. Fused deposition modeling (FDM), also known as *fused layer modeling* (FLM), is the most notable process that makes use of only one material. A subcategory of this process is *multijet* or *polyjet modeling*, which can be used to create parts with graded properties. Here, the nozzle generally has two degrees of freedom (in the *X* and *Y* directions), whereas the entire platform can be moved in the *Z* direction. Thus 3D parts can be manufactured in this way. The adhesive bond between the extruded beads is formed during the cooling process. The bead structure results in a relatively low surface quality, as can be seen in *Figure 2.12*.

Advantages of Fused Deposition Modeling

- Good mechanical properties
- Compact system
- Can be used as an office system
- Possible to work with ABS
- With multiple-nozzle systems, multimaterial parts are easy to implement.

Figure 2.10: Model for a shoe–created with a 3D printer.

Figure 2.11: Model for a gear—unit created with a 3D printer.

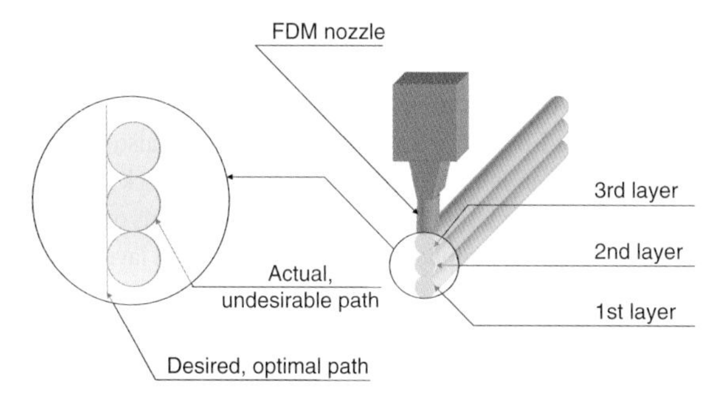

Figure 2.12: Component quality via FDM.

- Large number of system manufacturers
- Build-it-yourself systems are available.

Disadvantages of Fused Deposition Modeling
- Low surface quality
- Overhangs are hard to implement because no supporting material is present.
- Considerable work is required to remove the supports.

Stereolithography

Process Description
Stereolithography is the oldest of the additive manufacturing technologies. In this process, plastic parts are created by

Figure 2.13: Various components–created with FDM.

means of selective 3D polymerization of a photosensitive resin. Polymerization generally is performed using a UV laser in which the critical energy required to cure the part is reached only at the focal point of the laser beam. By lowering the platform, a new layer of liquid resin can be deposited on top of the already cured layer. By means of successive lowering and curing, a 3D part thus can be created.

Polymerization is defined here as a chain reaction in which unsaturated molecules are bonded to form macromolecules. This can be broken down into four steps:
- Chain initiation, or the primary reaction
- Chain propagation
- Chain termination
- Chain transfer (branching of a molecule chain)

The materials used in stereolithography have to respond to UV radiation and reach chain termination very quickly so that the resin is cured only in the illuminated areas. In order to give the parts their final strength, the construction process itself is often followed by a curing process in a UV cabinet.

Stereolithography parts are used above all as concept models or functional models during the product-emergence process. Another area of application for stereolithography is the creation of master patterns for casting under vacuum and investment casting.

One further development of stereolithography is micro-stereolithography, which can be used to produce very complex geometries that at the same time also have fine details. In this process, the resin is not cured point by point using a laser, but rather an entire layer is cured over its whole area by means of a digital light processing (DLP) chip.

Advantages of Stereolithography

- Easy to produce complex, thin-walled structures
- Practically no thermal stress, thanks to the low laser power (generally <1 W)
- High-precision parts can be produced.

Disadvantages of Stereolithography

- Relatively quick aging process of the materials owing to the UV component of natural daylight
- Changing resins is complex, time-consuming, and expensive.
- Overhanging areas of parts require support structures (*Figure 2.14*).

Other Processes

Mask Sintering

Mask sintering (MS) is very similar to SLS. Here too, energy is applied to a powdered feedstock material in order to fuse it. Unlike with laser sintering, mask sintering does not involve a single laser beam that is deflected by means of a scanner but rather illumination of a layer over a large area via a mask. The mask is printed for each layer in such a way that the energy emitted by a UV source is reflected onto the powder in the areas that are to be consolidated. The mask consists of a mirror that is printed with ceramic powder for each layer. Illuminating a complete layer over a large area greatly reduces the construction time per layer.

Digital Light Processing

Digital light processing (DLP) uses a similar process sequence to that of stereolithography, but through the use of a special chip, it is possible to illuminate and consolidate an entire layer at a time. The DLP projector is controlled by converting the construction data into a bitmap format and projecting them onto the construction plane as a mask via the mirror unit. As in conventional stereolithography, construction in a fluid means that support structures are necessary. This process is especially used in the additive manufacturing of microcomponents.

Laminated-Object Manufacturing

In laminated-object manufacturing (LOM), also called *laminated-layer manufacturing* (LLM), the feedstock is in the form of sheets of material, for example, plastic films or sheets of paper. These are applied to each other layer on layer, bonded, and then cut along the component contours. Both laser

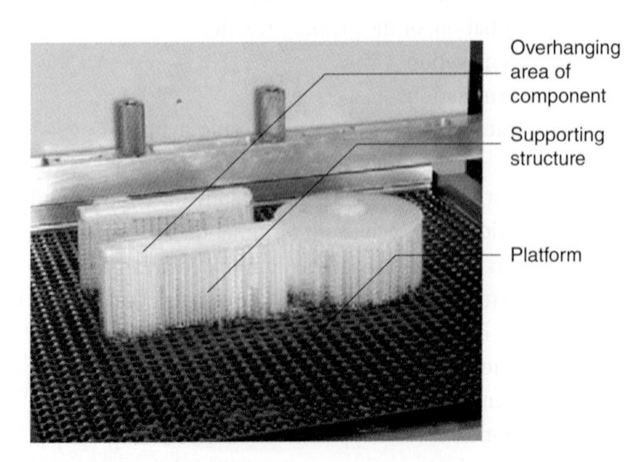

Figure 2.14: Component created using stereolithography with supporting structures.

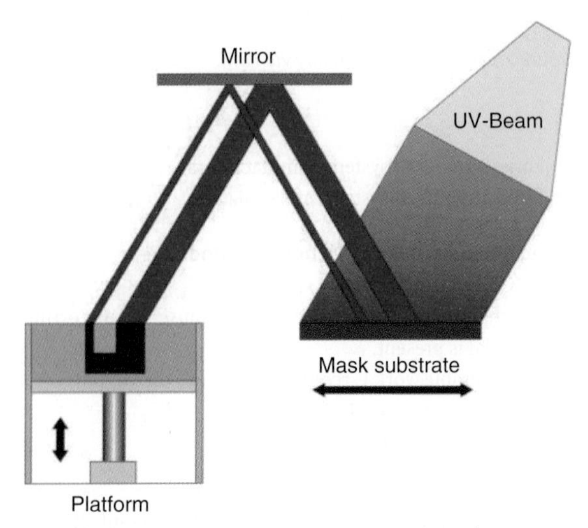

Figure 2.15: Principle of mask sintering (Sinter Mask).

systems and conventional cutting tools such as rollers are used for this. The completed block is removed from the system, and the sections that are not part of the component are removed. It is also possible to work with metallic materials. Thanks to the thinness of each individual layer, high surface quality can be achieved.

Laser Cladding

In laser cladding, a laser beam is used to create a localized weld pool on the surface of a metallic workpiece. The metallic feedstock material (generally in the form of a powder or wire) is applied to this weld pool by means of a feeder unit. By moving the weld pool over the surface of the material, it is possible to create a bead-shaped line. To protect the fused material against oxidation, the process generally takes place in an inert atmosphere. Thus 3D material solids can be built up by superimposing a number of individual layers. Components built up using this additive process have a comparable density to that of components created from the same material using conventional processes. Components created using laser cladding generally have a relatively rough structure that is very similar to the structure of a casting. Furthermore, as yet only low surface quality can be obtained using this process.

2.6 Summary

Additive manufacturing processes can be used to produce prototypes, casting molds, and end products. In all cases, the part being produced has to be available as a 3D CAD model. Depending on the part's intended purpose, a wide range of different layer manufacturing processes can be employed. In addition to the processes mentioned, a number of others exist that differ only slightly from the ones already described.

Current R&D work in the field of additive manufacturing processes is focused on a wide range of aspects. These include speeding up the production process, for example, through higher laser powers and thus faster deposition rates. In addition, further postprocessing steps have to be performed after the additive manufacturing process in order to achieve adequate component quality. This means that R&D work is centering on both the robustness of the process and ensuring component quality. The constant expansion of the range of materials that can be processed means that more and more areas of application are being opened up using additive manufacturing processes.

Additive Manufacturing Processes

Important points to remember:

1. **Rapid technologies** include both additive and subtractive manufacturing processes. Their goal is to produce parts or tools quickly.
2. **Additive manufacturing processes** join amorphous materials or materials with a neutral form layer on layer to form a physical workpiece directly from the 3D CAD data.
3. In order to implement additive manufacturing processes, the following requirements must be met:
   a. Data models for the parts being produced must be available on 3D CAD systems.
   b. During process planning, the solid/surface model must be broken down into individual layers.
   c. A process-specific NC program has to be generated.
4. Parts created using additive processes differ greatly as to their consistency, precision, and surface depending on the specific process used.
5. A large number of additive processes are used in industry today. The product-emergence process phase plays an important role here.
6. The additive manufacturing processes in use today can be classified based on two aspects:
   a. According to the **feedstock material**—powdered granulates, liquid-plastic resins, or solid materials
   b. According to the **shaping method**—directly 3D or by superimposing 2D individual layers
7. While the first process to be employed industrially was stereolithography, today a large number of additional commercial processes exist.
8. All currently used commercial processes operate two-dimensionally.
9. The five most important processes at present are
   a. **Beam melting** of fully functional metallic parts and prototypes.
   b. **Laser sintering** of both powdered plastics and single/double-component metals.
   c. **3D printing**, that is, the powdered feedstock material is consolidated via the targeted introduction of a binder liquid.
   d. **Fused deposition modeling,** which heats thermoplastic wires in a nozzle and fuses the material in the shape of a bead.
   e. **Stereolithography**, that is, layer-by-layer polymerization of liquid resin.
10. **Rapid prototyping** is defined as the quick production of illustrative objects or models that provide only limited functions compared with the subsequent component.
11. **Rapid tooling** brings together processes that can be used to create tools (e.g., for casting) in order to produce components in the genuine material.
12. **Rapid manufacturing** is used to describe manufacturing processes that allow the direct production of fully functional parts (end products).

3. Flexible Manufacturing Systems

Flexible manufacturing systems differ as widely as the manufacturing problems they are intended to solve. There are no limits to the possible combinations of machines, workpiece-transport systems, and control systems.

3.1 Definition

The term *flexible manufacturing system* (FMS) refers to a group of numerically controlled machine tools that are connected by a common workpiece-transport system and a central control system. Various dissimilar (complementary) or similar (alternative) computer numerical control (CNC) machine tools perform all the necessary machining operations on workpieces from a group of related parts. The workpieces pass through the manufacturing process automatically. In other words, the sequence of machining operations is not interrupted by manual intervention, changeovers, or reclamping tasks. These systems thus make it possible to continue working through break times. At the end of a shift, operation can be tapered off with reduced personnel (*Figure 3.1*).

Highly automated systems even make it possible to incorporate material inventory, clamping devices, quality control, and tool management into the flow of machining operations and information. Assembly areas likewise can be integrated.

The use of numerically controlled machine tools makes it easy to continually adapt to changes in the design and machining operations as these changes are introduced into the system. Flexible manufacturing systems do not require minimum lot sizes but also can process single parts and small lots without stopping for changeovers. This requires suitable CNC programs, tools, and clamping devices. It is not necessary to combine lots to achieve larger quantities. This means that the amount of capital tied up in stock can be kept low, thus reducing inventory costs.

Flexible manufacturing systems are not limited to the machining of prismatic workpieces. They also can be used for turning, sheet-metal machining (*Figure 3.2*), or other methods. Aside from the various machine tools, this also requires a variety of transport systems. Individual or multiple prismatic workpieces are usually clamped and then transported on pallets. Turned parts are grouped in suitable bins holding relatively large quantities. Machining centers use pallet changers, but in turning applications, these are replaced by automatic handling devices, for example, robots or gantries. The handling device removes one part at a time from the bin or feeder mechanism and then carries it to the chuck. A double gripper exchanges the machined workpiece for an unmachined piece and deposits the machined part in a finished part bin.

The general **objective** for the use of flexible manufacturing systems is to

- Produce a variety of workpieces
- By a variety of machining methods
- In any desired sequence
- In lots of various sizes
- In a fully automatic manner, without manual intervention
- And in a profitable manner

3.2 Flexible Manufacturing Islands

The concept of the *flexible manufacturing island* is mentioned here for the sole purpose of correcting original errors in its definition. This term refers to a special form of shop organization that can be used with great flexibility. This concept has nothing to do with the flexible manufacturing systems discussed in the rest of this chapter!

Today, the term *flexible manufacturing island* is understood to mean a clearly delineated shop area with various conventional and CNC machine tools and other devices. This equipment is intended to perform all the necessary tasks on a limited selection of workpieces. An essential component of the flexible manufacturing island is the spatial and organizational grouping of the machines and production facilities that allow these parts to be machined as completely as possible. The people employed in such a shop area work very independently in formulating plans, making decisions, and

Figure 3.1: FFS 630 flexible manufacturing system for the machining of engine parts at DaimlerChrysler AG, Untertürkheim plant. It consists of two CWK 630 dynamic machining centers and two HEC 630 take-five centers with a linear pallet system on two levels and separate clamping stations. (Image courtesy of Heckert.)

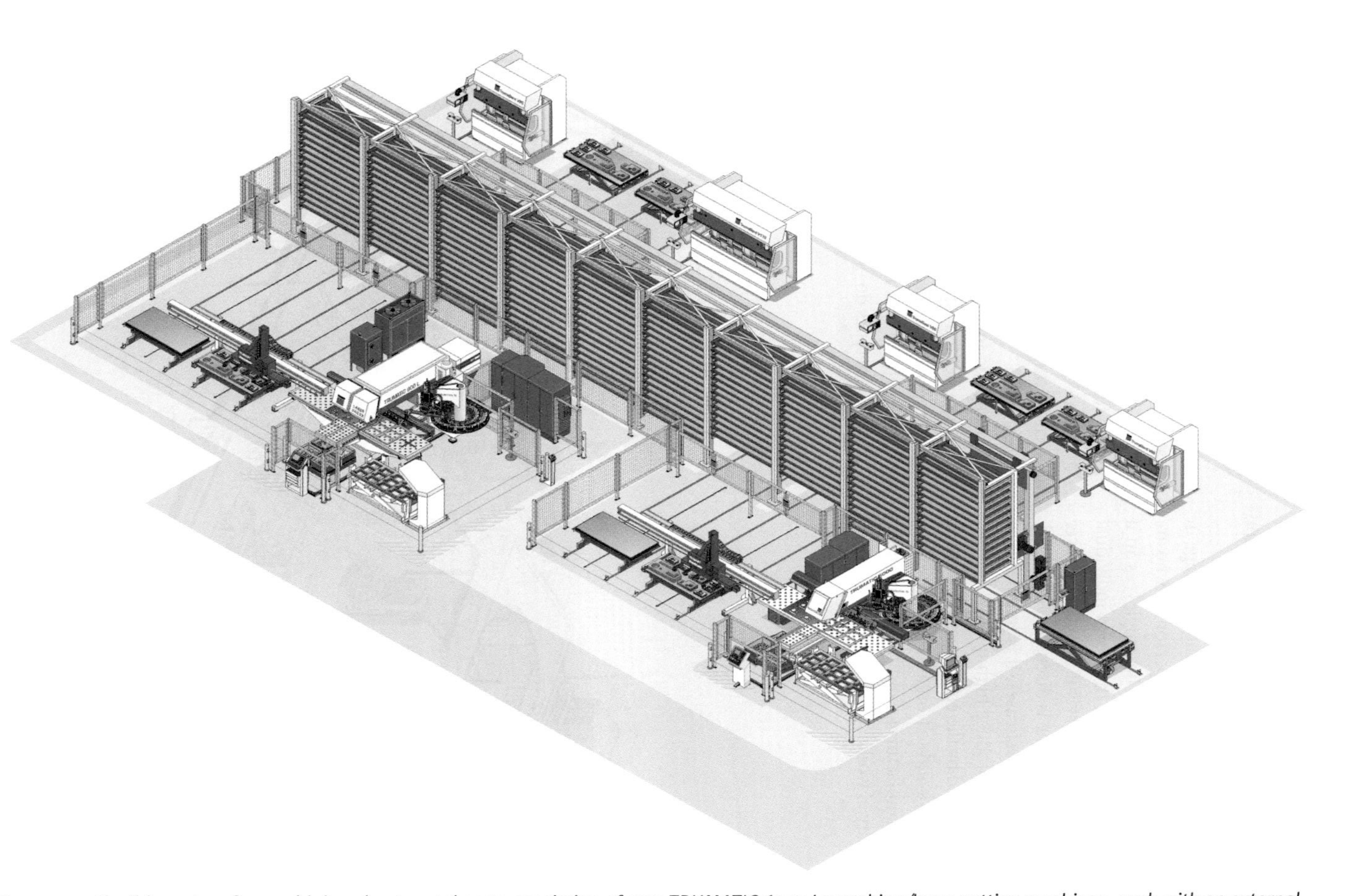

Figure 3.2: Flexible system for machining sheet-metal parts consisting of: two TRUMATIC 6000 L punching/laser-cutting machines, each with an external tool magazine; TRUMATOOL automatic loading and unloading system; TRUMALIFT SheetMaster; TRUMASORT sorting systems for finished workpieces; TRUMAGRIP residual skeleton-removal system; a centralized storage system; and a TRUMABEND press brake with linked transport vehicles. (Image courtesy of TRUMPF; www.trumpf.com.)

monitoring the tasks that are performed there. The flexible manufacturing island dispenses with rigid work assignments, thus expanding the range of availability and tasks for each individual worker.

Manufacturing islands can be used to great advantage in production processes that require flexible personnel who are capable of working wherever needed. The employees in a flexible manufacturing island organize their own individual operations, usually without any supervisor in authority over them. All work that needs to be done is discussed, managed, and assigned by the group. In most cases, it is important to meet specified deadlines and produce high quality. Flexible working time models arranged by the employees themselves are of benefit here. Teamwork and self-organization reinforce the concept of flexible manufacturing and help to improve motivation and the quality of work.

3.3 Flexible Manufacturing Cells

(Figures 3.3 through 3.5)

This term refers to a standalone CNC machine—usually a machining center, turning center, or other CNC machine—that is equipped with extra automation devices for unmanned operation of a limited duration. This requires the following features (→ *Figure 3.5*):

- An adequate **supply of parts** in the form of loaded pallets or single-part storage (for single-shift operation, for example)
- Automatic **loading** of the machine from the workpiece magazine and the return of each machined part to its appropriate storage location
- An expanded **tool magazine** so that frequent changes in workpiece types do not constantly require tool changes

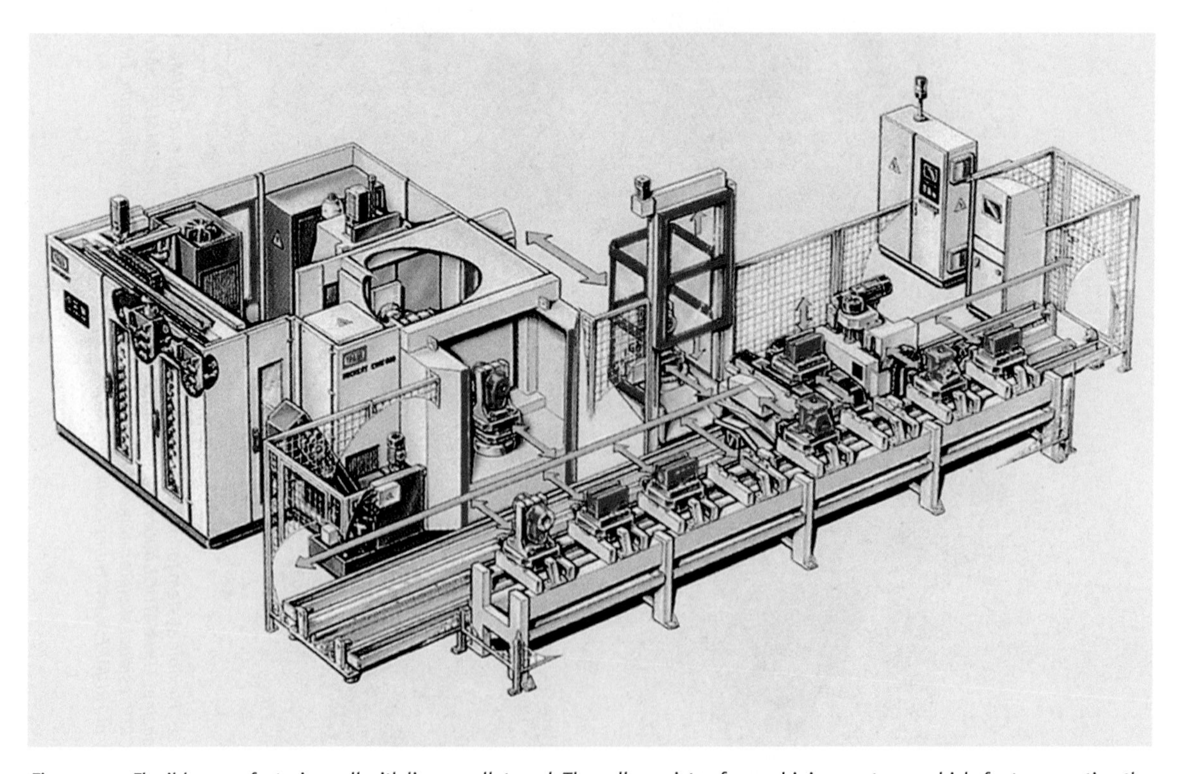

Figure 3.3: Flexible manufacturing cell with linear pallet pool. The cell consists of a machining center, a vehicle for transporting the pallets, multiple clamping and storage locations for the pallets, and a cell control system. (HECKERT.)

Figure 3.4: Flexible manufacturing cell with circular pallet pool for eight pallets and a pallet changer integrated into the machine. (Hüller Hille.)

- Automatic **tool changing** with **monitoring devices** to check for breakage and wear, as well as automatic retrieval of replacement tools
- **Dimensional monitoring** of the machined workpieces (e.g., by means of measuring probes and corresponding analysis software), making it possible to automatically readjust compensation values or automatically shut down the machine when tolerances are exceeded (breakage monitoring)
- Automatic **shutdown** of the machine after processing the entire supply of parts or when an error message is issued

If the third shift operates without human operators, the pallets for it are ordinarily loaded and unloaded manually during the first and second shifts.

The required **storage capacity of the workpiece magazine** is determined primarily by the machining time required for each workpiece. At an average machining time of 30 minutes, 16 pallets will be sufficient for eight hours of production. In some plants, the machines are not used to the fullest extent during one of the shifts. In our example, 10 pallets would be sufficient for an output of approximately 60 percent. This saves both pallets and pallet locations and helps to prevent tool breakage. The use of clamping systems with multiple fixtures increases the machining time of the pallets, thus reducing the number of pallet changes required.

When parts are fed individually, the machining time should not drop below three minutes so that approximately 160 parts may be prepared for an eight-hour shift.

Shorter machining times require large pallet pools or workpiece magazines and entail higher costs that jeopardize the profitability of the operation. In order to "get by" with a small number of similar clamping devices, it should be possible to machine various types of workpieces. This requires a larger amount of program memory in the CNC system. Direct numerical control (DNC) operation also

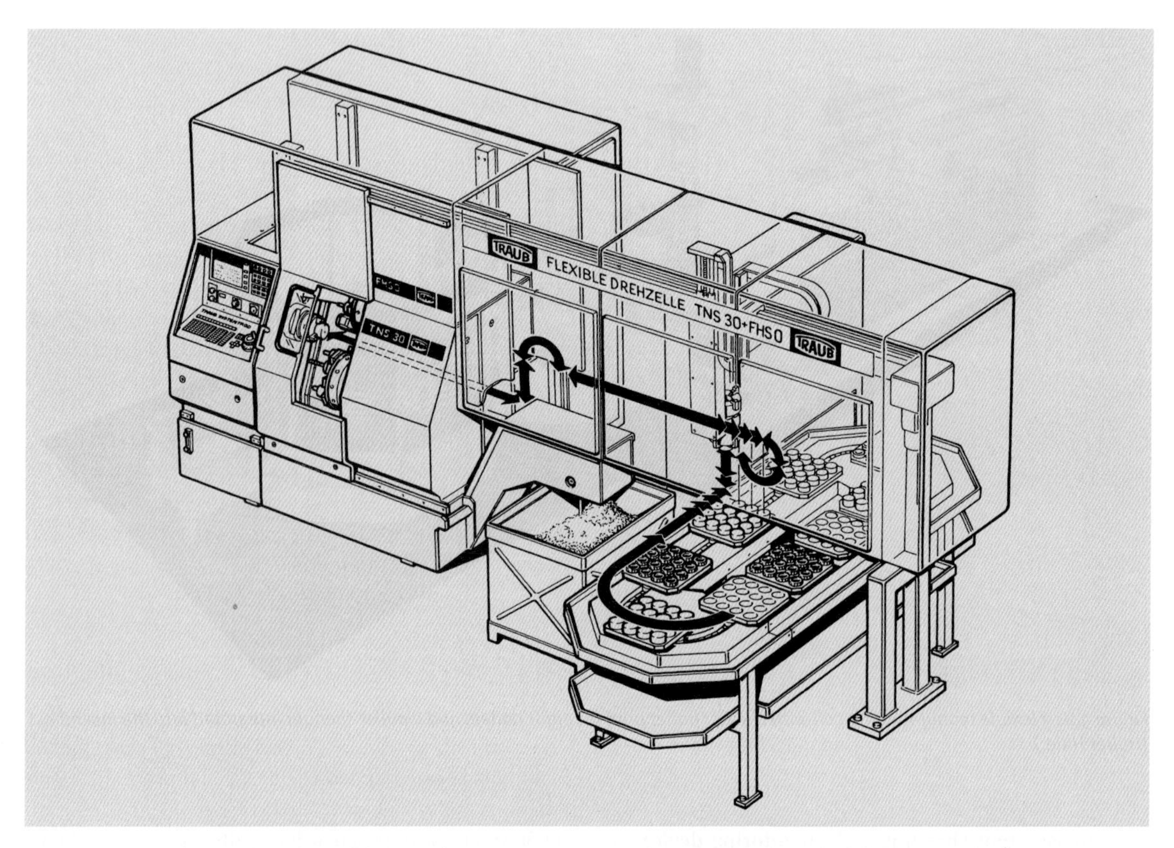

Figure 3.5: Flexible manufacturing cell for turned parts. A gantry-type robot assumes the task of changing workpieces. It takes the blanks from one pallet and deposits the finished workpieces on another pallet. (TRAUB.)

allows the central management of the CNC programs and prevents CNC programs with the same name but different contents from existing.

When machining is finished or interrupted, it may be necessary to modify the pallet coding accordingly and/or to issue a message to the **transport-control system** informing it that this pallet is now being conveyed to a different machine. This also can be accomplished by means of suitable management software in the CNC system, allowing the user to read the status (e.g., completed, interrupted, or unmachined) of each individual workpiece from the CNC screen on the following morning.

Occasionally, two machines of this type will be combined to form a machining unit called a *double cell.*

3.4 Technical Characteristics of Flexible Manufacturing Systems *(Figure 3.6)*

In order to satisfy the stringent demands as regards automation, flexible manufacturing systems must display the following technical characteristics for the machining of prismatic parts:

- Multiple **CNC machines that are suitable for use in flexible manufacturing systems**—usually flexible manufacturing cells—the size and quantity of which are appropriate for the quantity and range of parts being machined
- A sufficient **supply of workpieces** in order to ensure that the system can operate automatically and with a

Production features
Monocyclic production
Independent of lot size
Self-replacing machines
Self-complementing machines
Flexible automation

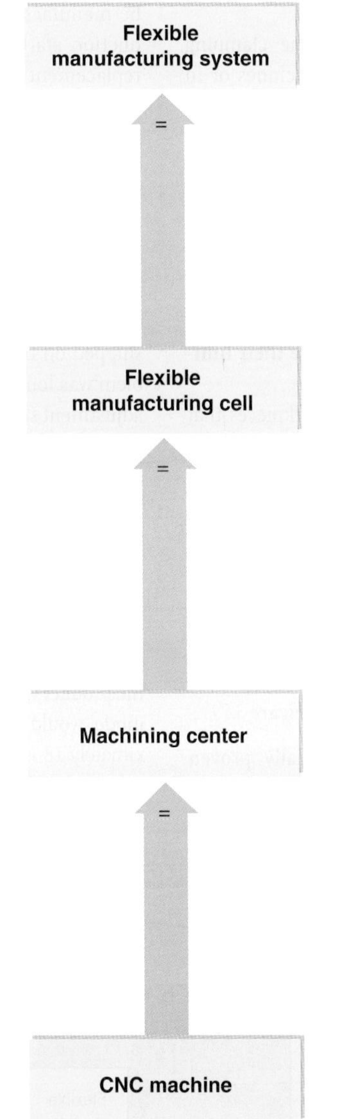

Technical features
+ Multiple-machine concept
+ Workpiece-transport system
+ Linking with respect to workpieces
+ Tool logistics
+ Automatic production control via host computer and DNC

Limited variety of parts
Medium lot sizes
Mixed production
Continued operation during breaks
Production tapers off in third shift
Shift operation requiring minimal operator intervention
Work changing without setup

+ Workpiece magazine or pallet pool
+ Expanded tool magazine
+ Loading and unloading systems
+ Connection to a computer
+ Monitoring devices
+ Integrated measuring system

Small to medium lot sizes
Multiple repetitions per year
Frequently changing variety of parts
Shop-floor organization

+ Automatic workpiece changing
+ Automatic tool changing
+ Multisided machining (fourth NC axis)
+ Expanded program memory
+ Automatic program call-up

Standalone machines
Single-part or series production
Manual workpiece changing
Operation requiring intensive operator intervention

Three NC axes
Manual tool changing
Manual program call-up
Punched-tape or simple DNC operation

Figure 3.6: Expansion in three stages from the NC machine to FMS.

minimum of personnel or without human operators for a limited time but in any case for as long as possible

- An automatic **system for conveying and changing workpieces** that manages and transports the parts from the time the unmachined parts are clamped until the machined parts are unclamped
- A DNC system for automatic **management and preparation of the CNC programs** and the correction values for tools and fixtures
- Automatic **tool supply** with management of all tool data and correction values from the presetting device to the machine and back again

- Automatic **chip disposal** for each machine
- Automatic **cleaning** of the workpiece, the clamping device, and the pallet in the processing machines or in separate washing machines, followed by forced-air drying
- **Host computers, measuring stations, central monitoring, manufacturing-data acquisition/production-data acquisition, and error diagnosis systems** can be installed as needed and specified.
- Parts machining with high process reliability

Despite their great technical sophistication, **currently installed flexible manufacturing systems have their limitations**, particularly in regard to

- Size, weight, shape, and material of the workpieces that can be machined
- Type of machining operation to be performed (three, four, or five sides; inclined surfaces or drilled holes; technologies)
- System output (quantity produced per hour)
- Type and quantity of the available special tools
- Accuracy of the machined workpieces

3.5 Application Criteria for Flexible Manufacturing Systems *(Figure 3.7)*

Flexible manufacturing systems have especially proven their worth in situations involving products that can only be manufactured in **small to medium quantities** at production startup and in the subsequent manufacturing of replacement parts.

Two **typical examples** are described here: A plant is manufacturing pneumatic cylinders in four different diameters. The customer can choose the length in millimeter increments ranging between certain minimum and maximum values. From the very start, it has proven too costly to maintain an inventory of all possible designs. It was necessary to develop a system that would allow production to fill orders within 24 hours of receipt so that the product could be shipped on the following day. An ideal solution to this problem was found: The cylinder body, piston rod, and tension-adjustment screws are now combined, manufactured, tested, and picked on an order-by-order basis. The pistons, covers, and nuts are provided as standard series parts at the assembly stations. In the meantime, the plant has successfully applied this principle to other products, too.

Another plant had to prepare a series production process for a new product that was still in development. Relatively large production quantities were expected immediately after the product was introduced on the market, but many variations of the product would be required. It also was known that the part model would change considerably during the period of approximately 18 months that it would take for the machines to be delivered. Manufacturing on single-purpose machines or

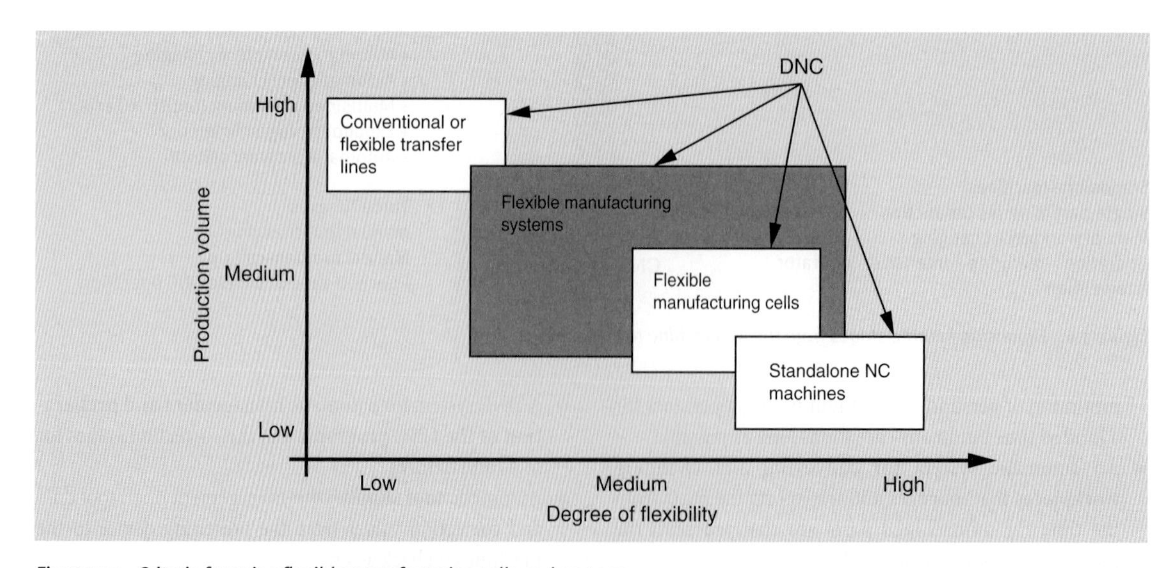

Figure 3.7: Criteria for using flexible manufacturing cells and systems.

transfer lines was ruled out from the beginning because it would have been necessary constantly adapting the machines to the changing shapes and machining methods. The solution was a flexible manufacturing system consisting of multiple standard machining centers. This allowed design modifications and variations to be accommodated without difficulty. Transfer lines and rotary index machines were installed to produce the required quantities, which increased rapidly after the introductory phase. Large-scale series production of the standard parts then was performed on these transfer lines and rotary index machines. The FMS continued to be used to its full potential for a number of years to produce special designs in small quantities.

This area of applicability of flexible manufacturing systems—ranging between single machines and transfer lines—can be seen clearly in *Figure 3.7*. This range has widened considerably after years of positive experiences with flexible manufacturing systems.

The **planning** of a flexible manufacturing system begins with an analysis of the parts that may need to be produced based on their sizes, weights, material composition, required quantities, lot sizes, and range of variations. This analysis will point to the necessary machining methods, the number of tools required, and the machining times. The type, number, and size of the required machine tools can be determined on this basis. The clamping method, the sequence of machining operations, and the number of CNC axes that will need to be controlled are determined at the same time.

Each flexible manufacturing system should be custom tailored for the user while making maximum possible use of standard components. Not just the parts currently being planned for production but also the user's future projections and production strategies should be taken into consideration.

The integration of existing CNC machines into a flexible manufacturing system is possible, but in the interest of a standardized, uncomplicated approach, it should not be treated as an absolute requirement. On the other hand, conventional, manually operated, or mechanically programmed machines cannot be integrated into an FMS. This is prevented by the lack of pallet-changing devices for such machines, as well as the rigid method of programming used with them, which does not provide for automatic modifications.

Under certain conditions, however, it may be practical to use special single-purpose NC machines, such as drill-head changers, face-milling machines, or special machining units.

Therefore, flexible manufacturing systems are not new machines but a combination of components that are already available: machining systems, automation systems, and information systems.

3.6 Manufacturing Principles

(Figures 3.8 and 3.9)

Various basic options exist for machining groups of related parts on CNC machines. These are

- **Complementary machining operations performed consecutively on multiple machines in multiple setups** (→ *Figure 3.8a*). This requires a certain amount of space between the machines for the purpose of intermediate storage of the semifinished parts. It also entails repeated clamping and unclamping of the parts, which impairs accuracy. There is also the risk that the breakdown of a single machine could bring the entire production operation to a halt.
- **Complementary machining operations, cycle-driven and performed consecutively on multiple CNC machines in a single setup** (→ *Figure 3.8b*). This is a possible solution when many tools are required. This solution also eliminates the need for pallet park stations or workpiece storage locations between the machines. Complementary machining ($A+B+C+D$) requires that the CNC programs be broken down into multiple individual programs with standard machining times in order to prevent idle time on individual machines. When a machine breaks down, however, there is the same risk as depicted in *Figure 3.8a*.
- **Final machining in a single setup on one or two machines.** This can involve complementary machining ($AB+CD$) or complete machining ($ABCD$). This makes it possible to achieve greater accuracy and reduce idle time. Without automatic workpiece transport, workpiece storage locations would be required at each machine (→ *Figure 3.8c*). This is the typical area of application of standalone CNC machines—usually machining centers, double cells, or turning cells. Unnecessary downtime can be prevented by providing automatic workpiece changers at each machine.
- **Final machining in an FMS with automatic workpiece transport** (→ *Figure 3.8d*). This is suitable for the simultaneous production of multiple (and possibly dissimilar)

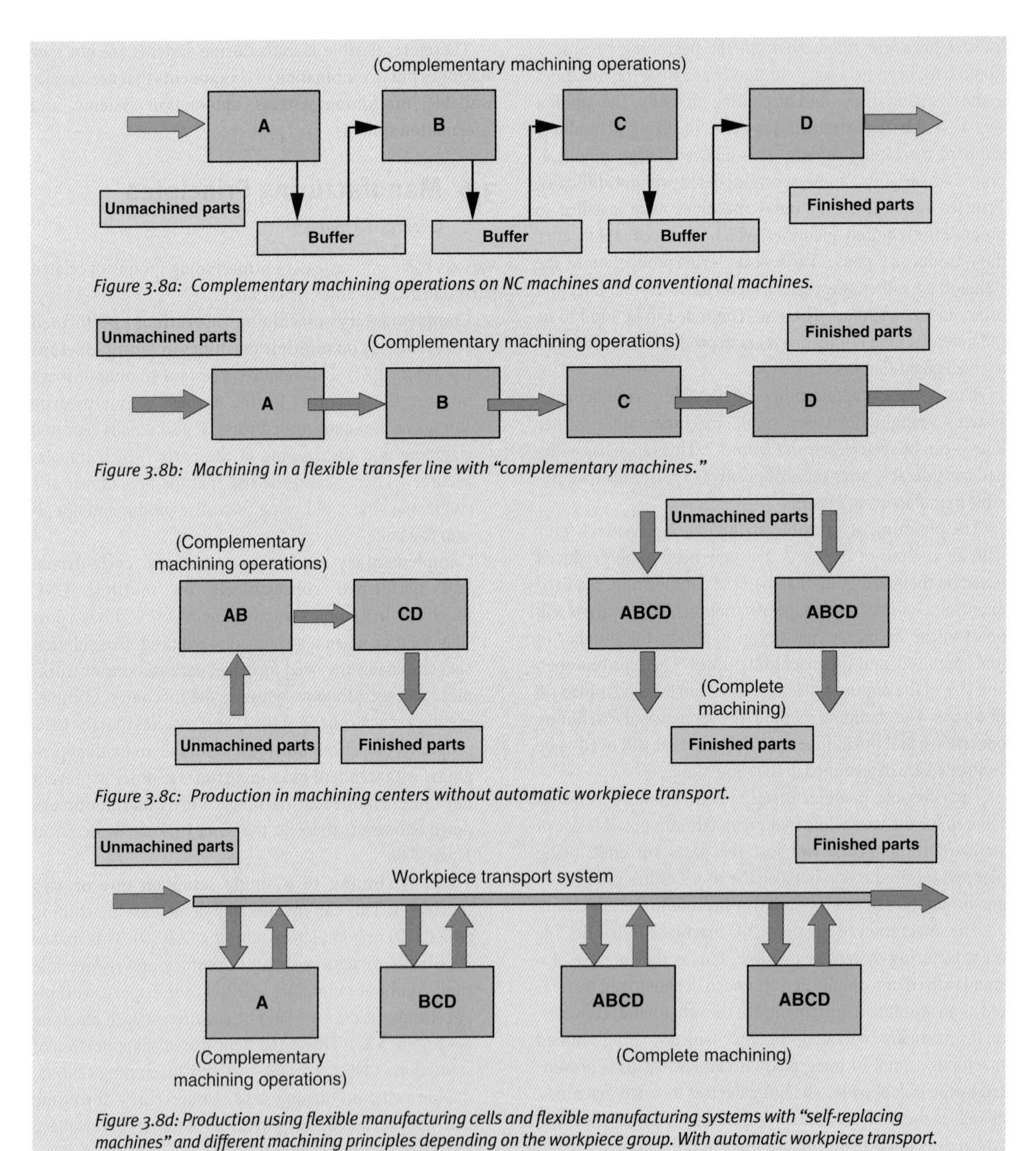

Figure 3.8a: Complementary machining operations on NC machines and conventional machines.

Figure 3.8b: Machining in a flexible transfer line with "complementary machines."

Figure 3.8c: Production in machining centers without automatic workpiece transport.

Figure 3.8d: Production using flexible manufacturing cells and flexible manufacturing systems with "self-replacing machines" and different machining principles depending on the workpiece group. With automatic workpiece transport.

Figure 3.8: Manufacturing principles with unlinked and linked machines.
The letters ABCD refer to the different machining operations that must be performed on the workpieces, such as milling, drilling, reaming, and tapping.

workpieces in any desired lot size. It is important for the appropriate CNC program, tools, and clamping devices to be present each time the workpiece is changed.

- The supply of workpieces is prepared for the machining operation and then provided at multiple shared pallet park stations for easy access. This approach is suitable for either complementary machining on two or more machines ($A + BCD$) or complete machining ($ABCD$).
- Most modern flexible manufacturing systems are based on the principle represented in *Figure 3.8d*.

3.7 Machine Selection and Layout

Machine tools are selected based on the size of the workpiece and the machining operations that must be performed. The FMS can comprise either universal machines (e.g., machining centers or flexible manufacturing cells), single-purpose machines (e.g., multiple-spindle drill-head changers or milling units), or special machines depending on the selected manufacturing principle (*Figure 3.8*). Sometimes it is also necessary to combine machines from different manufacturers. A prerequisite for this is that all the machines are able use the same tool holders, pallet changers, and table heights.

Another prerequisite is that all the machines are equipped with numerical controls that are suitable for use in flexible manufacturing systems.

An experienced manufacturer should be assigned the task of **planning and designing the overall system.** This manufacturer also will assign the orders to the subcontractors and ensure that standard interfaces are used. This keeps the responsibility for the proper long-term functioning of the overall system in a single set of experienced hands.

The general contractor also should be responsible for planning the washing machines, measuring machines, and reversible clamping devices that will be integrated into the system, as well as the transport system, including the pallets, the setup stations, and the clamping devices. An **FMS task group** drawn from within the buyer's own company should work together very closely with the individual manufacturers. This task group should detect planning errors in a timely fashion and should use a high-performance simulation system to carry out a very close examination of the operating sequence that eventually will be used. The procurement of a suitable CNC programming system and the programming of the parts to be manufactured also can begin at this time. The flexible manufacturing system should be able to start production immediately after it has been installed.

Based on the **experience gathered so far**, it seems advisable to use standardized machines and controllers to the greatest possible extent and to avoid ordering any special custom solutions from manufacturers. As more and more dissimilar machine types are integrated into the system, it becomes increasingly difficult to continue production when a machine breaks down. As a rule, the machines that are still functioning properly should be able to take over the work of the machine that is down without great difficulty so that production can continue (even if output may suffer somewhat). The principle of the **self-replacing machine** is extremely helpful in this case.

As far as the required degree of **flexibility** is concerned, no machine should be designed to be used exclusively with one special workpiece. Each machine in the flexible manufacturing system must offer universal (i.e., flexible) applicability, requiring only a change in tools and CNC programs. This is the only easy way to adapt manufacturing operations to changing market demands and design modifications. Any special machines that are present should not be allowed to turn into bottlenecks that are difficult to get around. This will also allow the system to be expanded more easily and affordably later on.

3.8 Workpiece-Transport Systems

(Figures 3.9 through 3.11)

The planning of a flexible manufacturing system usually begins with a basic determination of the most suitable workpiece-transport system. The number and types of machines to be installed in the FMS are determined, and then the machine layout and its connection to the transport system are planned in detail. In many cases, the amount of floor space that will be available for setting up the machines is already known. It is usually too small to allow much leeway for experimenting with different layouts. The workpiece-transport system then will be determined by this available floor space as well. The choices include

- **Linear, rail-guided systems** with one or two vehicles (→ *Figure 3.9*)
- **Roller conveyors or double-belt systems** with multiple pallets in circulation (→ *Figure 3.10*)
- **Floor-mounted systems** with multiple inductively guided vehicles (automated guided vehicles, abbreviated AGV) (→ *Figures 3.11 and 3.12*)

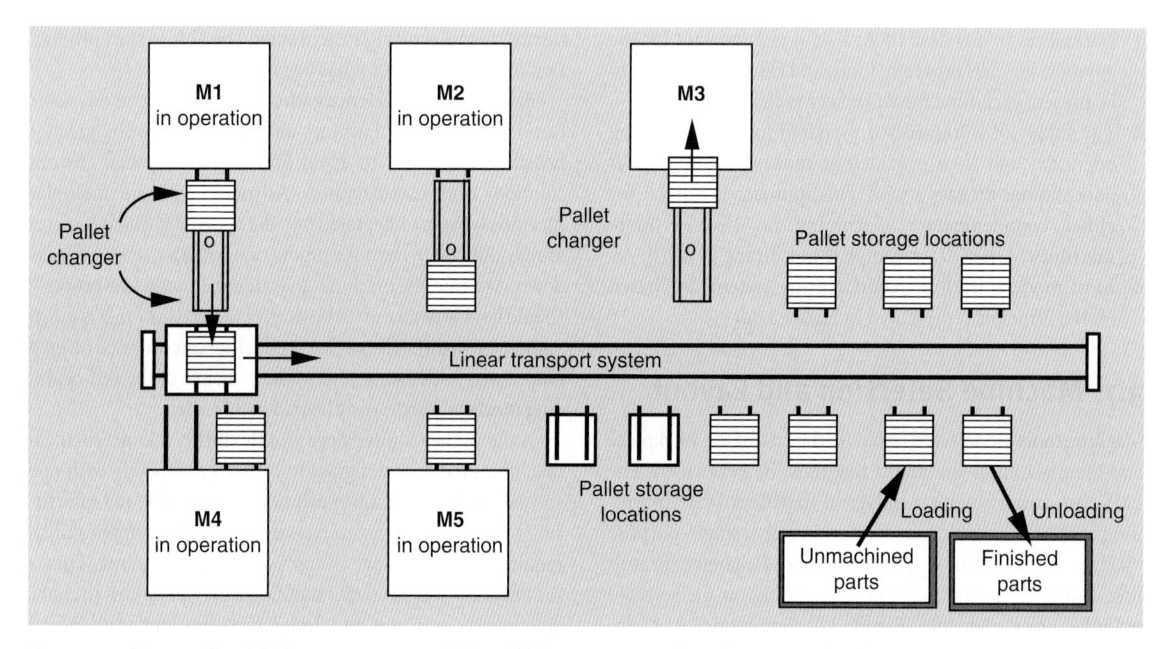

Figure 3.9: Linear, rail-guided transport system with parallel arrangement of machines on both sides of the conveyor track.
Characteristics: The vehicles feature one transport location.
If necessary, a vehicle can be equipped with two transport locations, or a second vehicle can be added, and the transport control can be modified accordingly.
The prepared pallets are parked at storage locations and picked up by the vehicle when needed.
Machines 1 to 3 are equipped with pallet changers, Machine 4 has one fixed loading station and one fixed unloading station.
Machine 5 is equipped with only one changing station.
This gives rise to various strategies for executing the changes.

The automatic conveying of workpieces from the clamping station to the machines (and back again) represents an important component of a flexible manufacturing system. This is usually accomplished with (standardized) pallets that assume the task of transporting the parts, as well as inserting and removing them at the machines. The clamping devices, which are fastened to the pallets, serve to hold the workpieces precisely in predetermined positions. The number of pallets in an FMS is limited by the number of pallet-transport locations and part stations plus the number of machining stations. It is possible to make use of the third dimension with multilevel shelf systems and suitable storage and retrieval machines. Because shelf magazines also can be constructed with two sides, the number of storage locations in the shelf magazine can be very large.

The **clamping and unclamping** of the parts in the fixtures is usually performed manually. Relatively small parts can be clamped in multiple fixtures in order to reduce the overall machining time. This method also can be used to combine the first and second setups of the parts in a single fixture should this prove advantageous. Hydraulic clamping fixtures are also seeing increased use, for example, to standardize the clamping operations and keep them consistent.

A coding device on the pallet or the fixture allows the **recognition and identification** of the workpiece that is currently clamped on a particular pallet. This device is "read" electronically before processing begins on the machine, and the code is compared with the CNC program that has been prepared for the machine. The "go-ahead" for machining is contingent on the results of this comparison. This device generally can be omitted because the transport control reliably can assume the tasks of management, error-free conveying, and identification at the machines.

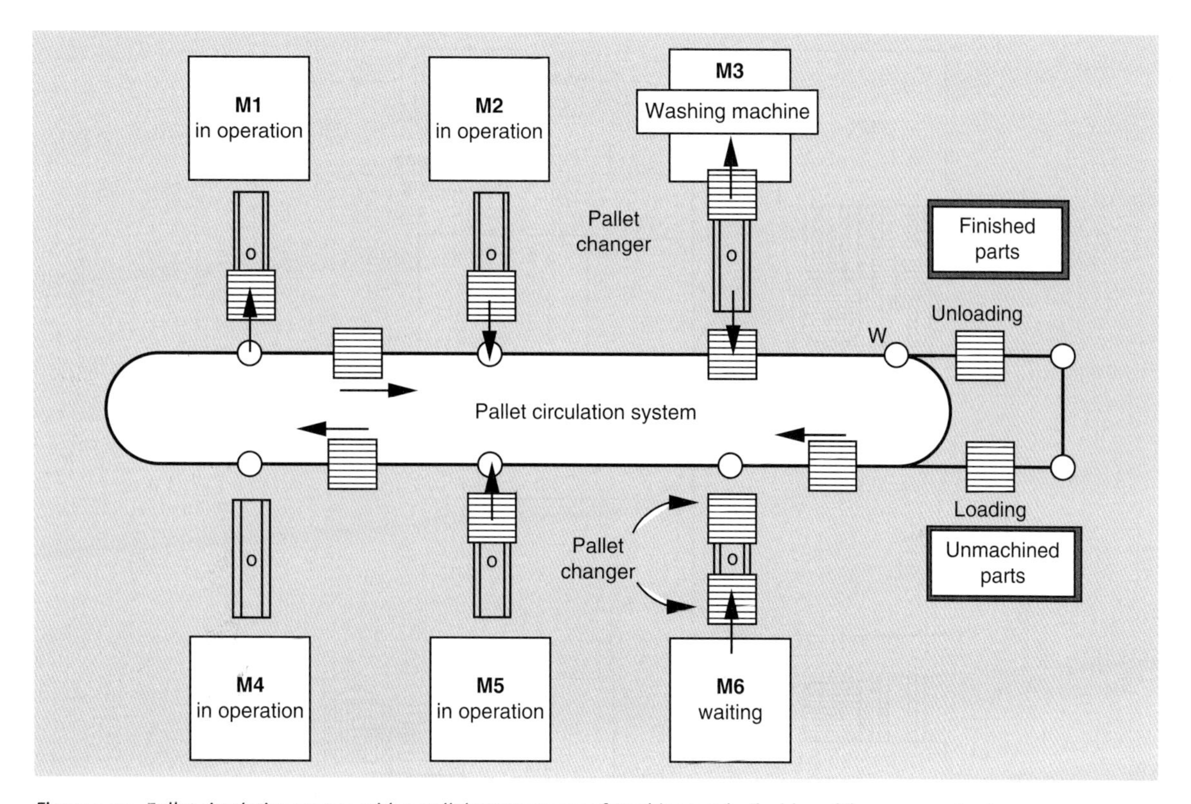

Figure 3.10: Pallet circulation system with parallel arrangement of machines on both sides of the conveyor track.
Characteristics: No additional pallet-storage locations are provided. The unmachined workpieces circulate in the transport system until the loading location of a pallet changer is free.
The pallet changers at the machines also serve as loading and unloading buffers, holding one pallet that is ready for machining and one pallet that has already been machined, respectively. After exiting the machines, the machined pallets are automatically brought to the washing machine and then to the loading and unloading stations on branch W.

From both technical and economic standpoints, the transport system represents an essential part of the overall system. It is therefore very important to devote sufficient attention to planning the sequence of functions so that unnecessary corrections, costs, and travel can be avoided.

Transport systems must satisfy **stringent requirements,** such as the following:

- Additional machines.
- Maintenance and repairs should cause as little down time as possible.
- Ideally, this system should be capable of transporting the exchanged tools to and from the machines, eliminating the need for any additional tool-transport system.

Setup stations, loading stations, and unloading stations are among the elementary components of an FMS. Ergonomic solutions take the operators' needs into account and make their work easier. The pallets/fixtures/workpieces also have to be easily accessible, for example, by means of automatic lowering, tilting, and rotating. Access to the workpieces via platforms or ladders represents an accident risk and should be avoided.

Selection of the Transport System

Various transport systems are available. The proper selection depends on the workpiece dimensions and weights, as well as the machine layout.

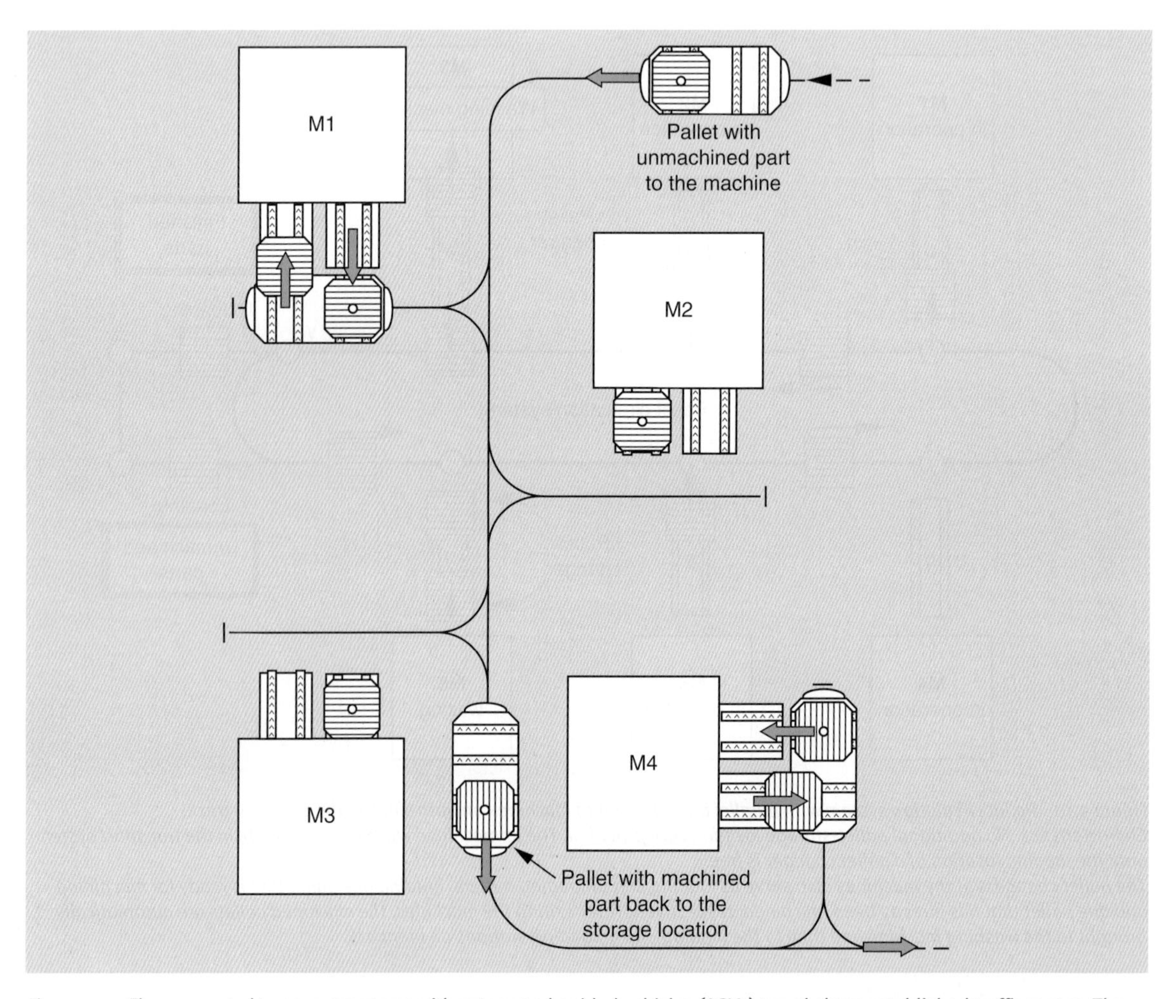

Figure 3.11: Floor-mounted transport systems with automated guided vehicles (AGVs) travel along established traffic routes. The vehicles are radio controlled with a wire embedded in the floor.
Characteristics: The machines can be distributed across the available floor space in any desired manner. The vehicles travel in "one-way traffic" from the material storage location to the machines and back. There they are unloaded and then loaded with new workpieces. The speed of travel is restricted for reasons of safety.

Linear, rail-guided transport systems are used most commonly (→ *Figures 3.9 and 3.13*). These systems save space and offer a high speed of travel. Each of their vehicles can be equipped with one or two pallet locations, as desired. They also make it possible to arrange the machines on either one side or both sides of the track. The track can be extended easily for the purpose of expanding the system at a later time. The control logistics are simple and trouble-free.

Pallet-circulation systems using **double-belt conveyor lines** are familiar from the field of automatic assembly technology. These systems are basically suitable for this application as well, but only for workpieces that are not too large or heavy. In a system of this type, the transport pallets carry workpieces that are clamped on fixtures and then transfer these workpieces to the machines. The pallets wait outside the machine, then receive the machined workpieces, and carry them onward. In processing that involves multiple machines, the pallets sometimes contain part-specific destination codes indicating that the stations that still have to be visited. In this situation, each workpiece must be returned to the pallet on which it was originally carried. Where large pallets are involved, expensive transfer stations are required at the machines.

In mass-produced **roller conveyor** systems, the pallets are transported on powered roller tracks by means of friction. Multiple track segments can be assembled to give the overall track the desired form. The conveyor height and track width can be adjusted as needed. The load-carrying capacity can be as high as 750 kg/m. Conveyor speeds can be selected from a range of about 1 to 12 m/min. The drive system consists of a geared electric motor. Such systems are very seldom used.

Roller conveyors using roller chains are similar. In this case, roller chains move the pallets. Diverting or pivoting devices change the transport direction at the corners of the track.

However, **underfloor drag-chain conveyors** (which are familiar on assembly lines) are now obsolete owing to the excessive cost of installing and repairing them and their great susceptibility to breakdown.

Driverless carts (aka automated guided vehicles [AGVs]) of various sizes have proven their value in applications involving relatively large pallets or heavy workpieces (→ *Figure 3.12*).

Figure 3.12: Automated guided vehicle for one pallet in an FMS.

Operating under radio control, these vehicles follow a wire embedded in the floor. Rack-switching features and queue stations also can be built into these systems. The vehicles travel along existing traffic routes and can reach almost any desired location in the shop area. Existing machines can remain in place on their expensive foundations, and relatively remote material- and tool-storage facilities also can be integrated into the material flow. For reasons of safety, the speed of these vehicles is considerably slower than the speeds attained in linear, rail-guided transport systems. For this reason, it is usually necessary to have several vehicles underway at the same time.

In turning cells, parts are usually inserted and removed with the help of **handling units, gantry-type robots, or floor-mounted robots.** Another alternative is the use of machines that feature a vertical design and do not require separate handling units.

Sequence of Transport-System Functions

The core of a flexible manufacturing system is a transport system that functions intelligently, thus providing transport control. In contrast to systems involving circulating pallets, flexible manufacturing systems that are linear, rail-guided, and equipped with carts require central control. The selected transport strategy must be adapted to the conditions that currently exist in the system. In other words, the logistics change accordingly as the task at hand changes from startup to normal operation to changeover and finally to operation under reduced loads (as production tapers off).

Let us first examine the sequence of functions in the normal operation of a flexible manufacturing system (*Figure 3.13*). Several different workpieces are ready to be picked up from the pallet-park stations. The machines are working, and the transport vehicle is positioned to wait for its next task. There is one machine operator stationed at each of the two clamping stations. This operator will remove the machined workpieces from the clamping devices and then perform the clamping of new, unmachined parts.

All machines include two pallet-transfer stations. The station on the left is intended to receive the unmachined workpiece as it arrives. The one on the right holds the machined workpiece that is ready to be picked up. Alternatively, revolving pallet changers are used. Each machine in the FMS can be programmed with the CNC program for the machining of all the workpieces in the system and is also equipped with the tools necessary for machining them. A washing machine (M4) and a measuring machine (M5) are also integrated into the system.

The automatic sequence is performed according to the following **strategy:**

1. Machine 2 has finished its machining operation and has parked the pallet at the changing station on the right. This machine also has taken the pallet that was waiting at the station on the left so that it can begin machining this workpiece.
2. Two messages are issued to the transport control:
   a. Bring a new pallet.
   b. Pick up the finished pallet.
3. The transport control "knows"
   a. Which workpiece has to be conveyed to M2 (this was entered and stored before machining began).
   b. At which park station this workpiece can be found (the transport control has already parked this pallet there and saved the station number).
4. The transport vehicle first travels at the appropriate park station, where it docks and receives the pallet.
5. The vehicle travels to M2 and transfers the pallet to the unoccupied changing station on the left.
6. The vehicle then takes the machined pallet from the changing station on the right and carries it to the washing machine. All workpieces must be washed and dried at this location. If the loading station for M4 is occupied, the vehicle will temporarily park the pallet at an available storage location.
7. At M4, the vehicle takes a clean pallet from the changing station on the right (if one is available) and carries it to the measuring machine. If a measured workpiece is present, there at the unloading station the vehicle will transport it to an open clamping station. There it will be manually unclamped and deposited with the finished parts, provided that the measuring machine has issued an "Okay" message. After an unmachined part has been clamped in place, the operator presses a button or key indicating that it is ready to be picked up.
8. As soon as the vehicle is free, it picks up the pallet and parks it at an unoccupied storage location.

These operations are repeated continuously in a chaotic sequence. Each motion is recorded in a table within the control system. The table is preserved in the event of a power

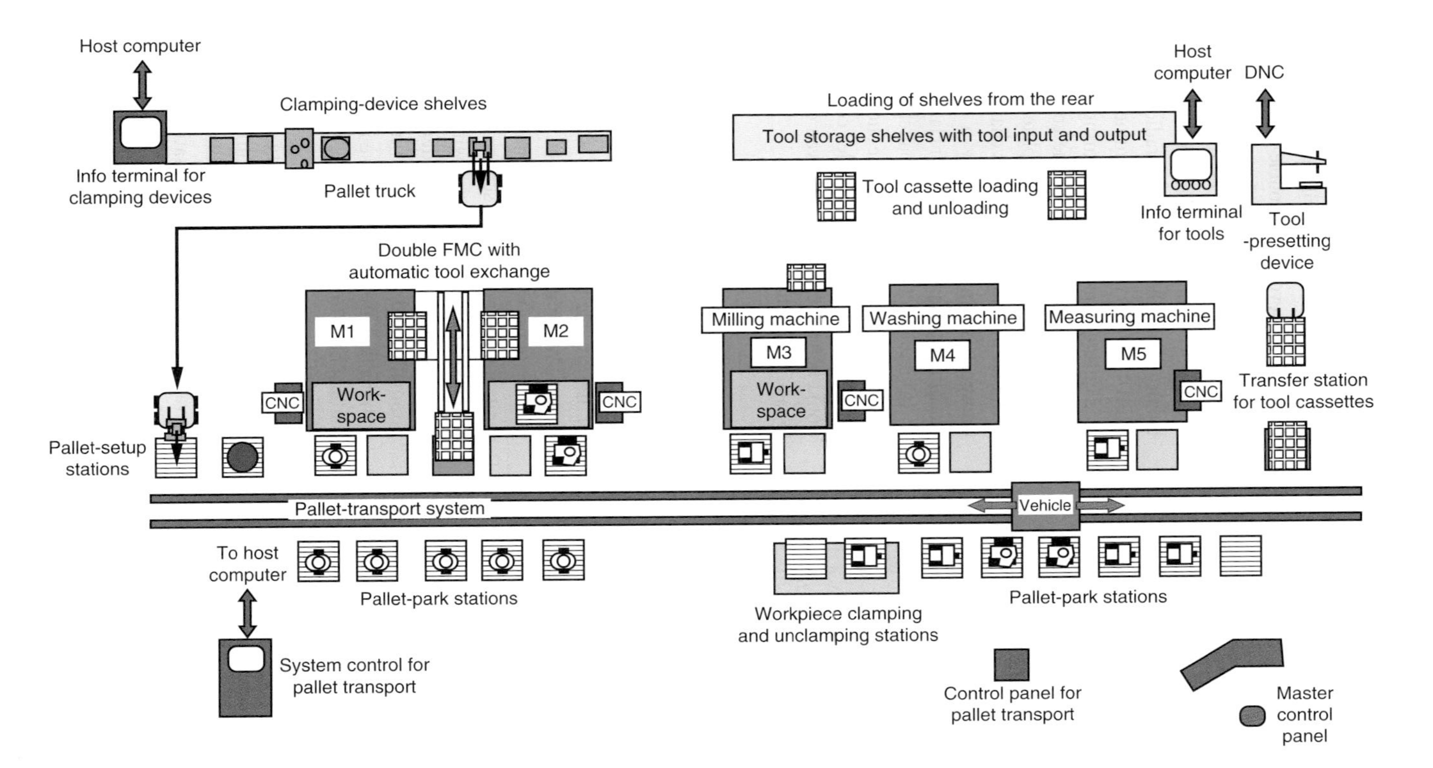

Figure 3.13: Typical layout of a flexible manufacturing system.
Characteristics: The machines are arranged on one side of the transport system, allowing easy access for maintenance and repair work.
The pallet-park stations are located on the other side of the track so that only short distances need to be traveled when the machine. Assignments are optimized.
The flexible double cell is supplied with tool-changing cassettes. A gantry-type robot performs tool handling at both machines (M1 and M2).

failure, allowing a trouble-free restart when power is restored. Pallets that have been parked temporarily are automatically fetched by the vehicle as soon as the loading station at the destination machine is free.

So much for a normal, trouble-free process. In actual practice, the transport-control system is constantly receiving new signals from the individual stations, indicating that workpieces should be picked up from them or delivered to them. These orders are stored, prioritized, and executed according to a predetermined strategy and the programmed priority. Without a doubt, the first priority is the operation of the machines. In other words, new workpieces must be delivered to the machines, and their unloading stations must be emptied.

If a finished workpiece is ready to be picked up at an unloading station and the vehicle is free, then this particular order can be executed (even if there is currently no workpiece to bring to the loading station). When we add to this the orders for tool exchange and pallet changeover, we discover additional logistical tasks. The pallets that require changeover must be removed from circulation. Pallets with replacement tools must be brought to the changing stations that are provided for this purpose and then picked up from these stations again, etc.

When a single machine or the entire system is being emptied, the logistics change again: Workpieces are picked up from the machines but no longer brought to them. However, the washing and measuring machines must continue to operate.

As the **operation of the system is winding down** and unmachined parts are no longer available for a given machine, the machine is shut down. For example, the machine may *time out* after failing to find a pallet at the loading station for a specified period of time. Once all the workpieces have been machined, washed, and measured, the entire system shuts down.

From this it is clear that when working with relatively large systems, it is absolutely essential to perform simulations at an early stage. This will prevent unpleasant surprises and costly modifications later on. Strategies then are modified, conveyor speeds are raised to the limit, and other remedial measures are identified. **Vehicles with two storage locations** reduce idle time because they allow two pallets to be exchanged simultaneously in a single docking and positioning operation. Another alternative is to employ a second vehicle and then coordinate the travel commands issued to the two vehicles so that collisions are avoided.

Transport-System Control

The control system is matched to the transport system. Completely different control principles therefore must be applied to roller conveyors and double-belt conveyors than to rail-guided or AGV systems. With circulating pallets, the conveyor track receives the workpieces and must automatically bring them to one or more specified machines. This is accomplished by means of pallet codes using radiofrequency identification (RFID) systems that are programmed accordingly. These are read during transport, thus ensuring that each freely circulating pallet is transported to the correct individual station. When machining has been completed and the pallet is exiting each machine, the pallet code is changed to indicate the next transport destination (possibly a machine tool, washing machine, measuring machine, or clamping station).

In circulation systems with pallet stowage space and multiple clamping cubes, experience shows that a transport control without unambiguous pallet coding and management would be very expensive and even overstrained with regard to safety and reliability.

3.9 CNC Systems Suitable for Flexible Manufacturing

High-performance CNC systems are an essential requirement for the trouble-free operation of a flexible manufacturing system. While it is true that the first flexible manufacturing systems were installed before CNC systems were available, the power and scope of function of these early systems cannot be compared with the stringent demands that must be satisfied by a modern FMS. In order to allow automatic and sometimes unmanned operation, CNC systems that are suitable for use in flexible manufacturing systems have to offer various special functions. Examples of such functions include

- A large **data-storage** capacity, allowing temporary independence from the DNC computer. A large amount of storage capacity is also required for tool compensation values, zero-point compensation values, compensation for differences in clamping adjustments, and tool data management. However, a large storage capacity is no problem with today's control technology.
- Powerful **management** of the stored data so that the entire set of data can be updated, displayed, checked, and

corrected at any time. It should be possible to perform these actions at either the CNC system or a central host computer.

- If **multiple CNC programs** are stored in the CNC system, each program must be capable of being retrieved individually by an external command and started automatically when operation is ready to begin.
- Uninterrupted operation requires **tool management within the CNC system** for replacement tools and alternate tools, including tool-life monitoring and automatic assignment of the respective compensation values.
- There must be a provision for assigning **various compensation values** to each tool depending on the program. This will make it possible to use the tolerance range that is specified for a given machining operation.
- There must be provisions for inspecting the machined workpieces by means of **measuring probes** and the special measuring programs that are stored in the CNC system. An "Okay" or "Not okay" signal must be issued, as determined by the results of the measurement and contained in the measurement records.
- Another essential requirement is a powerful **data interface** (e.g., Ethernet) that provides a connection to the DNC system and allows bidirectional data transfer between the DNC computer and the CNC system.
- **Pallet-management** or **workpiece-management** functions must be included in the transport control system so that processing priorities (i.e., the sequence of pallet retrieval) can be defined. A status display must be provided to indicate the general status of all workpieces (machined/unmachined/machining interrupted/etc.).
- Automatic **manufacturing data acquisition** and **production data acquisition** (MDA/PDA) within the CNC system involves the recording of all problems that occur during the unmanned shift, reporting of interrupted machining operations, and the constant provision of information regarding the static utilization rate of the machine.
- There must be provisions for activating the reduced feed rates that are required for nighttime operations at **reduced stock-removal rates.**
- When individual components of the FMS break down, **emergency strategies** must be in place so that emergency operations can be maintained with the system components that remain intact.
- In the event of a problem, a message is automatically sent via SMS to a defined mobile telephone and as an e-mail to the user.

3.10 Host Computers in Flexible Manufacturing Systems

(Figure 3.14)

The task of the host computer is higher-level control and monitoring of the individual components in an FMS. Today, an industrial desktop PC or laptop with a standard operating system and additional software for the specific host-computer functions is an adequate solution.

Depending on their scope and the range of tasks involved, FMS process-control functions can include the following tasks: visualization of the overall system, controlling background processes, updating the databases, fault messages, and DNC functions. Other functions include tool management with a link to tool presetting and integration into the PPS.

Data networking takes place via field bus systems and the input/output (I/O) level of the PC networks by means of Ethernet and transmission protocols via TCP/IP.

In principle, it is possible to design an FMS to work without a host computer. This essentially applies to automatic manufacturing processes, that is, pallet delivery, machining, and pallet parking. This can be accomplished using automatic workpiece transport control. All other organizational tasks connected with the FMS then have to be scheduled, monitored, and executed promptly by manual means so that waiting times are not caused by missing workpieces, tools, clamping devices, or other potential sources of problems.

This essentially defines the important **coordination tasks** that are transferred to a higher-level host computer. These are determined by the system's layout and degree of automation. They may contain the following functions, for example (→ *Figures 3.13 and 3.14*):

- Receipt of **production orders** (including production quantities and deadlines) from the production-planning system and monitoring of deadlines by means of feedback messages
- **Machine assignments** in normal operation, taking into account the current tooling; that is, when jobs change, which machines would require the fewest tools to be replaced?

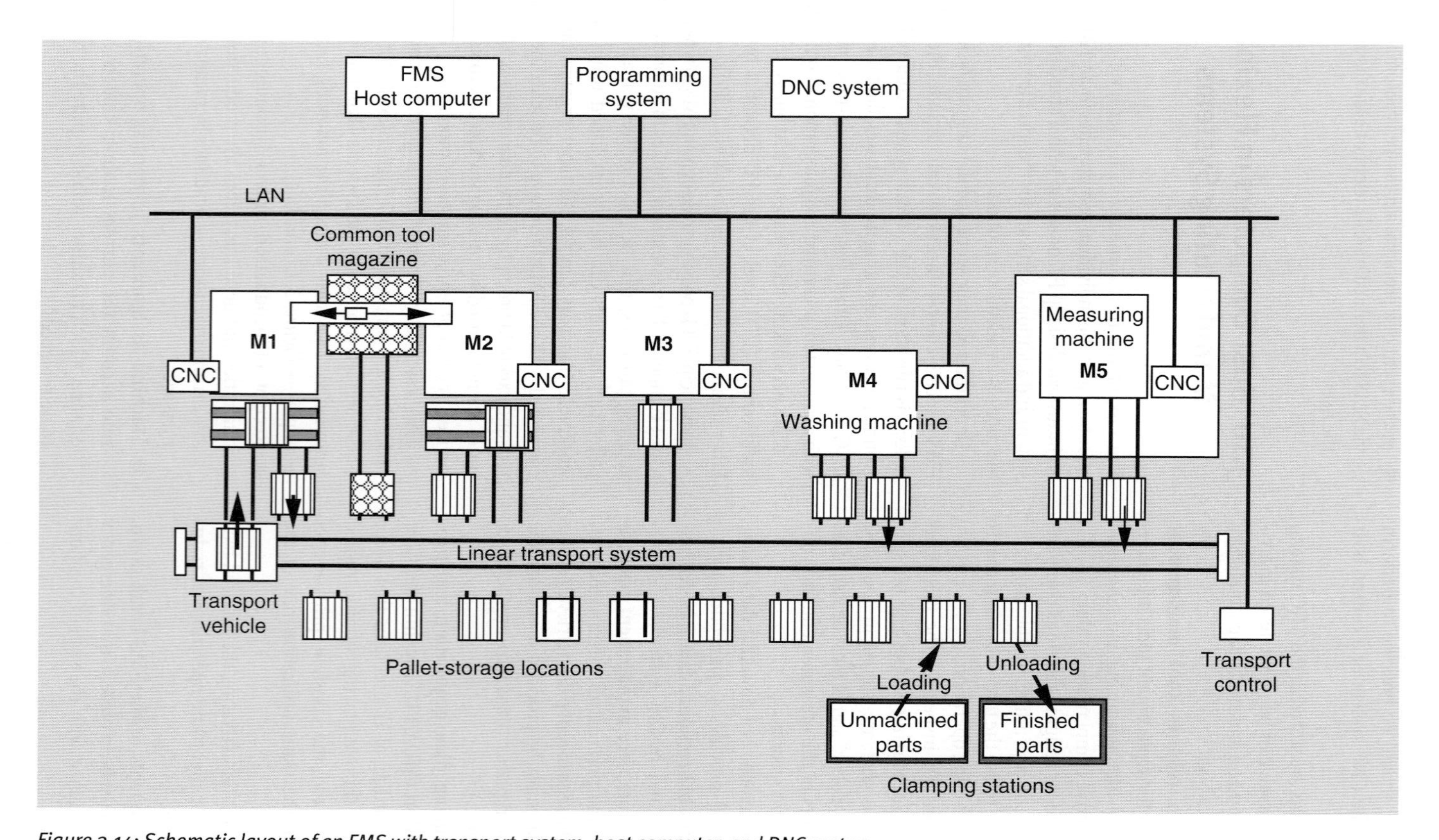

Figure 3.14: Schematic layout of an FMS with transport system, host computer, and DNC system.
There is free data access from each machine to each computer to allow retrieval of CNC programs, setup drawings, assignment plans, machining notes, tool exchange lists, status reports, advance planning, etc.
The transport-control system simultaneously serves as a central system controller. The above-mentioned types of information can be retrieved and displayed here as well.

- Machine assignments for **rush orders** but without complete removal of the currently running orders from production
- Preparation of **pallets** for changeovers to different clamping devices (e.g., quantity, deadlines); preparation of clamping devices
- Preparation of **unmachined parts** (after the production-planning system has already determined that enough unmachined parts are available)
- Retrieval of the necessary **tools** (i.e., custom, special, or series-produced tools), including **tool data** and their availability at the machines
- Issuing of a **tool-difference list** as output to the shop, indicating the tool exchanges that must be made for each machine
- Instructions to the DNC computer to make the appropriate **CNC programs** ready for retrieval
- Messages to the **transport-control system** regarding the assignment of workpieces to the individual machines, identification codes, etc.
- Preparation of the associated **measuring programs** for a measuring machine built into the system or for the machine's internal inspection function
- Information provided to **shop personnel** regarding changes in production and the necessary preparations, status messages, and alternative strategies to be adopted in the event of machine breakdowns, deadlines, production quantities, etc.
- **Visualization of the operating sequence** on the host computer with advance warning of foreseeable problems, etc.

At this point, it is usually possible to begin fully automatic production of the new workpieces.

When a machine temporarily breaks down, the host computer adopts a prepared failure strategy to divert the workpieces to other machines (provided that this is compatible with the production planning system and possible from a technical standpoint). The decision then rests with the personnel.

Another task is the **central monitoring of the system** and its status via analysis of **manufacturing data-acquisition and production data-acquisition** data. The following **feedback messages** are important to this monitoring activity:

- Machine ready for operation/running/waiting/malfunctioning
- Machine under repairs/time-related information/reason
- Machining enabled/suspended
- Pallet missing/in transit/ready/being machined
- Number of parts produced per lot
- Scrapped parts/reason
- Transport system ready for operation/malfunctioning

These and other types of data provide the information needed for informative status reports and valuable statistics from which tendencies and weaknesses also can be clearly derived.

3.11 Economic Advantages of Flexible Manufacturing Systems *(Table 3.1)*

Flexible manufacturing systems offer certain economic advantages over alternative methods of manufacturing. In single-part or small-scale series production these advantages result from exploitation of various system properties.

1. **More complete utilization of the means of production with respect to both time and technology.** This is accomplished by means of greater automation, by continuing operation through break times, and by tapering off operations in the second or third shifts, shifts with a minimum of personnel or without human operators or on weekends.
2. **Increased productivity** through fast, interrupted adjustments to changing production tasks.
3. **Reduced need for production area** owing to elimination of intermediate storage areas and working surfaces at the machines, for example, use of high-bay warehouses.
4. The ability to modify the CNC programs and thus **adapt** to geometric modifications of the workpieces and technological changes in the machining process.
5. **Fast response to market changes** by means of flexible priority changes in production.
6. The **ability to expand and adapt** to new tasks or greater production quantities at a later time.

Individual, unlinked CNC machines and flexible manufacturing cells are used primarily to produce small or medium quantities of workpieces. However, **lots ranging from extremely small to medium size can be produced more profitably by linking multiple CNC machines.**

Program Runtimes Minus Downtimes	Formula	Hours/year	Days Remaining	% of hours
Theoretical Utilization Time	**365 days × 24 hours**	**8,760**	**365**	**100%**
- Saturdays and Sundays	52 weeks × 2 days × 24 hours	−2,496	261	−28%
- Holidays	8 days × 24 hours	−192	253	−3%
Third shift on 253 days	253 days × 8 hours	−2,024	253	−23%
- Absenteeism (average)	52 weeks × 1.5 hours	−78	253	−1%
- Organizational problems	253 days × 1.5 hours	−380	253	−4%
- Changing workpieces and orders	253 days × 4 × 0.5 hour	−506	253	−6%
- Replacement of worn tools	20 tools/day × 2.5 min/tool × 253 days	−211		
Total Downtime		**−5,888**		**−67%**
Available Program Runtime:	**8,760-5,886**	**2,874**	**253**	**33%**
Gain in Utilization in FMS				
+ 6 hours tapering off in third shift	253 days × 6 hours	1,518	253	17%
+ Continued operation during break times	1 hours/shift × 2 ×253	506	253	6%
+ 50% less absenteeism	4	0	253	0
+ 60% fewer organizational problems	380 hours × 60 %	228	253	3%
+ No interruption when changing orders		506	253	6%
+ No interruption when changing tools		211	253	2%
Gain in Utilization:		**+3,009**		**+34%**
Program Runtime Per Year:		**5,883**	**253**	**67%**

Table 3.1: Loss of Utilization Without Automation and Gain in Utilization Through Flexible Automation (Theoretical Numeric Values); Saturday Operations Would Bring an Additional 14 Percent Gain in Utilization Time

This is accomplished primarily by avoiding machine waiting times and using the extra time that is freed up through the use of an improved, forward-looking organization.

If clamping devices and tools are prepared first and the CNC programs are loaded into the CNC machines, these machines can be used to a considerably greater degree in an FMS than as individual, unlinked CNC machines.

There is no generally accepted scale for evaluating the **degree of flexibility.** In other words, it is impossible to measure flexibility in absolute terms. It is only possible to compare the numbers resulting from different approaches with the same manufacturing task.

Examples of these numeric measures include

- Reduction of the throughput times for the parts
- Reduction of the inventory of semifinished parts
- Reduction of the inventory of finished parts by gearing production to specific orders
- Reduction of changeover costs, expressed as the time required to change over to a new manufacturing task
- The number of possible changeovers per hour without machine down time
- The ratio of productive time to changeover time
- The range of finished parts to be produced by the FMS

- More complete utilization of the machines through the use of auxiliary automation equipment

After gaining some initial experience in operating a new flexible manufacturing system, it is often possible to further optimize the system on the basis of the employees' suggestions.

3.12 Problems and Risks in the Design of Flexible Manufacturing Systems *(Figures 3.15 through 3.17)*

The number of options provided by flexible manufacturing systems—and the problems encountered in implementing these systems—are undoubtedly too great to be described completely in this book. Overall planning must take into account the fact that the resulting FMS will have to produce the parts flexibly and at costs that are in line with market conditions. The degree of system automation exerts a considerable influence here. Too much automation raises the investment costs very quickly, as can be seen in the example of a turning machine (→ *Figure 3.15*). In such cases, it is worth considering whether two machines with a lesser degree of automation would not be more profitable.

Experience has shown that another problem is the maximum number of tools that can be stored at the machines. To avoid having to equip all machines with expensive, oversized tool magazines, the design of the parts first must be oriented toward the use of standard tools. If the number of tools remains excessive despite this approach, the best solution

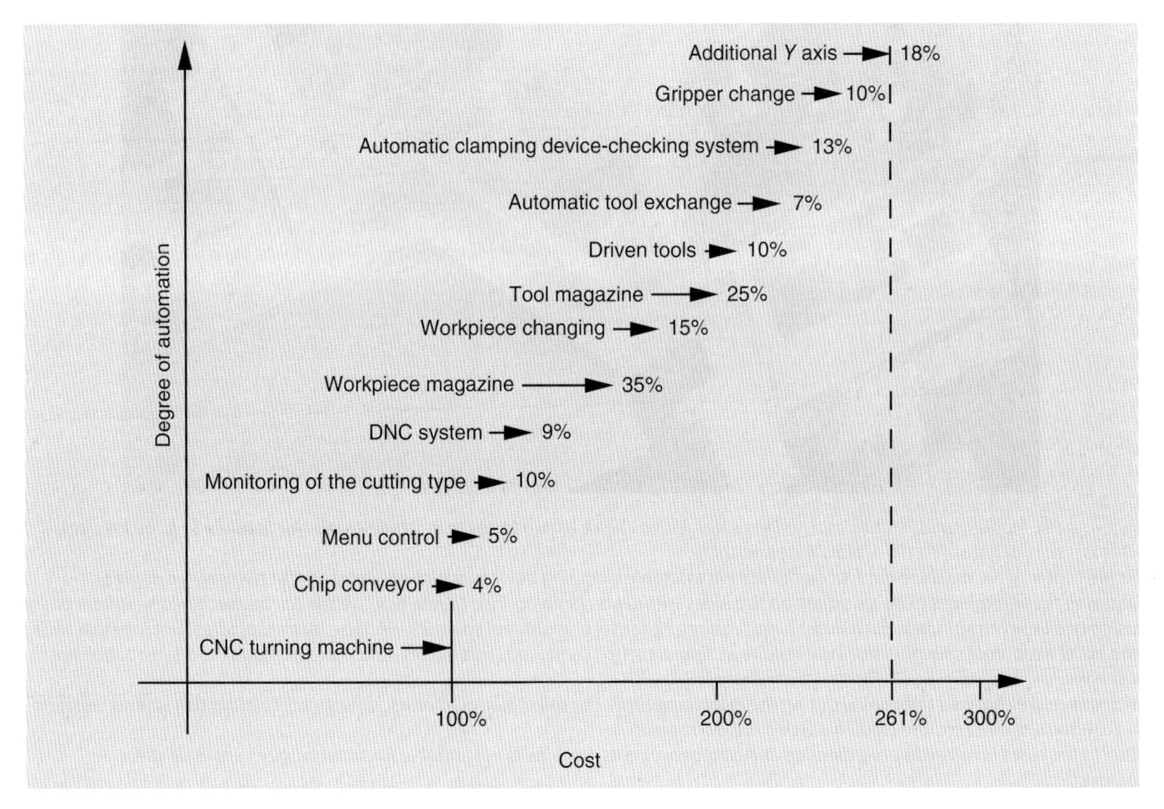

Figure 3.15: Additional costs for upgrading a CNC turning machine to a flexible manufacturing cell.
This may require upgrades that are more or less expensive. It must be determined in each individual case whether these additional costs are worthwhile, or if it might be more cost-effective to use two machines with a lesser degree of automation.
The percentages shown above are intended only as approximate values and include both hardware and software components.

Figure 3.16: Modular multistation machining center with a choice of two to seven machining units for medium to large lot sizes. This is a special type of flexible rotary index machine.
At the center of the machine are four workpieces in rotatable clamping devices, which are indexed four times by 90 degrees, allowing the workpieces to be machined on five sides without reclamping. Two to four workpieces can be machined simultaneously using two to seven spindles. Each turret head features either six or eight tool spindles. Because all the spindles turn together with the turret head, tool changing does not require a "Spindle stop" command. This makes it possible to change tools in approximately 1 second.
Workpiece are changed (manually) during the machining time. The roller guides, motors, drives, recirculating ball screws, position-measuring systems, etc. are located outside the workspace.
Due to the costs involved in programming and changeover, this machine is not suitable for small lot sizes or single-piece production.

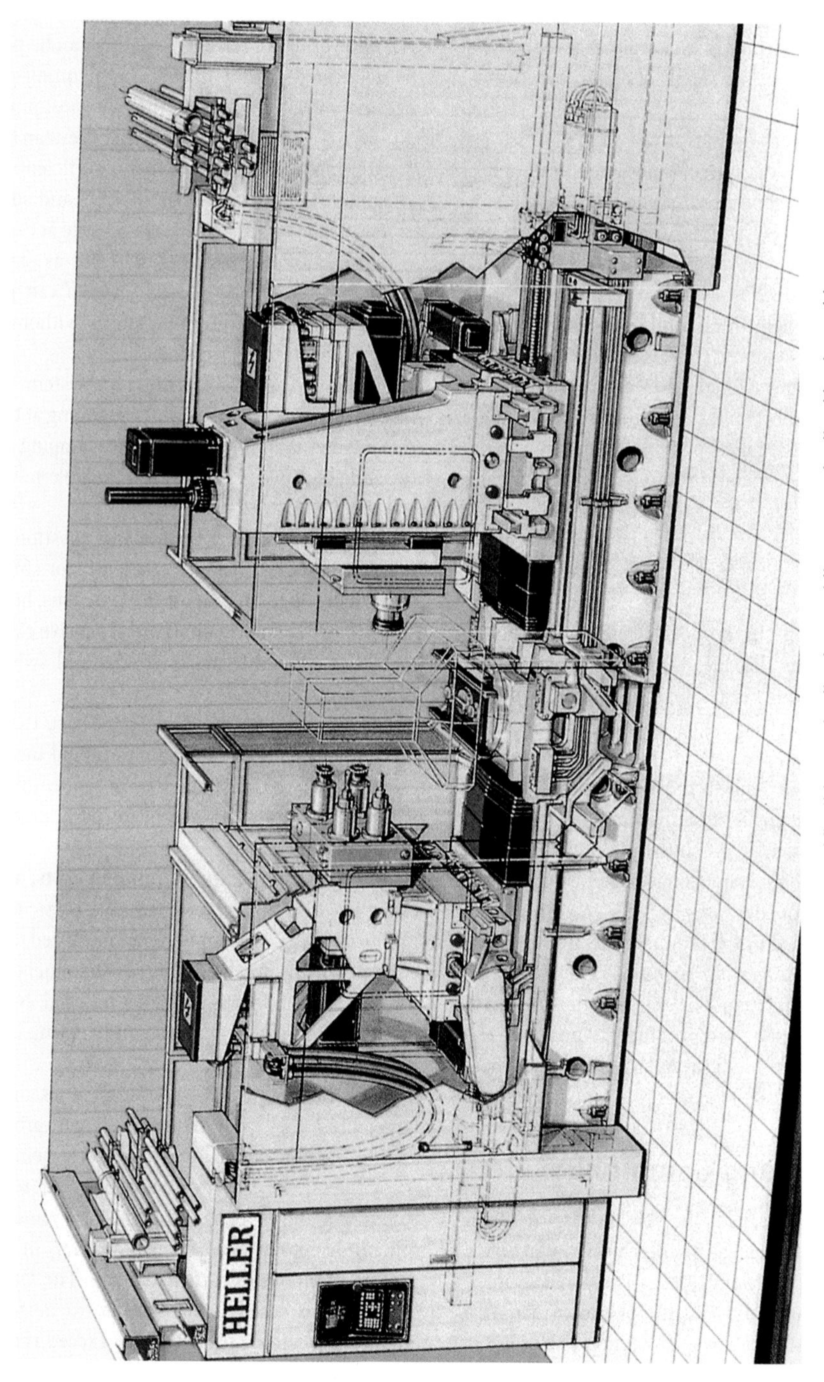

Figure 3.17: Structural units for flexible transfer lines. Use of flexible standardized assemblies on standardized basic machines. Customer-specific production lines can be assembled in this manner.

may be a central tool magazine for automatic tool exchange. When machining is performed on materials that reduce tool life to a few minutes, a constant supply of replacements must be ensured.

Temporary emergency operation of an FMS is also very difficult to implement. If the system is designed to use a host computer, its extensive set of coordinating tasks cannot simply be adopted from other systems, and the personnel certainly will not be sufficiently prepared to perform manual control of the missing functions.

The risks also include the high **total investment costs** of flexible manufacturing systems. This results in a great financial burden in the form of **fixed costs and depreciation,** which have a negative effect on the operating result when the capacity is not used to the fullest possible extent. In order to prevent this, substantial expenditures must be made as early as the planning phase. The productivity of an FMS must be calculated in such a way that the increased hourly cost of operating the machines pays off in higher output, thus keeping unit costs down.

Some of the arguments in favor of flexible manufacturing systems involve benefits that are difficult to prove from a commercial standpoint, for example, reduction of work-in-progress (WIP) inventory and regular inventory. Other arguments are based on market-related benefits. It is important not to assign too much value to these arguments. All too often, flexible manufacturing systems have been saddled with manufacturing tasks for which they were not designed originally. Of course, some systems have been able to prove their flexibility in this manner.

As planners are guided by the requirements in each given situation, it is not uncommon for them to wind up with an entirely different type of system. For example, flexible rotary index machines can be used most profitably to produce relatively small workpieces in medium to large lot sizes, where they offer many advantages (→ *Figures 3.16 and 3.17*).

3.13 Flexibility and Complexity

Flexible manufacturing means being able to adapt to changing situations without difficulty. In a flexible manufacturing system, this means that the overall system is tailored as precisely as possible to the range of products that is to be manufactured and that the system allows production changeovers without difficulty and with minimal time expenditure.

Although in principle any part can be produced using an FMS, the group of related parts that can be produced should be delineated already during the planning phase. It is also necessary to check whether the four most important preconditions have been met and whether they can be maintained:

1. A group of related parts that is sufficiently large, with an appropriate "degree of similarity" and sufficient quantities to allow the machines to operate at capacity
2. A selection of machines that are as standardized and interchangeable as possible, which can perform all the necessary machining operations without manual intervention or postprocessing
3. A suitable workpiece-transport system with automatic pallet changing and/or parts changing at the machines
4. Suitable clamping devices (The clamping of smaller parts using multiple fixtures is also possible.)

This, in turn, gives rise to the question of how to use multiple-spindle machines efficiently for simultaneous production of two to four identical parts. This, however, requires larger, more complex, and more expensive clamping devices, possibly with additional rotational and swivel motions for five-axis machining.

If you expand the group of related parts that can be produced too widely, then the planned capital investment will be exceeded very quickly, and the subsequent production costs will be too high—which will have a particularly serious effect on the "simple" parts.

Product complexity, equipment costs, and production costs are directly related. To save costs, more and more production facilities are being modernized, automated, and equipped with computer control. Personnel costs and inventories of finished parts are being reduced. Nevertheless, the products still may be too costly if planning is performed on too lavish a scale.

If production and assembly facilities are planned too large, then they also require a larger production space, adding further to the cost of the parts being produced. In many cases, the system complexity that results from the technical planning phase is not necessary and should be examined for appropriate savings potentials in the course of a second, economically oriented phase. The reasons for excessively large solutions generally are to be found in overly ambitious planning goals that far exceed the solutions that are actually necessary or else in particular design engineers who do not restrict themselves to the **limited, carefully selected assortment of tools.** This leads, for example, to machines with overly large tool magazines and excessively complex production equipment. Therefore, every machining production operation should have its tools strictly subdivided into standard, series-produced, and special tools, with hourly rates staggered accordingly.

Flexibility in production is easier to achieve if the parts being produced do not pose any particular problems and if adequate attention has been paid to the subsequent manufacturing process, including all conceivable sources of problems. This is why it is in the interests of all those responsible for the product—from the manager to the design engineer and the service technician—to make the overall system as simple and straightforward as possible. This makes it easier to achieve profitable, flexible production.

Used properly, manufacturing flexibility should lead to lower overall costs. This can be achieved through a number of factors, such as

- Appropriate organization
- Reduction of production and delivery times
- Adherence to deadlines
- Reduction or elimination of inventories
- Simultaneous production of different parts
- Production of different lot sizes
- Changing production priorities for rush orders
- Avoiding rejects and rework
- Consistently high production accuracy
- High production quality across the board
- Training and instruction of personnel concerning the capabilities and limitations of the system

In flexible manufacturing systems, this applies to **all lot sizes.** These factors add up to overall cost savings thanks to the capability for quick, interruption-free changeover of the FMS to different parts.

3.14 Simulation of Flexible Manufacturing Systems

For the planning and use of flexible manufacturing systems, simulations are used that have entirely different tasks and focuses than simulations used for CNC programs. In the simulation of flexible manufacturing systems, the emphasis is on the response of the overall system to changing requirements, with the main purpose being to determine its cost-effectiveness. The cost-effectiveness of an FMS does not depend, however, on the individual machines but rather on **the interaction of all FMS components.**

However, the ongoing changes in the workpieces, parts mix, and lot sizes also require adjustment and generate unforeseeable changes. Therefore, it is virtually impossible to make reliable statements as to the future behavior of the system or to predict the effects of changes in parameters without a realistic simulation of the entire manufacturing situation.

As the size of the system and the investment increases, so does the scope of these unforeseeable risks, thus leading to doubts on the part of the buyer. For this reason, vendors and suppliers of flexible manufacturing systems use computer-assisted simulation already at the beginning of the planning phase. This is done by entering the system layout and the data for the system into a computer program. It is then possible to change various details and test their specific effects on the production process. Bottlenecks or overdimensioning can be detected very quickly this way.

The **system planner** first uses **static analyses** based on data provided by the customer to make an initial statement about the number of machining stations required. He or she then determines the suitable transport equipment, the number of clamping devices and pallets, and the number of setup and storage locations.

Given the complexity of such systems, these static analyses are not sufficient by themselves but rather form the basis for the subsequent **dynamic simulation.** This dynamic simulation is more flexible, less expensive, and faster than the method that was used previously, which was to set up the entire system on a smaller scale using a technically sophisticated "toy construction set" and to use this setup to perform tests.

The money spent for a simulation will pay for itself many times over and thus can be recommended to any buyer of an FMS. The simulation may show that a completely different system concept would be much more advantageous and would guarantee the required cost-effectiveness and profitability.

In most cases, the most important goal of simulation during the planning phase is to optimize the investment costs. The results of the various simulation experiments are shown on the screen in color graphics.

However, the **system user** also can use the same simulation technology later to check how the system would respond to new conditions. He or she can make reliable predictions, detect bottlenecks, and plan ahead. Changes in the range of parts or new orders can be "played through," as can the effects of technical or organizational problems.

With simulation, the user can determine the following, among other things:

- The degree of utilization of the machines with different product mixes
- The optimal lot sizes
- The degree of utilization of the washing machine and measuring machine
- The degree of utilization and bottlenecks in the transport system
- The number of workpiece carriers and clamping devices required
- The number of tools required in the magazine
- The effects of shorter or longer malfunctions
- The manufacturing throughput times per workpiece
- Compliance with deadlines
- The activities of the system supervisor

Unlike **planning simulation,** which has to deal with defined values and various system configurations, **user simulation** assumes an online connection to the manufacturing host computer. This means that the simulation system can receive the relevant data directly from the process and can work in a process-oriented manner. The results of the simulation then can be incorporated directly in the production planning.

Comparable simulation programs are also available for robots and assembly systems, although these have different tasks and emphases.

System-Dependent Parameter Values

Analysis of a computer model based on **different FMS configurations** is performed by entering system-dependent parameter values such as

- Number and arrangement of the machines
- Number of locations in the tool magazine and in the tool-storage facility
- Tool-changing time (chip-to-chip)
- Strategy and time requirements for the tool exchange
- Workpiece- and tool-transport system
- Number of transport locations in the vehicle
- Average velocity of the transport system
- Conveyor tracks
- Transfer time of the transport system
- Number of pallets in the system
- Number and variety of clamping devices
- Changeover times for changes in production

After this, it is possible to enter the **production-related parameter values,** such as

- Production quantities and lot sizes
- Number of different workpieces
- Number of tools required per workpiece
- Required system tools, special tools, or custom tools
- Required machining operations and machining times (from the CNC programs)
- Necessary clamping positions (three, four, five, or six-sided machining)
- Changeover times

Based on these specifications, the simulation system calculates the results of changes to the production program or product mix, such as

- Utilization and idle times of the FMS components
- Overcapacity, undercapacity, and remaining capacity
- Bottlenecks in the system
- Throughput times and end dates
- Effects of rush orders and changed production priorities
- Possible emergency strategies in case of problems
- Requirements for tools and replacement tools
- Costs, utilization times, and changeover and idle times
- Personnel requirements

3.15 Production Planning System (PPS)

One of the main tasks of a production-related simulation system is to ensure that a cost-optimized production process is possible despite the large number of different manufacturing tasks. There are, however, a number of preconditions that must be met that are not taken into account by simulation systems with a strictly manufacturing technology orientation. This is the task of the PPS.

The **strategic goal of a PPS is** to control and optimize the order throughput time from the receipt of an order to shipment in such a way that the demanded delivery dates and costs of the orders being filled will be maintained. Given the generally very complex interrelationships of these tests, especially in the case of flexible manufacturing systems, a PPS is an absolute requirement for subsequent smooth operation.

A PPS has fundamentally three different areas of application:

1. The **planning functions,** in order to plan all orders, administer them, and initiate their production on schedule.
2. The **control functions,** in order to issue the production orders in accordance with the available manufacturing capacities and stocks of material and to monitor compliance with deadlines.
3. **Supporting** the sales department by forecasting future availability of capacity, time, and materials. In this way, the sales department will be able to make binding offers without disrupting or otherwise limiting current planning.

In an FMS one can assume that for the most part the production of standard products in variable lot sizes is involved and that customer-specific modifications only have to be taken into account in isolated cases.

In some cases, the function of the PPS may be to control the throughput times of orders in such a way that delivery times are as short as possible without altering ongoing production. Nevertheless, owing to the competitive environment, many companies are forced to promise shorter delivery times and therefore have to integrate rush orders into their plans. This requires them to change their existing production plans. Such demands cannot be met in the long run without the help of a high-performance PPS. All this goes to show that a PPS has to be able to deal with various priorities, thus requiring a great deal of flexibility. To do this, the system has to have comprehensive data that are updated on a continual basis.

These data make it possible to identify in detail the inventories, schedules for incoming goods, material requirements, utilization of production capacity, costs, and so on. This information then can be incorporated into planning in such a way that the current production costs and the additional costs for special customer requests are always available to the sales department. Prices, discounts, and commissions can be calculated much more quickly and realistically than with manual preliminary costing.

If one assumes that all the necessary parameter values are available, such as bills of materials and production times per workpiece, then the PPS can provide the following data:

- Material and time requirements
- Material and capacity planning
- Production costs
- Purchasing control
- Shop-floor control
- Assembly times
- Fault messages and repair times
- Cost calculation

Also included are data interfaces for **additional functions,** such as

- Quality control with statistics
- Production of repeat parts
- Prognoses
- Order management
- Account management
- Bookkeeping for salaries and wages
- And so on

The PPS should be able to calculate the production costs and delivery times for outgoing orders based on these data. There is an additional function that makes it possible to determine the latest possible start of production based on the specified delivery date and to use that as a basis for the incoming delivery deadlines for materials and purchased parts. And this all should be possible very quickly!

The machine assignments resulting from the PPS are followed by a simulation, which shows whether the machine capacity is sufficient, whether other orders would run into problems, or how bottlenecks could be eliminated or bypassed by means of short-term replanning. This wide range of functions means that a PPS is truly valuable as a planning tool—as long as it is continually supplied with current data and is used properly.

3.16 Summary *(Figure 3.18)*

The use of powerful CAD systems, coupled with the ability to make direct use of CAD-generated data in the manufacturing process, is allowing new products to be developed with increasing speed. The short life span of modern products no longer allows a company the comfort of installing relatively

WALDRICH COBURG

in an assembly shop at

ABB

for the machining of turbine components

Overall length: ~122.5 m
Overall width: ~16.5 m
Overall height: ~11.2 m

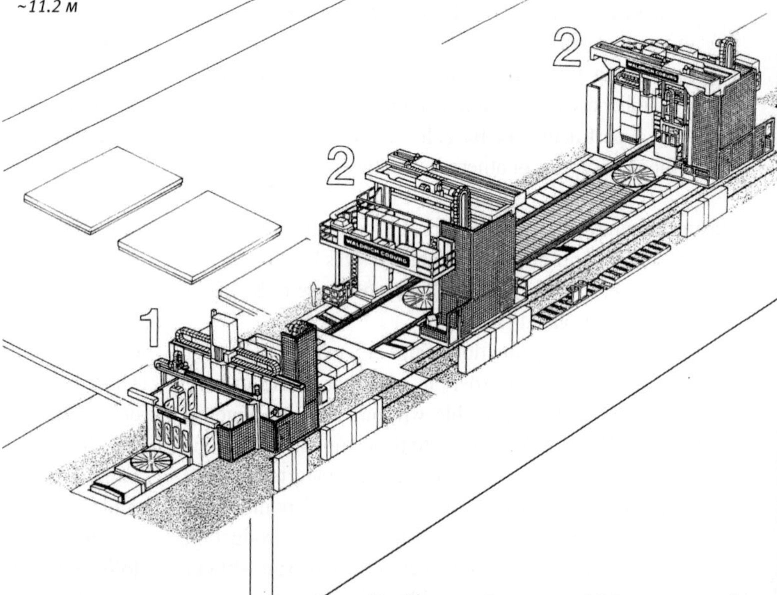

Machine 1: *Gantry machining center with integrated vertical rotary table*
Table width 2.5 m
Faceplate 2.5 m
Milling output 95 kW

Machine 2: *Gantry machining center, consisting of:*
Two gantries with one clamping plate and
Two vertical rotary tables
Clamping plate width 3.5 m
Clamping plate length 19 m
Gantry movement range 23.5 m each
Faceplate diameter 3 m
Vertical clearance 4 m

WALDRICH COBURG

Werkzeugmaschinenfabrik
Adolf Waldrich Coburg GmbH & Co
Hahnweg 116 · D-96450 Coburg
Telefon 09561-650 · Telex 663225
Telefax 09561-60500

Figure 3.18: Flexible manufacturing system for large turbine and generator components consisting of five gantry machining centers for milling, drilling, and turning. (Image courtesy of WALDRICH COBURG.)

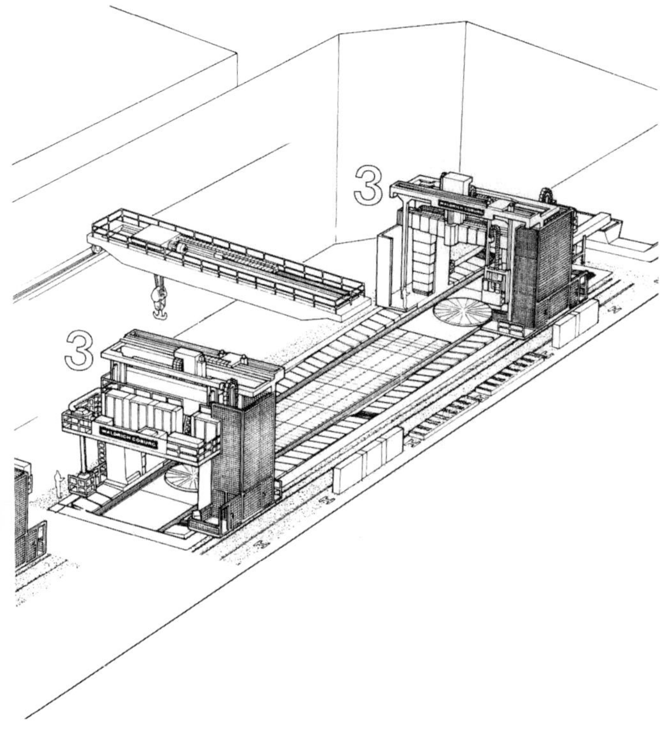

Machine 3: Gantry machining center, consisting of:
Two gantries with one clamping plate and
Two vertical rotary tables
Clamping plate width 5 m
Clamping plate length 18 m
Gantry movement range 31 m each
Faceplate diameter 5 m
Vertical clearance 5.5 m
Horizontal clearance 6.5 m
Support extension distance 3 m

Controllers: Sinumerik 880 M

Explanations regarding the illustration:

Design
This FMS consist of five large machines, and replaces an original set of 18 different machine tools for milling, drilling, and carousel turning. The gantry machining centers are used to manufacture housings for gas turbines, steam turbines, and generators. Complete machining of these very large workpieces is performed with great precision and only a few setups.

The combination of milling machines and vertical turning machines allows flexible, continuous utilization of the manufacturing system.
The throughput times for large workpieces have been reduced by more than 50 percent. The usual setup, transport, and waiting times have been almost completely eliminated. Production costs have been reduced by up to 60 percent through the use of state-of-the-art technologies.

High flexibility
Twenty-two automatically changeable spindle units (for turning and milling, with horizontal, vertical, and fork heads with B and C axes) are used on the five milling supports with no accuracy limitation.
If all gantry machining centers have an identical machine concept, then operating personnel and CNC programs can be interchanged as desired. This allows flexible production planning and the desired freedom as regards assignment of the individual gantry machining centers.

Manufacturing control technology
All machines are provided with the manufacturing data by a host computer via DNC. The host computer also plans, in accordance with the production sequence, the requirements and the operation times of the spindle units at the individual gantries.
On the basis of the CNC programs, the host computer generates the tool-exchange lists that the operator uses to place the required tools in the chain magazine of the machining center and to remove tools that are no longer needed.

inflexible production equipment, secure in the knowledge that it will be able to use this equipment in mass production for several years without making any significant changes to it. The variety of manufactured parts also has increased steadily, and the buyer expects the system manufacturer to adapt to this increased variety. Because production is changing so rapidly from one type of part to another, there is also a risk that bloated inventories will very quickly turn into expensive, unmarketable scrap.

<tx>For these reasons, planning efforts are increasingly centered around the demand for automated solutions for manufacturing small and medium-sized lots. However, because there is no single manufacturing system that is ideal in all cases, profitability can be optimized only by using solutions that are specially designed for the requirements at hand. A sufficient number and variety of machines and automation devices are now available for this purpose. These can be combined to form many different variations. Most system concepts also allow incremental expansion so that the high investment costs can be distributed across several years. The experience gained in operating these systems makes it easier to prove their profitability.

The use of flexible manufacturing systems requires a thorough analysis of the manufacturing task at hand. It is also necessary to specify the main objectives and to consider growth rates and future changes. Numerous manufacturers now possess the necessary experience and offer their support *and advice* in accordance with this experience. This makes it quite possible to keep the risk within acceptable limits.

One objective of flexible manufacturing systems is **profitability in manufacturing extremely small to medium-sized lots.** This can be achieved only by exploiting the theoretical properties of such systems. These properties are discussed below.

Improved utilization of the available machine-hours can be achieved by

- Continuing operations through break times
- Tapering off operations in the second or third shift
- Reducing machine downtimes that arise for organizational reasons
- Changing to a new production order without interrupting operation
- Manufacturing by automated methods with reduced personnel

The attainment of the most important objectives must be monitored constantly from the very beginning. These concerns must take priority over all other technical planning concepts.

- Does the FMS fulfill the technical requirements that have been defined in writing?
- Is there assurance that the many types of manufacturing-related data will be used automatically and consistently (i.e., CAD/CAM and tool data)?
- What is the limit on financial investments and the cost per machine-hour that will guarantee profitability?
- Will the system operate profitably despite incomplete utilization of the machines?
- After making all the necessary compromises in the planning phase, will the system still be flexible, productive, and profitable?
- Or would other manufacturing concepts be significantly more suitable and cost-effective?

Experience has shown that the best ideas for a new system will not emerge until the planning work is being carried out. The stringent requirements imposed on flexible manufacturing systems are obviously being met in many cases; otherwise, we would not be seeing more and more of such systems being installed.

Flexible Manufacturing Systems

Important points to remember:

1. **Flexible manufacturing cell:** A CNC machine, generally a machining center, with a limited supply of parts that are processed consecutively. As a rule without DNC or host computer, provided that the program memory within the CNC system can store the necessary programs.
2. **Flexible manufacturing island:** Shop area with all the machines and devices necessary to perform all the required machining operations on a limited selection of workpieces.
3. **Flexible transfer line:** Workpiece-based linkage of a number of CNC machines using the linear principle; that is, all the parts pass through the individual stations and are processed by means of various sequential programs.
4. **Flexible manufacturing system:** A group of CNC machines (statistically at least 6 to 10) that are connected with each other via a common workpiece-transport system and a centralized control system. The machines perform all the required machining functions on a limited range of parts without any need for manual interventions to support or interrupt the automatic sequence.
5. **In general:** In an FMS, various production facilities are connected with each other via a common control and transports system in such a way that it is possible to produce all the following in a cost-effective manner:
   - Various workpieces . . .
   - By a variety of machining methods . . .
   - In any desired sequence . . .
   - In lots of various sizes . . .
   - In a fully automatic manner without manual intervention.
6. Flexible manufacturing systems do not contain any new machine concepts but rather represent a **combination of already existing components:**
   - A number of interchangeable or complementary CNC machines
   - + A workpiece-transport system
   - + A loading and unloading system for pallets and workpieces
   - + Monitoring devices for the overall system
   - + Removal systems for chips and coolant
   - + A host computer with system-specific software
   - = A flexible manufacturing system.
7. Flexible manufacturing systems represent a **combination** of communication, automation, and computing capability.
8. **Transfer lines** are much more productive than flexible manufacturing systems but are unfortunately **not a viable alternative.**
9. The **simulation** of flexible manufacturing systems has different functions and emphases than the simulation of NC programs. For flexible manufacturing systems, the main priorities are the behavior of the system as a whole and cost-effectiveness under changing conditions.
10. Besides the design of the system itself, the most important goal of dynamic simulation during the planning phase is optimization of investment costs.
11. It is also possible for the subsequent operator of the FMS to use the simulation to determine the behavior of the system in the event of changed production requirements, faults, or the failure of individual components.

4. Industrial Robots and Handling

Dipl.-Wirtsch.-Ing. (FH) Christian Schmid

Industrial robots have similar characteristics and properties as those of computer numerical control (CNC) machines but very different kinematic systems and special, flexible controllers. Their areas of application, programming, and special requirements are likewise very different. In less than 40 years, they have initiated major changes in the industrial environment.

4.1 Introduction

While the use of CNC technology has speeded up machining times in metal-working machines in industrial production to a maximum extent by means of appropriate equipment and programming, the use of robots has required completely new manufacturing concepts. CNC machine tools now have reached a high state of development worldwide. The scope for reducing machining time is very limited. For this reason, increased attention is being paid to reducing nonproductive times and to manufacturing processes that can be performed efficiently using industrial robots.

In industrial applications, industrial robots work cooperatively with other devices and machines in the production process. This also can be referred to as a *manufacturing cell.* Therefore, robots should not be viewed in isolation but rather as part of an interactive system comprising many components in the manufacturing cell. This becomes evident, for example, when the robot controller exchanges signals and/or data with sensors, transfer equipment, machine tools, or other industrial robots.

Today, industrial robots have become an essential fixture not only in the automotive sector but also in any number of other manufacturing industries. Robots were used even at the beginning of the 1970s for simple handling tasks and in particular for spot welding and seam welding. The main reason for this was increased efficiency, such as reduced processing times, improved product quality and reduced costs.

Even technologically problematic processes such as the automatic assembly of engine components, the assembly of chassis components or transmissions, or even the processing of natural materials such as wood, leather and textiles, or elastic, yielding materials in general, have been automated through the use of robots. Progress in the field of sensors has made possible new applications in the fields of quality assurance and inspection.

Advances in microelectronics and in control and drive technology in the mid-1970s laid the basic foundations for industrial robot technology. Starting from robots with hydraulic drives, whose characteristics as regards precision and dynamic performance soon demonstrated their shortcomings and limitations, the next major step was to develop robots that made use of electric drives. The first industrial applications of robots involved operations that were easy to implement. As the capabilities of industrial robots improved and suitable tools for joining and processing tasks were developed, new process-related areas of application began to open up. Eventually, more demanding requirements for the dynamic performance and precision of robots emerged, that necessitated robot controllers to come up with a new level of performance and special functions for processing tasks. The result was industrial robots provided with an increasing degree of "integrated intelligence."

4.2 Definition: What Is an Industrial Robot?

Previously, there was absolutely no uniform international definition for this term. Therefore, the term *industrial robot* has been defined in the ISO/TR 8373 Standard. This reads as follows:

Manipulation Industrial Robot:

An automatically controlled, reprogrammable, multi-purpose, manipulating machine with several degrees of

freedom, which may be either fixed in place or mobile for use in industrial applications.

The individual terms in this definition are also defined as follows:

Reprogrammable: Whose programmed motions or auxiliary functions may be changed without physical alterations.

Multipurpose: Can be adapted to a different application with physical alterations.

Physical alterations: Means alterations of the mechanical structure or control system except for changing programming cassettes, read-only memories (ROMs), etc.

European Standard EN775 defines *industrial robots* as follows:

A robot is an automatically controlled, reprogrammable, multipurpose, manipulating machine with several degrees of freedom, which may be either fixed in place or mobile for use in industrial applications.

To sum up, the following criteria must be met: Industrial robots

- Are freely programmable
- Are servo controlled
- Have at least three CNC axes
- Have grippers and tools
- Are designed for handling and processing tasks

4.3 Structure of Industrial Robots

The task of industrial robots is to pick up, hold, and guide workpieces and/or tools. A number of different components are involved in initiating and executing the motions that are necessary for this. Each of these subsystems (*Table 4.1*) contributes to the robotic system's solution of the motion and holding task.

The most important components of a robot are

- The mechanical frame, including the gear units
- The actuators for motions within the workspace
- The sensors for detecting the axis positions and ambient conditions
- The robot controller

4.4 Mechanical Elements/Kinematics

The mechanical structure of a robot is described in terms of its kinematic system. The following criteria are important here:

- The form of motion of the axes (translational or rotational)
- The number and arrangement of the axes, for example, their sequence and positions
- The length of the translational axes
- The shape of the working envelope

In general, industrial robots have the structure (*Figure 4.1*) of an open kinematic chain with no branches, whose elements or lever arms are linked to each other in pairs via joints. Each joint has exactly one degree of freedom, either rotational or translational. One end of the kinematic chain is joined to the robot's base, which serves as the basis for the spatially fixed reference coordinate system; the free end serves as a mount for the end effector (gripper or tool).

Each robot has an associated work envelope that defines the set of the positions that can be reached. A distinction can be made between various basic types of work envelope depending on the structure of the robot as defined by the arrangement of its joints. If, for example, you take a robot arm with the joint arrangement RRT (*Figure 4.2*), then the first joint performs a rotational motion that the second joint expands to create a toroidal surface. This is contracted or expanded by the linear joint, thus forming the complete work envelope of the robot. Common basic forms of work envelopes are

- Cuboids (robots of type TTT)
- Cylinders (e.g., RTT)
- Spheres (e.g., RRR, RRT)

The toroidal form is not very common with most robots and generally most resembles a sphere. The function of the robot's wrist joint is to achieve any desired orientation at the required position. Quite understandably, it has shorter elements than the arm, and it also has not linear joints. The typical structures for wrists are TRT and TRR. The last element of the wrist is connected to a flange to which the effector is coupled.

A distinction is made between various types of industrial robots (*Figure 4.3*).

Subsystem	Characteristics and Subfunctions
Mechanical structure	1. Structure of the motion subsystems 2. Definition of the degrees of freedom and of the working space 3. Securing the position and orientation of the handling objects
Kinematic system	1. Spatial relationships of the individual joints of the robot arm and end effector 2. Temporal relationships between the motion axes and the motion of the end effector
Axis control and drives	1. Closed-loop control of the dynamic drive processes 2. Feeding the drive energy to the axis drives of the motion subsystems 3. Generating the motion of individual axes
End-effectors	1. Gripping and handling of product parts (joining, screw connections, checking, etc.) 2. Processing of workpieces with tools (welding, deburring, grinding, painting, etc.)
Sensors and sensor systems	1. Detection of the internal states of the manipulator and end effector (position, velocity, forces, torques) 2. Detecting the states of the handling objects and the environment 3. Measurement of physical variables 4. Identifying and determining the status of workpieces and interactions 5. Analysis of situations and scenarios in the environment
Control	1. Control, monitoring of motion and handling sequences, and movement tasks 2. Synchronization and adaptation of the manipulator in the handling process 3. Avoiding and eliminating conflict situations
Programming	1. Creating control programs (using software systems such as compilers, interpreters, simulators, etc.) 2. Interactive/automatic planning of the robot task
Computer	1. Execution of the computer processes (program development, processing of sensor data, data transformation) 2. Implementation of human-machine communication 3. Global control and monitoring of flexible manufacturing systems and machines (including industrial robots)

Table 4.1: Components of a Robotic System (WAR 90)

4.5 Gripper or Effector

An effector is the part of the robot that performs that actual handling operation. It is fastened to the robot's wrist and connected to power and control cables to enable it to grip, hold, and transport workpieces and to position them in the desired location. Depending on the specific handling task, sensors, assembly tools, or additional electrical cables (e.g., for welding current) may be used. Painting robots may use hoses to deliver the paint.

A total of three main axes are necessary to move to each point in the three space coordinates plus three additional "orientation axes" for the gripper to bring the workpiece into the desired spatial orientation via rotational, tilting, and swivel motions.

Given the diverse applications of industrial robots, a broad spectrum of grippers has to be available for all sizes, constructions, functions, and applications. A distinction is made between the following grippers, subdivided according to their functional principle:

- Mechanical grippers, such as parallel, pincer, and three-point grippers
- Pneumatic grippers, such as flexing fingers, perforation grippers, and pin grippers
- Suction grippers (especially suitable for workpieces with smooth surfaces)
- Magnetic grippers (only suitable for paramagnetic materials)
- Adhesive grippers (seldom used)
- Needle grippers (e.g., for textiles, leather, etc.)

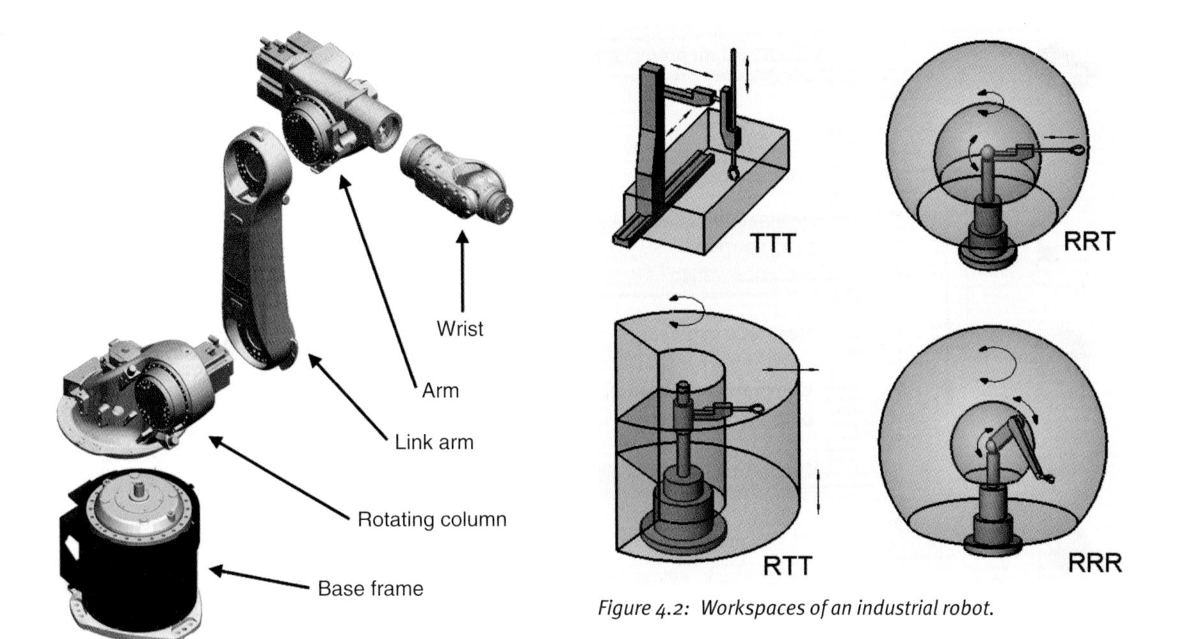

Figure 4.1: Structure of an industrial robot.

Figure 4.2: Workspaces of an industrial robot.

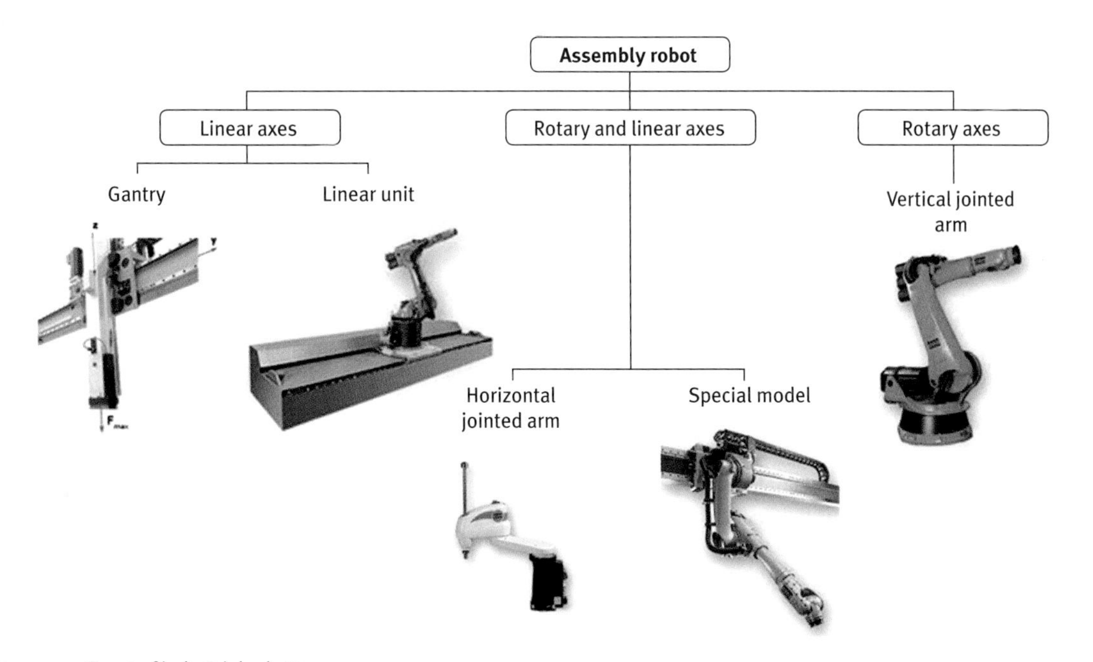

Figure 4.3: Types of industrial robots.

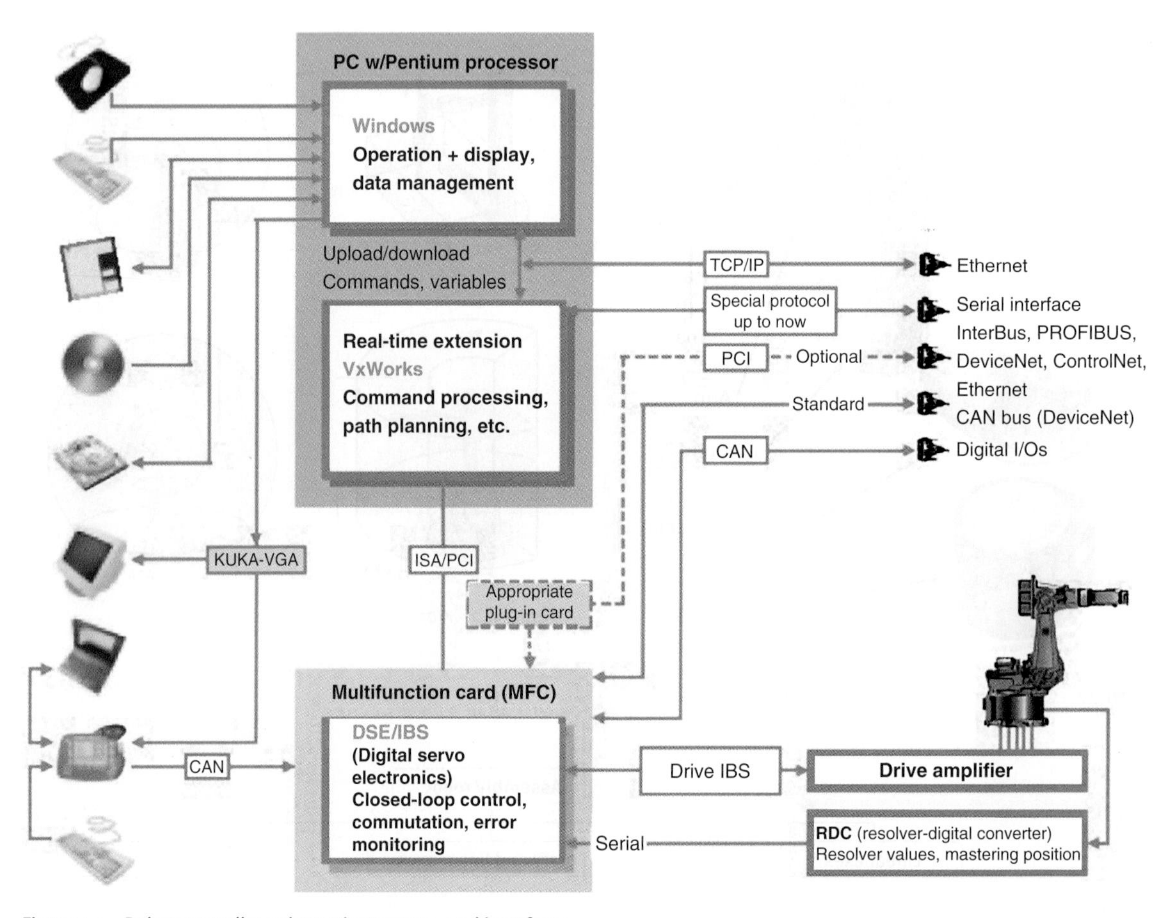

Figure 4.4: Robot controller, schematic structure, and interfaces.

Special **gripper-changing systems** contribute significantly to the flexible use of industrial robots. Especially when assembling small lots, grippers and assembly tools have to be exchanged frequently because in such cases the robot often has to perform a number of different operations one after the other. Here, the changing system provides additional flexibility if the tool change takes place automatically.

The important factor here is not just a secure mechanical coupling but also the continuous flow of power and information. To ensure this, plug connections for electrical and pneumatic lines are provided in the interchangeable flanges. Spot-welding guns that require cooling also have to have a cooling-water connection.

4.6 Controllers *(Figure 4.4)*

The intelligence of an industrial robot, and thus its ability to act flexibly, resides in the robot controller. All the necessary input data, such as the path, velocity, or operator interventions, are processed in the controller. As output data, they affect the robot drives or robot tools in accordance with the specified logic.

Today, most robot controllers feature a human-robot interface in the form of a programming system that runs on a standard PC. It is also possible to use custom-tailored control units for production operations or universally applicable teach pendants.

Functions of the Robot Controller

The robot controller contains all the components and functions that are necessary for operation, control, programming, and monitoring of a robot. The robot controller can perform an extremely wide range of functions within a robotic cell:

- Controlling the robot's motions
- Influencing the process components in the system
- Influencing the conveyor and in-feed components
- Controlling the gripper functions
- Receiving and evaluating sensor signals
- Receiving and processing process information for influencing the process
- Diagnostic functions for error detection on the robot or in the process
- Providing support to the operator
- Supporting the programmer when setting up the automation cell

In order to perform all these functions, the robot controller needs not just a high-performance processing unit but also interfaces with the robot, the periphery, the process, and the operator.

Operator Control and Programming Device

Various components are available for operator control and programming of robots. Normally, robots are operated and programmed using a teach pendant (hand-held programming unit). The teach pendant (*Figure 4.5*) can be used to input all the necessary commands and to activate all the functions of the robot controller.

Figure 4.5: Control panel.

Processing Unit

The processing unit consists of a main computer that carries out the overall coordination within the robot controller. Subordinate to this main computer are the axis-control computers, which perform closed-loop position-control and monitoring functions for the robot axes. Today, the main computers of robot controllers are equipped with high-performance multicore processors. These improve the performance of the controller and allow it to carry out other automation functions in addition to controlling the robot.

Power Unit

The power unit of the robot controller supplies the necessary power for the electronic units and the drives/servo controllers. These are the most important components of the power unit.

As a rule, the servo controllers are supplied with power from a direct-current (dc) circuit that, in turn, draws its power from a rectifier that is operated off the power mains. The rectifier is connected to the mains via a transformer. Besides adapting the voltage, the transformer also filters out high-frequency disturbances that otherwise would be transmitted from the network to the robot controller or also in the opposite direction from the servo controllers to the power mains.

Today, the connection between the servo amplifiers and the robot controller generally is implemented in the form of a bus system. This can be used not only to transmit the control commands to the servo unit but also to transmit the feedback messages and fault states to the robot controller. This makes possible centralized diagnostic functions for these system components as well.

Setting and adjustment of the servo units to their respective robot axes are performed in a similar manner. The necessary parameters for an optimal control response are sent to the servo controllers via the bus. This means that when maintenance is performed, it is no longer necessary to make any adjustments on the servo controllers.

The power unit of the robot controller as a whole, and in particular, the ballast circuit, is a source of heat loss. For this reason, the control cabinet generally is equipped with a cooling system.

Drives

The purpose of drives is to move each axis in closed-loop control and to hold each in position after movement. Drives have to satisfy very high dynamic requirements owing to the major changes in the dynamic behavior of the robot depending on the weight of the workpiece or the greater or lesser degree of extension within the work envelope.

Axis Control

The power unit of the robot controller supplies the necessary power for the electronic units and the drives/servo controllers. The servo controllers are the most important components of the power unit. Based on the motion commands from the robot controller, the servo amplifiers control the individual axes of the robot in such a way that the programmed motion sequence is followed exactly. As a rule, the servo controllers are supplied with power from a dc circuit that, in turn, draws its power from a rectifier that is operated off the power mains. The rectifier is connected to the mains via a transformer. Besides adapting the voltage, the transformer also filters out high-frequency disturbances that otherwise would be transmitted from the network to the robot controller or also in the opposite direction from the servo controllers to the power mains.

Measuring Systems

The purpose of measuring systems is to measure the position and/or angle of all axes and the movement speed and acceleration of the individual axes. Although these generally make use of incremental measuring systems, in some cases absolute measuring systems are essential. This is necessary, for example, in welding robots in order to recognize the positions of all the axes immediately after a power failure.

Safety Functions

The safety-related functions of robot controllers include, in part, the following:

- Hardware and software limit switches for all axes
- Light barriers for immediate switch-off to protect personnel
- Key-operated operating-mode switch
- Monitoring of the switching systems for enabling switches and emergency-stop switches
- Monitoring of speeds in setup mode
- Switch-on diagnosis for all non-solid-state switching operations that perform safety functions (e.g., switching drives off, etc.)
- Monitoring of voltage and temperature

The functions just mentioned are subject to standards and regulations with the primary purpose of protecting personnel. These standards and regulations are currently undergoing major revisions. The next section describes the "safe robot" technology, the direction in which this technology is moving, and what is possible at the current time.

4.7 Safe Robot Technology

Robots and Operators Working Together Closely—Safe Operation *(Figure 4.6)*

According to the current state of the art, for safety reasons, industrial robots can only be operated in secured work areas. People are always prevented from getting near the robot by protective fences or other complicated safety facilities. In the production sequence, the robot is controlled and monitored by a higher-level PLC, and only a minority of robots is additionally guided by external sensors. In any case, in production operation, robot motion is only possible when it is ensured that no operator is present in the robot's work envelope. Through the use of "safe robot" technology, it is now possible to overcome this separation between the operator and the robot for many applications. Besides being able to dispense with expensive safety systems, it is also possible to implement new system concepts, especially in the field of assembly technology. The result is an ideal combination of the strengths of humans (sensory capabilities) and robots (working performance). This makes it possible to find solutions for automation functions that previously had no cost-effective implementation using more cost-effective partial automation.

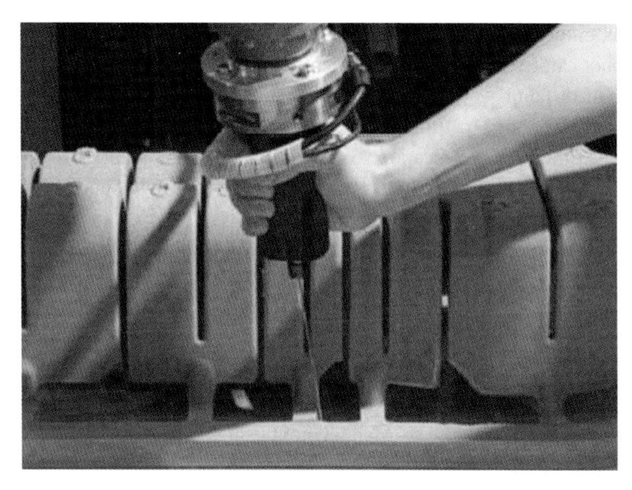

Figure 4.6: Design study for manual handhold grip to control the robot.

Using the "safe robot" technology, a monitoring technology with redundant, two-channel monitoring directly on the robot, the decision is now made locally as to whether or not the robot is allowed to move into a blocked area. The current position of all axes is recorded permanently and continuously compared with the configured limit values within a few milliseconds. Because this solution does not require any mechanical cams, there can, of course, also be a larger number of work envelopes or safety zones. If a zone is violated, the robot can immediately initiate the necessary stop itself. This significantly reduces the stopping distances in the event of a fault while also reducing the required floor space.

The "Human-Guided Robot"—Safe Handling

As described earlier, until now, robots could only be operated "behind bars." It was inconceivable that an operator could work in the area of or even next to a robot. These safety requirements explain the low level of automation in automotive industry final assembly lines and also explain the slow rate of progress in the fields of mobile robotics and service robotics. The operator can move the robot using a joystick that is fastened directly to the robot or the robot tool—without any program or additional PLC! This way it is also possible to teach points and paths without having to write any program text.

The preconditions for this kind of cooperation between humans and robots are reliable limitation and monitoring of the velocity of the robot system (Cartesian), reliable monitoring of the velocities of each individual robot axis (axis-specified), and a three-position enabling switch on the joystick.

Semi-Autonomous Systems—New Areas for Robotics

Today, simple manipulators are used for many feeding and joining operations in assembly applications. For large components and high requirements as to the positioning accuracy, in some cases two operators are needed to operate a single manipulator. Although the manipulator does compensate for the weight of the parts being moved, it provides only limited assistance to the operators when it comes to applying the joining force. The greatest disadvantage of manipulators, however, is the fact that the operator is totally bound to the device.

With the human-guided robot, it is quite different. In this application, the robot fetches the part being installed automatically under program control, for example, from a pallet, and brings it to a location close to the installation position. This takes place in a secured work envelope. The operator then takes over control of the robot and performs only the critical joining operation. The power of the robot and the acute senses of the human supplement each other to form a highly efficient system. After the operator has guided the robot back to a secure working position, the robot can fetch the next part in automatic mode. This "division of labor" between the operator and the robot makes it possible for one operator to tend two assembly stations at the same time.

Differentiated Degrees of Automation

For the first time in the history of robotics, it is now possible to implement production operations with a differentiated degree of automation. While previously there were only the two extremes of automating a production operation completely (with the robot "behind bars") or performing it 100 percent manually (with no use of robots), today it is possible to employ automation in a customized, cost-optimized manner. The human-guided mode of operation has broken down the barriers between the human and the machine (→ *Figure 4.6*).

4.8 Programming

When it comes to the cost-effectiveness of a robotic system, the amount of time required to work out error-free application programs that are also suitable for the specific application is particularly important. *Programming* means creating a program and inputting it into the robot controller. The program itself is the sequence of all information that the handling system requires to execute a motion or work cycle. The fact that the analysis of the operation to be performed, the operating conditions, and the range of motions of robots all place different requirements on the programming means that there are also different programming methods. Every robot controller can process programs in at least one programming language.

Program Contents

An application program has to contain everything that is necessary for the complete work sequence of the industrial robot in the technological process. The individual components of this are described in *Table 4.2*.

Programming Methods

The creation of a robot program can be broken down into motion programming and sequence programming. Motion programming involves the definition of points along a path/motion segments. Sequence programming involves linkages between motion segments, defining process parameters, times, wait positions, velocities, accelerations, and communication with peripheral equipment. The programming methods for industrial robots can be subdivided as follows:

Online methods *(on-line, direct programming; close-to-process programming):*

- Programming in teaching mode (moving to specific points)
- Playback methods (moving along a path; the robot arm is guided directly by the operator)
- Manual input using buttons or switches (obsolete)

Offline methods *(off-line, indirect programming; programming remote from the process):*

- Textual programming using robot languages.
- Explicit programming languages. They specify the operation to be performed in elementary steps. Each motion element has to be described in an instruction. The programming is robot-oriented.
- Implicit programming languages. The task to be performed is formulated globally. The robot generates the individual actions itself, which requires a certain level of "intelligence." The programming is task-oriented.
- Programming in teaching mode with a "phantom arm."
- Interactive programming on the screen with graphic assistance.
- Acoustic programming (input using natural language with speech recognition).

Hybrid Programming Methods *(combined methods that bring together the advantages of online and offline methods):*

The geometric instructions (paths and positions) are programmed using teaching, whereas the sequence, control, monitoring, and communication instructions are input in the form of a code or language.

Automatic Program Generation

Once a target state has been defined, the program required to achieve it is generated automatically by a system. This requires the use of a program solution that breaks the task

Component	Possible Contents	Features for Users
Program sequence	Commands for motions Commands for gripper actuation Commands for process communication and for main computer command links Command sequence	Necessary components of the handling program
Preparatory functions	Command values for position Command values for orientation Special point patterns Work positioning or intermediate positioning	Free design and correction via the handling program
Motion conditions	Motion velocity Response in the approach phase Response during positioning Motion dependence in the axes Interpolation conditions	Ensuring stable motion response Optimization of motion Verification for special motion paths
Logical decisions	Programs with variable structure Program run depends on process signals and sensor signals	Need for complex operator control and assembly tasks Program selection
Monitoring/diagnosis	Function monitoring Process monitoring Response to faults Actions for remote servicing	Satisfaction of requirements for reliability and safety High-quality service support

Table 4.2: Basic Components of Handling Programs (HES 96)

down into subtasks, on the basis of which it plans the actions that have to be programmed and generates the program. Such approaches are already in use today for the automatic generation of path and process data, for example, in milling or adhesive bonding. In this manner, a robot program can be generated directly from CAD and/or CAM data.

4.9 Sensors

The purpose of sensors is to detect and respond to disturbances such as changes in position, deviations from the pattern, or other extraneous interference. The fact that robots are being used more and more in manufacturing applications can be attributed to, among other things, the development of suitable sensors. Sensors serve to detect the internal state of a robotic system, the current interactions between the end effector and the environment, and the external conditions in the area where the robotic system is being operated. Sensor functions are based on the conversion of a physical phenomenon present at the input (e.g., pressure, force, contact, or motion) into a quantitative electrical dimension. This electrical variable is digitalized and then evaluated by a sensor processor or by the robot controller. The current state of the robot and its environment are recorded at discrete intervals (e.g., measurement of the opening width of the gripper jaws) in order to ensure that the actions are performed properly or in order to record parameters directly from the environment so as to affect subsequent operations (e.g., identification of parts or distances). Further operations are controlled based on the detected status.

During operational execution of accessory measurements, it is possible to implement sensor-monitored and/or sensor-controlled actions. Sensor-monitored actions are executed as long as the measured variables exceed the specified limit values. In the case of sensor-controlled actions, the specifications for execution of the operation may be corrected to ensure that the specified limit values are always observed.

		Touch Sensors				Noncontact Sensors									
Principle		**Touch**				**Electrical**		**Optical/visual**					**Acoustic**		
Sensor type		Mechanical sensor	Strain gauges	Load cell (piezoelectric)	Pressure-sensitive plastic structures	Inductive proximity switch	Capacitive proximity switch	Light barrier	Reflection sensor	Laser scanner	Infrared sensor	Video system	Sonar ultrasonic barrier	Sonar	Ultrasound array
Signal type	**Digital**	X			X	X	X	X	X	X	X	X	X		
	Analog	X	X	X								X		X	X

Table 4.3: Technical Sensors (According to Hesse)

In other words, sensors are technical devices (measuring elements) that give robots a limited sensory capacity. They collect information about properties, states, or actions and make them available in the form of electrical signals. The actual primary conversion element that receives a nonelectrical measured value and forwards it as an electrical signal is also referred to as an *elementary sensor*.

4.10 Application Examples for Industrial Robots

KUKA.CNC is the new generation of controllers for direct CNC program code processing on a KUKA robot. With this, it is possible to execute CNC programs according to DIN 66025 directly by means of the robot controller. The complete scope of the standard code can be interpreted by the controller and implemented by the robot (G functions, M/H/T functions, local and global subprograms, control-block structures, loops, etc.). These options mean that the **areas of application** of industrial robots can be extended as follows:

- Milling of formed parts made of soft to medium-hard materials such as wood, plastic, aluminum, composite materials, etc.
- Polishing and grinding of formed parts
- Coating and surface treatment of complex component surfaces (*Figure 4.7*)
- Trimming and cutting of complex components and component contours
- Laser, plasma, and water-jet cutting of complex components

The direct integration of the CNC kernel into the robot controller turns the robot into a machine tool with an open kinematic system that combines the advantages of an industrial robot (i.e., large work envelope, high flexibility, low investment costs, six-axis processing, etc.) with the advantages of a CNC control system (i.e., G-code programming, CNC user interface, tool-radius compensation, looking ahead by many points, expanded spline path planning, convenient tool management, etc.).

Integration of the CNC kernel also makes it possible to process the CNC program directly on the robot controller. As a result, it is no longer necessary to perform any complicated conversion of CNC programs into robot programs. This makes it much easier to use industrial robots for typical machining processes. Therefore, both CAD/CAM programmers and CNC machine operators can use their existing know-how to program and operate industrial robots.

Figure 4.7: Milling cell in model construction.

Figure 4.8: Cockpit assembly in the automotive industry.

CNC-Robot Machining and Machine-Tool Automation

In addition to the fundamental capability to execute appropriate programs via the CNC kernel that is integrated into the robot controller, KUKA still provides the full range of functions of a robot controller. Thus it is possible to switch between CNC operation and conventional robot operation for various applications. This means that it is possible to select the ideal control, programming, and operator control environment for the process in question.

The CNC mode provides major advantages, especially in continuous-path processes. Processing programs with a large number of path points can be executed more precisely and with shorter cycle times via the CNC kernel and its subfunctions. These improvements in precision and cycle time are based most of all on the expanded path-planning function in the CNC kernel. By planning the path via looking ahead by more than 500 points, it is possible for the robot to maintain a constant velocity in the machining process and to plan acceleration/deceleration ramps in an optimal manner. But the continuous-path accuracy is also improved by means of various spline interpolations in the CNC kernel. Akima and B-spline calculations ensure that the robot moves along its programmed path as precisely as possible. These typical CNC functionalities are combined with the functions of a robot controller, thus producing optimal machining results. This includes, for example, robot compensation calculations, elasticity models, and so on.

But the advent of G-code programming also has redefined the use of robots in the context of CNC machine tools. In automated machining systems, it is now possible to program and operate not just the machine tool but also a robot per DIN 66025 code. This makes it easier to integrate robots and provides significant benefits, especially to the machine operator. From now on, operators no longer have to have robot-specific programming knowledge; instead, they can also operate the robot using their existing knowledge of CNC technology.

A special case in automated machining is machining parts that are handled by the robot. The combination of machining and handling makes this process variant especially efficient. Because the robot grips the components, moves them to a machining station, lets the machining be performed there, and then hands them over to a storage system, the process automation does not need any additional transport or handling kinematics. In such cases, the machine tool and automation system merge into a unit that provides a significant improvement in cost efficiency.

CNC programming opens up a wide variety of programming systems for the field of robotics. The market for CAD/CAM systems offers appropriate special solutions for a wide range of machining applications. This means that robot programming and the use of robots will see even more diverse employment in the future. An increasing number of CAD/CAM manufacturers have recognized these developments and therefore now equip their software with appropriate modules for robot programming and simulation.

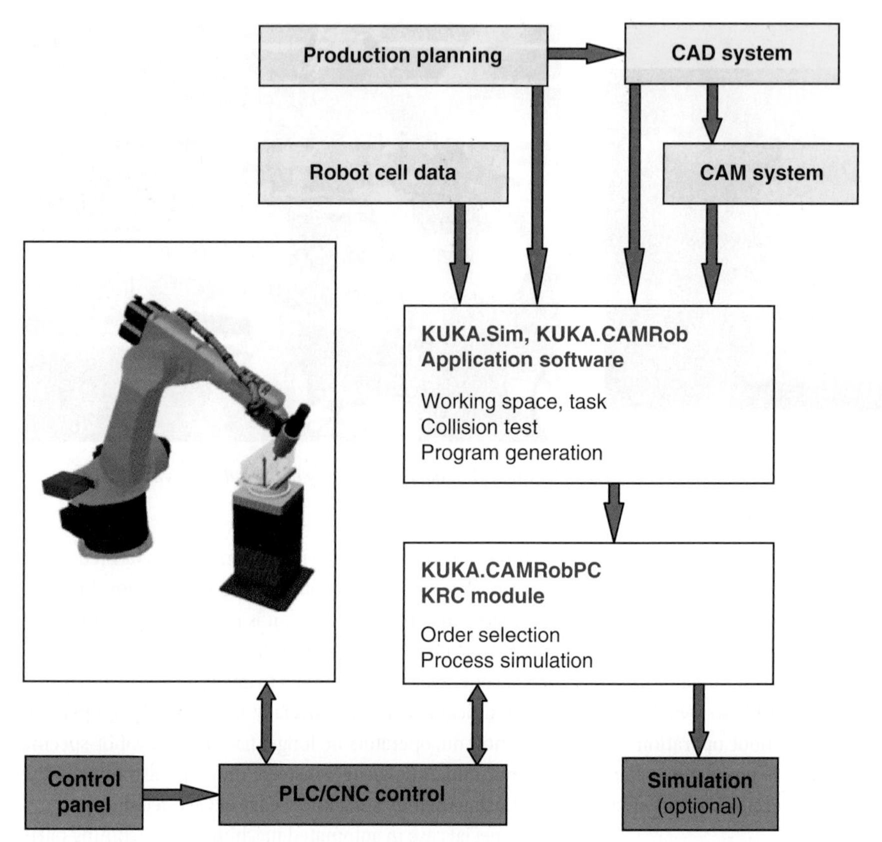

Figure 4.9: Programming concept of offline programming for KUKA robots.

4.11 Application Criteria for Industrial Robots *(Figures 4.7 to 4.14)*

Among the important reasons for employing flexible automation with robots are the following:

- An increasing variety of product types and variants based on market requirements
- Shorter production run times with more frequent changes of models
- Reduced quantities with greater conversion times and low utilization
- Conventional operating equipment
- Increased requirements for product quality
- Excessive throughput times with excessively large inventories and too much capital commitment
- Stressful and boring work that is hazardous to the operators' health
- Pressures to reduce the costs of products and increased product differentiation

The use of industrial robots depends, among other things, on the external conditions but also on economical considerations:

- **Environmental conditions:** Clean room, deep-freeze, toxic fumes, heat
- **Occupational safety:** Hazardous materials, weight, noise, etc.
- **Quality requirements in production:** Precision, zero-error production
- **Labor costs:** Increasing labor costs have an increasingly negative effect on competitiveness in industrialized nations.

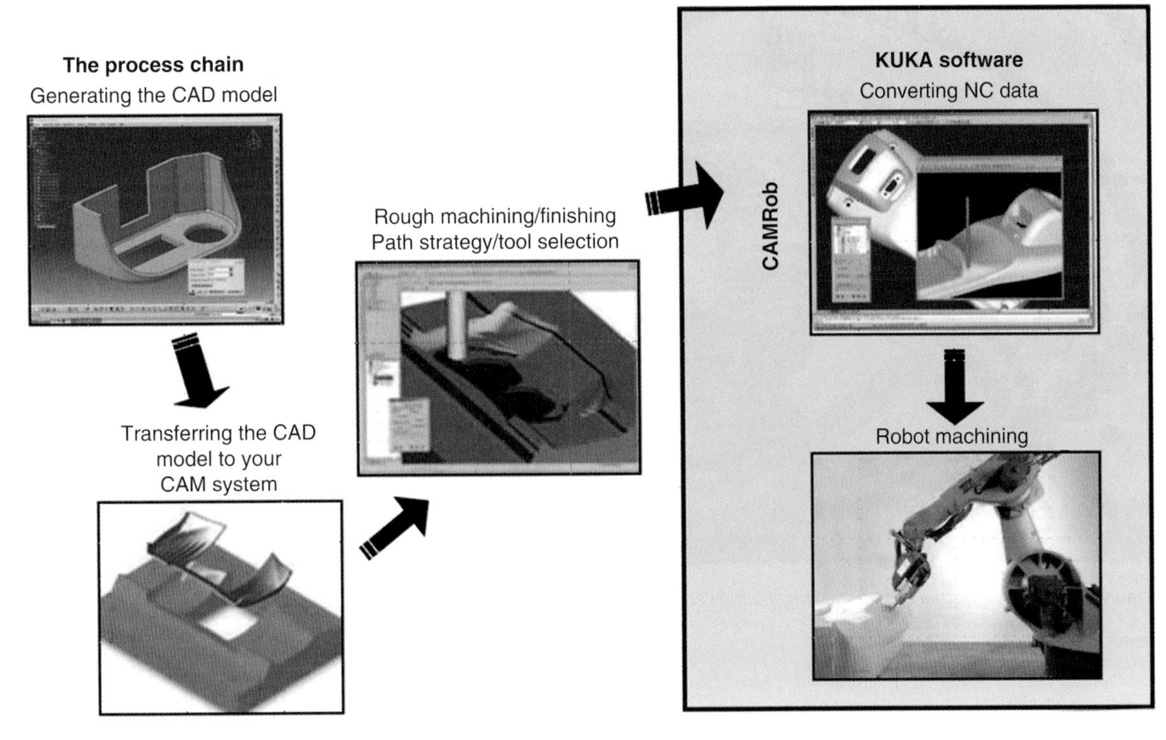

Figure 4.10: The process chain for creating a machining program.

- **Flexibility:** The use of robots increases flexibility in production.

The following criteria are important when selecting a robot:

- The general task at the place of use
- The payload, its point of application, and moment of inertia
- The work envelope within which the robot should move
- The process speed and/or the cycle time of the processes
- The required precision relative to the continuous-path characteristics or position
- The type of controller, including special requirements on the part of the systems integrator
- The ease with which the robotic system can be integrated into the surrounding industrial environment

These potential applications for industrial robots provide the following advantages:

- **High availability:** Unrestricted production at all times.
- **Optimized utilization of cycle times:** Utilization of non-productive times in the production process for additional tasks such as postprocessing of parts.
- **Increased output:** Robots can work at a continuously high speed with consistently high quality, resulting in a lower reject rate and greater production throughput.
- **Time savings:** Optimal positioning of the robot at the machines leads to shorter cycle times.
- **Cost savings:** High availability in combination with optimal production quality and short cycle times lead to a quick return on investment (ROI) and competitive production.
- **Fast:** Higher processing speeds than with manual processing with consistently high quality.

Figure 4.11: Automatic tool changing in machining centers.

Figure 4.12: Grinding and polishing of implants.

Figure 4.13: Milling of stone Basin.

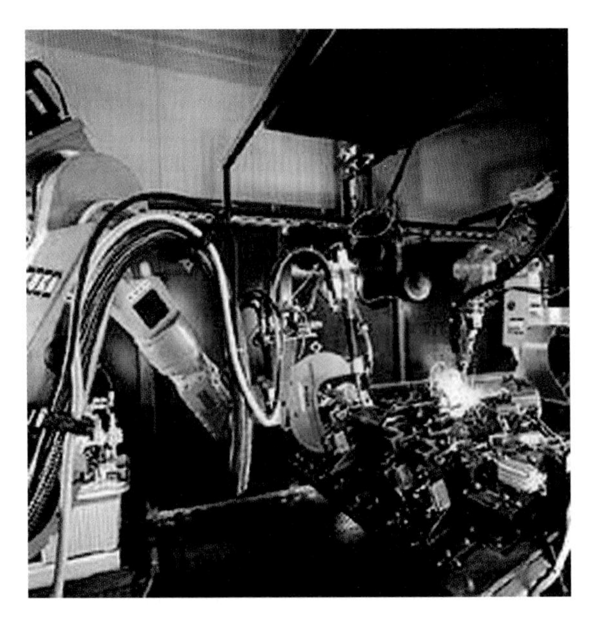

Figure 4.14: Arc welding.

- **Flexible:** Simple programming makes it possible to adapt flexibly to changing production processes and seasonal fluctuations.
- **Precise:** High positioning accuracy, even in continuous operation, leads to increased product quality and a lower reject rate.
- **An attractive long-term investment:** Industrial robots have a very long service life (15 years on average).

4.12 Comparison of Industrial Robots and CNC Machines

In their technical characteristics, industrial robots are typical CNC machines—yet at the same time very different in some major details. CNC machines have a number of fixed planes of motion based on linear main axes and one or two rotational or swivel axes. Given their design, these motion axes are very robust, which also allows high loading of the tools. This is especially important when machining metals. This robustness also allows very high dynamic accuracy during the machining process. The repeatability is also very high. For this reason, this approach is used primarily for the machining of metals with various types of machines.

In contrast, series-produced industrial robots should be versatile and able to perform many degrees of motion. This is achieved by means of a cascaded structure of up to seven swivel and rotary axes. As a result, their precision is limited. They are thus used mainly for the handling of workpieces or tools. In addition, they also have automatically changeable grippers or tools when necessary. For more demanding tasks, the robot is additionally equipped with special sensors or video cameras. This means, for example, that robots can detect various workpieces and their positions, grip them correctly, and adapt the machining to the varying workpiece positions.

Metal-cutting **CNC machines** can be subdivided according to their main areas of application, such as turning, milling, grinding, nibbling and punching, hybrid machines, as well as machining centers, for example, for drilling and milling. **Like CNC machines, robots** have to be designed appropriately for their specific area of application. Therefore, a distinction is made between assembly robots, handling robots, mounting robots, fitting robots, painting robots, welding robots, and deburring robots. A particular distinction is made based on the size and load-bearing capacity of the robot.

A **combination of a CNC machine with a robot** is very frequently encountered. In such cases, a special robot integrated into the machines takes tools from the magazine, exchanges them for the tools in the spindle, and returns the previous tools to the magazine. A series-produced robot is often sufficient for tool changing on turning machines. Robots are also often used for large but relatively lightweight workpieces, for example, for the handing of pressed parts for automobile bodies.

On the other hand, when it comes to the automatic changing of large, heavy workpieces or clamping fixtures at CNC machines, instead of robots, the use of pallets of various sizes with mechanical pallet changers is preferred.

There are no major differences as regards **controllers.** The number of CNC axes that have to be controlled and interpolated is also machine-specific. The controllers for CNC machines have significantly more commands and machining cycles (fixed machining subprograms) and generally also have an integrated, machine-specific programming system with graphical support for programming on the machine.

Programming of CNC machines is generally performed using CAD/CAM systems, followed by verification using simulations. The on-site programming in teaching mode that is common with robots is used only in exceptional cases for machine tools.

The **machining programs** for CNC machines can change a number of times every day in order to produce various parts. However, the sequential program for a robot is generally executed according to a single pattern that is defined once and then retained. This may, however, be modified in a limited manner depending on the process.

The areas of use and application for CNC machines and robots thus are extremely varied, as has already been described. The required flexibility for both robots and CNC machines is mainly provided by the CNC. For CNC machines, the main priority is "flexible reconfiguration," whereas for industrial robots and assembly robots, the main priority is "flexible conversion," that is, reusability in a different production location.

4.13 Summary and Outlook

The increasing global integration of economic activities and the associated employment of industrial products and systems require standardization of the interfaces and also of the communication between robots and assembly systems. The goal is to standardize these on a worldwide level and to make them simple to use. This means that the user can make investments with confidence and gives the manufacturer a stable environment for new development projects.

Without communication between the individual controllers in a system, modern automation technology cannot achieve much. Especially with open bus systems such as Ethernet and TCP/IP, it is possible to implement open- and closed-loop control of complete production systems in real time regardless of their location. There is a definite trend among robot manufacturers to equip and connect their own controllers with technologies that have already been in use in the computer industry for a long time.

The implementation of more and more software resources in the robot itself is opening up greater opportunities as regards maintenance. If previously a maintenance process was initiated by stopping the system or cyclically on a preventive basis, in the future the individual robots will interpret the counter values, measuring points, and other statistical data itself and will automatically report that it requires maintenance once particular states or events take place. Even in the context of service, it is possible to conceive of scenarios in which the robot is connected directly with the manufacturer's system via web portals in order to allow not only remote diagnosis but also ordering of spare parts. During evaluation of the system, various databases in the robot are accessed. These contain data regarding the maintenance history, the load profile of the drive train, and case-related solutions as an aid for interpretation. Connection to a service platform thus provides access to a broad range of experience. Maintenance is initiated automatically at the optimal time once the criteria are fulfilled. At the same time, this is also integrated with the production planning and control processing, the spare-parts system, and other relevant departments. This means that machine availability can reach a previously unheard-of level.

The new communication hierarchies and communication mechanisms ultimately also have an effect on quality management. In the future, the time-consuming use of hardcopy process information and product information, which is often associated with greater costs and more frequent errors, will be replaced by an automated process. Measurement data are interpreted based on continuous measurement of the product data and process data and on comparison of the measured values with reference models. The measurement results that have already been interpreted are then stored online in a central database. This means that the process according to which each product is created and the associated quality assurance data can be documented individually. By integrating the competence of the process supplier, it is then possible to implement a continuous process-control/trouble-shooting system to achieve optimal production output.

Industrial Robots and Handling

Important points to remember:

1. Definition of industrial robots: A robot is an automatically controlled, reprogrammable, multipurpose manipulating machine with several degrees of freedom that may be either fixed in place or mobile for use in industrial applications. Industrial robots
   - Are freely programmable
   - Are servo controlled
   - Have at least three NC axes
   - Have grippers and tools
   - Are designed for handling and processing tasks
2. Construction of industrial robots: Industrial robots comprise the following subsystems:
   - The mechanical structure
   - The kinematic system
   - Axis control and drives
   - End effectors
   - Sensors and sensor systems
   - A controller
   - Programming
   - A computer
3. Mechanical elements/kinematics: Industrial robots have the structure of an open kinematic chain with no branches whose elements or lever arms are linked to each other in pairs via joints. Each joint has either a rotational or translational degree of freedom.
4. Grippers/end effectors: A distinction is made between the following grippers, subdivided according to their functional principle: Mechanical grippers, pneumatic grippers, suction grippers, magnetic grippers, adhesive grippers, and needle grippers.
5. The controller: The robot controller can perform an extremely wide range of functions within a robotic cell:
   - Controlling the robot's motions
   - Influencing the process components in the system
   - Influencing the conveyor and in-feed components
   - Controlling the gripper functions
   - Receiving and evaluating sensor signals
   - Receiving and processing process information for influencing the process
   - Diagnostic functions for error detection on the robot or in the process
   - Providing support to the operator
   - Supporting the programmer when setting up the automation cell
6. A distinction is made between the following programming methods:
   - Online methods
   - Offline methods
   - Hybrid programming methods
   - Automatic program generation
7. Application criteria for industrial robots: The use of industrial robots depends, among other things, on the external conditions but also on **economical considerations:**
   - Environmental conditions
   - Occupational safety
   - Quality requirements in production
   - Labor costs
   - Flexibility

Part 4

Tooling Systems for Computer Numerical Control Machines

1. Tooling Systems

In order to be suitable for use in a computer numerical control (CNC) machine tool, a tooling system must be universal and able to be used flexibly for all kinds of machining tasks, very stiff, capable of achieving a high stock-removal rate, and yet still remain within an acceptable cost framework. It is equally important for the system to be able to scan, update, and manage the tool data in an integrated, closed process.

1.1 Introduction

The productivity and reliability of the cutting tools used in a CNC machine tool will have a decisive influence on the productivity of the machine itself. The productivity of the tool may be defined in terms of attainable cutting data, such as the cutting speed, the feed rate (or simply feed), the cross-sectional area or volume of the chips removed, or the tool life in terms of time, travel, or quantity.

The **reliability of tools** has both technological and geometric aspects. Both are critical to the profitability of the automatic machining of workpieces. **Technological reliability** indicates the degree to which the performance data are reproducible. In other words, it must be possible to determine in advance the cutting parameters at which the tool will not break and how many workpieces can be machined at an acceptable level of quality before the tool must be replaced.

The very principle that underlies the system of numerically controlled tool motions makes it absolutely necessary for the tool to possess **geometric reliability.** The CNC system processes not only the target dimensions of the workpiece but also the specified dimensions of the tool. If the actual tool dimensions deviate from the stored specifications, the machined workpieces may be unacceptable. Therefore, a tooling system for CNC machine tools must ensure that the cutting-edge position (which determines the final dimensions of the workpiece) does not stray from its intended position during the machining process. The system also must ensure that the desired position is resumed with great accuracy after the tool has been changed. Although modern controls can calculate tool compensation without difficulty, this always requires measurement of the actual dimensions of the tool. This is costly and time consuming to perform during the machining process and therefore should be restricted to rare cases, for example, when creating "fits" to close tolerances.

Aside from the geometric and technological aspects of reliability, the **flexibility of the tooling system** is very important to many operators of machining centers. After all, it is often necessary to change practically overnight to an entirely different type of workpiece (and to obtain the special tools required by this new workpiece). This situation is especially common in job shops. The logical solution to this problem is the **modular tooling system,** which allows the most suitable tools for the job (e.g., boring bars) to be assembled from standard elements.

Aside from the productivity and reliability of the tools, **the tooling system** must be **easy to handle.** This will allow the user to take advantage of the benefit of CNC technology, namely, the ability to adapt quickly to changing machining tasks. From a theoretical standpoint, the workpiece and tool can be regarded as interacting with each other. In practical applications, too, the cutting edge of the tool will indicate how productively—and therefore, how *profitably*—the workpiece can be machined by the overall **system consisting of the machine tool, control, and cutting tool.**

1.2 Tool Systems

Tool systems represent the connecting link between the cutting edge (where the chips are produced and the cutting forces take effect) and the machine tool, which absorbs the forces and simultaneously guides the motions between the workpiece and the cutting edge of the tool.

The design of the tool system is determined primarily by the machining process (e.g., drilling, milling, or turning) and the design characteristics of the machine tool on which it will be used.

Tool Systems for Drilling and Milling

The CNC machining center was developed from the classic boring-milling machine, in combination with numerical control technology. Its distinguishing characteristic is that tool changes are performed automatically. Even when the tools used in a boring-milling machine are changed manually in a particular case (for economic reasons, for example), the elements still should be suitable for a future upgrade to fully automatic handling.

The essential elements include

- **The tool holder,** which is used to insert the tool into the main spindle
- **The tool** itself, which is attached to the holder and measured off-line
- **The tool magazine,** in which the various tools required for machining are stored
- **A tool-changing device,** which transfers the tools (including their respective holders) back and forth between the working position and their magazine locations

Chapter 1 of Part 3, Computer Numerical Control Machine Tools, describes, among other things, tool systems in CNC machining centers.

The **tool holder** is the most important element of a tool system. Despite international efforts at standardization, there are different systems of tool holders. The differences pertain especially to the gripper location slots for automatic tool changing and to the devices used for pulling the tools into the spindle. These latter components include drawbolts and collets. The taper has been standardized and can be described by the nominal sizes 40, 45, 50, or 60, for example. The design of the end of the taper varies according to the tool-insertion system and is frequently interchangeable. *Figure 1.1* shows a series of holders used in the machining centers of various manufacturers.

In order for cutting tools to be exchanged quickly and safely on a given machining center, uniform tool holders are required for each tool. Their design is essentially determined by

- The **taper** in the machine spindle that holds the tool
- The **gripper location slots** for the automatic tool changer, including a reference mark for tool orientation
- The **type of clamping system** with which the tool holder is held positively in the working spindle so that the cutting forces a**re transmitted** securely

In order to hold down the costs of these tool holders when several different machining centers are used, the same type of tool holder should be used for all machines.

In the case of tool systems for machining centers, tapers according to DIN 69871 (ISO 7388) have become increasingly popular. *Figure 1.2* shows holders with a hollow shank (front) and with a taper.

The effort to create a uniform interface that also will satisfy requirements for high stiffness and precision and is also suitable for high speeds has been concluded successfully with the standardization of the HSK interface in DIN 69893. The international standard ISO 12164 was approved in 2001.

Tool Systems for CNC Turning Machines

CNC turning machines make use of two kinds of tooling systems with fundamentally different tool carriers:

- Indexing tool turrets
- Tool magazines combined with a tool-changing device

Each system has its advantages:

- Because of their short indexing times, **tool turrets** allow tools to be changed quickly.
- **Tool magazines** make it possible to store a larger number of tools without introducing a risk of collision within the work area of the CNC turning machine.

In both cases, the tool shanks are usually clamped in tool blocks, which, in turn, are held in indexed positions on the tool carrier. These tool blocks correspond to the tool holders used in machining centers and also come in two standard designs:

- With cylindrical shank
- With square blocks

For CNC turning machines, the interface between the tool carrier and the tool is governed by DIN 69880. In recent years, however, the joint between the exchangeable cutting head and the base holder has developed in a

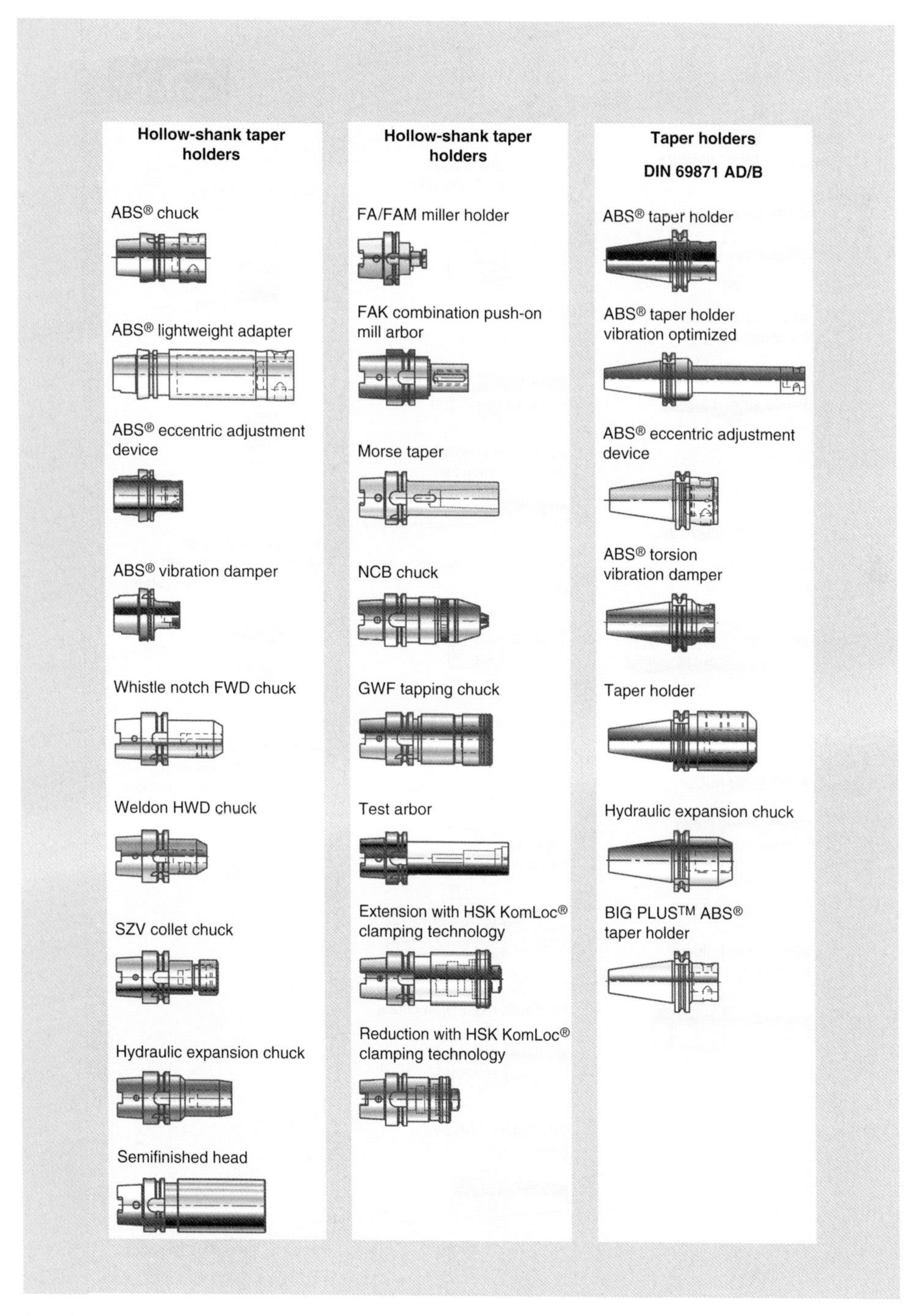

Figure 1.1: Overview of tool holders for drilling and milling tools.

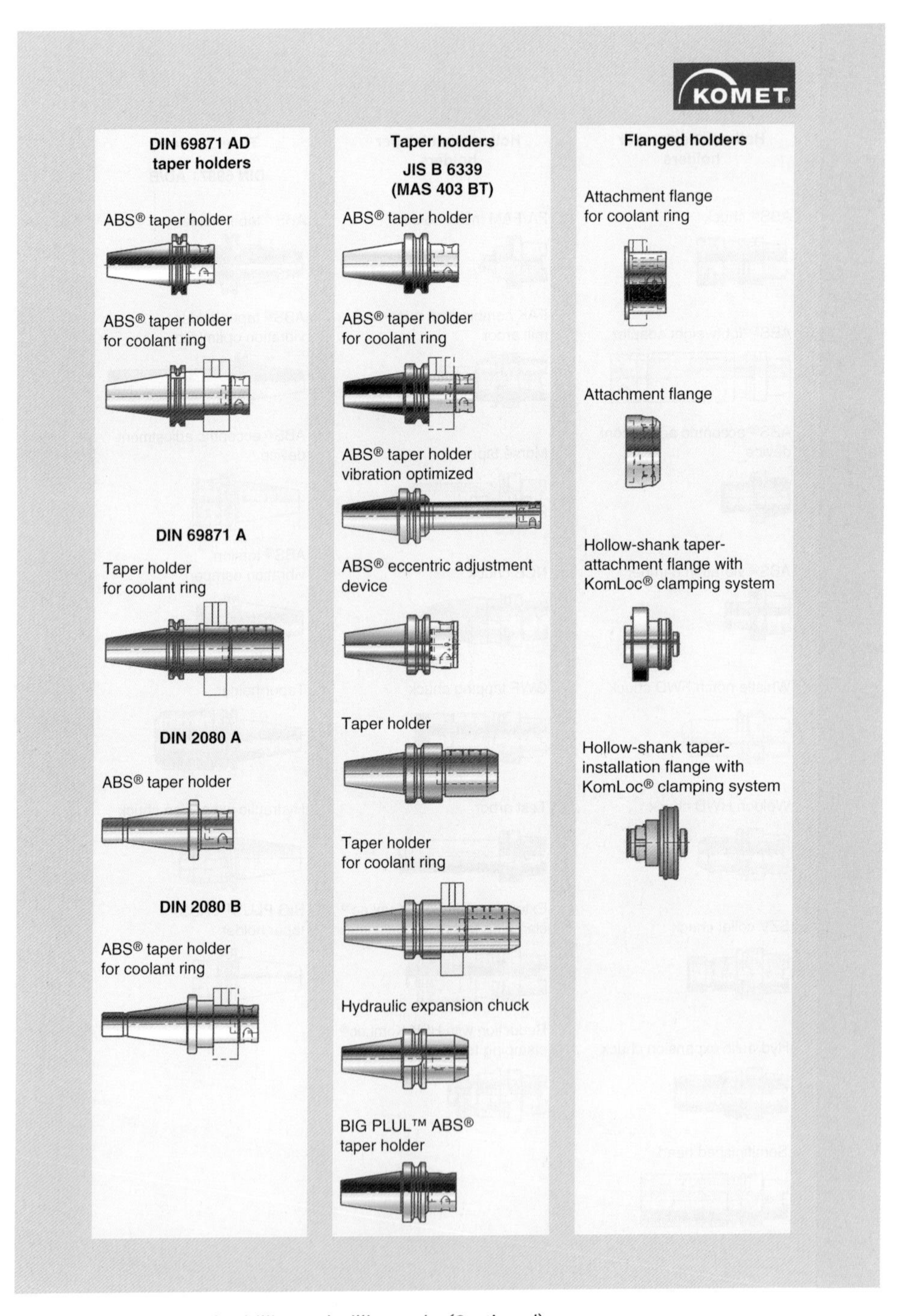

Figure 1.1: Overview of tool holders for drilling and milling tools. (Continued)

Figure 1.2: Tool holders according to DIN 69871, Part 1.

number of different directions. In the past, a number of company-specific systems have been used, for example, those from Hertel, Sandvik, and Widia. These systems achieved only limited acceptance, however. The brands currently established on the market include Coromant Capto, Kennametal Widia, Komet, and ISCAR, especially with regard to systems that are suitable for both rotating and stationary tools (*Figures 1.3 and 1.4*).

However, the standardized HSK interface offers the user an optimal solution. For the first time, the user is able to equip the production system with a **single** uniform tool system, that is, a machining center and turning machines. Standardization means that the user is not dependent on any specific supplier.

Tool Turret Designs (*Figure 1.5*)

Aside from certain standard tool turret designs that are also manufactured by various suppliers, many manufacturers of NC turning machines have developed systems that are specially tailored to the work area and overall concepts of specific machines. The standard types of construction include

- Star turrets
- Disc turrets
- Drum turrets

However, most of the machine-specific tool turret designs also can be traced back to one of these principles of tool arrangement (*Figure 1.6*).

Tool Magazines on CNC Turning Machines

Tool magazines are used less commonly than turrets on CNC turning machines because the tool-changing device is usually more costly than the indexing mechanism that is used with a turret. The principal arguments for a magazine-based solution are therefore the **reduced risk of collision** and the ability to insert a **greater number of tools** without manual intervention. The principle of using magazines on CNC turning machines has received an added boost from tool-system developments that provide for only the tool head and insert (and not the entire tool block) to be interchanged. This type of design allows a greater number of tool inserts to be stored in a relatively small space. It also allows these inserts to be kept available for a longer period of machining time, provided that the appropriate automatic tool-changing devices are present.

Driven Tools (Live Tooling) for CNC Turning-Milling Centers

Many turned parts require a finishing operation that is impossible to perform on most turning machines. Examples include eccentric, axial, or radial drilling and milling operations, such as milling a longitudinal slot on the outside diameter of a part, milling a transverse slot on the end face, drilling axial or radial holes with or without screw threads, or milling a polygonal section on the end face of a workpiece.

A modern CNC turning machine can be used to perform these tasks, provided that it features a numerically controlled main spindle (*C* axis) and live tooling in the turret head. Depending on the tools that are provided, it may be possible to use this type of turning center for turning, drilling, reaming, countersinking, milling, or tapping.

Several tool manufacturers offer axial and radial tool heads for this area of application. These tool heads feature a special coupling that allows them to be driven in the tool turret. As far as the CNC system is concerned, it is not sufficient

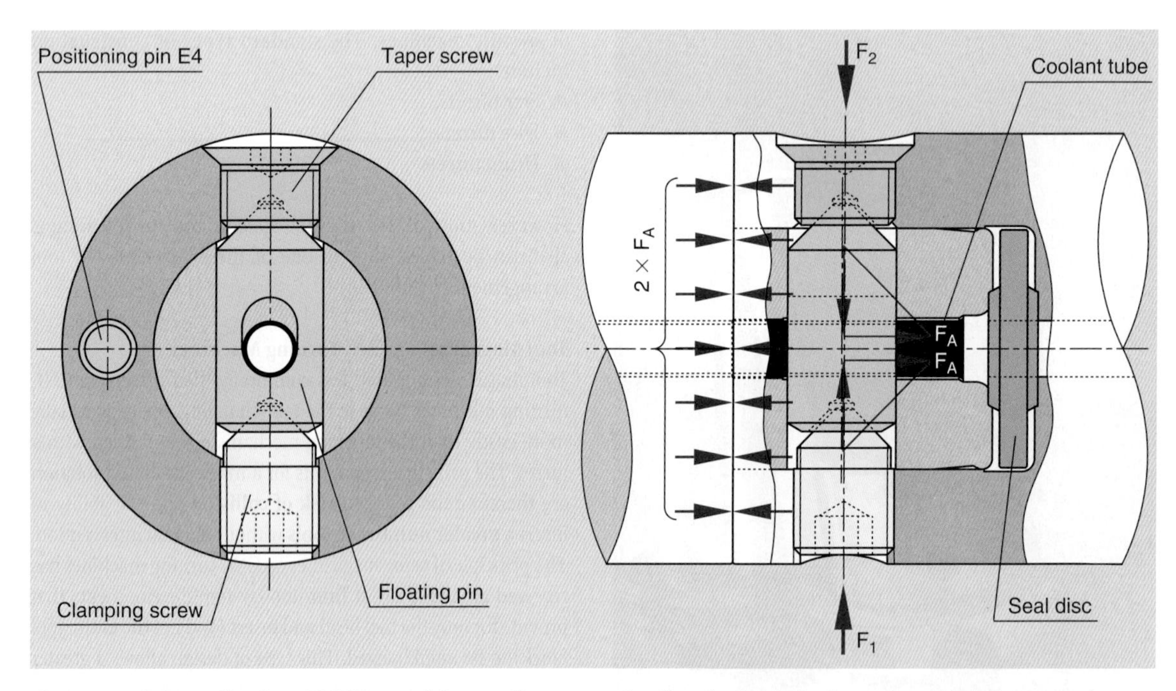

Figure 1.3: ABS coupling from KOMET, a modular coupling system for all tools, that is, both rotating and stationary tools. Tool holders equipped with the ABS system can be used to hold tools, extensions, or adapters.

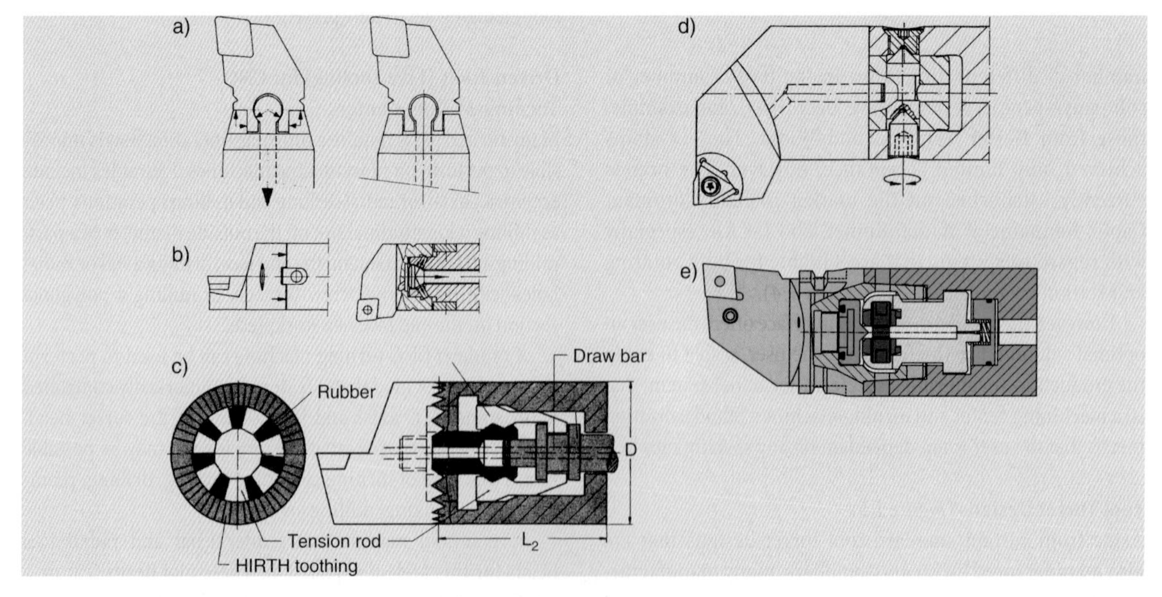

Figure 1.4: Systems for turning tools with exchangeable insert holders: (a) block tools (Sandvik Coromant); (b) multiflex (Krupp Widia); (c) FTX (Hertel); (d) ABS system (Komet); (e) hollow-shank taper (with KomLoc from Komet).

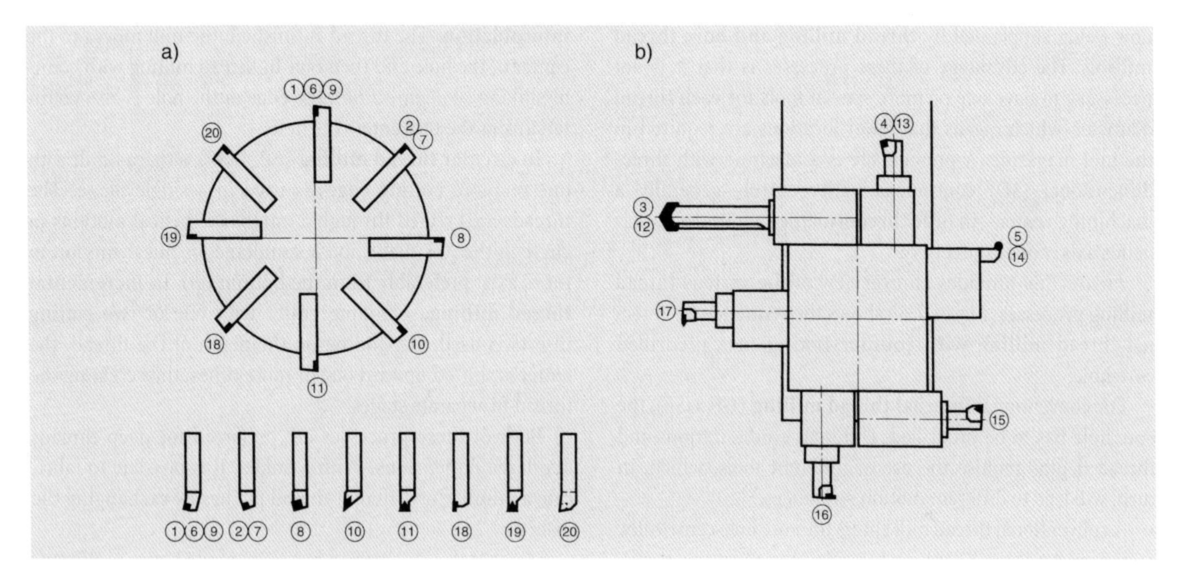

Figure 1.5: Tool turrets for NC lathes: (a) disc turret; (b) star turret (universal turret).

to merely provide a third axis for the spindle. A special method of programming is also required.

Drilling operations require spindle programming in terms of degrees and minutes. In contrast, milling operations require programming to be performed in rectangular coordinates, in the same manner as on a milling machine:

- *Z/C* for machining on the periphery
- *X/C* for machining on the end face

The special CNC software performs the complex interpolation of the axis motions as a function of the diameter.

Tool Selection

Numerically controlled machine tools open up new possibilities for the sequences of operations used in machining workpieces. In some cases, for example, the goal is to perform a complete machining process on a single machine using fewer setups than in conventional production. Tool selection also must be directed toward this goal. In other words, **tools for many different types of operations** must be combined on a single CNC machine. This would seem to increase the number of tools needed at each workstation and across the entire spectrum of different workpieces. However, numerical control also makes it possible to produce a great variety of workpiece contours by moving standard tools along controlled paths, thus eliminating the need for special profiling tools. This, in turn, reduces the required number of tools.

Selection of the proper CNC tools is not determined by individual contour elements or workpieces but by the entire spectrum of parts. **A few versatile tools of high quality** will always represent a better solution than a multitude of special tools tailored to geometrically unique applications. But technological optimization is important; the use of **indexable inserts** makes it easier to adapt the cutting material and geometry of the tool insert to the material, stability, and structure of the workpiece. Here it is important to pay special attention to **chip flow.** The relatively high rate of stock removal and the widely varying directions in which the tools are inserted and moved frequently will require special measures for rapid chip removal. Problems of this type are often solved by providing an internal **coolant supply** or building coolant lines and connections into the tool blocks and adapters. These solutions do not disturb the automatic sequence of tool-changing operations.

Example: Internal Thread Milling (→ *Figure 1.6*)

To create internal threads on CNC machines, the established method of tapping with or without a compensating chuck is now being supplanted by **thread milling and bore thread milling**. The advantage of these processes is that it is not necessary to have one or more special tools for each thread diameter, which means that fewer locations are required in the tool magazine. A prerequisite is a machine with three-dimensional (3D) continuous path control—generally a machining center—in order to control the required simultaneous axis motions precisely.

Figure 1.6a provides an overview of the various thread milling processes. *Figure 1.6b* shows the sequence for internal thread **milling with countersinking** in a predrilled core hole.

For conventional **internal thread milling** (*GF, GSF*), the core hole has to be predrilled. Different configurations and thread depths require the use of different tools, which, in turn, will lead to different motion sequences.

A conventional thread mill enters the core hole centrically, moves to the thread contour in a circular approach curve, and then upward by one thread pitch in a single 360° **helical interpolation.** The thread is finished; the mill moves to the center of the hole and then out. In thread milling with countersinking (→ *Figure 1.6b*), the edge of the hole is also countersunk as the tool enters the hole.

In **circular thread milling** (*EP, WSP*), a thread mill with one or more cutting edges is used in a single plane. The threads are created through a number of helical motions (a circle in the *X/Y* plane and a simultaneous linear motion in the *Z* axis, preferably from the bottom up). In **incremental thread milling,** a milling cutter with one or two cutting inserts is used. Depending on the depth of the thread, the cutter is shifted upward one or more times, thus creating the thread in separate stages.

Both of these processes are preferred for deep threads and large dimensions. With a holder, it is possible to fabricate a number of different thread pitches by exchanging the inserts.

In general, a distinction has to be made between **dimension-specific thread mills** and **non-dimension-specific**

Area:	Internal thread milling					Bore thread milling	
Preparation:	Drilling the core hole			(countersinking) except for thread milling with countersinking		None	
Process:	Thread milling with solid carbide tools			Thread milling with drilling inserts		Bore thread milling	Circular bore thread milling
Type:	GF	Conical thread milling	GSF	EP	WSP	BGF	ZBGF
Process-principles:							

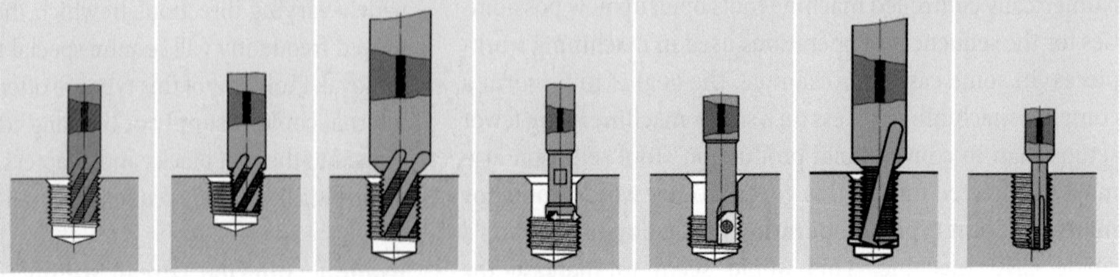

Figure 1.6a: The various internal thread milling processes.

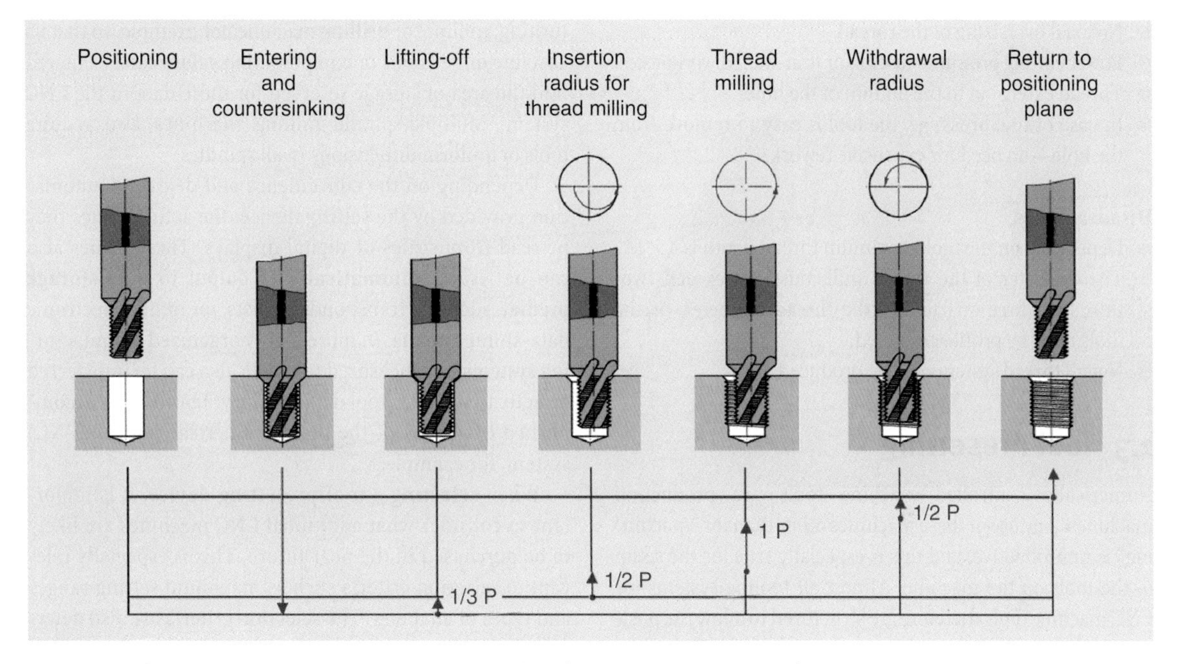

Figure 1.6b: Sequence for internal thread countersink milling (with a predrilled core hole).

thread mills. Dimension-specific tools are designed for a particular range of thread dimensions, and the sizes of threads that can be produced are fixed. With non-dimension-specific tools it is possible to produce any desired thread diameter for a specified pitch, provided that the ratio between the diameter of the mill and the diameter of the thread exceeds a certain minimum value.

With non-dimension-specific thread mills it is important to note that there is a specific **minimum ratio between the mill diameter and the thread diameter** that has to be observed. For standard metric threads, a ratio of 2:3 can be used as a guide value, and for metric fine threads, a ratio of 3:4 is used. The reason for this relationship is the profile distortion that arises during the thread-milling process. During this process, a linear thread profile moves in a helical interpolation, which distorts the thread profile in the pitch. This distortion must not exceed certain tolerances if a true-to-gauge thread is to be produced.

Unlike with ordinary thread milling, in which the core hole has to be present, the **bore thread milling process** (→ *Figure 1.6a, BGF, ZBGF*) is performed without premachining. The tool penetrates into the solid material, drills the core hole, and while it is being withdrawn, it mills the thread in a helical interpolation (*XY* circular and *Z* linear). This process is only used for short-chipping materials such as gray cast iron.

In **circular bore thread milling** (*ZBGF*), the helical interpolation is initiated before the tool contacts the solid material, at which point it mills the thread **from the top down.** Once the programmed thread depth is reached, the tool moves to the center of the thread and out.

The **approach radius** can involve either a 90° or 180° motion. The purpose of this approach motion is to minimize the angle of contact of the tool to prevent tool breakage. The more robust the tool, the shorter the approach arc can be.

General Advantages of Thread Milling

- Machining practically independent of diameter in dimensions and tolerance.
- A single tool for both right- and left-hand threads.
- No chip problems thanks to small, "comma-shaped chips" that are easy to remove from the hole.
- No change in direction of rotation of the spindle.

- No axial oversizing of the thread.
- Lower cutting pressure—useful for thin-walled workpieces.
- Thread extension to the bottom of the hole.
- In case of tool breakage, the tool is easy to remove from the hole—no need for expensive rework.

Disadvantages

- Depending on the tool, maximum thread depth is $4 \times D$.
- The diameter of the thread mill must not exceed two-thirds or three-quarters of the thread diameter, or the tools must be profile corrected.
- Not all thread systems can be produced.

1.3 Tool Presetting

Numerically controlled machine tools are production machines. Any use of these machines other than for "machining" is unproductive, and this is especially true for the setup of the tools on the machine. Almost all tooling systems for CNC machine tools therefore are structured to allow the positions of the cutting edges of the tools to be adjusted to a predetermined dimension or to allow the actual dimension to be determined. This is accomplished at a location separate from the machine tool by the use of **presetting devices.** These presetting devices are equipped with appropriate adapters that hold the tools just as they are held on the machine. The positions of the cutting edges are measured by optical methods using a microscope with an eyepiece or ground glass or by the use of touch probes (*Figure 1.7*).

The objective of tool **presetting** is to adjust the main edge and secondary edges of a cutting tool in relation to a fixed point in the tool holder (*Figure 1.8a*). It is important to perform this adjustment with the greatest possible precision in accordance with the presetting specifications. The tool must be designed to allow presetting.

Tool presetting is necessary, for example, in the manufacturing of workpieces with tools of firmly established dimensions. Even boring bars and other boring tools must be adjusted precisely with respect to their diameters, but the CNC system can compensate for deviations in the lengths of these tools.

The objective of **tool measurement** is to precisely determine the actual distance from the cutting edge of the tool to a fixed point within the tool shank (*Figure 1.8b*). A tool is measured before being used in a numerically controlled turning, milling, or drilling machine, for example, so that its absolute dimensions or compensation values can be entered into the area of storage reserved for these data in the CNC system. Multiple-spindle milling machines also require tools of uniform dimensions in all spindles.

Depending on the convenience and degree of automation provided by the setting device, the actual values may be read from scales or digital displays. These values also can be issued **automatically** as output to **data-storage media,** such as transponders, lists, or other electronic data-storage media. In more highly organized manufacturing systems, the measured tool data also can be transferred directly to the CNC tool data memory. This may be accomplished by means of the direct numerical control (DNC) system, for example.

When **selecting a tool-presetting device,** it is important to consider what additional CNC machines are likely to be purchased in the near future. This is especially relevant to selection criteria such as maximum setting ranges and types of adapters. The selection criteria are also determined by the general organization of tool distribution within a company: Are tools distributed from a central tool crib or assigned to a single CNC machine or a single group of machines? There are many different technical, organizational, and economic viewpoints to consider here, and each may suggest a different "ideal" solution for the company. Before making a decision, therefore, it is especially important to look for opportunities to study alternative solutions that have been attempted by other companies who are willing to discuss their own problems and experiences in this area. As long as the unique conditions that prevail within a given company are taken into account, practical examples are often more valuable than general theories. In any event, the use of CNC machine tools always requires systematic, careful planning of tool organization and presetting.

1.4 In-House Tool Catalogs *(Table 1.1)*

The dimensions and technical data pertaining to the use of tools in CNC machines are incorporated into the CNC part program. Differences in dimensions mean higher costs in the areas of programming and tool presetting. These differences also result in greater investments in tools and tool holders.

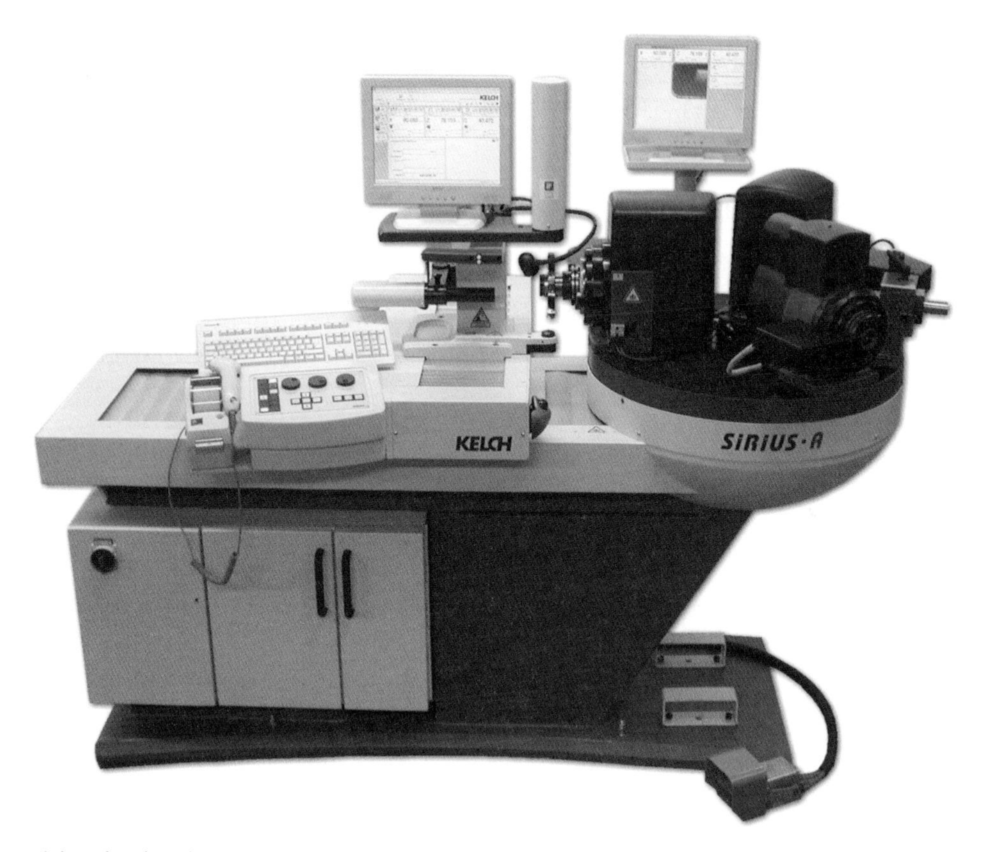

Figure 1.7: High-end tool-setting and measurement device with CCD camera and CNC-controlled axes, suitable for fully automatic measurement with all kinds of tools for turning, drilling, milling, etc. With automatic saving of data and transmission via direct numerical control (DNC) to the connected CNC machines. In the case of tools with an integrated memory chip, there is a choice of transmitting all the manufacturing-related data or assigning them to tool numbers using a separate tool data computer. This computer also performs external tool data management for all the tools, such as compensation values, remaining tool life, and collision data. It is also possible to call up and save the manufacturer's tool data directly via an Internat connection. If multiple presetting devices are connected via a network, then tool data and measuring programs can be used via a common multiuser database. The illustration also shows the reader unit for radiofrequency indentification (RFID) systems, a hand scanner for reading barcodes, and an additional monitor solely for depicting the inserts. (Courtesy of Kelch, www.kelch.de.)

Standardization is aimed at performing similar tasks with tools of the same type, but this requires a systematic analysis of the tools and applications. This is best accomplished in the form of a company-specific tool catalog. Today, preference is given to interlinked electronic systems that form a continuous data network: presetting device → tool-management computer → tool-storage facility → CNC machine → tool-storage facility and so on. In addition to depicting the tools in the correct scale, such systems also manage the tool type, the adjustment dimensions, recommended cutting parameters for specific materials, remaining tool life, weight class, and so on. The data output from the tool manufacturer's electronic catalogs are automatically applied.

A code number provided for purposes of classification will help to organize the tooling system and the tool combinations that are designed and used. The code may include the machining method, the type of NC machine, the type of tool holder, the tool block design, the type of

Computerized Tool Catalogs Should Meet a Number of Minimum Requirements:
• Data and graphics for tools should be adopted directly from all manufacturers and suppliers.
• It should be easy to find and select tool on the screen.
• Use of company-specific classifications and numbering systems
• Fields must be freely available for each tool for personal comments.
• The user must be able to register his or her own special tools.
• In addition to the manufacturer's data, the software also should contain a comprehensive catalog of non-manufacturer-specific standard tools.
• Putting together complete tools. Components from various manufacturers can be combined to create complete tools. Graphics, dimensions, and bills of materials are determined automatically. Suitable collets, drill-chuck inserts, and return pins are automatically added in the background during the assembly process.
• Acceptance and saving of complete tools; if broken down into separate parts, automatic updating of the component catalog.
• Tooling sheet and bill of materials with dimensioned graphical depiction, as well as data for the necessary tool compensations.
• Transmission of the setting and compensation values to the CNC system via DNC.
• To aid NC programming, it must be possible to transfer cutting parameters, types of machining, geometries, and limit values automatically.
• Tool management and tool information, that is, what tool is located where?
• Creation of a complete tool list with all data for command position input for presetting.

Table 1.1: Requirements for Digital Tool Catalogs

insert holder, and the type of indexable insert used. First and foremost, a code number of this type must be compatible with the company's own internal numbering system. The code in *Figure 1.9* therefore is intended only to serve as an example of a code that could be applied to turning tools.

Creating a Tool Catalog *(Figures 1.10 to 1.16)*

There is a new approach to providing tool data and graphics in digital form. In this approach, a tool catalog is created with the help of a so-called generator. This functions first as a source of data for the tools and second as a software tool for generation of data and graphics, and saving them directly to a database. The geometric parameters for creating true-to-scale two-dimensional (2D) and 3D graphics for use in simulations are also stored.

It is important to be able to output the graphics in all the typical formats. This provides better integration into various computer-aided design (CAD)/computer-aided manufacturing (CAM) and simulation systems.

The generator includes standard tools from various manufacturers, classified by the type of machining. The parameters can be accessed and modified, which makes it

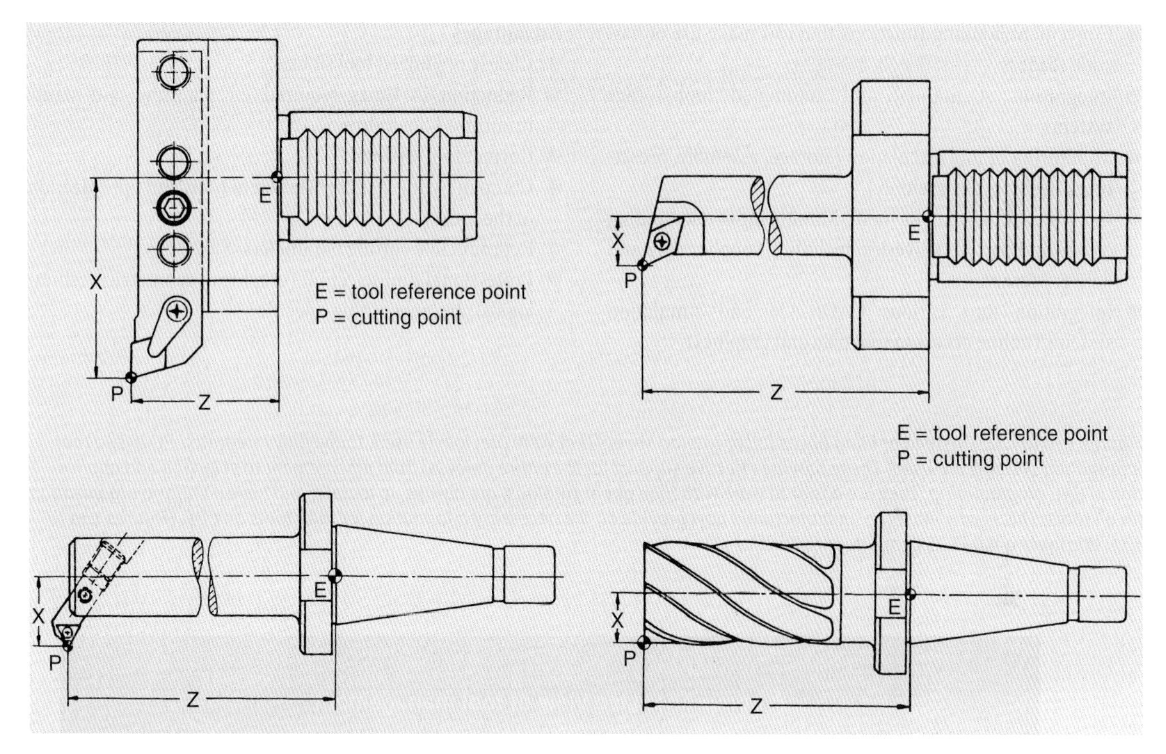

Figure 1.8a: Adjustable tools.

Figure 1.8b: Tool data that have to be measured.

simple to generate semistandard tools and tools from different manufacturers. The data and graphics are created automatically in a user-oriented graphical class/group structure.

TDMeasy has been developed to meet the special needs of small and medium-sized companies. It contains the same data structure as *TDM V4*; presentation and handling require just a few screen masks.

Functions

- Organization of tools, fixtures, measuring and inspection equipment, setup and clamping equipment
- Tool management structured by tool components, complete tools, and tool lists
- Organization of tool storage: Management of the tool storage and tools in use, new and used tools

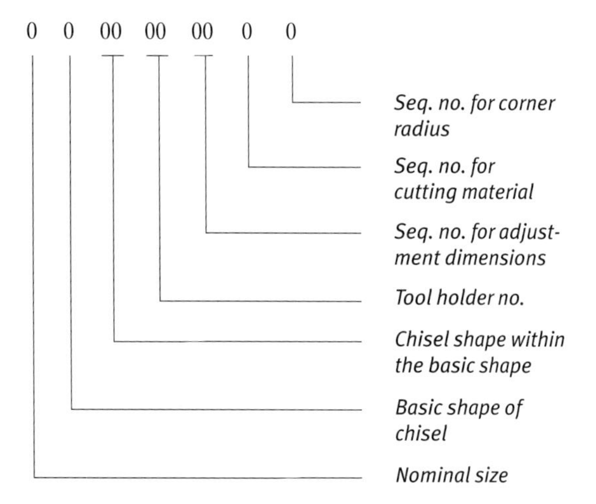

Figure 1.9: Examples of an in-house numeric codes for identifying turning tools.

- Convenient booking functions that can make use of bar-code readers
- Integration of manual and automated tool-storage systems
- Monitoring of tool circulation (storage, assembly, presetting, machine, disassembly)
- Integrated order management: Monitoring of outstanding orders and partial deliveries, including supplier data and delivery terms
- Integration into various CAD/CAM and simulation systems (online access to tool data and graphics)

Advantages

- Clearly organized tool stocks
- Reduction in times required for planning and setup; minimized downtime
- Correct tool settings
- Company-wide standardization of tools and optimization of the tool stock
- Reproducible cutting parameters
- Professional tool selection (graphical, geometric, technological, and machining-specific)

Figures 1.10 to 1.14: Instead of filing boxes full of catalog sheets that were previously used, today fully integrated PC-based tool-information systems are used. These provide effective support for the entire process, from procurement to selection and application in CNC programming. They are easier to work with than paper catalogs, are always up to date, and have a uniform organization for all tools. This is why most tool manufacturers now provide all the necessary information for their tools on CDs. (Figures 1.10 to 1.14: WinTool from DATOS Computer AG, Zurich.)

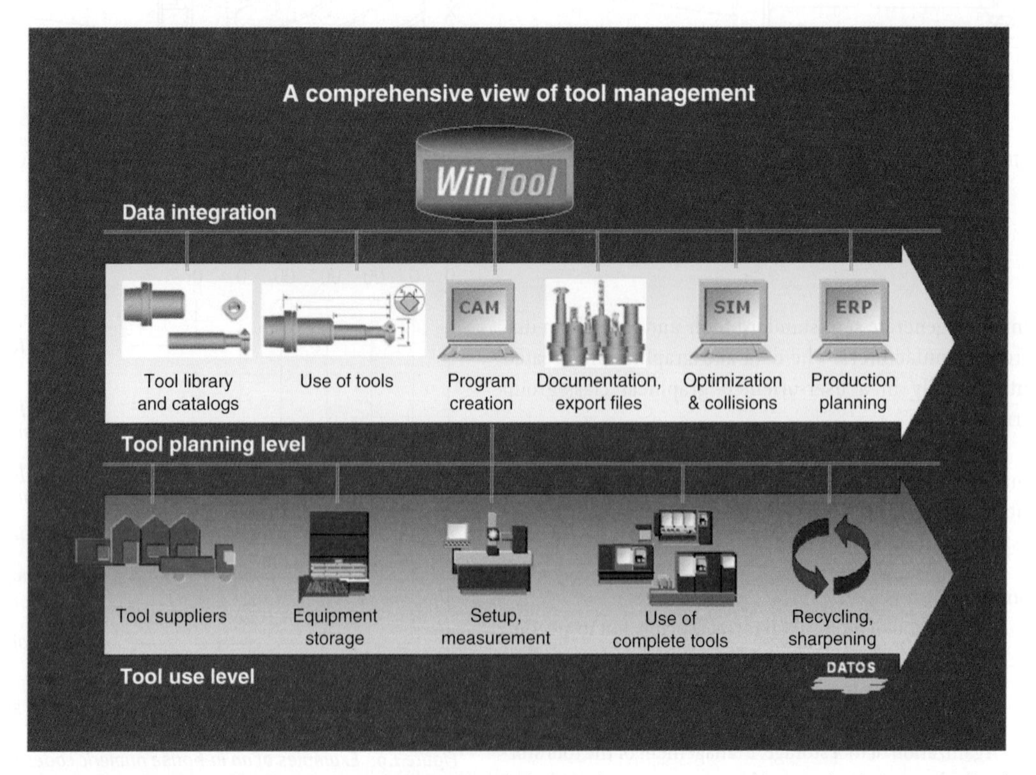

Figure 1.10: Overview of tool management. Process analysis in tool management and integration into a centralized tool database.

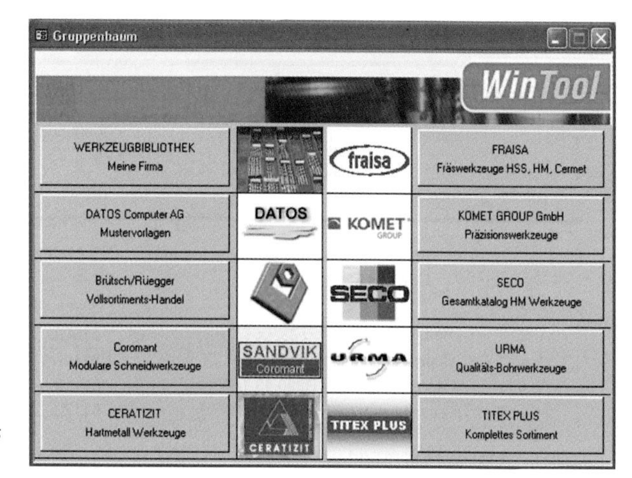

Figure 1.11: Clear database structure.
A standardized database structure for each tool type allows data exchange with CAM systems or presetting devices.

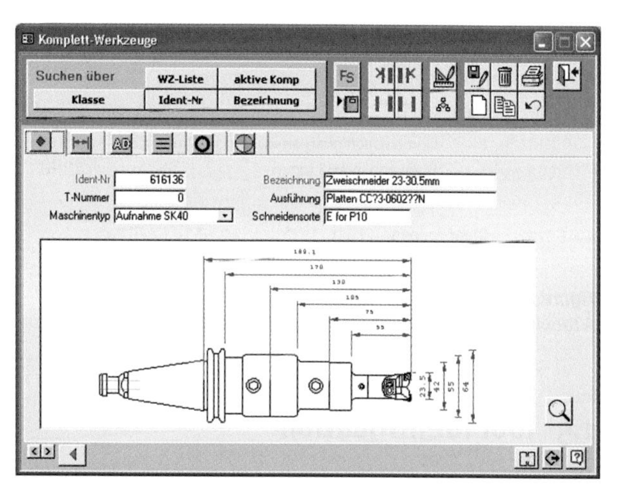

Figure 1.12: A complete tool. Complete tools with precise adjustment instructions are specified for production.

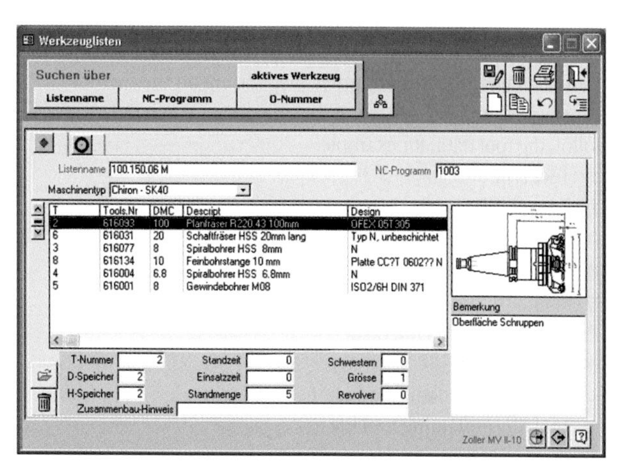

Figure 1.13: A list of the required tools is generated automatically for each CNC program.

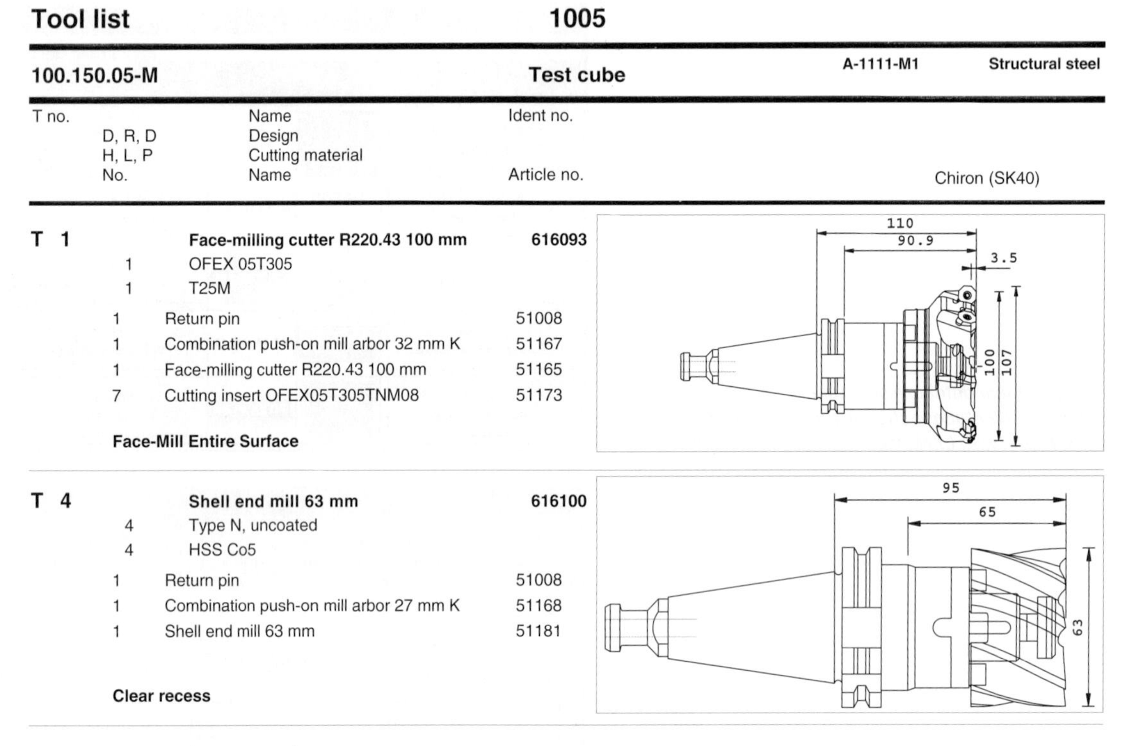

Tool list **1005**

100.150.05-M		Test cube	A-1111-M1	Structural steel

T no.		Name	Ident no.	
	D, R, D	Design		
	H, L, P	Cutting material		
	No.	Name	Article no.	Chiron (SK40)
T 1		**Face-milling cutter R220.43 100 mm**	**616093**	
	1	OFEX 05T305		
	1	T25M		
	1	Return pin	51008	
	1	Combination push-on mill arbor 32 mm K	51167	
	1	Face-milling cutter R220.43 100 mm	51165	
	7	Cutting insert OFEX05T305TNM08	51173	
		Face-Mill Entire Surface		
T 4		**Shell end mill 63 mm**	**616100**	
	4	Type N, uncoated		
	4	HSS Co5		
	1	Return pin	51008	
	1	Combination push-on mill arbor 27 mm K	51168	
	1	Shell end mill 63 mm	51181	
		Clear recess		

Figure 1.14: Tool list (abbreviated).
A tool list can be output as a setting sheet for the machine operator, for example.

1.5 Tool Identification

Tool management is intended not only to provide reliable identification of the tools themselves, but it also must fulfill the important task of providing the data that pertain to each tool in an unambiguous manner. Depending on the memory capacity of the CNC system, it may be necessary to enter the following tool data, for example:

- Tool type
- Tool number
- Substitute tool
- Magazine location
- Standard tool/series-produced tool/special tool
- Drill head/surfacing head
- Tool weight
- Maximum feed and torque
- Tool life/remaining tool life
- Early-warning limit before expiration of tool life
- "Tool broken/defective" flag
- Fixed location/variable location
- Tool radius 1/2
- Tool nose radius
- Collision radius 1/2
- Tool length 1/2
- Collision length 1/2
- Special tool code (customer-specific)
- Wear compensation 1/2
- "Tool deactivated" flag
- Fault code (reason for deactivating the tool)
- Assignment to machine
- Most recent use in machine, etc.

As the memory capacity of CNC systems continues to increase, there will be a demand for encoding additional

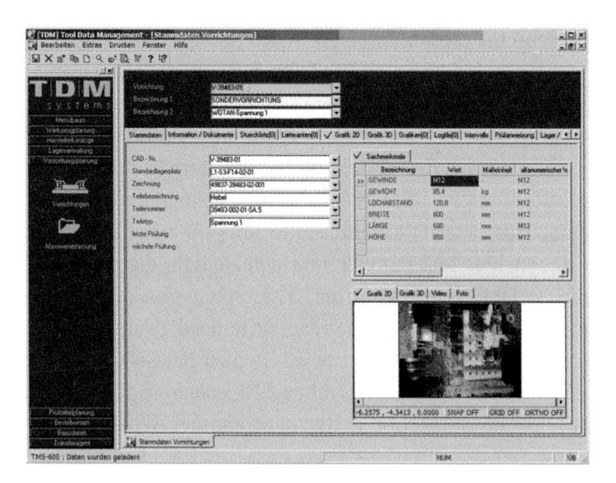

Figure 1.15: Digital equipment management systems also can be used to manage fixtures. (TDM Systems, www.tdmsystems.com.)

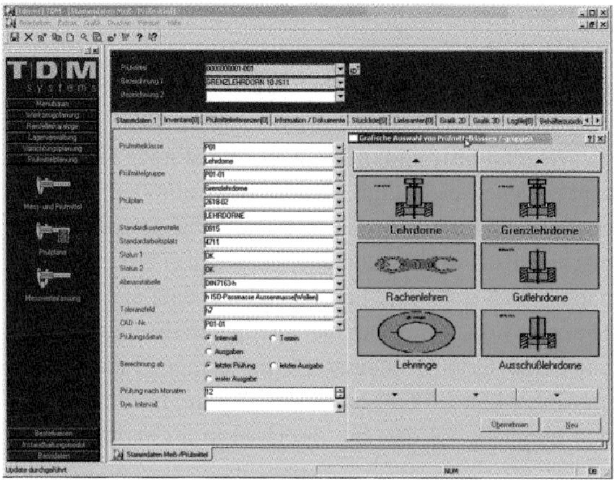

Figure 1.16: Reliable monitoring of measurement and inspection equipment is essential to ensure product quality. (TDM Systems, www.tdmsystems.com.)

types of identifying data. But even this list makes the following requirements clear:

1. The system must allow for automatic data input and output. Manual input is not acceptable because of the time it requires and the possibility of error.
2. Simple, mechanical methods of tool encoding (e.g., encoding rings) are not adequate.
3. The data must be stored in a manner that prevents them from being lost or assigned to the wrong tools.
4. Data input, processing, and output must be possible at multiple locations within the company.
5. The data must be managed within the CNC system after they have been entered only once in order to save time.
6. The identification system must be applicable to different types of tools.

These conditions are currently satisfied most effectively by **electronic tool-identification systems.** The most important component of these systems is the electronic data-storage chip, which is affixed to the tool and can be read with a special "read head" (*Figure 1.17*).

Function/Principle

In earlier systems, data were exchanged between the data-storage chip and the electronics by means of contacts. Contact wear and contamination occasionally caused read errors. Inductive devices that function without contacts are available today. These devices read the data with much greater accuracy, expressed as *data-read integrity.*

Figure 1.17: Two fixed and one rotating tool with an integrated data carrier.

These devices employ two different principles:

- **The read-only system** (*Figure 1.18*)
- **The read/write system** (*Figure 1.19*)

The **read-only system** employs data-storage media with a preassigned eight-digit identification number. The read heads in the tool crib, the tool-presetting device, and the machine itself work in conjunction with a central tool computer, which stores and manages all tool data in a database. The code-storage medium merely supplies the identification number to the tool computer. The tool computer then assigns previously entered tool-related data to the fixed identification numbers. All data are displayed in a clearly arranged form on the computer screen. When the tool is placed in the tool magazine, the read head determines the identification number, and the data thus are supplied automatically to the CNC system.

The **read/write system** uses data-storage media with greater storage capacities and can store up to 511 bytes of tool data. This capacity is sufficient to store the most important data, such as tool number, tool type, length, diameter, tool life, weight class, and so on. These data can be updated, modified, and read by the read/write head at any time. In other words, the tool carries all its own data with it at all times. When it is inserted into a CNC machine, it therefore requires no connection to the tool computer. When the tool

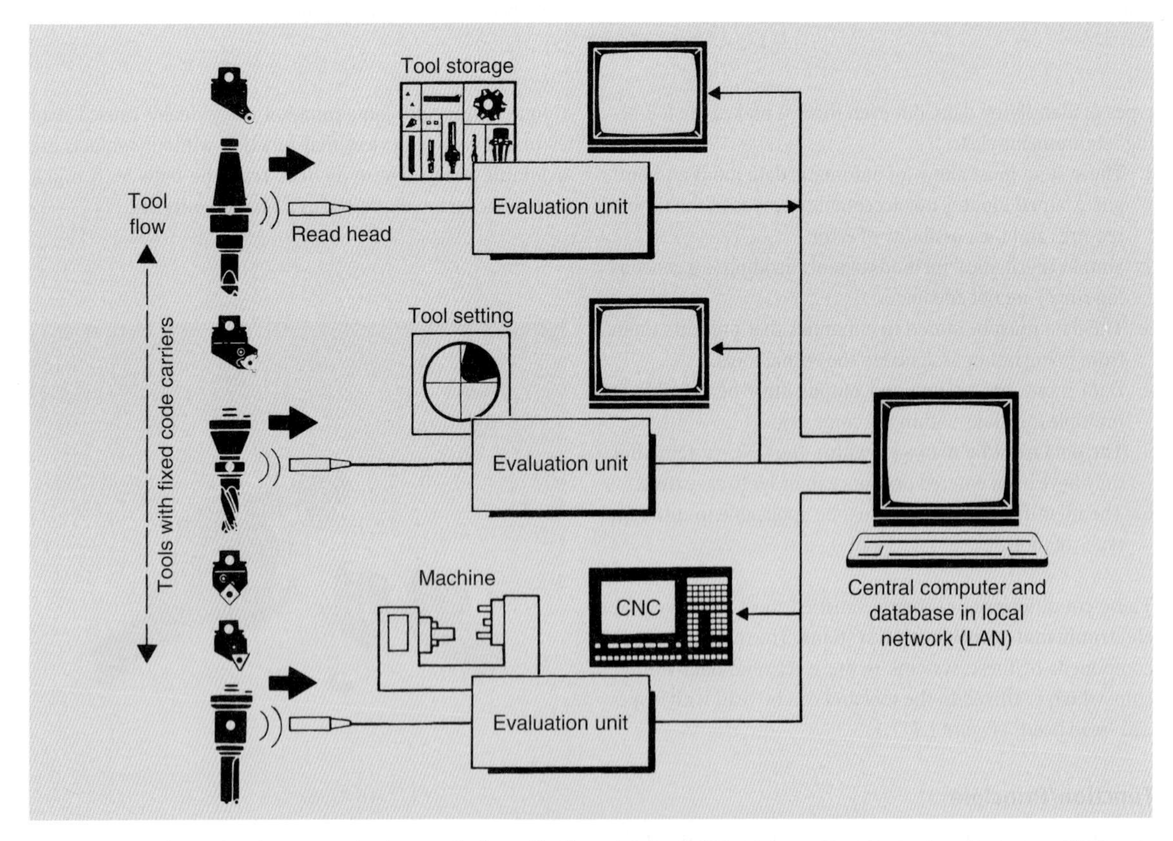

Figure 1.18: In the read-only system, the data carrier has a fixed number, and all the data assigned to each tool are stored in the central computer under this number and can be called up from there.

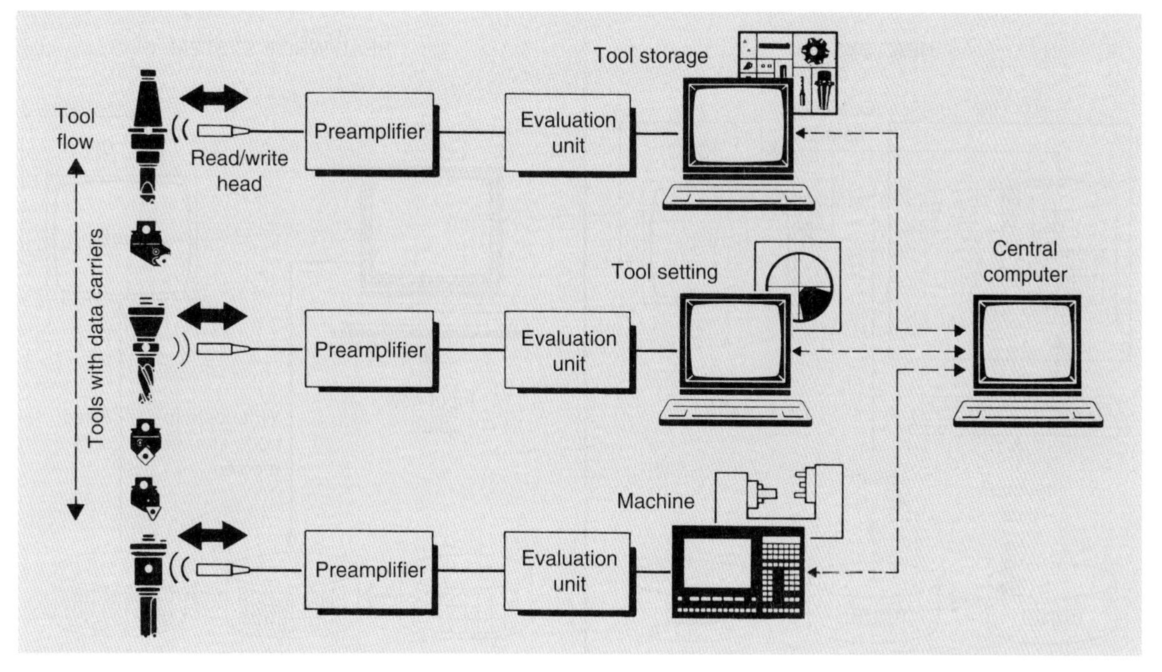

Figure 1.19: In the read/write system, the tool number and all the associated tool data are programmed in the data carrier and can be read directly.

is removed from the machine, the data contained in its data-storage medium (e.g., remaining tool life, wear compensation, etc.) are updated automatically.

If the CNC system is connected to a DNC computer, the data also can be passed across this connection to the tool computer if this becomes necessary for management-related purposes.

Components of a Tool-Identification System

An electronic tool-identification system as described earlier consists of the following components:

- The **code-storage devices**, also called *chips*, with fixed or modifiable coding
- The **read heads** or **read/write heads** with **preamplifiers**
- The **reading station**, which works in conjunction with the read heads and passes the identification number along to a computer or CNC system
- A **signal processor** for read/write systems, with outputs (RS232, V24) that are suitable for connection to a PC or a CNC system
- A **tool computer** for storing and managing the tool data
- Appropriate **software** for data storage, data management, data exchange, and clearly understandable displays on formatted screens

Technical data pertaining to the scanning distance, read time, programming time, write cycles, power supply, and so on can be obtained from the manufacturers of these systems.

Organizational Advantages of Electronic Tool-Identification Systems *(Figure 1.20)*

In view of the amounts of data that are required by sophisticated tool-management systems used with highly automated CNC machines, electronic systems offer tremendous advantages, such as

- **Automatic data flow** between the tool-presetting device, the tool, the tool computer, the CNC system, and the user
- Increased **integrity of data exchange** (because input errors are avoided) and additional monitoring of random read and write errors

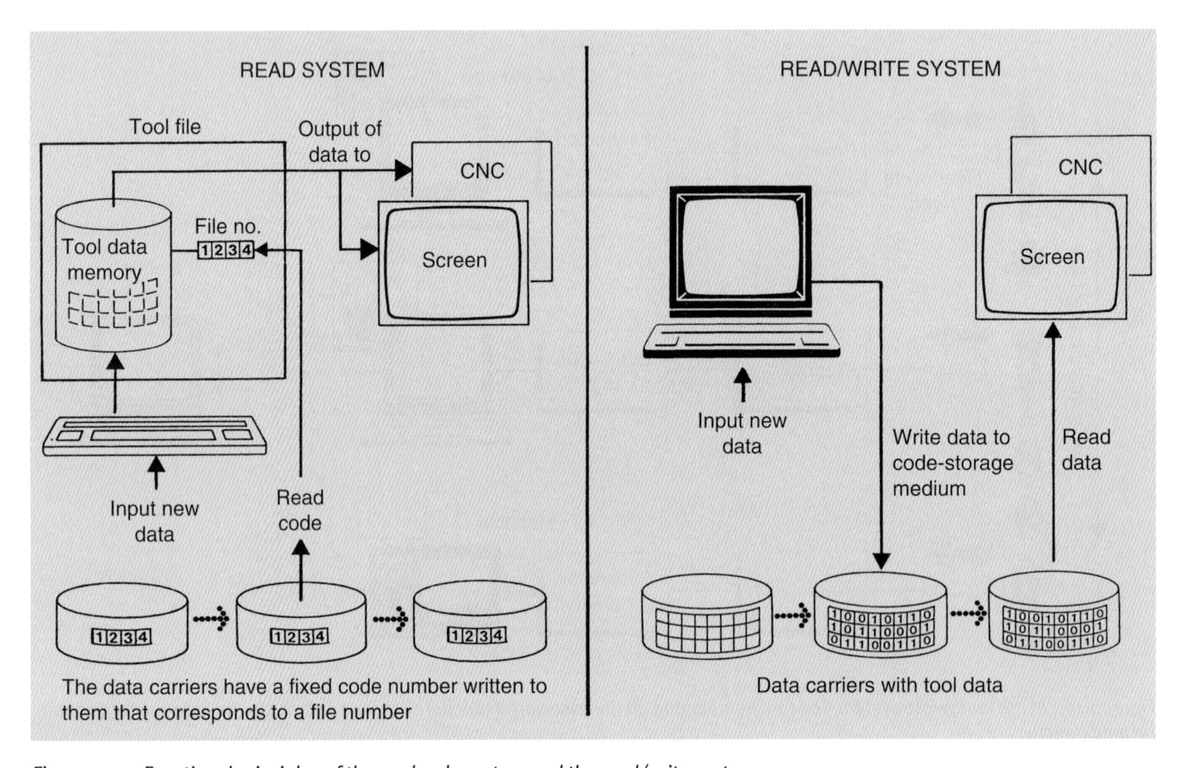

Figure 1.20: Functional principles of the read-only system and the read/write system.

- **Shorter setup times** on the machine
- Better **utilization of tool life**
- **Increased efficiency** in tool storage and tool presetting
- **Elimination of tool data cards** in the shop
- Improvement and automation of **tool statistics**
- **Support for workers** involved in assembly, measurement, and inspection of tools
- **Improved tool-management** capabilities

Tool Identification and Data Management with RFID

In computer-supported manufacturing, the flow of materials and the flow of information are inseparably linked. This applies to pallets, clamping devices, and workpieces but especially to tools that are used often in different locations with different data. Therefore, mechanical coding systems and barcode labels have proved to be unusable for automatic tool data management. In their place, systems working by means of induction, abbreviated as RFID, have become widely accepted. This principle and the associated components are rugged, insensitive to harsh environmental conditions, and provide reliable data integrity.

The use of RFID ensures that the locations of tools can be determined at all times, prevents damage to machines caused by the wrong tool data, and allows tools to be used for their entire service life without repeated measurement. This applies during use in the machine, during transport, and in the tool-storage facility, thus creating a closed circuit.

When the tools are sorted into the magazine, the tool data are automatically read, transmitted to the CNC memory, and assigned sequentially to the current magazine location during operation. No further read operations are necessary during subsequent automatic changing of tools from the magazine to the spindle and back. This leads to shorter changing times.

What Does RFID Mean?

The abbreviation RFID stands for "radiofrequency identification device." This is an automatic, electronic **identification technology** that is being used increasingly for noncontact identification of objects, goods, persons, animals, in-process control, the tracking of goods or flows of goods, in access-control systems, and many other areas of use. A number of different types of data carriers and read heads are available for this purpose.

An RFID system contains the following components:

- A **transponder** (also called an *RFID label*, *RFID chip*, *RFID tag*, or *radio tag*) as the data carrier
- A **send/receive unit** (also called a *reader*, *read/write head*, or *interrogator*) for communication with the transponder
- A **signal processor** that controls the bidirectional data transfer between the read/write head and the transponder and buffers the data. This is connected to
- A **computer system** for processing and managing the data

For various fields of application, it is possible to use special transponders that can be written to or read from in the high- or low-frequency range depending on the range that is required. The data exchange is noncontact by means of an inductive, that is, electromagnetic read/write system. Depending on the range required, the transponders can function either without their own power supply (passive transponders) or with their own power supplies provided by batteries (active transponders).

Transponders are available in various designs, for example, as adhesive labels, buttons, chip cards, pins, screws, or a plaque. The housings are hermetically sealed, extremely rugged, and resistant to impacts, vibration, pressure, chemicals, and high or low temperatures. For the identification of tools, the transponders are inserted in a borehole in the tool holder and bonded in place. The dimensions and position of the borehole in the holders are standardized (ISO 14443, ISO 18000-4, ISO 10536, ISO 15693, etc.) (*Figure 1.21*).

Fixed-programmed (read-only memory [ROM]) transponders can be written to only once. Recordable EEPROMs allow the stored information to be overwritten more than 1 million times. This is more than adequate because tools generally do not have to be rewritten to very often. The number of read cycles is unlimited.

These data carriers are used in all types of tool holder. They can even be used in up-to-date holders for shrink tools. The larger variants shown in the figure were mostly used earlier, before standardization. Above all, the version with a thread had the advantage that it was possible to exchange the data carrier without destroying it. In today's standard version, the data carrier is bonded in place. This means that it can no longer be exchanged without destroying it, but it is actually not necessary to do so because the data carriers and tool holder remain associated with each other for their entire service life.

Another area of application for these data carriers is the field of assembly technology. These data carriers are always

Figure 1.21: Transponders, data carriers for industrial use in tools.
Dimensions: 10 × 4.5 mm, protection classification IP 67. Capacity: 511 bytes or 2,047 bytes. Type: EEPROM, read and write. (Balluff.)

necessary when a number of small tool carriers are used in a fully automated assembly line.

In practice, it is important in many applications that the data no longer have to be acquired optically (i.e., using scanner systems, hand scanners, etc.), but instead the data are transmitted to the read/write system by means of radio waves, which only takes a fraction of a second.

Functional Principle of RFID

Data transmission between the transponder and reader/receiver unit takes place by means of electromagnetic waves. At low frequencies, this occurs inductively via a near field and, at higher frequencies via an electromagnetic far field. The distance over which an RFID transponder can be read can vary between a few centimeters and more than a kilometer depending on its design (passive/active), the frequency band being used, the transmission power, and environmental influences.

The **reader** generates an electromagnetic field that is received by the antenna of the RFID transponder. An induced current is generated as soon as the antenna coil comes near the electromagnetic field. This current activates the microchip in the RFID tag. In passive tags, the induced current also charges a capacitor that supplies power to the chip for a short period. In active tags, this is done by the built-in battery.

Once the microchip is activated, it can receive commands from the reader. It transmits its serial number and other data being scanned by the reader to the signal processor (*Figure 1.22*) by modulating a response to the field transmitted by the reader.

Example: The passive transponders manufactured by Balluff for tool identification (type BIS M) are first scanned in the low-frequency (LF) range at 70 kHz to provide the power supply and then the data transmission takes place at 455 kHz. Recordable transponders have a memory capacity of 511 bytes or, in rare cases, up to 2,047 bytes.

According to current experience, 511 bytes are sufficient to store the tool data for 95 percent of all applications.

DIN 69873 defines the dimensions of the data carriers: 10 mm in diameter and 4.5 mm high.

Memory Subdivision (*Table 1.2*)

The data carrier memory can be subdivided internally into block sizes of 32 and 64 bytes (also called the *size of a page*).

Memory size up to 1,023 bytes = 32 bytes per block
Memory size up to 2,047 bytes = 64 bytes per block

Figure 1.22: Evaluation unit interface with the machine controller. Interface: Choice of serial, parallel, InterBus, PROFIBUS, or DeviceNet.

Technical Advantages of RFID

The advantages of RFID compared with barcode systems include

- Noncontact identification is possible (even without visual contact).
- Penetrates various materials, such as cardboard, wood, oils, etc.
- Memory can be read and written to as desired.
- Identification and reading occur in less than a second.
- Resistant to environmental influences.
- Shape and size of the transponder can be adapted as needed.
- Transponders can be completely integrated into the product.
- High security thanks to copy protection and encryption.
- The RFID chip is a permanent data-storage medium that can be used to store all the product data. No redundant database is necessary to obtain initial information.
- Objects equipped with RFID can be scanned more than 20 times more quickly than those with barcodes.
- RFID tags can be read even when heavily fouled.
- The placement and orientation of the object being scanned are much simpler than with barcodes. It is sufficient for the object to be within the reading range of the scanner.

Transponder	Read Time (ms)
Data Carrier with 32 Bytes per Block	
From 0 to 3	14
For each additional byte	3.5
From 0 to 31	112
Data Carrier with 64 Bytes per Block	
From 0 to 3	14
For each additional byte	3.5
From 0 to 64	224

Table 1.2: RFID Read Times in Dynamic Operation: The Times Indicated Apply from the Time the Data Carrier Is Detected (If the data carrier has not been detected yet, 30 ms has to be added for the buildup of energy until the data carrier is detected.)

1.6 Tool Management

This refers to the acquisition, organization, and management of tools and all the current data associated with them. Many companies are unaware of the economic advantages provided by closed, clearly organized, and seamless computerized tool-management systems. For this reason, we will now go into some of the benefits of *closed data circuits* in greater detail.

Tool Management in Production Operations

(Figures 1.23 and 1.24)

Each complete tool that is used on a CNC machine must be planned and made available at the right time. If, for example, a CNC program on a machining center requires 30 tools, then about 100 individual tool components have to be set up. Good organization of tools therefore is essential to be able to operate efficiently and without errors.

In production operations with large setup requirements—for example, where there is a large number of machines, a highly varied range of products, or small lot sizes—tool management becomes a serious challenge. This is why such shops are making increasing use of computerized tool- and equipment-management systems.

But it is not just logistical requirements that are important in tool management but also technical requirements: The tools have to be prepared in such a way that the CNC program runs smoothly. Problems often occur in production processes where there is a division of labor, for example, when a programmer, a purchaser, a tool presetter, and a machine operator all work on the same production order. It is very easy here for errors in communication and organization to occur. The consequences of tool errors can be quite costly, especially when they become apparent only on the machine itself. Machine downtime, tool breakage, collisions, or even damage to the machine are all possible.

Such problems can be avoided effectively by means of good documentation and management of the tools. This includes information on the following:

- CNC program and machine
- List of the complete tools
- Bill of materials for tool components
- Quantities
- Storage locations
- Technology of components
- Graphics of complete tools
- Collision dimensions
- Measurement instructions
- Possibly a solid model (3D depiction)
- Possibly manufacturer information, order numbers, or standards

A centralized tool database is essential to ensure an efficient, error-free tool process in manufacturing plants. Employees from all departments can always access the current tool data via the company network so that everyone is working with the same data. Supplementary information or lessons learned from working with the tools can be inserted

Figure 1.23: Genius 3s universal tool-measuring machine with SX loading magazine—for fully automated, unmanned, noncontact measurement and inspection of ground or eroded tools. The measuring system has five CNC axes and two fully automated image-processing systems. The loading magazine can be used during nonproductive times to store, measure, and log up to 75 tools. Interfaces such as NUM and the network can be used to transfer command or compensation data. (Courtesy of ZOLLER.)

by any employee authorized to do so. For example, tested cutting parameters for tools can be collected systematically as part of a continuous-improvement process.

The tool database encompasses all the technical and logistical tool information. The structure of the database has to comply with industry standards and the requirements of the tool manufacturer and be compatible with CAM and enterprise resource planning (ERP) solutions. This is necessary so as to be able to manage and exchange the data efficiently.

An integrated tool-management system means that tool orders can be determined based on the tools used in the CNC programs and the order data in the production-planning system (PPS). Selection of tools and the tools kept available can be planned exactly. This makes it possible to have a lean, low-cost tool-storage facility without having to risk supply bottlenecks.

Optimal, trouble-free tool data management is possible only if all the tool manufacturers in question supply their tool data in a suitable digital form.

Tool Management in CNC Systems

Every program is based on certain assumptions about the dimensions of the tools, regardless of whether the program is created manually or automatically. If the actual tool data deviate from the assumed values, the resulting parts will display dimensional deviations that may be so severe that the parts must be scrapped. Great importance therefore is assigned to the area of tooling, including the entry and management of the associated compensation data.

The tool data managed and processed in the CNC system include, for example, the following:

- **Tool length compensation,** that is, compensation for the difference between the assumed and actual tool lengths (in the *Z* direction)
- **Tool diameter compensation,** also known as *cutter-radius compensation*, which is used to automatically calculate cutter centerline paths with constant offsets for any cutter diameter. For nontangential transitions, this logic calculates suitable transition radii at corners along the

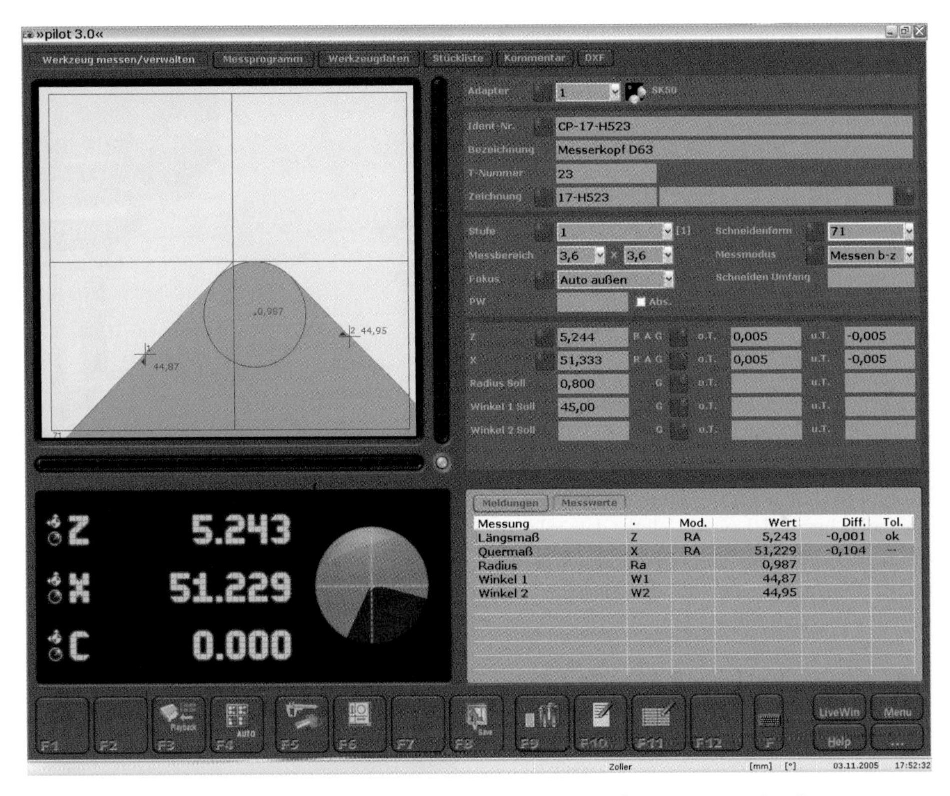

Figure 1.24: Electronics with Pilot 3.0 software for measurement, adjustment, and management of tools. Graphical user interface employs the latest research in the field of human-machine interfaces (HMI, University of Kaiserslautern). Automatic measurement with micrometer accuracy using ZOLLER image-processing technology and with no action required by the operator. (Courtesy of ZOLLER.)

outside contour, as well as points of intersection on the inside contour. The advent of "true" cutter-path compensation made it possible to use the workpiece data as program data.

- **Tool wear compensation,** providing automatic or manual compensation for the amount of wear detected on the cutting edge of the tool without requiring any change to the value that was entered originally (additive compensation).

These types of tool compensation were already provided by earlier generations of numerical controls and thus are well known. However, newer CNC systems offer additional, lesser-known functions and capabilities that allow fully automatic operation of flexible manufacturing cells and systems.

These include

- **Tool-life monitoring,** which involves the continuous tracking of the usage time for each tool in the magazine. When this usage time reaches the specified limit value, a replacement tool is retrieved.
- The automatic **alternate tool-management** feature keeps a replacement tool on hand in the magazine for this purpose. This logic manages up to nine identical replacement tools for each tool, all stored under the same tool number. As the service life of the currently active tool expires, the next replacement tool is inserted in its place.
- After the used tool has been returned to the magazine, the CNC system "hides" it electronically and then activates the alternate tool the next time the program calls for it to be retrieved.

- The conditions of all tools in the magazine can be checked on the screen at any time.
- The limited number of tool locations in a magazine gives rise to another management problem: the prompt insertion of new tools to replace tools that are "used up" or no longer needed in the magazine *without* stopping the machine.
- This is the task of **tool-management software.** This software looks ahead to find sections of the program that do not involve any programmed tool changes and allows manual or automatic replacement of tools in the magazine to occur during these periods. As the next programmed tool change approaches, the procedure is interrupted early enough to make the magazine available for changing the tool to be used in the main spindle.

Newer machine concepts employ suitable controlled devices or conventional robots for this kind of tool replacement. As a new tool is inserted, however, all the data related to compensation and tool life also must be read, stored at the correct location, and managed automatically. This is a task involving a great deal of responsibility.

1.7 Tool Measurement and Monitoring with Laser Systems *(Figure 1.25)*

Figure 1.25: Measuring a drill with the LaserControl system. Thanks to intelligent electronics and measuring software, it is no problem today to perform measurements even when the tool is surrounded by coolant. (Blum-Novotest, www.blum-novotest.com.)

A laser system for tool monitoring is essentially a high-precision light barrier whose output signals can be used for measured-value acquisition by any CNC system or PLC. Using the visible laser beam, the adjustment mechanisms, and the adjustment aid integrated into the receiver, it is possible to align the system quickly in the machine without any additional equipment. It also has to be designed for use in the harsh conditions in machining centers, which is why the optics of the transmitter and receiver are protected against chips and coolants by a pneumatically operated seal and sealing air.

The laser system must be able to work together with the CNC system to measure the length and diameter of tools in the machine at micrometer accuracy and to monitor even the smallest tools for tool breakage throughout the entire working range of the machine. The basic version can check tools with a diameter of 1 mm at a distance of up to 20 m. Using focused systems, it is even possible to monitor tools with diameters of less than 0.05 mm or measure their length, diameter, and concentricity. By **measuring the tool** in the workspace at its rated speed, it is also possible to detect clamping errors, at the same time determining the effective compensation values for length and radius.

Breakage Monitoring

A classical method for breakage monitoring is checking the tool geometry. Mechanical measuring probes, often called *load cells*, have been available for a long time now. These have several disadvantages that make them unsuitable for universal laser systems that are intended for a truly wide range of applications.

The main requirements are

- Checking rotating tools
- Avoiding physical contact during measurement

- Measuring even the smallest tools
- Reducing the measuring time

An optimal solution was found subsequently by installing a light barrier in the workspace of the machine in such a way that it is in range of all the CNC axes. It is advantageous to use a laser system with a visible laser beam.

Tool-breakage monitoring for drills, countersinks, reamers, screw taps, and so on involves checking the length only. This can take place while the tool is oriented or rotating. The system provides the following options:

- Passing through the laser beam perpendicular to the tool axis; an intact tool interrupts the beam, generating a signal.
- Passing through the laser beam lengthways to the tool. With this method it is possible to monitor tools even under extreme ambient conditions, such as while coolant is still passing through the interior of the tool. This requires that the precise position of the laser beam be known.

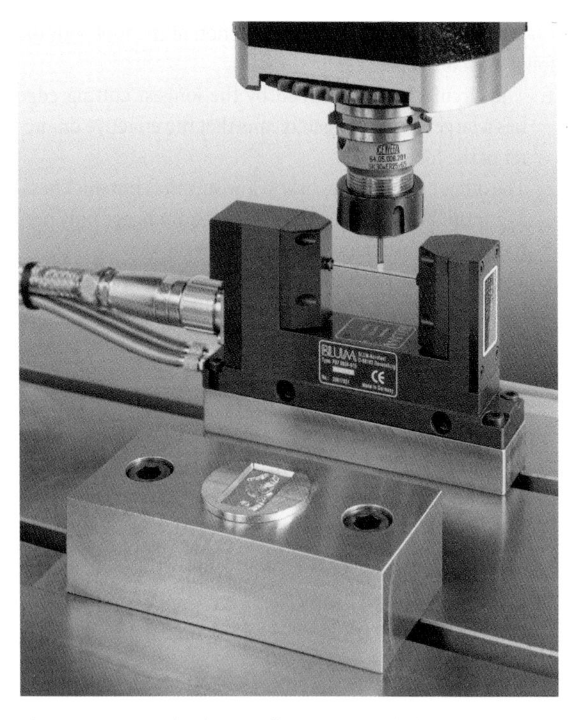

Figure 1.26: Monitoring a milling cutter with the LaserControl Mini system. Each individual cutting edge is checked for breakage or wear in an extremely short time. (Blum-Novotest, www.blum-novotest.com.)

Monitoring of Individual Cutting Edges *(Figure 1.26)*

In addition to relatively simple length monitoring, thanks to their precision lasers, systems also can detect even the smallest flaws in cutting edges, including the wear dimension. In monitoring of individual cutting edges, the system checks whether each individual cutting edge is within a specified tolerance value. If the system has a high scanning rate, this check can be performed practically at the rated speed of the machining process. With this method, it is possible to detect four different fault situations:

- Cutting-edge breakage
- Tool wear
- Built-up edge
- Clamping errors

Tool Measurement in the Workspace

If the laser system with a high repetition accuracy of the switching point is used in a modern CNC machine tool, then a tool setting accuracy of a few micrometers can be achieved. Tools whose length and diameter are unknown can be measured with great accuracy over a measuring cycle at the rated speed of the spindle. The value being measured (i.e., tool length or diameter) is saved in the tool-compensation memory under the current tool number.

Temperature Compensation

To ensure dimensional accuracy of workpieces over a longer period, the thermal elongation of the CNC axes can be determined and compensated for very quickly.

High-Speed Cutting

High-speed cutting (HSC) is, among other things, characterized mainly by high spindle speeds and highly dynamic drives. Spindle speeds of up to 60,000 rpm require balanced tool holders and perfect placement of the tools in the spindle.

Clamping errors cause eccentric rotation of the tool with the following consequences:

1. The circular path described by the longest cutting edge has a larger radius, which means that the effective diameter of the tool is larger.
2. The other cutting edges have a low specific chip thickness, as a result of which they do not engage or engage only partially. This drastically increases tool wear.
3. The transverse force created by eccentric rotation of a tool can destroy the tool and the spindle bearing.

Special measurement cycles for fast laser measuring systems make it possible to detect the longest cutting edge at the rated speed, thus allowing determination of the effective length and radius compensation. It is also possible to check whether the eccentricity is within a programmable tolerance value.

Additional Capabilities *(Figure 1.27)*

The laser measuring process provides a number of other options, including determining the radius and center of ball-nose cutters or checking the radii and chamfer of form cutters. All these functions assume that the CNC system is equipped with the appropriate software.

1.8 Summary

Tools for CNC machines are an integral part of the system that comprises the machine tool, the workpiece, and the numerical control. These tools must be manufactured with such great precision that they can be guaranteed to be interchangeable. Furthermore, it must be possible to determine their optimal performance parameters (i.e., cutting values and tool life) in advance. Finally, these tools must lend themselves to fast, precise handling so that they can be interchanged as dictated by the programmed sequence and tool wear. However, these features of a high-quality CNC tooling system can take effect only when tool selection and preparation are performed carefully and systematically. The general prerequisites for this are off-line tool presetting and the cataloging of tools. When in doubt—or when experience is lacking—advice should be obtained from a well-known tool manufacturer or an experienced user.

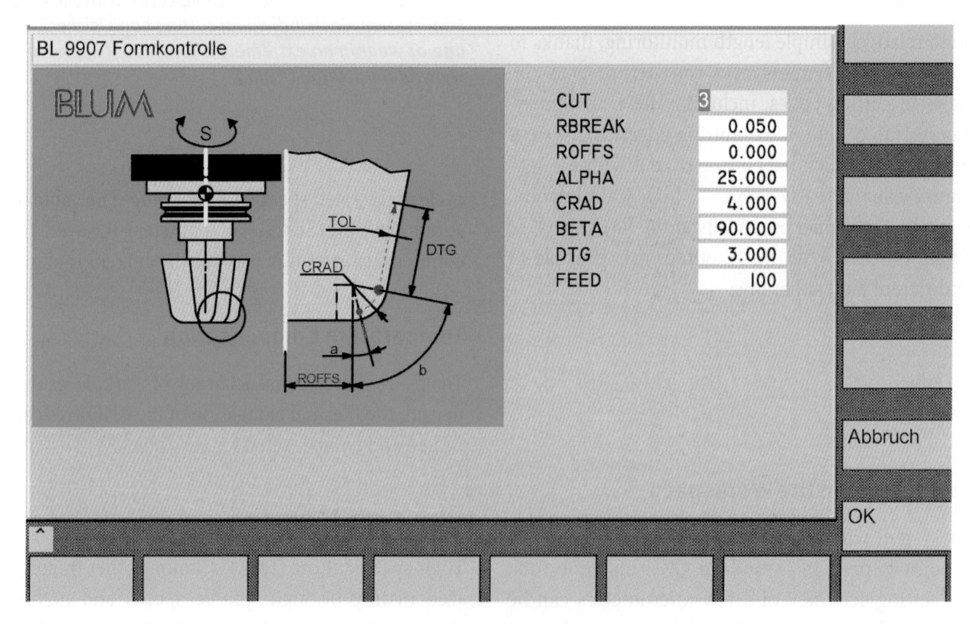

Figure 1.27: Input menu for shape monitoring of milling cutters with rounded cutting edges using a Siemens 840D as an example. (Blum-Novotest, www.blum-novotest.com.)

Tooling Systems

Important points to remember:

1. A **tooling system** is the totality of all tools that are used in a manufacturing system and are integrated into the organization of tools. All tools and data are systematically registered under code numbers in a **tool card file** or computer file.
2. A tooling system for NC machines consists of the following components:
   - A tool holder for inserting the tools into the spindle
   - A tool magazine to hold the tools on the machine
   - The tools themselves, sometimes provided with either manual or automatic presetting capabilities
   - A tool-changing device, which transfers the tools (in their holders) from the magazine to the spindle and back again
3. The **tools, sleeves, and holders** must
   - Be as stiff as possible
   - Be suitable for automatic tool changing
   - Allow presetting
   - Allow encoding (in some cases)
4. In **programming systems** and **manual data-input controls,** all tool numbers, dimensions, and other technical data are stored in—and can be retrieved from—memory. This allows **tools** to be **selected automatically** to a great extent by specifying the spindle speed, cutting depth, and feed values.
5. Each machine generally requires **at least three sets of tools.** These are located
   - In **the magazine of the machine**
   - In an inspection facility, in inventory, or in the metrology room for purposes of measuring and/or presetting
   - In the preparation cart for the next type of workpiece
6. Modern CNC systems include software for **tool management in the machine.** The features of this software include
   - Replacement tools or alternate tools in the magazine
   - An area of storage reserved specially for compensation values representing the length, diameter, wear, and service life of each tool
   - Assignment of tool number and location number in the magazine (variable location encoding)
   - Oversized tools and reservation of empty locations adjacent to these tools (when applicable)
7. **The precondition** for seamless tool management is a closed data circuit; that is, all the tool data are managed and updated in the CNC system, and when the tools leave the machine, the tool data are transferred to the external tool computer or transponder.
8. **Location coding** is ordinarily used on turning machines. This requires the programmer to know the location of each tool.
9. Tool measurement and monitoring in the machine is performed primarily using laser systems.
10. Preference is given to RFID systems for automatic tool data scanning and management.

2. Close-to-Process Production Measurement Technology in Combination with Mechatronic Tool Systems

Dipl.-Ing. Alexander Blum, Dipl.-Ing. Jacek Kruszynski

Close-to-process measurement technology, tool measurements, and mechatronic tools are new methods for preventive quality assurance during the production process.

2.1 Introduction

In modern production processes with CNC machines, the measurement and monitoring of tools are part of the closed-loop control of process chains. The goal is **preventive quality assurance.** Tool measurement data or results of measurements on the workpiece are used as compensation values for parameters in the overall process. The machining processes are kept stable and the machining results constant by means of automatic tool compensation or, if necessary, by changing tools.

Depending on requirements, intelligent measurement strategies may be necessary. These, in turn, may require the use of new **tool-measurement systems.** This is complemented by measurements of the workpieces to determine the manufacturing results, which likewise are used as input parameters for the process. On the whole, *in-process*, *post-process*, and *close-to-process* measurement technologies provide a wide variety of options for acquiring process-related quality data.

2.2 Parallel Measurement Technologies

The **in-process measurement technology** operates inside the machine in parallel with the productive time, allowing extremely short response times. However, it has severe inherent limits owing to limited accessibility and vulnerability to failure. Well-known variants include output-monitoring systems in spindle drives or diameter-monitoring systems with measuring probes. **Postprocess measurement technology** also operates in parallel with the productive time but outside the machine. It can use a wide range of sensor types to perform practically any measurement task. Greater costs are involved, however, owing to linking, workpiece handling, measuring-tool handling, and setup work.

2.3 Close-to-Process Measurement in Idle Times

In **close-to-process measurement technology,** the measuring sensors are integrated into the machine, where they perform monitoring and measurement of tools and workpieces. Examples include workpiece-measuring probes or bore gauges that can be loaded in from the magazine or tool-measurement systems installed in the workspace. In workpiece-oriented close-to-process measurement technology, it is as if postprocess measurement technology were taking place within the machine tool. Because various machine components such as axes and clamping equipment are used while the measurement is being performed, the measurement process does result in idle times, but there is no investment required for separate systems. A major benefit in terms of precision is the fact that the clamping situation remains unchanged by the process. Moreover, the time difference between machining and measuring is minimized; that is, the measurements are performed in approximately the same thermal state as the machining. Thus close-to-process measurement creates a closed process chain in the machine tool that enables simple, timely correction of the process parameters, therefore stabilizing the process while maintaining high flexibility.

If production-measurement technology is employed close to the process within the machine tool, for example, by means of **workpiece-measuring probes,** then at least theoretically there remains an element of uncertainty because the same axis-measuring systems are used for both machining and the subsequent control measurements. System-related errors therefore will not be detected. Here, the machine tool is provided with the "objectivity" of postprocess measuring stations in the form of so-called **bore gauges** with an integrated measuring system, that is, a measuring system independent of the machine.

2.4 Close to the Process with Bore Gauges *(Figure 2.1)*

Bore gauges such as the compact BG40 from Blum are located in the tool magazine like any other tool. They are exchanged into the spindle for **machine-independent measurement in the workspace.** With a measuring time of significantly under 0.5 second, this type of measuring equipment is superior to any conventional workpiece-measuring probe. Their functional principle is based on measuring contacts with floating bearings; when these contacts are deflected, a drive needle is compressed, thus casting a larger or smaller shadow on a miniature light barrier inside the measuring head and therefore functioning as a linearized measuring system (*Figure 2.2*).

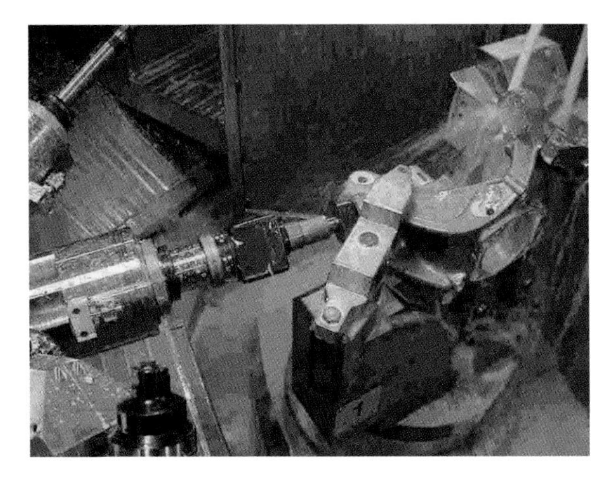

Figure 2.1: BG40 bore gauge in operation.

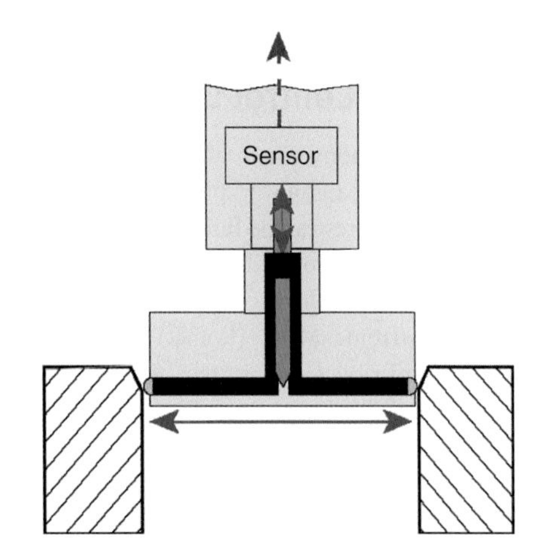

Figure 2.2: Measuring element with floating measuring contacts.

Such measuring elements can be used to measure the bore diameter independent of the spindle position. Another device variant (BG41 series) is equipped with individual elements and allows determination of the diameter, shape, and position (*Figure 2.3*). The diameter-specific measuring heads guarantee a repeatability of less than 1 μm in a measuring range of up to 400 μm and a resolution of 0.15 μm. Data are transmitted via infrared (IR) light from the up to eight measuring elements per measuring head to a receiver and from there to a simple interface connected with the machine controller. Power is supplied by a 9-V lithium battery with a capacity of up to 150,000 measurements.

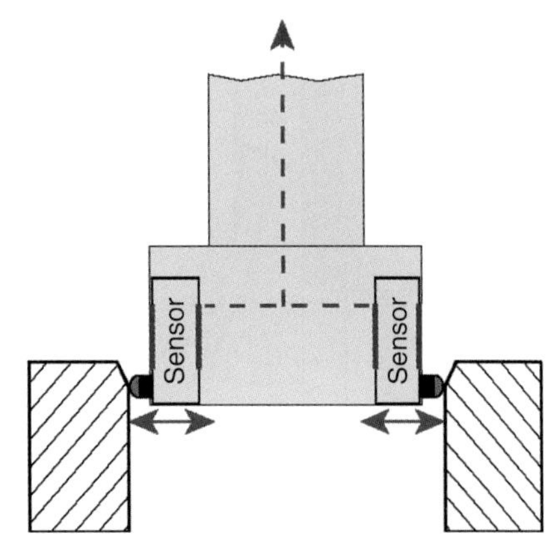

Figure 2.3: Measuring element with individual measuring units.

2.5 Actuator-Driven Tool Systems Close the Control Circuit

If dimensional deviations occur after verification of the machining result by means of workpiece measurement, then tool-changing procedures can be initiated or automatic tool compensation can be performed. In such cases, increasing use is being made of so-called actuator-driven tool systems, such as the KomTronic M042 (Komet) precision boring system or the KomTronic *U*-axis system, to close the control circuit in the manufacturing process (*Figure 2.4*).

2.6 Mechatronic Tool Systems

(Figures 2.5 through 2.8)

Mechatronic tool systems enable automatic adjustment of the cutting edge. The cutting-edge holders are monitored via sensors and equipped with their own drive in the tool head. Adjustment of the cutting edges can be performed statically before machining (M042) or dynamically during machining (*U*-axis tool system). Energy is transmitted inductively, that is, without contact. The stator coil located on the machine side is supplied with energy via a power cable. On the opposite side, on the tool head, there is a rotor head that supplies a stable direct-current (dc) voltage. The power supply can be either rotating or stationary. The same applies for the data transmission, where IR light guarantees extremely high transmission speeds and reliability. An IR transmission/reception module is installed in the machine room as a counterpart to the transmission/reception module installed as part of the tool head. The main field of application for precision-adjustment heads is compensation for wear on the cutting edges. *U*-axis tool systems represent an expanded form. Designed for the creation of freely programmable geometries, they allow adjustment of the cutting edge even during the machining process itself.

2.7 A Closed Process Chain *(Figure 2.9)*

An example of a closed process chain using the components just mentioned begins after rough machining with the use of an actuator-driven tool. After all the holes have been drill-finished, the tool is replaced with the bore gauge.

The bore dimension is checked quickly and independently of the machine in the last bore machined in the clamping operation.

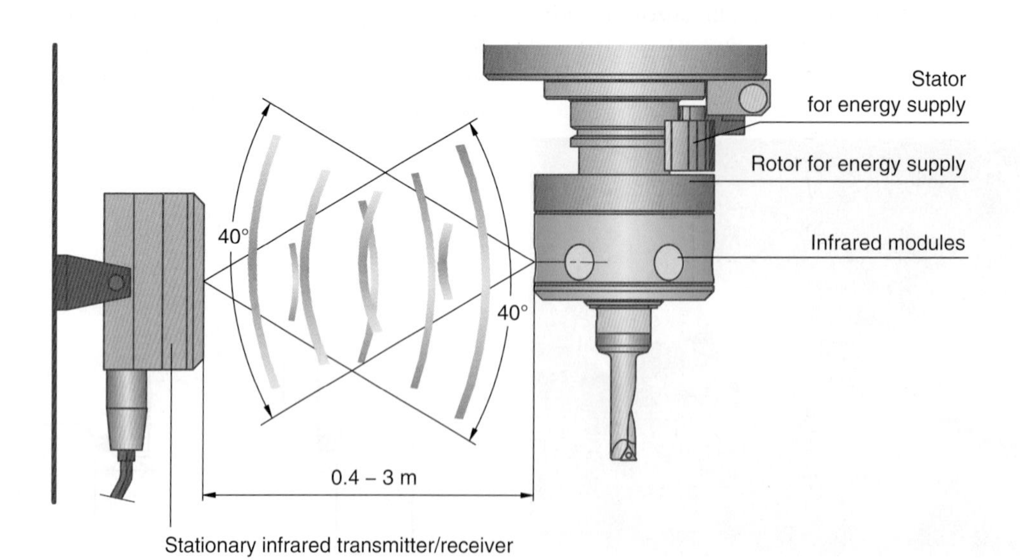

Figure 2.4: Principle of actuator-driven KomTronic® M042 precision boring system.

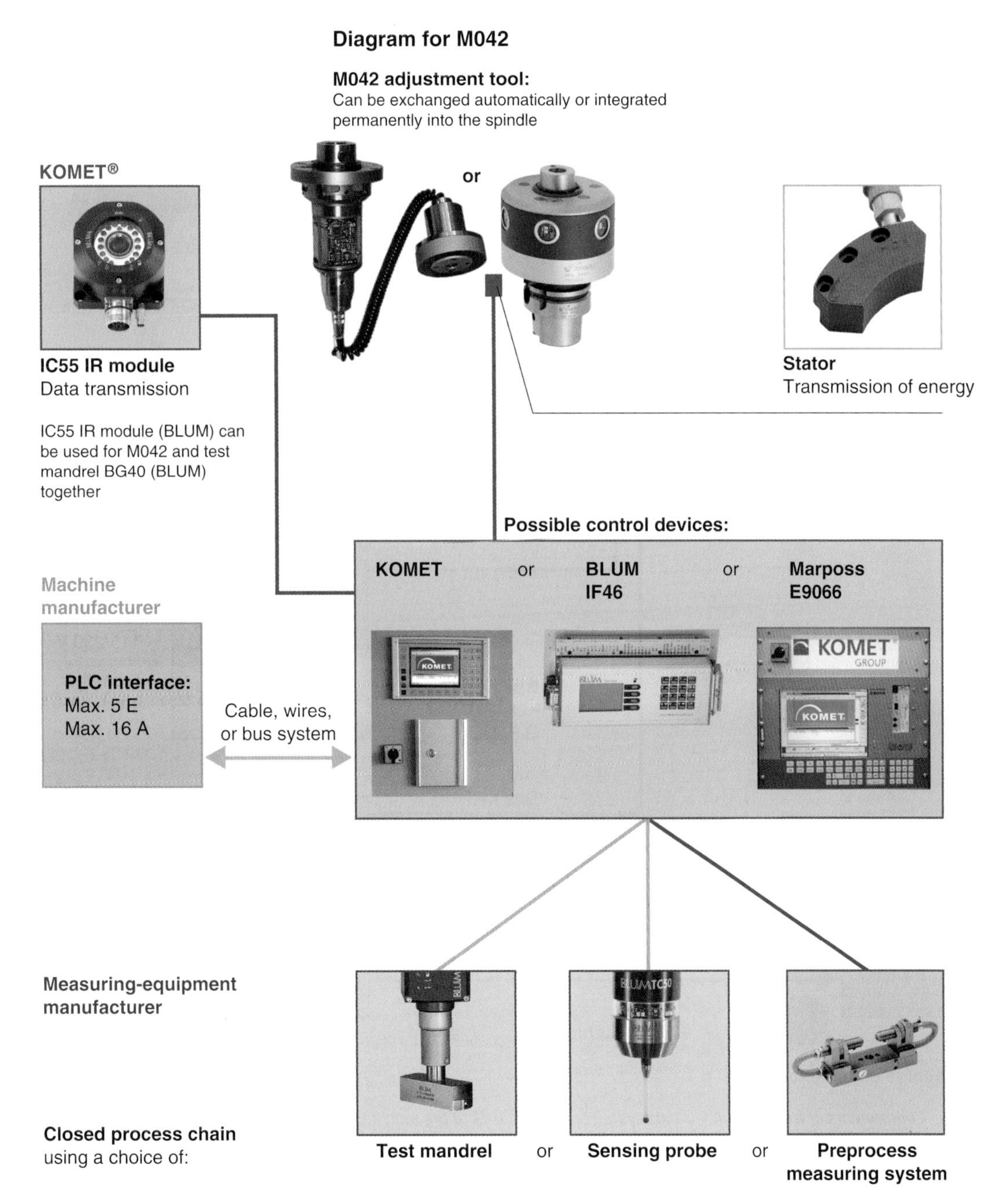

Figure 2.5: KomTronic® M042 precision boring system: connection to the machine controller of machining centers.

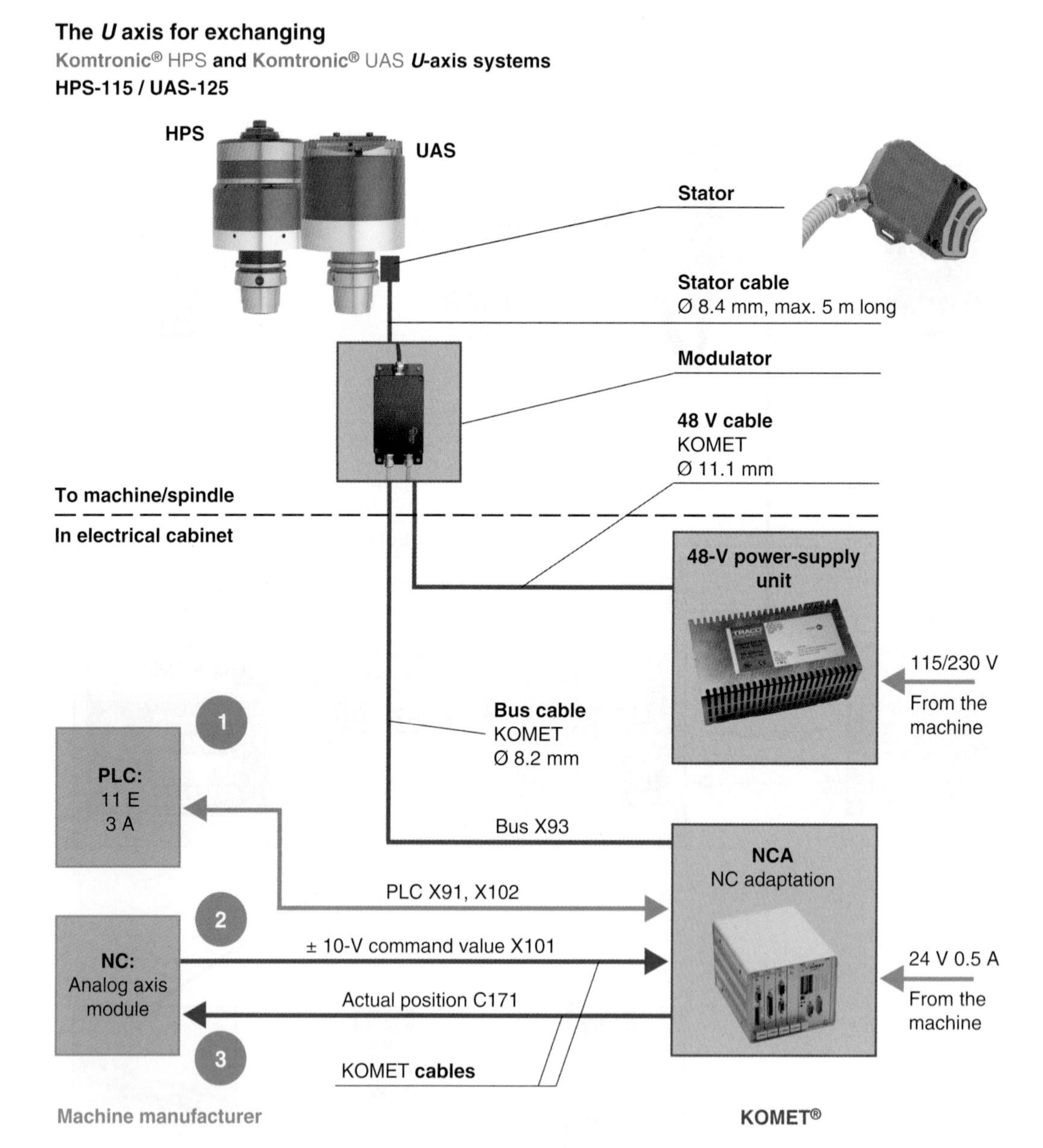

Figure 2.6: Components of a KomTronic® U-axis system and interfaces to the machine controller.

Figure 2.7: Machining of cylinder heads at a machining center with a KomTronic® M042 precision boring system.

Transmission of energy and data in the working space

The stator covers about 90° of the spindle flange. The arrangement of the stator element should be clarified based on the spindle drawings.

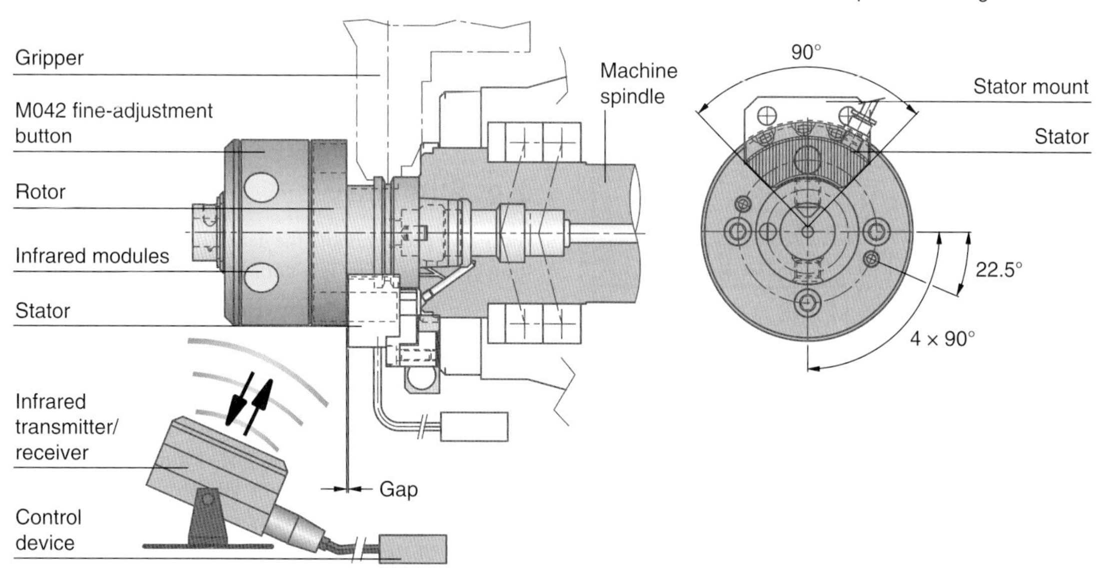

Figure 2.8: Actuator-driven M042 precision boring system (Komet). Static adjustment is performed before machining. Cutting-edge compensation is performed by means of a micrometer-accuracy measuring system on a slide.

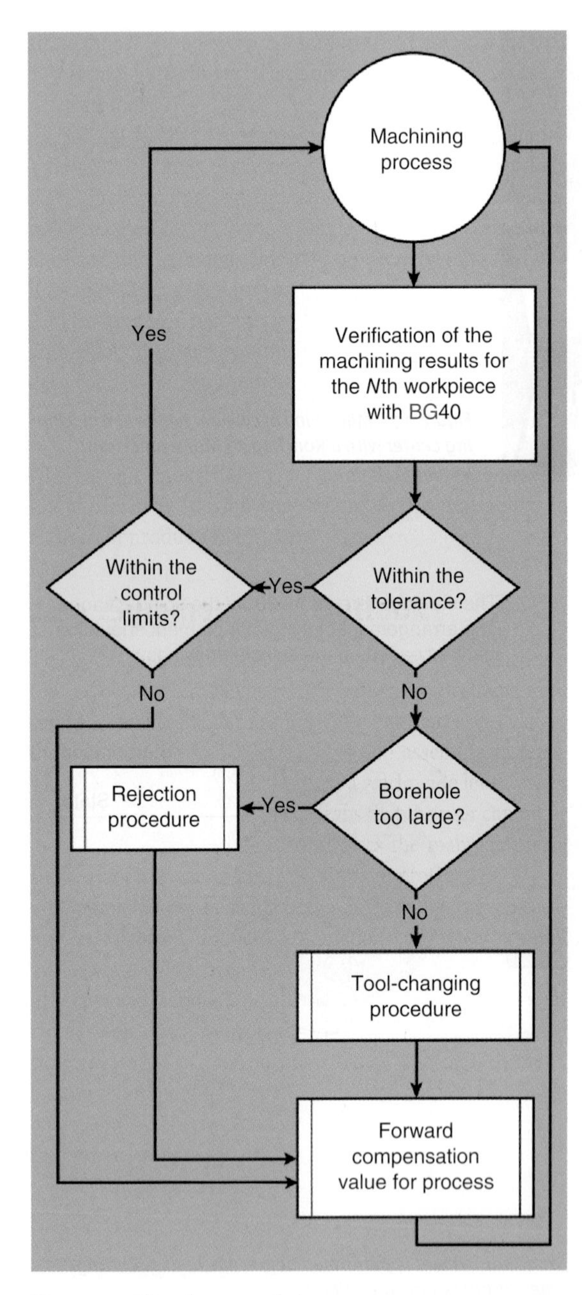

Figure 2.9: Closed process chain.

If the control limit (*UCL* or *LCL*) is exceeded, then a wear compensation also will be taken into account for the next workpiece (Figure 2.10).

This compensation must be determined in accordance with the usual rules, that is, from the rolling average with a damping factor. When according to empirical values the maximum value for wear compensation is reached, the tool is exchanged.

After it is exchanged, again the actuator-driven tool is adjusted to the compensated value. Any errors during tool changing can be compensated for through the use of a laser system for tool measurement.

If the measurement reveals spikes that lie outside the normal distribution, then it is advisable to interrupt production and alert the operating personnel. If only a simple cutting-edge breakage is present, then production can continue immediately once a new tool is provided.

2.8 Outlook

If a machine tool is equipped with a BG41 bore gauge with several sensors for detecting the cylindricity and circularity of a bore, as well as with a KomTronic *U*-axis system, today it is possible to compensate for tool deviations even during machining in a closed-process chain.

2.9 Summary

Given cost pressures and the demand for shorter production time per piece, there is a noticeable trend in contemporary manufacturing processes toward close-to-process measurement in the same setup as is used for machining. In machining centers, this is performed on a shop-specific basis through the use of measuring probes and appropriate measurement software. This is performed by calling up measuring cycles that have been prepared previously and saved in the CNC system and which are used to perform a set-point/actual-value comparison. Based on the results, an automatic check is performed as to whether the tolerance limits have been observed and whether it is possible to continue work, whether tool corrections have to be made, or whether a fault message is issued.

The use of close-to-process production-measurement technology is especially notable in highly productive operations, that is, where linked production lines are used. In such

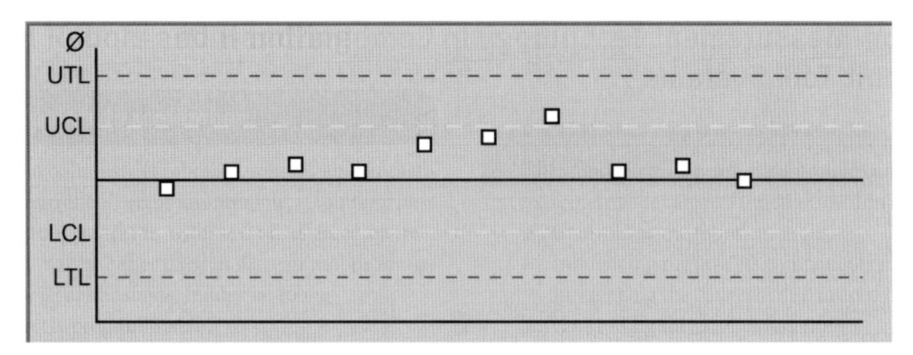

Figure 2.10: Wear compensation after exceeding the upper control limit.

cases, interruption of the manufacturing procedure owing to external measurement operations would cause time delays in all the upstream and downstream machining processes. But increasing use is being made of process-integrated measurement technology for monitoring measurements even in manufacturing situations where a postprocess measurement with subsequent final machining does not appear to be profitable for reasons of precision and cost-effectiveness. Here the industry prefers either production-measurement technology through the use of bore gauges in conjunction with adjustable actuator-driven tool systems or other automatically adjustable tool corrections in the CNC.

Both these techniques can be used together with an algorithm to eliminate systematic sources of problems close to the process without requiring any additional time for in-company transport for measurement and additional clamping and alignment operations. The user has to decide which of the described methods should be used based on the production situation, the workpiece, and economic factors.

Measurement in the machine is no substitute, however, for high-precision measuring machines in a metrology room if the manufacturing data for each workpiece or each manufacturing lot are required for reasons of precision (e.g., in the case of multiple fixtures).

Production-Measurement Technology in Combination with Mechatronic Tool Systems

Important points to remember:

1. Quality data relevant to the process can be acquired by means of **in-process, postprocess, or close-to-process measurement technologies.**
   - In-process measurement technology and postprocess measurement technology operate in parallel with productive time, whereas close-to-process measurement takes place in idle times.
2. Within the close-to-process production measurement technology, a distinction is made between **workpiece-oriented** and **tool-oriented** measurement technologies. Measurement and monitoring are performed within the machining space.
   - In **tool-related** measurement technology, the **cutting tools** are monitored by noncontact laser systems or tactile measurement systems before and/or after the machining process.
   - In **workpiece-oriented** measurement technology, the **workpiece** is measured in its original clamping situation in the machine tool.
   - Actual values are compared with set-point values. In the case of deviations that lie outside the tolerance limits, the stored tool parameters are corrected automatically (tool offset in the X, Y, or Z axis, tool length).
3. In **workpiece-oriented close-to-process measurement technology,** two types of measurement systems are dominant:
   - **Tool-measuring probes** in combination with the corresponding measurement software
   - **Bore gauges** for machine-independent measurement in the workspace
4. Through the use of **bore gauges,** it is possible to check the following characteristics:
   - The **bore diameter**, position independent of the spindle position
   - **Shape and position** of boreholes (e.g., circularity, cylindricity)
5. If dimensional deviations occur after checking of the measurement characteristics with the aid of the bore gauge, **automatic tool compensation** can be performed when **actuator-driven tool systems** are used.
6. **Bore gauges in combination with actuator-driven tool systems** thus enable manufacturing in a controllable, closed-process chain.
7. The time-related or economic **advantages of the individual measurement technologies** depend on the specific manufacturing process, the quantities produced, and the costs for measuring equipment and tools.
8. **Measurement in the machine** also requires the appropriate software in the CNC system in order to
   - Execute the various measurement cycles
   - Evaluate, display, save, and output the measurement data
9. Measurement in the machine is **not a substitute for postprocess measurement** on high-precision measuring machines in a metrology room if such measurement is necessary for reasons of precision or documentation of quality.

Part 5

Computer Numerical Control Programs and Programming

1. Computer Numerical Control Programs

To better understand numerical control (NC), it is useful to have some knowledge about the design and structure of NC programs. Such knowledge is essential when making manual program corrections on computer numerical control (CNC) machines.

Before delving too deeply into the abstractions of manual NC programming, it will be helpful to examine the current state of technology:

1. NC programs are no longer written in machine code (or *control code*) in the form described herein. This form of programming has been replaced by a variety of highly intelligent, user-friendly NC programming systems. These systems can be either integrated into the CNC system—in which case they are designed for specific machines (shop-floor programming)—or run on an external PC—in which case they are applicable to machines and controls from different manufacturers. Even subsequent geometric and technological modifications of finished programs can be inserted much more quickly by this method than with DIN code. DIN code is still the standard data-entry format and is used by almost all manufacturers of CNC systems. Therefore, it may be helpful to know about the structure and individual commands used in this code.
2. A new trend has appeared, prompted both by high-speed cutting machines, with their need for extremely large amounts of data in extremely short time periods, and the increased use of computer-aided design (CAD) systems and the direct use of the data formats generated by the CAD system. These involve either nonuniform rational-basis splines (NURBS) or Bezier formulas. This provides a number of advantages:

- A considerable reduction in the quantity of data
- Improved, smoother machine characteristics
- Higher speeds
- Greater accuracy
- Elimination of geometric data conversion by postprocessors, etc.

However, it is **not** possible to insert **manual modifications** or additions into these geometric formats at a later time. For a more detailed explanation, → *Part 2, Chapter 3, Spline Interpolation (NURBS)*.

1.1 Definition

A **program** consists of a sequence of instructions that cause a computer or an NC machine to perform a certain processing task. In the case of an NC machine, this is understood to mean the manufacture of a certain workpiece by means of relative motion between the tool and the workpiece, with the dimensional input specified directly in millimeters or inches. An **NC part program** of this type contains not only the dimensional data required for machining but also all additional M-functions and auxiliary commands. The result is that all the data needed for the fully automatic manufacturing of the workpiece are made available in the correct sequence.

1.2 Structure of NC Programs

Figure 1.1 shows the basic structure of an NC program. The content of the program consists of any desired number of **blocks** that describe the entire work process of the machine in a step-by-step fashion. Each block corresponds to a line in the NC program. The individual blocks can be numbered. This makes it easier to find a specific block (especially in the case of error messages) and can serve as a jump mark. Each block, in turn, consists of one or more words. In the word-address format that is used commonly today, the words are composed of address letters and numerical values. The **address** (*Table 1.1*) specifies the memory area that is intended to receive the following numerical value, that is, which group of functions is being referenced. As a general rule, each address is allowed to appear only once in a given block, but most control systems allow multiple G-functions and

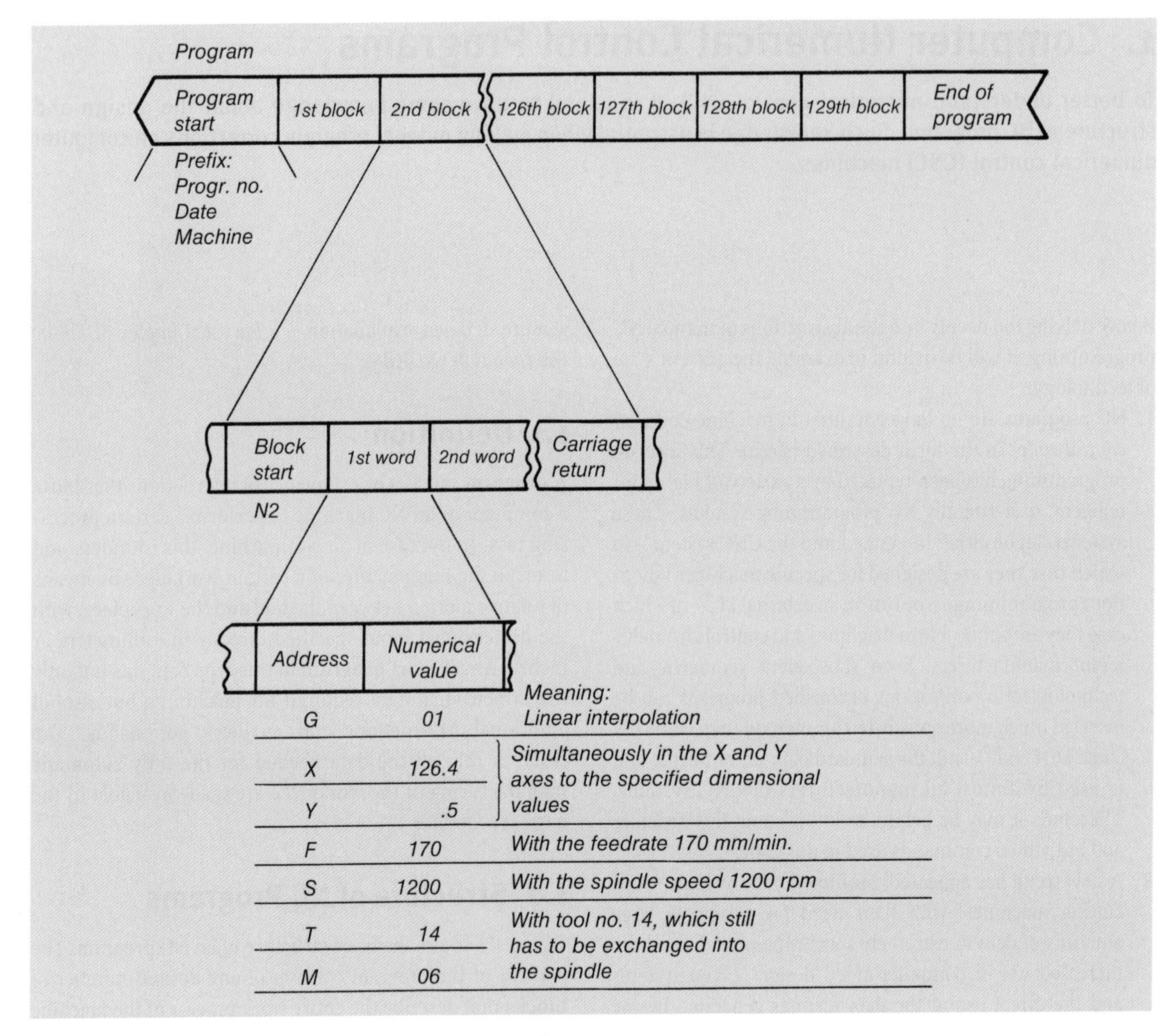

Figure 1.1: The basic structure of an NC program in address format.

M-functions per block, provided that these do not conflict or cancel one another.

A block can contain various **instructions.** A distinction is made here between

- **Geometric instructions,** which control the relative motion between the tool and the workpiece (addresses X, Y, Z, A, B, C, W, etc.)
- **Technological instructions,** which specify the feed rate (F), spindle speed (S), and tools (T)
- **Motion commands,** which determine the type of motion (G), for example, rapid traverse, linear interpolation, circular interpolation, or selection of planes
- **M-functions** for the selection of tools (T), index table positions (M), and coolant supply on/off (M)
- **Retrieval of compensation values** (H), for example, for tool-length compensation, cutter-diameter compensation, tool-nose compensation, or zero offsets (G)

Letter	Designation	Address for
A		*Angular dimension about the X axis*
B		*Angular dimension about the Y axis*
C		*Angular dimension about the Z axis*
D		*Angular dimension about the additional axis or unassigned*
E		*Angular dimension about the additional axis or unassigned (error code, etc.)*
F	*Feed rate*	*Also called rate of feed*
G	*Go*	*Preparatory functions*
H	*High*	*Tool-length compensation*
I		*Auxiliary parameter for circular interpolation or thread pitch parallel to the X axis*
J		*Auxiliary parameter for circular interpolation or thread pitch parallel to the Y axis*
K		*Auxiliary parameter for circular interpolation or thread pitch parallel to the Z axis*
L		*Unassigned*
M	*Miscellaneous*	*Machine commands, on/off commands, switching functions*
N	*Number*	*Block number*
O	*Offset*	*Avoid using tool offset parallel to an axis if possible*
P		*Third rapid traverse limitation*
Q		*Second rapid traverse limitation*
R	*Reference*	*First rapid traverse limitation or reference plane*
S	*Spindle rev.*	*Speed of main spindle*
T	*Tool number*	*Possibly with compensation value*
U		*Second axis parallel to the X axis*
V		*Second axis parallel to the Y axis*
W		*Second axis parallel to the Z axis*
X		*First main axis*
Y		*Second main axis*
Z		*Third main axis*

Table 1.1: Address Assignments According to DIN 66025

- **Cycle or subroutine calls** for program sections that recur frequently (P, Q)

The **numerical values** of the dimensional data define the target positions. The programming system should allow these values to be entered in **decimal-point notation.** In other words, it should allow all leading and trailing zeros to be omitted. This shortens the program to a considerable degree and helps to prevent errors. All numerical values without decimal places are placed in front of the decimal point, with decimal fractions coming after the decimal point.

Examples

$$X400 = X\,400.00\text{ mm}$$
$$X.23 = X\,0.230\text{ mm}$$
$$Z14.165 = Z\,14.165\text{ mm}$$

Finally, an additional distinction is drawn between main blocks and subordinate blocks as follows:

- **Main blocks** are characterized by the fact that all addresses are accompanied by the current numerical values. This makes it easier to resume execution of a long program after it has been interrupted. Main blocks can be marked by placing a colon in front of the *N* address or by assigning them block numbers that are multiples of 100 or 1,000.
- **Subordinate blocks** contain only the words whose values have changed.

Meanings of Commands

The meanings of the basic commands, their syntax, and the program structure are defined in the DIN 66025 standard. In addition, practically all control-system manufacturers provide specific, nonstandardized commands in their own syntax. These additional functions can be anything from functions for program execution (e.g., calculations, loops, program jumps) to special functions (e.g., working-area limits, programming of help systems, configuration commands). Generally, the scope and capabilities of these specific commands are considerably greater than the commands included in the DIN standard. Information on these functions can be derived from the applicable programming guides.

1.3 Program Structure, Syntax, and Semantics

The term **syntax** refers to the formal rules that govern the structure of instructions in a particular programming language. The *meanings* of the words are not established by rules of syntax but by **semantics.**

The combination of syntax and semantics determines the **program structure,** which consists of characters, words, and blocks, as well as the arrangement of these data on data-storage media.

The structure of a typical NC program for three-axis continuous path control, written in accordance with the Standard EIA RS 274 B, might appear as follows:

N4, G2, X ± 4.3, Y ± 4.3, Z ± 4.3, I4.3, J4.3, K4.3, F7, S4, T8, M2, $.

The meanings of the symbols are explained below:

N4 This represents the four-digit block number. Each program can be divided into a maximum of 9,999 blocks.

G2 This symbol represents the two-digit preparatory functions that can specify the type of interpolation, the cycle, the direction of tool-nose compensation, or the entry of geometric dimensions.

X ± 4.3, Y ± 4.3, Z ± 4.3 These are the dimensional data specifying four positions to the left of the decimal point and three positions to the right of it; that is, the maximum programmable value is 9999.999. The decimal point can be omitted or written in the normal fashion (decimal-point programming).

I4.3, J4.3, K4.3 These are the auxiliary parameters used to represent the arc center used in circular interpolation. A given block may contain only IJ, JK, or IK, which correspond to the interpolation planes *XY*, *YZ*, and *XZ*.

F7

F6.1 G94 = feed in mm/min
F4.3 G95 = feed in mm/revolution
F5.2 G04 = dwell in seconds
F7 G104 = dwell in revolutions

S4 This represents the four-digit spindle speed, specified directly in revolutions per minute.

T8 This is the eight-digit tool number, specified with or without the compensation number that is retrieved for this tool. Five- and six-digit tool numbers also can be programmed.

M2 These are the two-digit (maximum 99) miscellaneous functions or M-functions, for example, coolant on/off, tool changes, or direction of spindle rotation.

$ This is the end-of-block character.

1.4 On/Off Commands (M-Functions) *(Table 1.2)*

No switches are provided on the machine for turning the machine functions on and off. All these commands must be programmed. This is accomplished with the on/off commands, which are identified by the following address types:

S For the spindle speed.
T For tool selection (tool number).
M For all miscellaneous functions.
F For the feed rate

Some examples of blocks with these on/off commands are listed below:

N 10 S 1460 M 13 $	Step 10: Spindle speed 1,460 rpm, spindle rotating clockwise, coolant ON.
N 60 G 95 F 0.15 $	Step 60: Feed 0.15 mm/revolution.
N 140 T 17 M 06 $	Step 140: Insert tool number 17 into the spindle.
N 320 M 00 $	Step 320: Program interruption until a new START signal is given.
N 410 M 30 $	Step 410: Program end, spindle STOP, coolant OFF, rewind punch tape to program start.

On/off commands are stored just as if the switching were performed with ordinary switches. These commands can be altered or undone by **overwriting** them in the program. If it is practical, multiple on/off commands can be combined within a single block.

When using M-functions, it is important to bear in mind that some of them take effect immediately (i.e., when the block begins), whereas others only take effect at a later point, when the block has been processed completely. Which of these is the case should be defined in programming guide for the respective machine.

At this point, it would be advisable to take a closer look at the table of M-functions and practice programming a few commands.

1.5 Dimensional Data

Dimensional data are significant to the machine in three ways:

1. The **values** of these data determine the target positions for the motion.
2. The **sign** (positive or negative) of the data indicates the direction of motion or defines the quadrant.
3. The **sequence** of these data determines the sequence of motions in the program.

Dimensional data are coded with the address letters X, Y, Z, A, B, C, U, V, W, I, J, K, and R. In newer CNC systems, it is also possible to assign axis addresses containing more than one character (*Table 1.3*). In this case, the dimensional data have an equal sign (=) placed in front of them. The advantage of this is that it is much easier to read programs for machines with a large number of axes (example: X1 = 123.000 X2 = 234.500 X3 = ?, etc.).

Dimensional data can be specified in the drawings in the form of **absolute dimensions** and/or **incremental dimensions** (*Figure 1.2*). The NC program therefore should accommodate both types of dimensional specifications. This will make it possible to enter the dimensions directly from the drawings.

An absolute dimension indicates the distance of a position from the program zero point, whereas an incremental dimension indicates the distance between the new position and the previous position. The codes **G91** and **G92** can be used to switch between absolute and incremental programming at will without losing the program zero point.

If the tool is intended to move through the sequence of seven positions specified in *Figure 1.3* and then return to the zero point, for example, the input data will differ in accordance with the type of dimensions being used (*Table 1.4*).

The **advantage of absolute programming** is that any subsequent change to a particular position will not affect any other geometric dimensions. In incremental programming, this change would require corrections to the programming of all subsequent positions as well. Interrupted programs also can be restarted more easily when absolute programming is used.

The following points can be regarded as the **advantages of incremental programming:**

1. The sum of all *X* dimensions and the sum of all *Y* dimensions must be zero in programs that return to the starting position. This makes it easy to check for programming errors. When absolute and incremental programming methods are mixed, however, this simple error check is no longer possible.
2. Subroutines, such as those used to create drilling patterns, grooves, fillets, and milling cycles, are very easy to copy and transfer to other positions.

Code	Function
M00	Program stop. Spindle, coolant, and feed OFF. Restart via “Start” button.
M01	Optional stop. Acts like M00 if “Optional Stop” switch is set to ON.
M02	Program END.
M03	Spindle ON, clockwise.
M04	Spindle ON, counterclockwise.
M05	Spindle STOP.
M06	Perform tool change.
M07	Coolant 2 ON.
M08	Coolant 1 ON.
M09	Coolant OFF.
M10	Clamping ON.
M11	Clamping OFF.
M13	Spindle ON, clockwise and coolant ON.
M14	Spindle ON, counterclockwise and coolant ON.
M19	Spindle STOP at specified angular position.
M30	Program end and reset to start of program
M31	Cancel lock-out
M40 – M45	Switch gear stage.
M50	Coolant 3 ON.
M51	Coolant 4 ON.
M60	Workpiece change.
M68	Clamp workpiece.
M69	Unclamp workpiece.
All M-functions not listed here are unassigned or freely available.	

Table 1.2: Switching Functions According to DIN 66025, Part 2

```
N1000 ZOTSEL (GT300-NPV.zot); Selection and path to zero-point offset table
;--------------------- Machining with right-side spindle -----------------
--
N1010 MainSp(S2)              ;Feed will affect spindle 2
N1020 SMX(S2=3000)            ;Maximum speed at G96 second spindle
N1030 G8(SHAPE80)
;----------------- Move chuck to starting position (necessary for PLC)
N1040 S2CLOSE=66              ;Close chuck S2, 66 = inside-outside clamping
;-----------------------------------------------------------------------
; Zero point G59 1 mm to right of the left-side end face
N1060 G0 G90 DIA G18(X,,Z) G53 G48 G90 X=260 Z=300 Z2=1 M205
1070 IF TARTTYPE$ = "CASTING" THEN
N1070 (MSG T3 Roughing tool right)
N1080 M6 T3                   ;Tool change
N1090 G0 G47 G96 G59 X100 Z-10 Z2=1 S2=200 M204 ED1
N1100 (MSG facing on spindle 2))
N1110 G0 X45 Z0 M8
N1120 G1 X10 F.17
N1130 X8 Z.2 F.1
N1140 X-.5
N1150 G0 X45 Z-10
N1160 G0 G53 G48 X=260 Z=300  ;Move away to changing position
1180 ENDIF
;-----------------------------------------------------------------------
N1170 (MSG T4 finishing tool right))
N1180 M6 T4                   ;Tool change
N1190 G0 G47 G59 G96 X=45 Z=-10 Z2=1 S2=220 M204 ED1 M8
N1200 (MSG turn first end on spindle 2)
; Contour turning with standard machining cycle
N1210 G171 (P DameKontur, CD2, LD1, CR0.5, CA0, CES1, UCV0)
N1220 G0 G53 G48 X=260 Z=300 Z2=1 M9
N1230 M30
```

Table 1.3: Example of Axis Addresses with Multiple Characters and Additional Explanations, Some of Which Appear on the Screen

1.6 Preparatory Functions (G-Functions) *(Table 1.5)*

The two-digit motion-related functions (G stands for "go") and the dimensional data (X, Y, Z, R, A, etc.) go hand-in-hand. The **G-functions** specify the computer program that will be used to process the subsequent dimensional data in the controller. The dimensional data indicate **where** the motion is going, whereas the motion-related functions indicate **how** they will get there. In fact, multiple G-functions are required in most cases to prepare the numerical control system to interpret the dimensional data. For this reason, the term *preparatory function* is also used. This makes it necessary to store multiple G-functions and keep them activated. Depending on the control system, it may be necessary to program these G-functions in multiple consecutive blocks, or it may be possible to place all the G-functions in one common block.

For the sake of a clearer structure, the complete set of G-functions is divided into three overall types and various smaller groups. G-functions are only allowed to overwrite other G-functions from the same group. To put it differently, only one G-function from each group can be active at any given time.

A distinction is made among **three types of preparatory functions.** The functions appearing in boldface below indicate

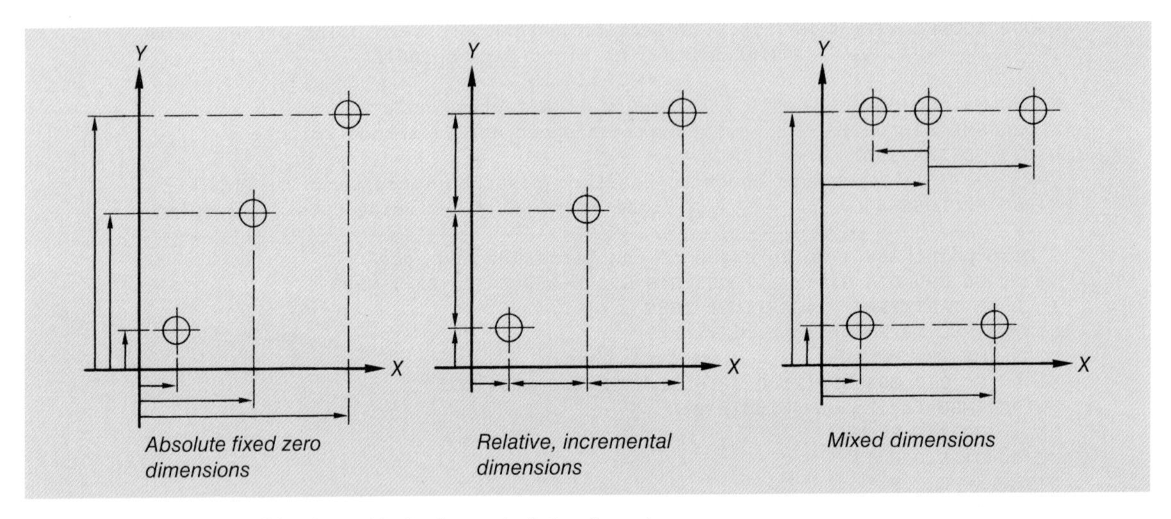

Figure 1.2: Dimensioning of drawings with absolute and relative dimensions.

the switching state. In some control systems, these do not need to be programmed in addition to the other functions.

- **Modal G-functions, that is, functions that remain in effect over several blocks,** for the following purposes, for example:
  - The type of interpolation — **G00,** G01, G02, G03, G06
  - Selection of planes — **G17,** G18, G19
  - Tool compensation — **G40,** G41, G42, G43, G44
  - Zero offset — G92, **G53**–G59
  - Positioning characteristics — G08, G09, G60, G61, G62
  - Work cycle — **G80**–G89
  - Dimensions — **G90,** G91
  - Feed specifications — G93, **G94,** G95
  - Entry of spindle speed — G96, G97
- **Preparatory functions that only remain in effect in a single block include**
  - Dwell time — G04 (in combination with F for the time specification)
  - Acceleration/deceleration — G08, G09
  - Tapping — G63
  - Reference-point offset — G92

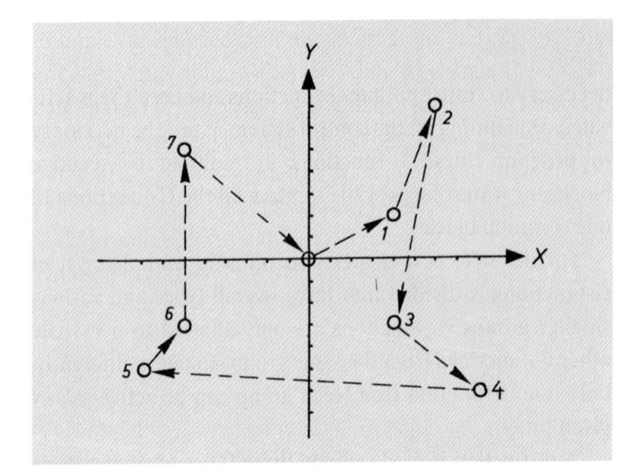

Figure 1.3: Drilling pattern.

Examples of the Effects of G-Functions

There are various ways to get from a starting point S to an end point E (*Figure 1.4*). These options are selected by choosing different G-functions:

First option: N100 G00 X200 Y140 $

Motion from S to E at rapid traverse speed, 140 mm at an angle of 45°, parallel to the *X* axis the rest of the way.

Second option: N200 G01 X200 Y140 F400 $

Motion in a straight line (linear interpolation) at a feed rate of 40 mm/min.

Third option: N300 G03 G17 X200 Y140 R205 F120 $

PT.	Absolute programming		Relative programming	
	X	Y	X	Y
1	4	2	+4	+2
2	6	7	+2	+5
3	4	−3	−2	−10
4	8	−6	+4	−3
5	−8	−5	−16	+1
6	−6	−3	+2	+2
7	−6	+5	0	+8
0	0	0	+6	−5
			Σ=0	Σ=0

Table 1.4: Path Dimension Table for the Drilling Pattern Shown in the Figure 1.3 Drilling Pattern with Absolute and Relative Programming

Motion in a counterclockwise circular path at a radius R of 20 mm from the arc center M_3.

Fourth option: N 400 G00 X200 $ and N401 Y140 $

Motion by in-feed at rapid traverse speed, first parallel to the X axis, then parallel to the Y axis.

Fifth option: N500 G02 X200 Y140 R–130 $

Motion in a clockwise circular path at radius R130 from the arc center M_5.

Preparatory Functions to Which No Fixed Meanings Are Assigned by Standards

The meanings of G-functions are defined in DIN 66025, Part 2, and should be uniform for products from all manufacturers of NC systems.

Examples

N 10 G81 $

From step 10 on: Drilling cycle G81 is retrieved. In other words, all Z dimensions are incremental dimensions. In-feed in the X and Y axes takes place at rapid traverse speed. The drill feed rate is specified in millimeters per minute.

N 40 G02 G17 X460 Y125 I116 J–84 $

Block 40: Clockwise circular motion until end point X460 Y125 is reached.

N70 G04 F10 $

Block 70: Dwell time of 10 seconds; that is, the spindle continues to turn while the feed motion of the axes is suspended for 10 seconds.

N 100 G17 G41 H11 T11 $

From block 100 onward: The compensation value that was entered at memory location H11 and is now retrieved from there takes effect in the XY plane. The offset is to the left when viewed in the direction of motion.

N 160 G54 $

Block 160: The first group of zero-point offsets for all axes is called up.

1.7 Cycles

To perform work processes that recur frequently, most numerical control systems contain a special type of subroutine with preprogrammed fixed cycles. Cycles are intended to help simplify programming and reduce program size by allowing each recurring sequence to be called up only once and then supplied with the necessary parameter values each time it runs.

A distinction is made between the following types of cycles:

Drilling cycles *(Table 1.6 and Figure 1.5).* For drilling, reaming, countersinking, and tapping. Defined in DIN 66025 (G80–G89) (For an example, → *Figure 1.6).*

Code	Function
G00	Positioning in rapid traverse, point-to-point control
G01	Linear interpolation
G02	Circular interpolation, clockwise
G03	Circular interpolation, counterclockwise
G04	Dwell time
G06	Parabolic interpolation
G09	Exact positioning
G17	Plane selection *XY* } Interpolation parameter for circular programming
G18	Plane selection *XZ* } Interpolation parameter for circular programming
G19	Plane selection *YZ* } Interpolation parameter for circular programming
G33	Thread cutting with constant pitch
G34	Thread cutting with increasing pitch
G35	Thread cutting with decreasing pitch
G40	Delete all called-up tool-compensation values
G41	Tool-radius compensation, offset to the left
G42	Tool-radius compensation, offset to the right
G43	Tool compensation, positive
G44	Tool compensation, negative
G53	Delete the called-up zero-point offset
G54–G59	Zero-point offset 1–6
G60*	Positioning tolerance 1
G61*	Positioning tolerance 2, also run loop
G62*	Quick positioning, rapid traverse only
G63	Set feed to 100%, for example, tapping
G70	Dimensions input in inches
G71	Dimensions input in millimeters
G73*	Programmed feed rate = axis feed rate
G74*	First and second axes move to reference point
G75*	Third and fourth axes move to reference point
G80	Delete the called-up cycles
G81–G89	Defined drilling cycles
G90	Absolute dimensions (fixed zero dimension)
G91	Relative dimensions (incremental dimension)
G92	Set programmed reference point offset/memory
G94	Feed rate in mm/min (or in/min)
G95	Feed rate in mm/revolution (or in/revolution)
G96	Constant cutting speed
G97	Spindle speed in rpm

* All the G-functions not listed here are not permanently assigned and are freely available.

Table 1.5: G-Functions According to DIN 66 025, Part 2

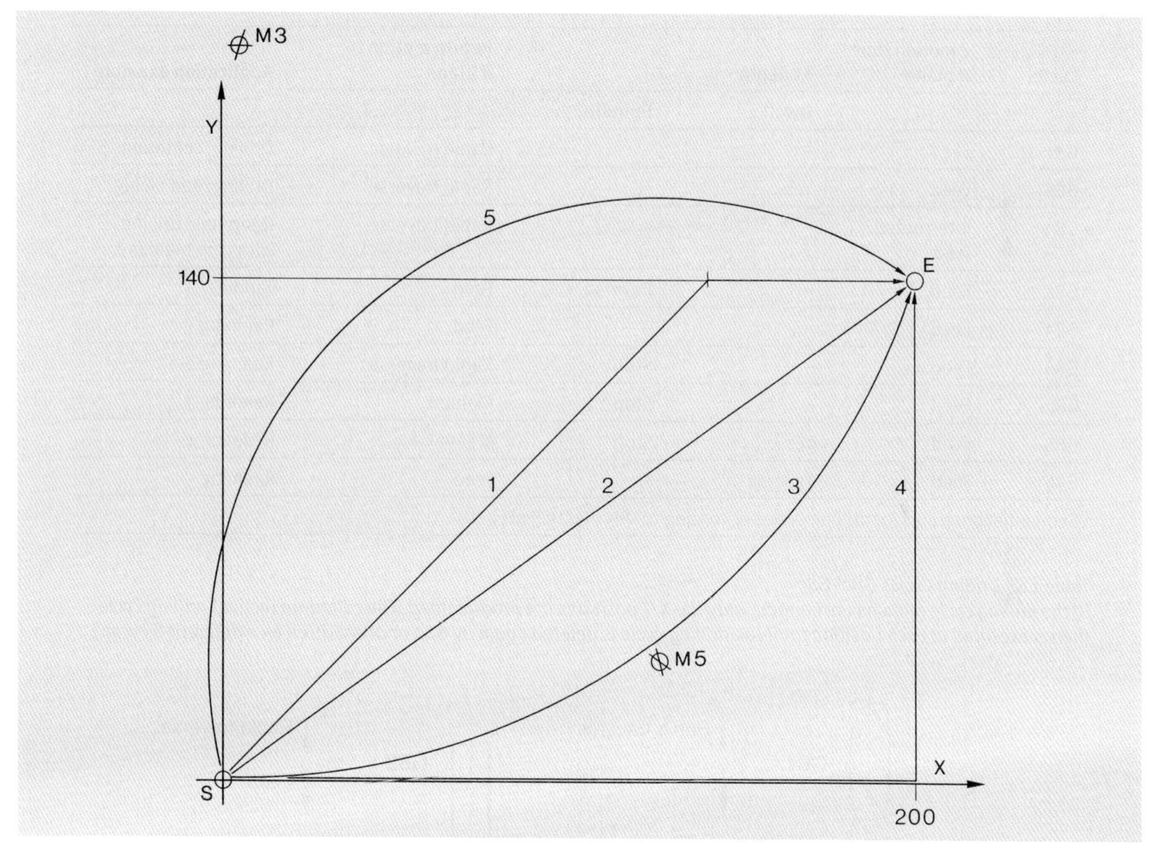

Figure 1.4: Using various G-functions on various paths to move from starting point S to end point E.

Milling cycles. For slot milling, pocket milling, reaming (by milling), thread milling, shank milling, etc. These cycles are not standardized but are designed differently by each controller manufacturer.

Turning cycles *(Figure 1.7).* For longitudinal cutting and facing, axially parallel or conical threading with automatic in-feed, as well as cycles for creating fillets and grooves with automatic selection of the number of cuts. This type of cycle also lacks standardization and varies from one controller manufacturer to another.

Variable cycles. These are also known as *subroutines* and are designed especially for each type of machine. Examples include the tool-changing cycle (M06) and the geometric cycles used to create hole circles, rows of holes, drilling of deep holes, milling of circular segments, pocket milling, etc.

1.8 Zero Points and Reference Points *(Figures 1.8 and 1.9)*

Zero points and various reference points are defined for each numerically controlled machine tool. It is generally easy to direct the motion to these points. The workpiece dimensions are programmed in reference to these points.

Cycle	*Z* motion from *R* plane	At depth		Return motion to *R* plane	Application example
		Dwell	Spindle		
G81	Feed	–	–	Rapid traverse	Drilling, centering
G82	Feed	Yes	–	Rapid traverse	Drilling, end facing
G83	Interrupted feed	–	–	Rapid traverse	Deep-hole drilling with chip breaking
G84	Feed	–	Reverse	Feed	Tapping
G85	Feed	–	–	Feed	Reaming 1
G86	Feed	–	Stop	Rapid traverse	Reaming 2
G87	Feed	–	Stop	Manual	Reaming 3
G88	Feed	Yes	Stop	Manual	Reaming 4
G89	Feed	Yes	–	Feed	Reaming 5
Sequential programs for drilling cycles according to DIN 66025, Part 2					

Table 1.6: Drilling Cycles G80–G89
[If the drilling cycle remains unchanged, only the X/Y positions are programmed. The called-up (active) drilling cycle then is executed at each position until the drilling cycle is deleted again by G80 or overwritten by a different G cycle.]

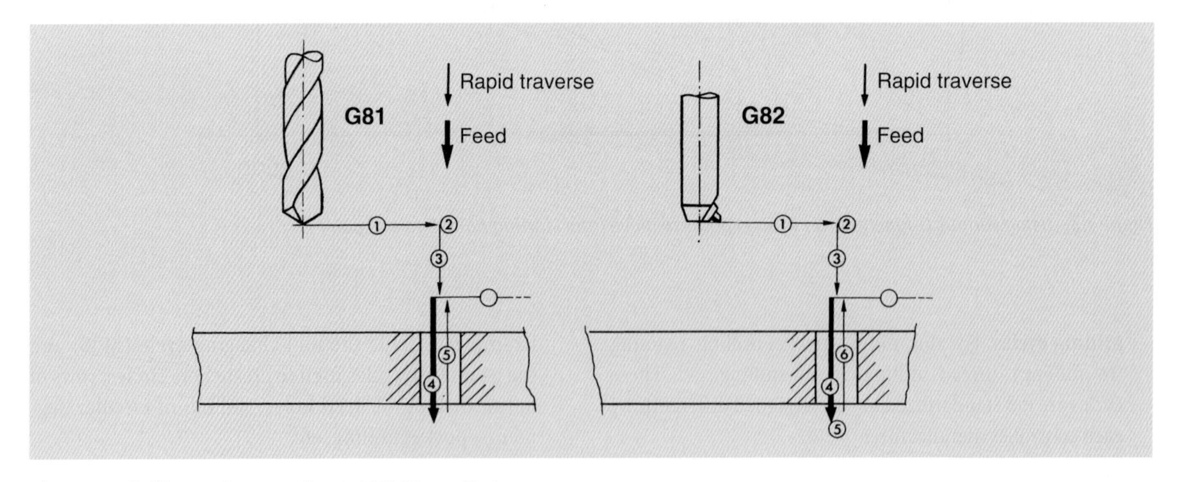

Figure 1.5: Drilling cycles according to DIN 66025, Part 2.

Machine Zero Point

The machine zero point cannot be shifted from its fixed location at the origin of the machine coordinate system. To be of practical use in the shop, the NC machine needs to have a zero point that can be referenced quickly and easily, for example, immediately after the machine is switched on. Ideally, the controller should be able to direct motion to this position in each axis automatically. It is also advantageous if this command can be issued either by the program or by pressing the corresponding key.

This zero point must be located with an accuracy of ±1 path increment. In applications involving resolvers and inductosyn

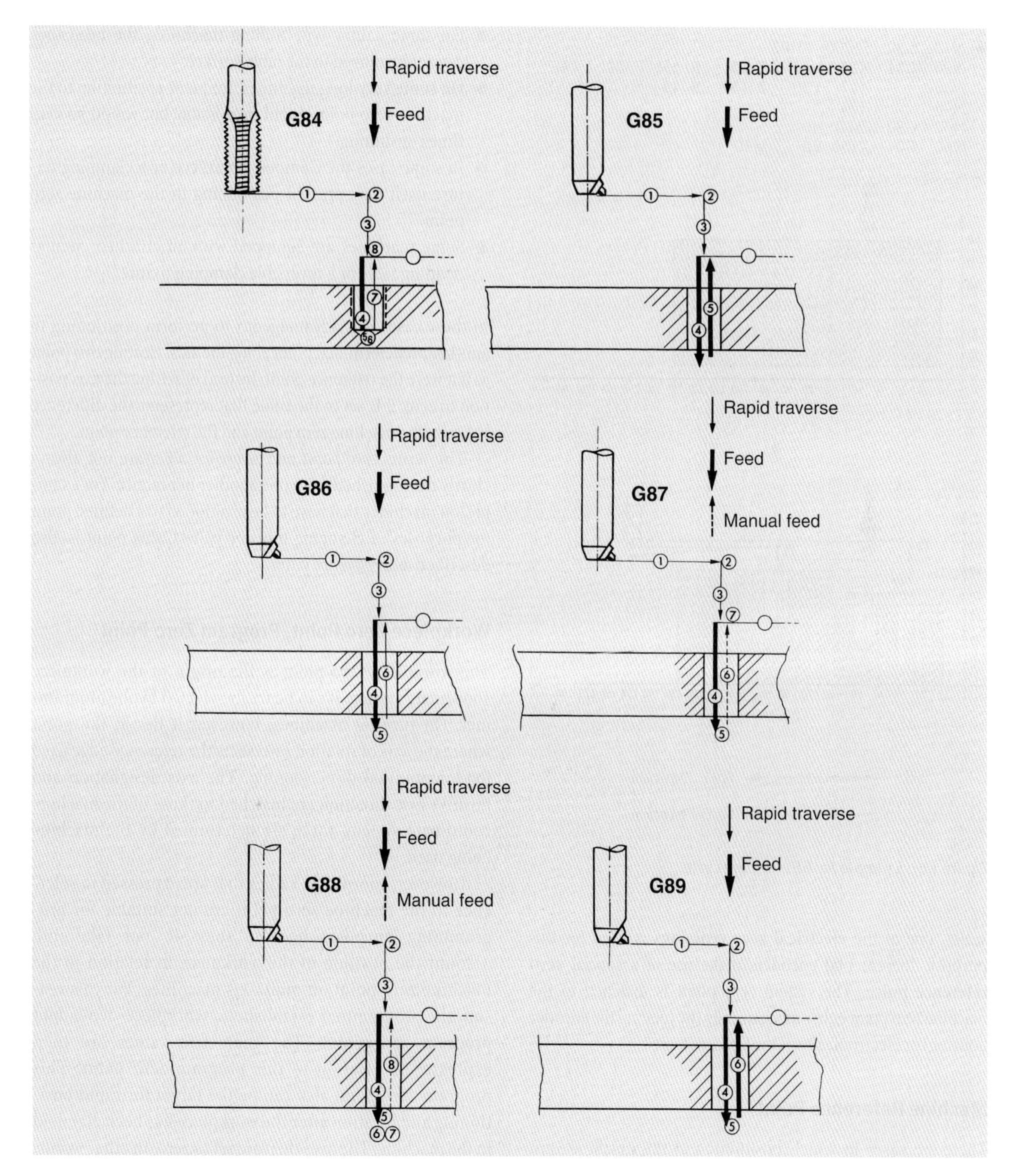

Figure 1.5: Drilling cycles according to DIN 66025, Part 2. (Continued)

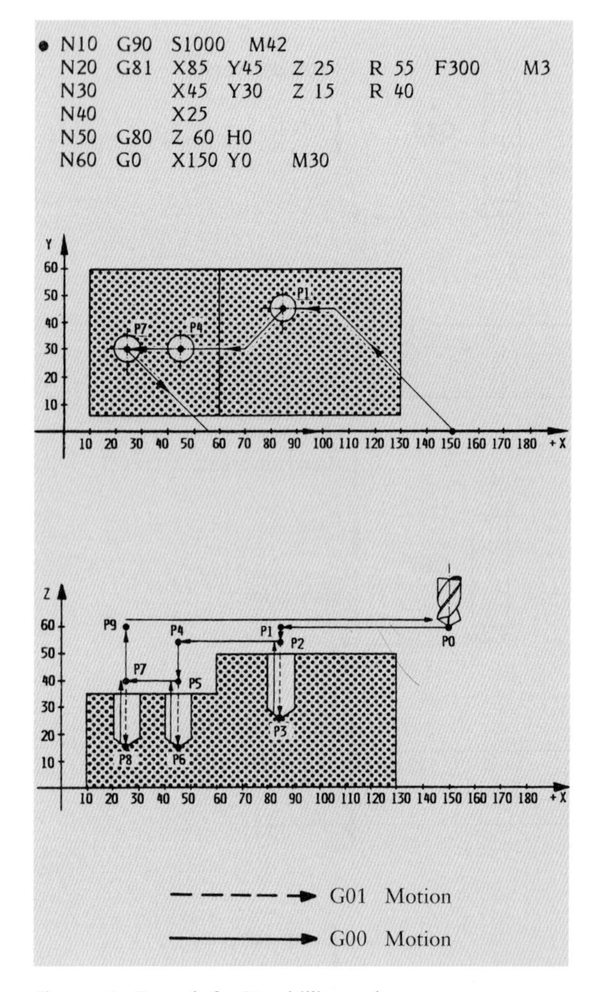
N10 G90 S1000 M42
N20 G81 X85 Y45 Z 25 R 55 F300 M3
N30 X45 Y30 Z 15 R 40
N40 X25
N50 G80 Z 60 H0
N60 G0 X150 Y0 M30
Y
Z
+X
P0
P1
P2
P3
P4
P5
P6
P7
P8
P9
G01 Motion
G00 Motion

Figure 1.6: Example for G81: drilling cycle.

scales, one of the **electrical zero crossings** is used for this purpose. Incremental encoders make use of a special **zero reference pulse.** The system zero point is matched to the machine zero point either mechanically (by rotating the encoder housing) or electronically in the CNC system.

Machine Reference Point

There are cases in which positioning at the machine zero point of a particular axis is not possible. Some examples of such cases are listed below:

- On some gantry-type milling machines, the headstock cannot be moved to the table surface.
- On some very long machines, frequent positioning to the machine zero point might be possible but would be very time consuming.
- In some cases the clamped workpiece or a clamping fixture itself may prevent positioning to the machine zero point.
- Some machines are equipped with an attached circular dividing table or a reversible clamping fixture.

In these cases it is advantageous to perform positioning to another predetermined point along the axis, treating this point as if it were the reference point. Instead of setting the axis position to zero, it is set to the value that represents the difference between the machine zero point and the reference point.

The terms *zero point* and *reference point* are not always clearly distinguished from one another in practice. For example, on machines that have to be positioned to a defined point for purposes of changing tools or pallets, this point is also described as a *reference point.*

Workpiece Zero Point, Program Zero Point

The workpiece zero point is the origin of the workpiece coordinate system. It can be freely selected by the programmer. On turning machines, however, it lies at the point where the axis of rotation intersects the reference edge used for longitudinal dimensioning. The axis designation and positive axis direction are matched to those of the machine coordinate system; these are determined by the machine being used.

However, coordinate values that are expressed in reference to the machine zero point are not suitable for programming purposes because they do not take into account the position of the workpiece in relation to the machine zero point on multiaxis machines. For this reason, the programmer establishes a workpiece-dependent program zero point. The geometric values are then expressed in relation to this program zero point. This program zero point also can be the target for rapid positioning and location after the workpiece has been clamped in the machine. The axes then are set to zero. In other words, the zero point is shifted from the machine zero point to the program zero point.

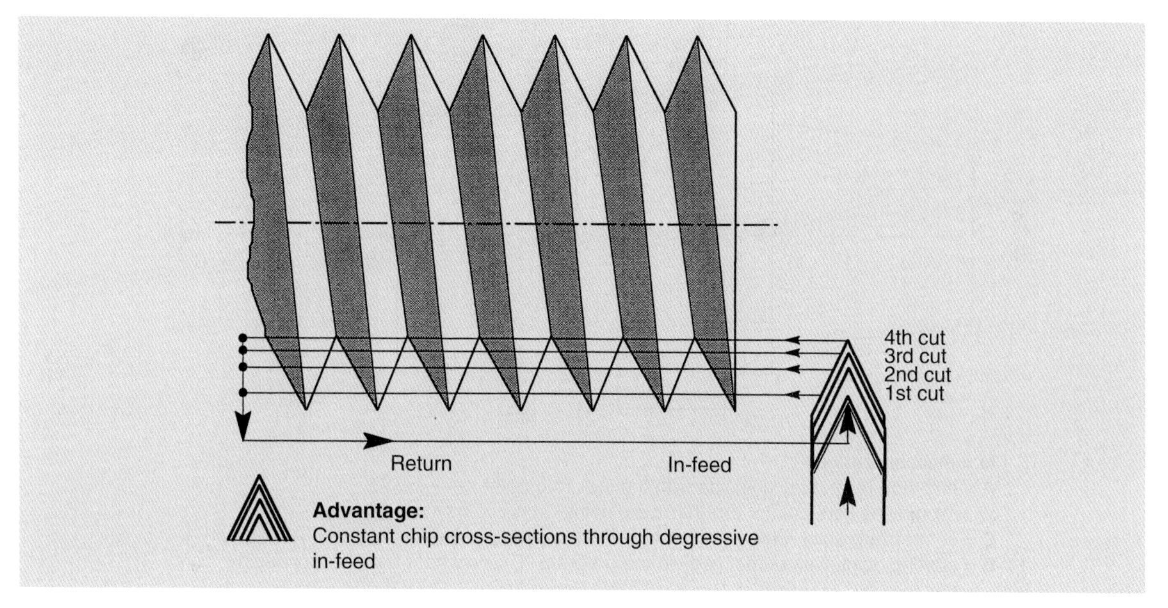

Figure 1.7: Thread cutting with degressive in-feed to keep the chip cross sections constant or reduce them with each cut.

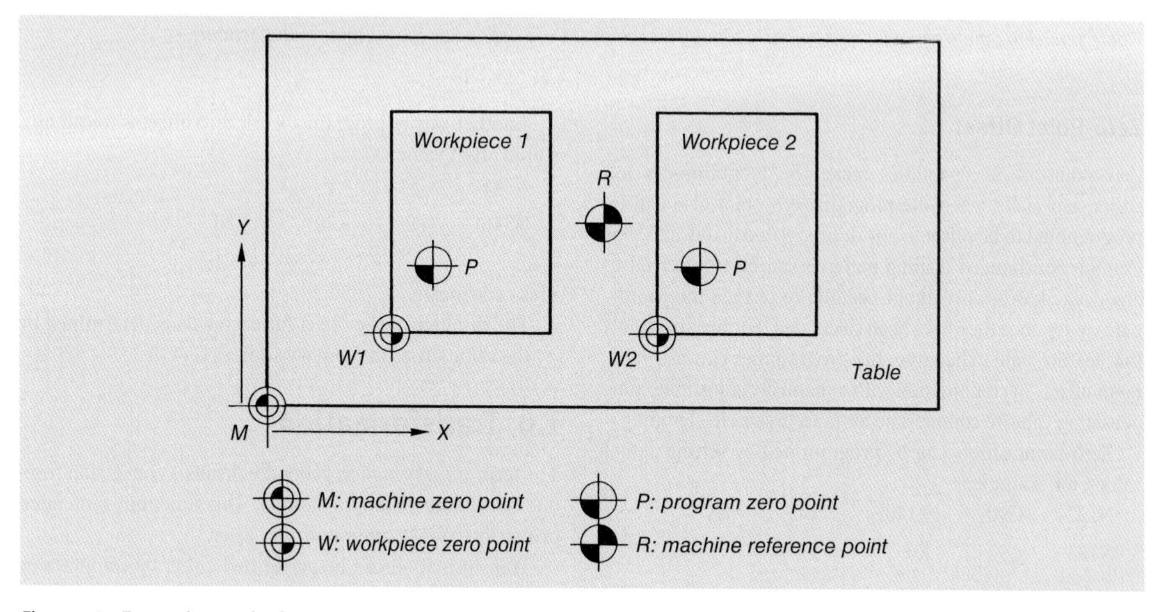

Figure 1.8: Zero points and reference points in drilling and milling machines (M = is defined by the machine manufacturer).

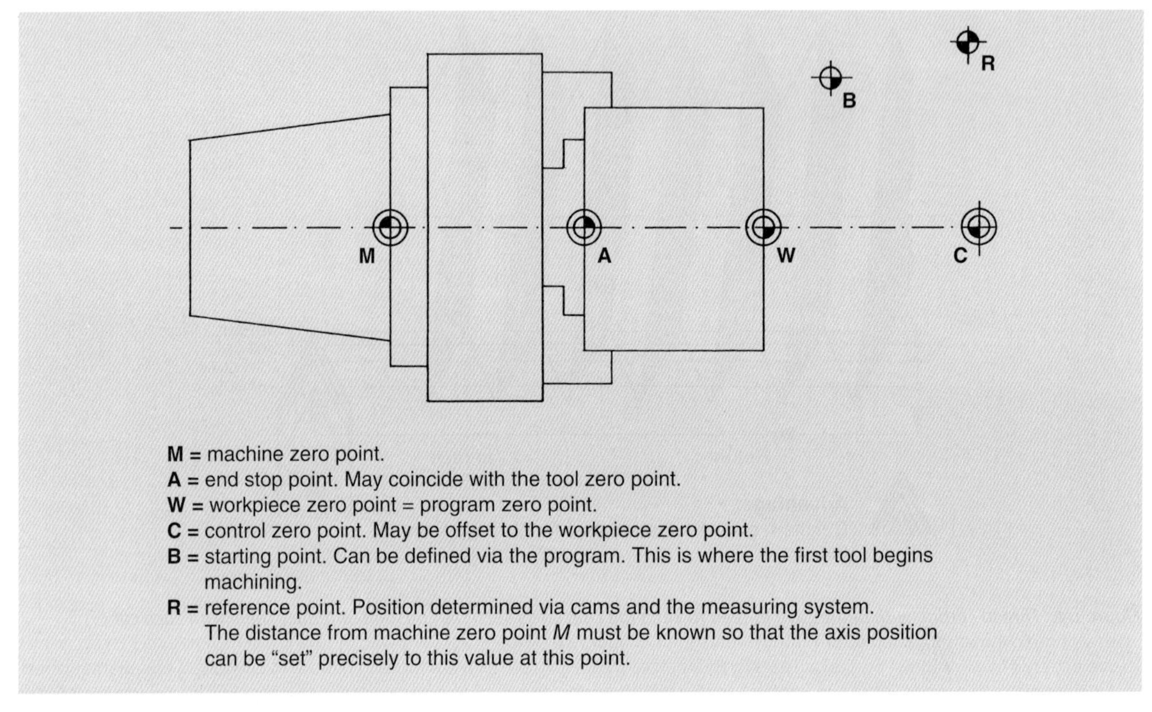

R
B
M
A
W
C
M = machine zero point.
A = end stop point. May coincide with the tool zero point.
W = workpiece zero point = program zero point.
C = control zero point. May be offset to the workpiece zero point.
B = starting point. Can be defined via the program. This is where the first tool begins machining.
R = reference point. Position determined via cams and the measuring system. The distance from machine zero point M must be known so that the axis position can be "set" precisely to this value at this point.

Figure 1.9: Zero points and reference points in turning machines (M = is defined by the machine manufacturer).

Zero-Point Offset

Zero-point-offset capabilities make the programmer's job easier, especially when the program zero point also can be programmed to be offset to any desired value in any axis. For example, unchanged drilling patterns can be transferred to any desired location without needing to change the coordinate values once they have been calculated. For machine tools that feature pallet changing, the programmer can use zero-point offsets in combination with a manually adjustable compensation value to compensate for inaccuracies in clamping.

Zero-point offsets can be programmed by setting actual values, for example:

```
N122    G90    X140.5    Y0      $
N123    G92    X0        Y100    $
```

This code means

The *X* axis moves to 140.5, where it is set to 0.

The *Y* axis moves to 0.0, where it is set to the dimension 100.

Another way to program a work zero offset is to call up a stored offset value:

```
N345    G54
N346    G0     X20    Y20
```

This code means

The distance of the offset for each axis is determined by the value stored at memory address G54.

1.9 Transformation

The term *transformation* generally means a conversion from one coordinate system to another. This function is intended primarily to make programming easier.

Transformation can be performed either by computer or in the CNC system. A distinction is made among the following transformation options:

Coordinate transformation. This may be needed, for example, to convert an NC program from Cartesian coordinates to polar coordinates or to the kinematics of a robot or a hexapod (conversion of space coordinates to machine coordinates).

Transformation of point patterns. This simplifies programming by allowing predefined programmed patterns of points or individual points (drilling patterns) to be shifted, rotated, tilted, connected, or mirrored by means of simple commands.

This also includes **compensation for inclined positions,** in which the workpiece does not lie in an axially parallel position on the table. This requires the NC program—which is defined in relation to the machine coordinates—to be converted to the axis coordinates.

World Coordinates or User Coordinates

The device-independent Cartesian coordinates that are used in computer-aided design (CAD) systems and serve as a reference system for programmers can be converted to device-specific coordinates (e.g., screen coordinates, machine coordinates, or robot coordinates) by means of coordinate transformation. For more on this, → *Part 2, Chapter 3, FRAME.*

1.10 Tool Compensation

Even when the part programs, fixtures, and tools are prepared correctly, it is necessary to implement tool compensation. Compensation is necessitated not only by deviations in the tools themselves but also by clamping tolerances, machine-related errors, and tool wear. In many cases the required compensation values can be determined only by metrological inspection of the workpiece. There are various types of compensation that differ widely from one another. These different types will be examined more closely below.

Tool-length compensation (*Figure 1.10*) is used to make allowances for the difference between the specified tool length (on which the program is based) and the actual tool length. Such differences may result from the regrinding of tools, for example. Either the magnitude of the difference or the absolute tool length itself is entered into the memory area reserved for compensation values. Compensation values can be retrieved in the program by referring to the address H. Another option is to retrieve these values along with the corresponding tool T by referring to the compensation number. The same applies to **cutter-offset compensation.** The functions G43/G44 are used to determine the compensation direction regardless of the quadrant in which machining is performed (*Figure 1.11, top*).

The task of **cutter-radius compensation** or **cutter-path compensation** is to calculate the necessary tool centerline path that will provide a constant offset from the programmed workpiece contour. This applies to turning as well as milling (*Figures 1.12 to 1.14*).

On machines with pallet-changing features, an axially parallel **offset** of the individual workpieces must be taken

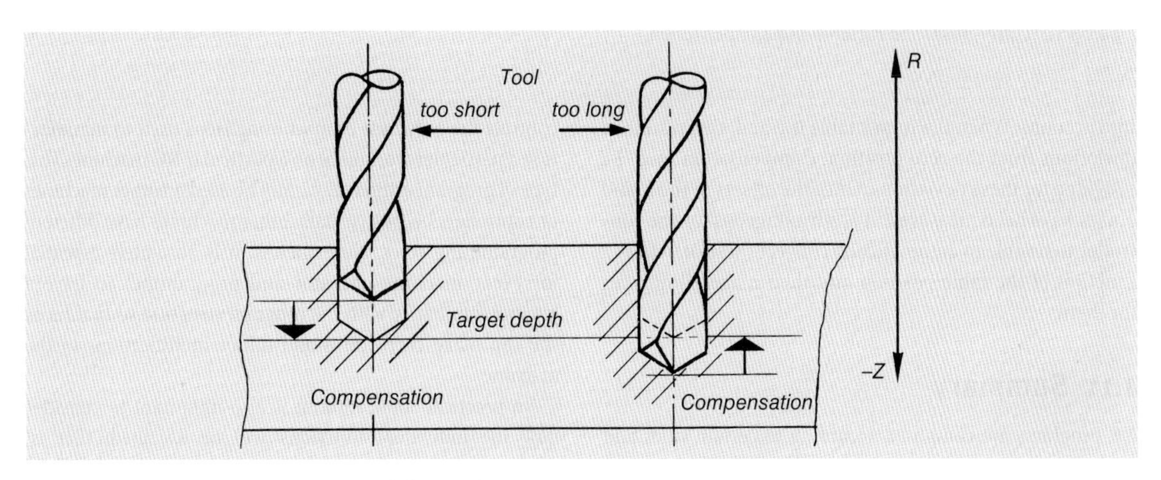

Figure 1.10: Tool-length compensation in drilling tools.

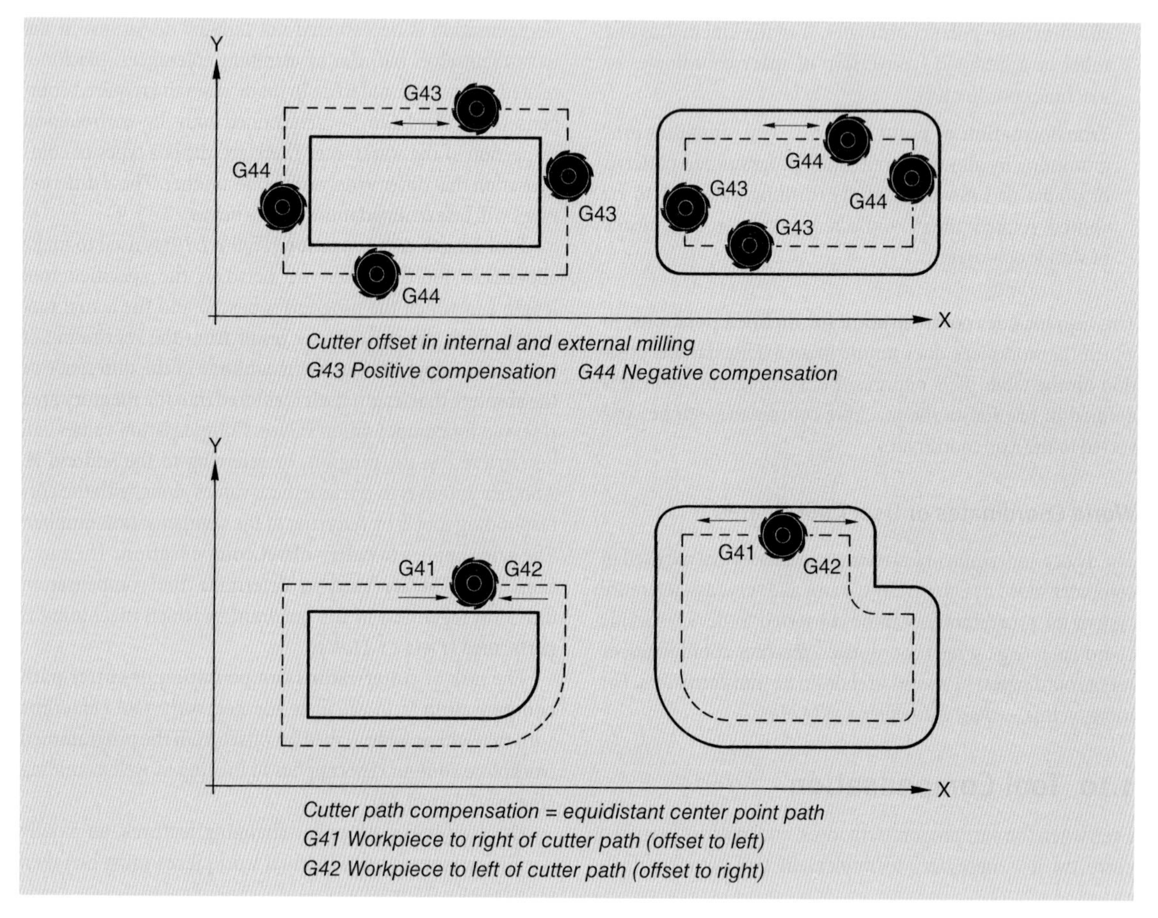

Figure 1.11: Cutter-offset and cutter-radius compensation.

into account. When the rotary table is tilted, this results in deviations from the zero position. Compensation must be provided for these deviations. Once the offsets in the *X* and *Y* axes have been measured, it is advantageous for the controller to be able to automatically convert these offsets as a function of the table position and then make allowances for them.

1.11 Summary

NC programs are composed according to certain rules and special instructions (the programming guide). The program controls the sequence of machining steps used to manufacture a particular workpiece on a particular NC machine. This type of program ordinarily cannot be used on other machines or controllers because usually other machines have different programming rules. Nevertheless, it is absolutely essential for every machine operator and programmer to have a general familiarity with the program structure so that he or she can make corrections and minor modifications to the machine.

In practical application, it is also important to consider how the individual functions will be executed, that is, whether they take effect at the beginning or end of the block.

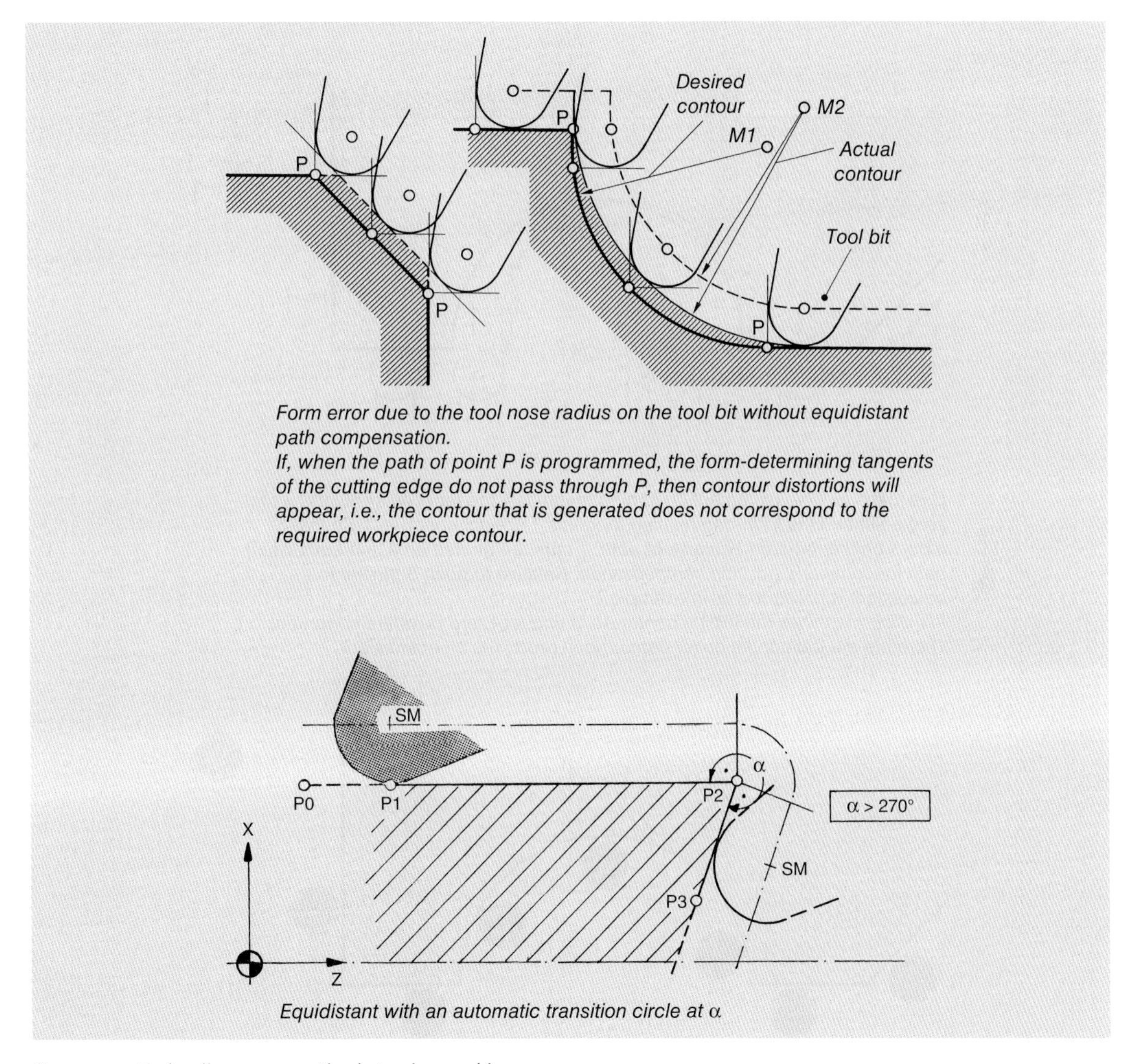
Desired
contour
M2
M1
Actual
contour
Tool bit
P
P
P
P
P
Form error due to the tool nose radius on the tool bit without equidistant path compensation.
If, when the path of point P is programmed, the form-determining tangents of the cutting edge do not pass through P, then contour distortions will appear, i.e., the contour that is generated does not correspond to the required workpiece contour.
SM
α
P0
P1
P2
α > 270°
SM
P3
X
Z
Equidistant with an automatic transition circle at α

Figure 1.12: Tool-radius compensation in turning machines.

This information can be obtained from the applicable programming guide.

The generally applicable rules for the structure of NC programs are defined in DIN 66025, Parts 1 to 3. By dividing each individual command (word) into an address letter and a numerical value, even large, complicated programs can be made clearly understandable. This often will make it easier to make the necessary manual corrections in the shop.

Frequently recurring program sequences (e.g., drilling, turning, and milling cycles) are stored as subroutines that can be called up and supplied with the required parameter

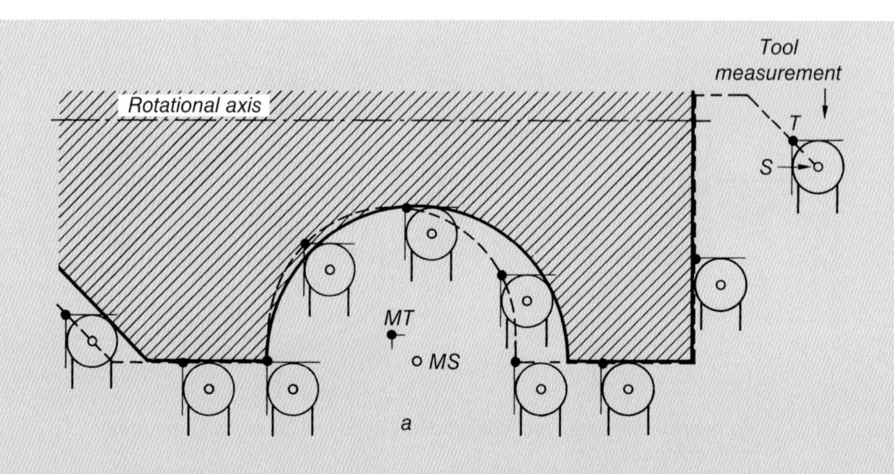

Due to the tool measurement on turning machines, the tool center point S = tool zero point and the form-determining tangents T are not congruent. So that S moves on the required equidistant path, T must be guided on a non-equidistant path for command position compensation. Relative to point S the same equidistant is produced as in example B.
In milling machines the milling center point and tool zero point are congruent. Therefore the path of the cutter center point produces an equidistant.

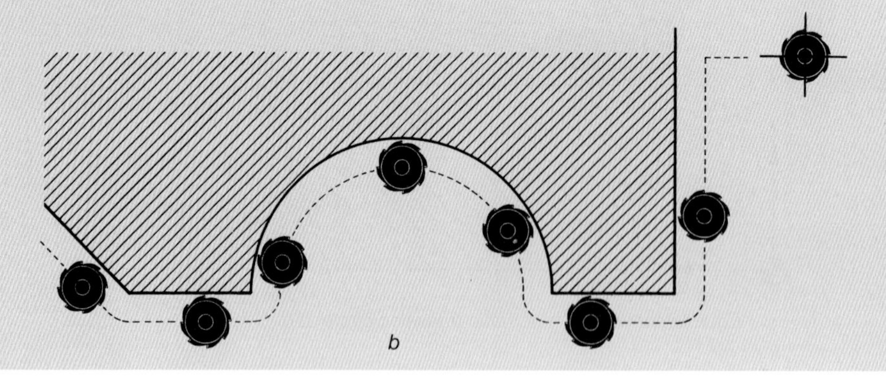

Figure 1.13: Difference between
(a) tool-nose-radius compensation in turning machines and (b) milling-cutter-radius compensation in milling machines. The workpiece contour is programmed; the transition radii and equidistants are generated automatically by the CNC system.

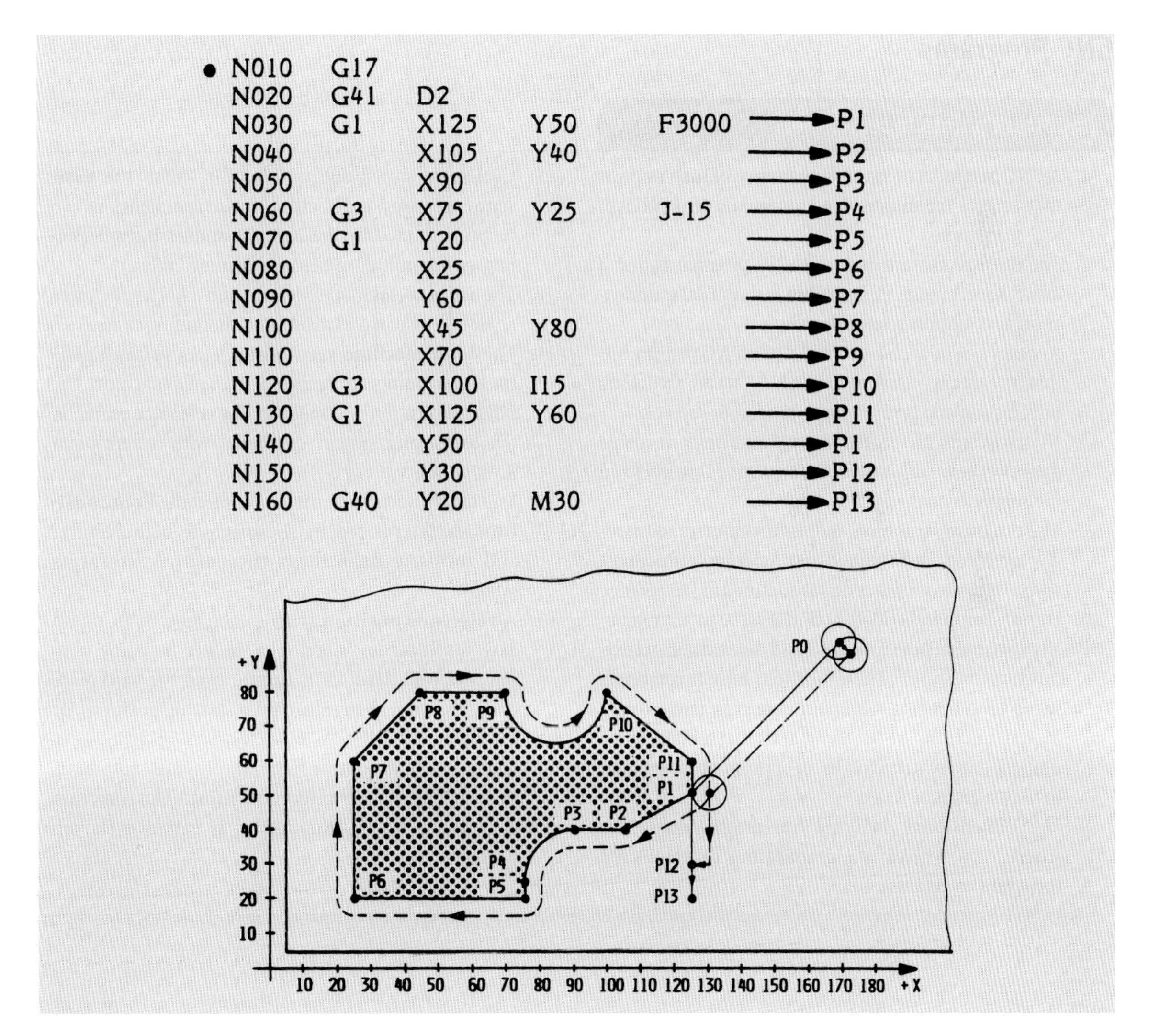

Figure 1.14: Programming example for G41: Tool radius offset to the left.

values, such as drill depth, retraction plane, or tool number. This reduces the size of the program and simplifies subsequent corrections.

The stored, retrievable compensation values that represent cutter radius, tool length, tool-nose radius, work zero offsets, and other machine-specific data are essential components of all NC programs. The options for creating NC programs are explained in *Chapter 2*.

CNC Programs

Important points to remember:

1. An NC program consists of a **sequence of instructions** that control the motions and automatic sequences of an NC machine.
2. The essential characteristic of an NC program is that it allows direct programming of **dimensional data** (dimensions) and **motion-related functions** (G-functions).
3. Another essential **characteristic of an NC program** is that it contains all dimensional data and switching information and is very easy to modify or replace.
4. NC programs are stored on automatically readable data-storage media or can be transferred directly from the computer to the CNC system.
5. The **program structure,** address assignment, entry of displacements, definition of the preparatory functions, and miscellaneous functions are defined in DIN 66025.
6. **On/off commands (M-functions)** serve to automatically switch machine functions and automation-related functions on and off. These functions include rotational speed, tool changing, workpiece changing, and so on.
7. **Modal functions** are the commands that remain active until they are switched off by an appropriate command in the program.
8. The **dimensional data** tell the programmed axes **where** to go, whereas the **preparatory functions** tell them **how** to get there.
9. **Cycles** are stored **subroutines for which the basic sequences** are fixed but the geometric dimensions can be programmed by the user. (Examples include drilling cycles, milling cycles, and turning cycles.)
10. The **use of cycles** makes NC programs shorter and more reliable. It also makes manual programming easier.
11. The term **machine zero point** refers to the rigidly fixed zero points along the NC axes of a machine. The physical position is established by reference marks in the measuring system combined with a zero-point limit switch.
12. The **machine zero point,** also called the **home position,** is determined by the machine manufacturer and precisely defined via the position-measuring system.
13. A **reference position** is an established point on an axis; its location in relation to the machine zero point is well defined. It can be used for workpiece changing, tool changing, or as a starting position, for example.
14. One particular control function provides automatic **positioning to the reference point.** This function can be activated by the user or as part of a startup program.

2. Programming of CNC Machines

The profitability of CNC machines depends largely on programming methods and the capabilities of the programming system. The faster an error-free program can be provided to the machine, the more effective and flexible NC manufacturing will be.

2.1 Definition of NC Programming

NC programming (a.k.a. *part programming*) is the creation of control data for the machining of a workpiece on a numerically controlled machine tool. This activity can be performed manually or by machine—in other words, with computer assistance. This requires a special type of programming software, which also offers interactive user guidance and graphic representation of the input values. The on-screen graphic/dynamic simulation of the machining operation offers one final opportunity for visual error checking before the program is released for use in actual machining.

2.2 Programming Methods *(Figures 2.1 to 2.4)*

The first decision to be made is **where programming is to be performed:** in the office or on the shop floor. Various **programming aids** and methods are available for each of these areas: a primarily machine-specific method of program creation or a universal programming system that may even make comprehensive use of workpiece data generated in the CAD system for NC programming. The machining process, that is, the cutting sequence, the use of tools, and the selection of machining parameters, is also performed by the NC programming system. Which of the programming methods is selected is determined automatically by the **programming means** used for that purpose.

Each user can identify the programming method that best suits the requirements imposed in his or her case. The basic options are compared in *Figures 2.1 and 2.2*, whereas their characteristics and features are summarized in *Figure 2.3*.

Manual Programming

This method requires no computer support. In this method, the programmer describes the machining task in controller-specific NC code, usually in accordance with the German standard DIN 66025. Before actual programming begins, however, it is necessary to create a tooling plan, a work plan, and a clamping plan (→ *Figure 2.4*).

Aside from pencil and paper, the only **aids** available to the programmer are his or her own experience, tables, pocket calculators, the programming guide, and a coding device for creating an automatically readable data-storage medium. If even such equipment is unavailable, the program has to be typed into CNC memory once it has been created. Manual programming therefore can be ruled out right from the start for applications involving three-dimensional (3D) machining or other highly complex machining.

An essential **characteristic of manual programming** is that the individual **tool motions** are programmed without checking for data, tools, or machining operations that have been entered incorrectly. If all tools have been used, the workpiece would have to conform to the drawing specifications. This can only be determined by testing the machining operation on the machine and correcting it in the program for subsequence workpieces. Some CNC systems offer the capability of testing finished-part programs in DIN code by graphic methods before processing begins. This is a very time-consuming and expensive process.

Today this method is used only with special machines in which the entire machining process can be defined by entering a few parameters into a special controller.

Teaching/Playback Method

Here the machine is positioned manually, and then the displayed final-position values are saved simply by pressing a key, or the entire motion sequence is saved. This method is used primarily with painting robots. With machine tools it is used only when exact definitions of the dimensions are possible only on the workpiece, for example, when machining

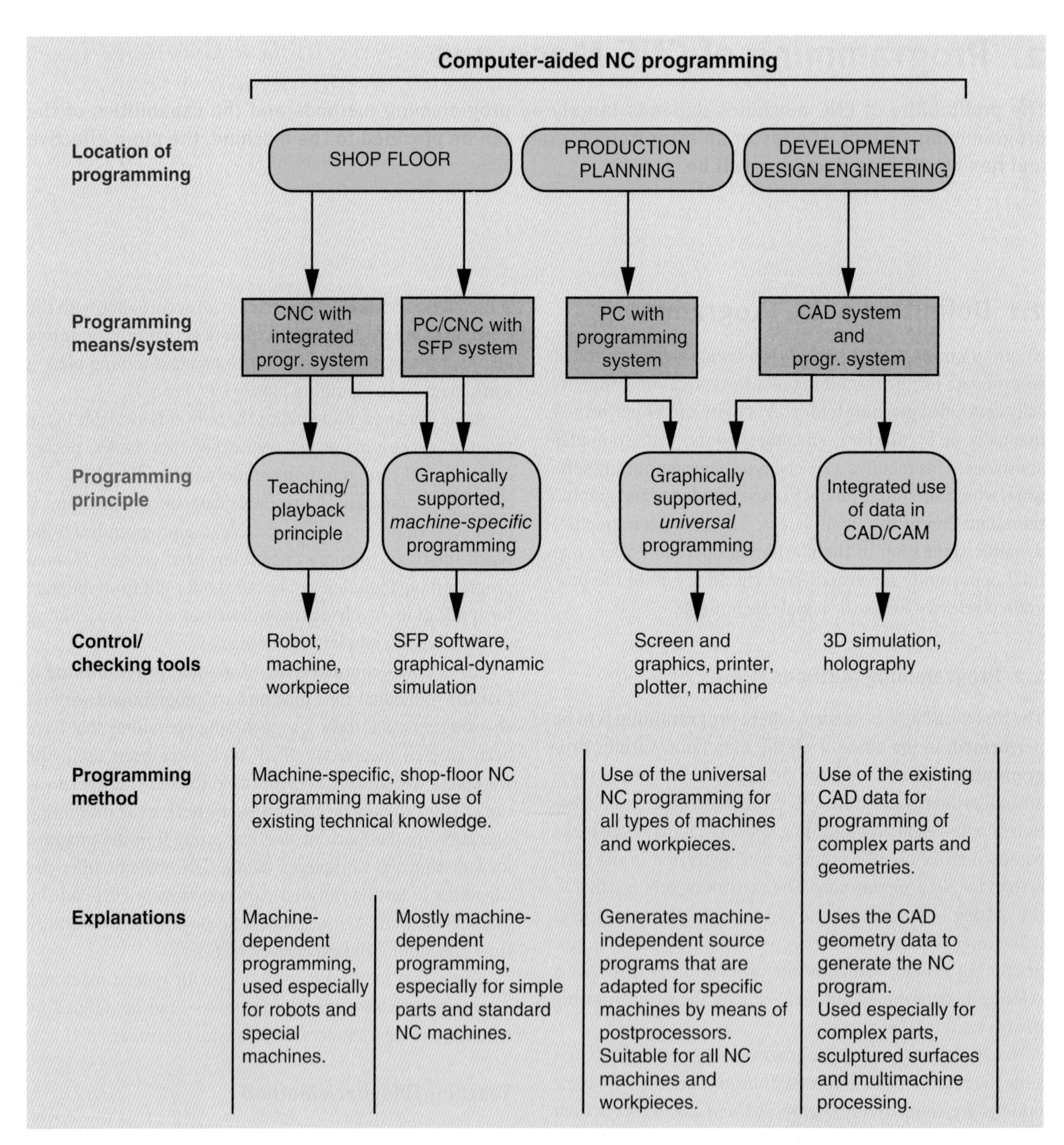

Figure 2.1: Interactions between programming means, programming principles, and programming methods in computer-aided programming.

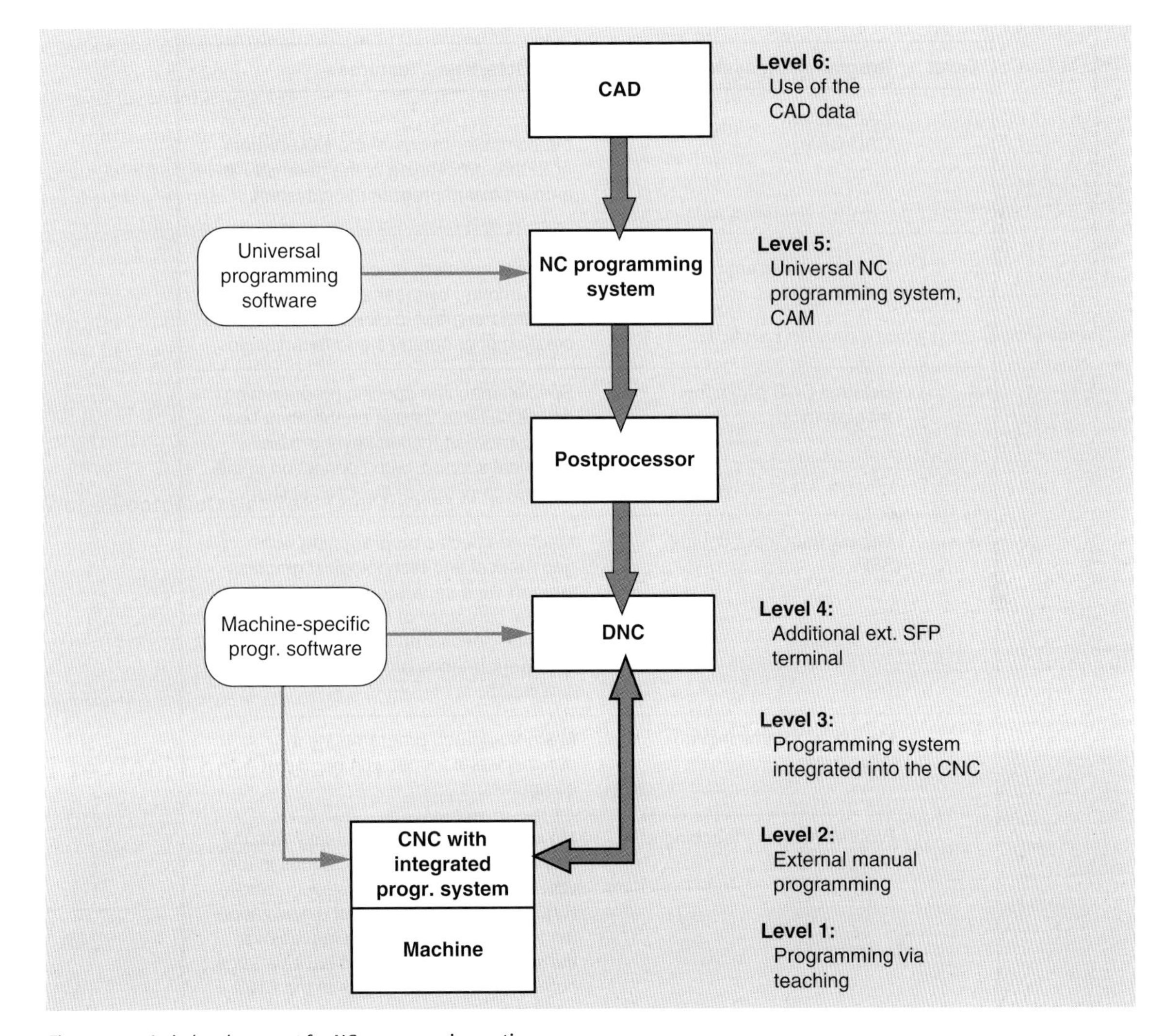

Figure 2.2: A six-level concept for NC programming options.

drilling patterns on boring machines. In this case, also, the time required for program input plays a subordinate role to the processing time.

In milling applications, a comparable method known as **digitizing** (**scanning**) has gained wider acceptance. Here the machine tool or measuring machine scans the surface of a workpiece model in a line-by-line fashion, continually storing the positions. These data are used later to mill an identical surface (**playback**). There are only a few controllers that also allow scale changes to the digitized model or the entry of tool-compensation data.

Manual Input Control

The processing power of modern CNC systems makes it possible to integrate a **machine-specific NC programming**

Level	Programming system	Characteristics, features
6	CAD system CAD/CAM	Use of workpiece geometries present in the computer to create the NC program or further processing of the geometry in a downstream programming system.
5	Universal programming system CAM	Problem-oriented, machine-independent programming by means of a programming language or graphics for support in creation of geometries and technologies.
4	External CNC panel for programming	Special, machine-specific programming device suitable for use on the shop floor with symbolic/function keyboard and graphical support, with connection to the internal computer of the CNC system.
3	Manual data input control SFP	Machine-specific programming with geometrical and technological graphics support via a separate computer or the internal computer of the CNC system. Machine-specific and therefore very high-performance programming and simulation.
2	Manual programming in DIN format	Machine-related programming in punched-tape format with no use of computers.
1	Programming via teaching	Moving to the position manually with the machine and saving the positions in the CNC system (primarily used with robots). (With a comparable level of digitalization the machine automatically moves over the surface of a model and saves a large number of individual points in the CNC system.)

Figure 2.3: General characteristics and features of NC programming principles.

system with graphics and interactive user guidance. The CNC system thus becomes a powerful manual data-input control and can be programmed by the user in the shop.

One **advantage** of manual programming is that it is **specially designed** for a specific machine as regards functionality and performance and its particular design characteristics. This is the only way to program turning machines with two heads, a *C* axis, and driven tools with manual data-input controls. Collision monitoring, wait conditions, and the distribution of tasks to the two heads can be determined by the controller much more quickly on a machine-specific basis than in a universal programming system, as well as taking them into account in the program structure in a much more systematic way. **Shop-floor programming** (SFP) has gained

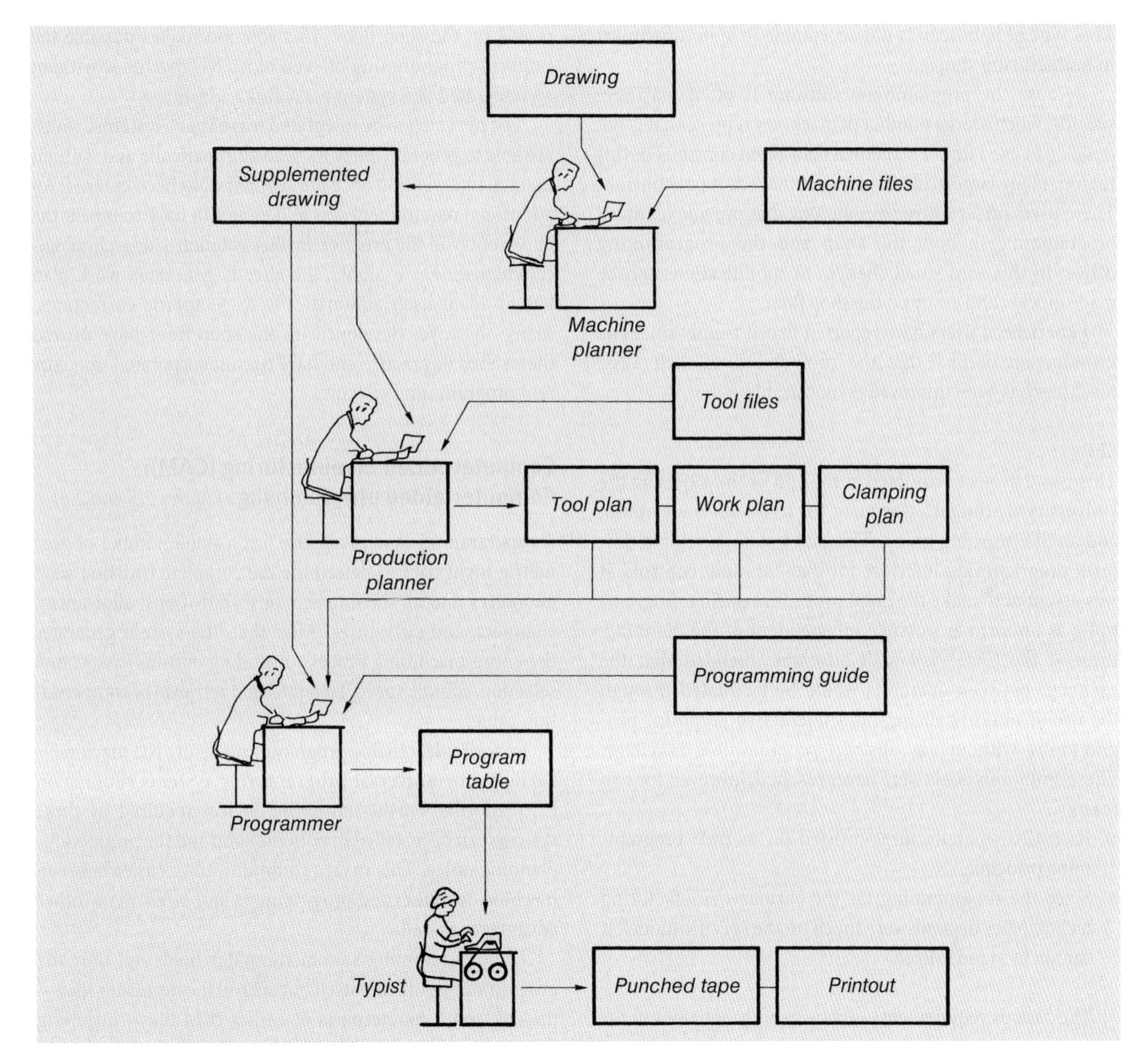

Figure 2.4: Principle of manual NC programming. All the manufacturing documents, the workpiece drawing, and the programming guide are needed in order to create the NC program and enter it in a list block by block. The NC program is created by writing or transcribing the program list, after which the NC program is saved.

wide acceptance in turning, milling, and nibbling/punching applications in particular.

Machine-specific manual data-input controls have the **disadvantage** that each new machine from a different manufacturer involves a different programming system. This makes it difficult to swap personnel and programs between machines and is not feasible in the long run. What is more, NC programs that are created cannot be used with already existing machines. This would require investment in yet another programming system, which would mean that the production-planning department and the shop floor would be using different systems.

This type of handicap is not acceptable in state-of-the-art manufacturing shops.

Because the programming software is often machine-specific, when a large number of machines is present, it is not possible to use a uniform manual data-input control. For this reason, many large-scale users reject manual input controls. These users prefer universal solutions that provide identical programming in both **the shop and the programming office.** In this way, small changes or modifications can be made quickly and easily on the shop floor.

Experienced users have reported that it is only since the introduction of SFP that the profitability of their CNC machines has been improved demonstrably.

SFP

This was a project promoted at the end of the 1970s by the University of Stuttgart, Germany, the IG Metall trade union, and an SFP working group. The goal was to develop a uniform programming interface for manual input controls. It was essential to make the input procedure during programming as uniform as possible independent of the manufacturer of the CNC system. It also was intended that the exchange between operators would be facilitated through the use of uniform dialogs and interactive graphical programming without any abstract programming language. These ambitious goals were never reached, however, for two reasons:

- Each CNC manufacturer insisted on its own programming principle.
- Since the programming for the various manufacturing technologies depends very much on the specific model, it cannot be standardized.

The various requirements become readily apparent if, for example, you check whether the programming of turned parts with internal and external machining, and possibly also with end-face and lateral surface machining, correspond to the programming input for the milling and drilling of cubical parts. This meant that a number of different SFP systems appeared, each of which was designed for a special type of machine.

The variety of SFP systems meant that there was no way to eliminate the variety of systems used on shop floors. The result of this was that companies with a large number of NC machines considered making production-planning systems usable on the shop floor. This solution makes possible the uniform programming of even older NC machines without an integrated SFP system—a definite advantage.

The next step to be integrated was **expert systems,** which are able to generate the work plan automatically and without manual intervention once the geometry has been entered. An additional requirement was added: not to have to repeat the entire entry in the event of smaller geometric or technological changes. For example, the already-generated work plan has to adapt itself automatically to geometric corrections. Many objectives developed on the shop floor have proved themselves in practice and have been incorporated into current programming systems.

Computer-Aided Manufacturing (CAM): Computer-aided programming *(Figures 2.5 and 2.6)*

A programming system can use just a small number of geometric input data to determine the complete **finished-part geometry** and all the intersections, transitions, allowances, chamfers, and curvatures. After that, the system generates the entire machining process, including cutting passes, tool selection, spindle speed, feed rate, and retrieval of compensation values.

Fast, simple, reliable programming of any NC machine is facilitated by **universal programming systems** running on PC platforms. Furthermore, the devices required by these systems can be installed in both the shop and the production-planning office. This makes communication easier between machine operators and programmers and offers many other advantages as well.

The programming system computer holds not only the programming software itself but also all the necessary files—the electronic counterparts of earlier card files—including the machine file, tool file, and fixture file. The required post-processors are also stored in this computer.

After programming has been completed and the machining operation has been simulated, all the production documents can be created as output for use in the shop. These documents include the program list, the program medium, the clamping plan, and the tooling plan (→ *Figure 2.5*). *Figure 2.6* shows the general principle and flow of information in computer-aided programming.

One **characteristic of all computer-aided programming systems** is that they involve the programming of the

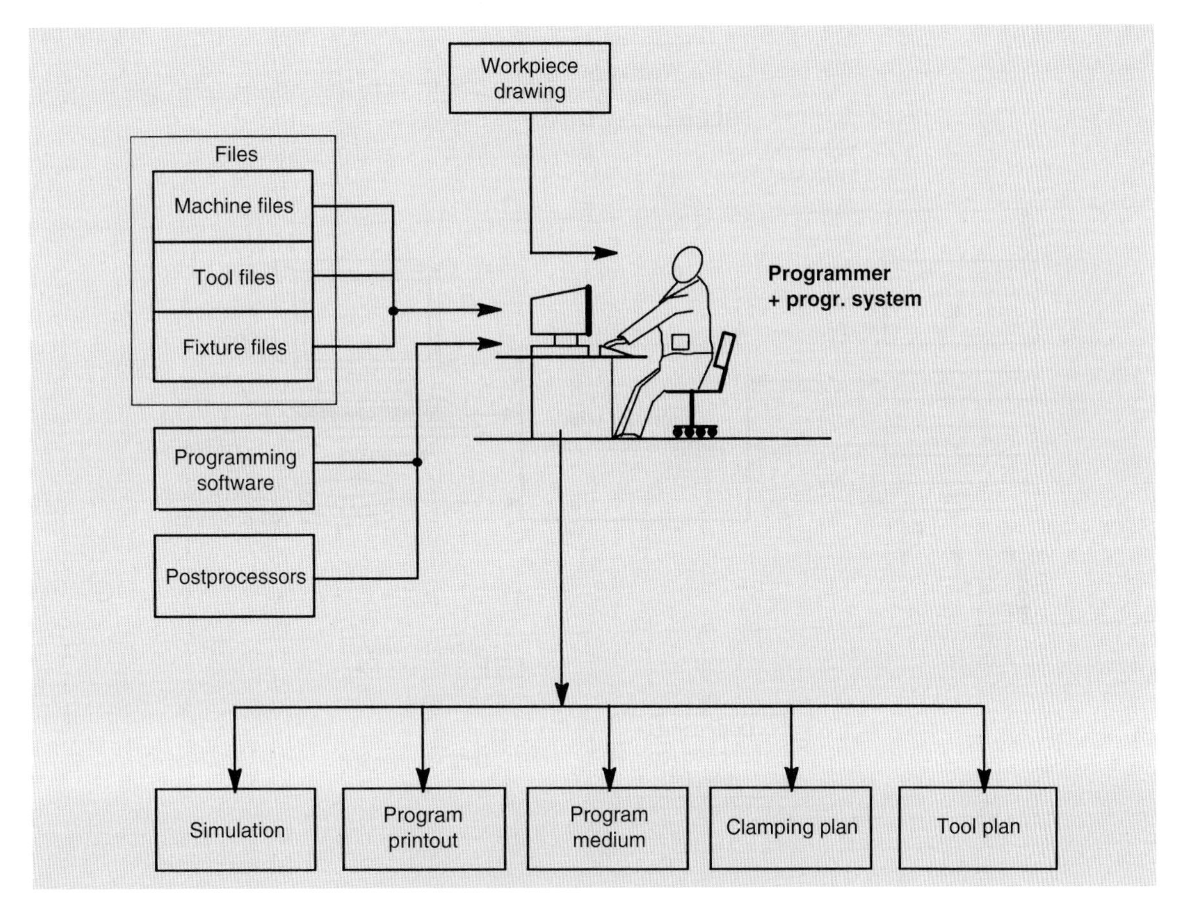

Figure 2.5: The principle of computer-aided NC programming.
All manufacturing documents and the NC program are created, checked, and saved using the computer.

exact contours and shapes of the **workpiece as described by the drawing,** instead programming the tool motions. To a great extent, the system automatically generates the selection of required tools and the machining process that culminates in the production of a finished workpiece. This eliminates the need for trigonometric auxiliary calculations and tedious paging through tables. It also increases the number of NC programs that can run in a given amount of time and makes programming more effective. Despite the additional costs for the programming system, this method ultimately is more economical than manual programming.

Postprocessors

When programming has been completed, the system creates a generalized part program, also known as a **source program** or **CLDATA file** (cutter line-data file). This can be tailored to any suitable NC machine. To this end, it has to be adapted to the machine tool on which it will run. This adaptation is accomplished in the computer by means of a machine/control-interface program called the **postprocessor.** A special postprocessor is required for each machine/control combination so that the part program can be produced in the specified output format. Some systems feature a **generalized**

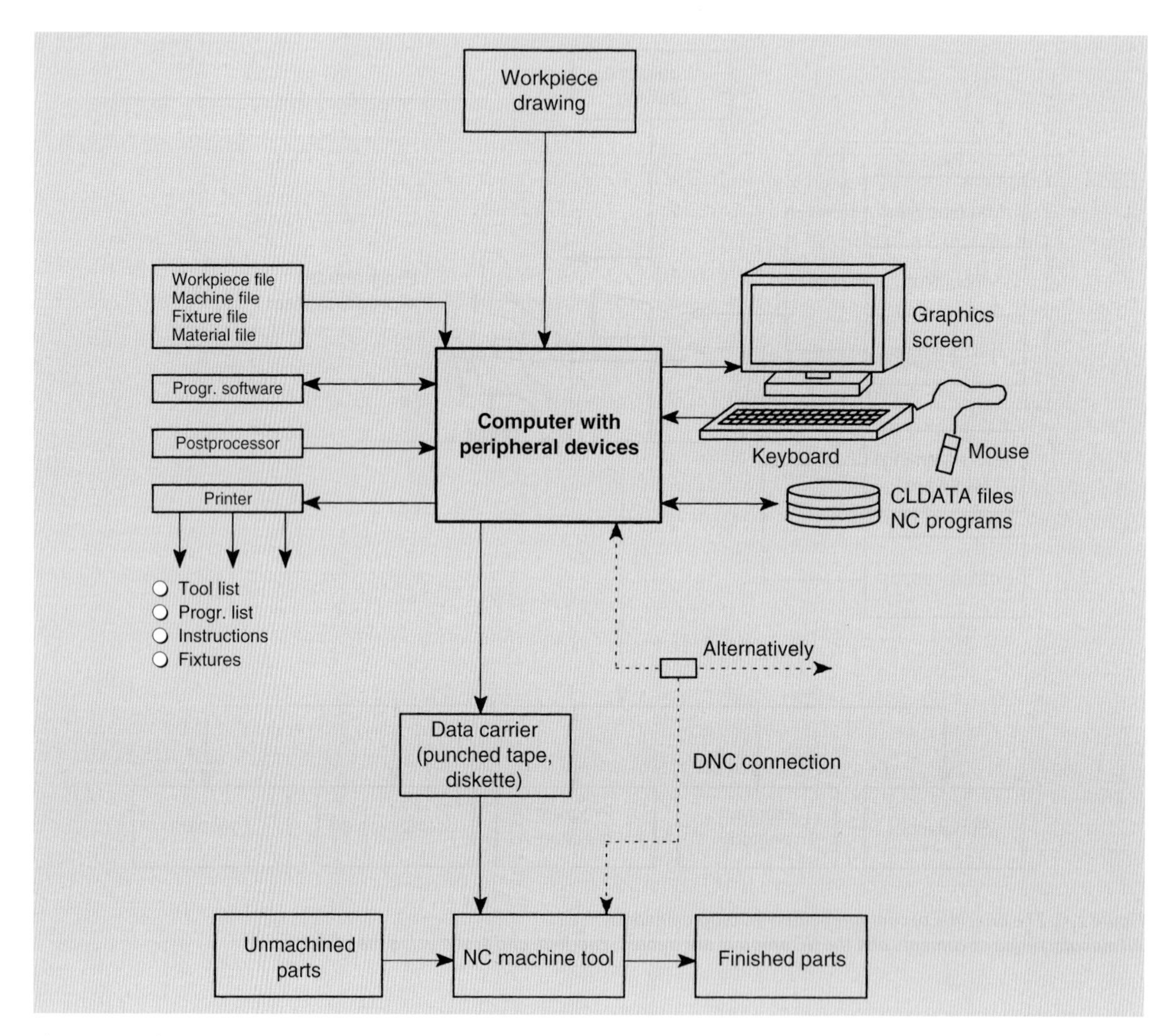

Figure 2.6: Information flow in computer-aided NC programming.

postprocessor that the user can modify for any machine. Other systems can operate without postprocessors.

If the programmer exceeds the specified limit values or commits **input errors** that go undetected by the programming system, they will be detected in the postprocessing stage at the latest. An error message then will be issued, accompanied by an indication of the source of the error. The postprocessor takes into account the size and kinematics of the machine and the limit values that govern the feed rate, rotational speed, and tools. It also issues the proper M-functions and assigns tools, compensation values, and reference points to one another. The result is an NC program that is immediately ready to run on a particular machine.

In earlier times, NC programming always was performed by means of abstract **NC programming languages** without interactive user guidance or graphic support. This type of programming has become technically obsolete and is no longer found.

Use of CAD Data

NC programming also must be regarded in light of the increased use of **computer-aided design (CAD) systems.** When the workpiece geometry data, which are contained in the CAD system anyway, are used directly in NC programming, it becomes possible to greatly reduce the time and expense required for creating production documents, as well as the likelihood that errors will be committed when defining the geometry.

This data integration requires standard **data interfaces** in the CAD system and the programming system, which make it easy to transfer the CAD data into the NC programming system. The NC programming system performs any compensation that is still necessary, supplements the technological data, selects the tools, and establishes the sequence of machining operations.

In this connection, it is important to point out that CAD systems are designed to perform not only tasks related to NC programming but also entirely unrelated tasks that exploit their impressive capabilities. The only situation in which CAD systems should not be used is when only part programming has to be performed. (→ *Part 6, Chapter 3*).

2.3 The NC Programmer

Experience has shown that skilled workers and technicians with a few years of occupational experience on conventional machines are best suited to do the work of NC programmers. They possess the **required technical knowledge** and can

- Read technical drawings and determine the necessary machining method on the basis of these drawings
- Determine the necessary machining operations
- Properly apply their technical knowledge of tools, materials (including cutting materials), and cutting values

Knowledge in the following areas is also required:

- The machine and its capabilities
- The programming guide for the respective machine
- The function of the commands
- The axis designations and motion commands
- The available automation systems, such as tool-changing systems and pallet-changing systems
- The positions of the individual zero points and reference points

Manual programming (→ *Figure 2.4*) imposes the greatest demands on the programmer. The programmer must be familiar with the details of each machine and each controller. He or she also must master a variety of trigonometric computations and be able to use them correctly. Herein lies the problem of manual programming. For this reason, the manual programmer can only take care of four or five NC machines at a time.

On the other hand, the programmer can work much more effectively when a suitable, universal, powerful **programming system** is provided (→ *Figure 2.5*). After a sufficient period of orientation, the programmer should be able to use this system to program various machining methods, for example, turning, milling, nibbling/punching, and wire EDM. This type of system even makes it possible to supply many different brands of machines and controllers with NC programs.

2.4 Graphics to Make Work Easier

(Figures 2.7 and 2.8)

On-screen graphics constitute a major aspect of computer-aided programming. First integrated into programming systems and then into CNC systems themselves, graphics have proven to be a significant factor in the quicker acceptance and more extensive use of NC technology.

Graphic support can be divided into the following types:

- **Input graphics** for geometric and technological data. This allows the programmer to visualize the unmachined part, the finished part, and the required tool.
- **Auxiliary graphics** for displaying drilling patterns, milling and turning cycles, clamping devices, tool geometries, and other functions that are specific to the machining operation being used.
- **Simulation graphics,** which provide a dynamic representation of the machining process when input has been completed (or even earlier). Any input or processing errors will be shown clearly in the simulation. These graphics must be good enough so that the programmer does not need to use the machine for purposes of error checking.

This type of graphic support for program entry is preferred by NC programmers because it corresponds most closely to practical shop operations. The ability to trace the step-by-step creation of the program corresponds perfectly to the activity

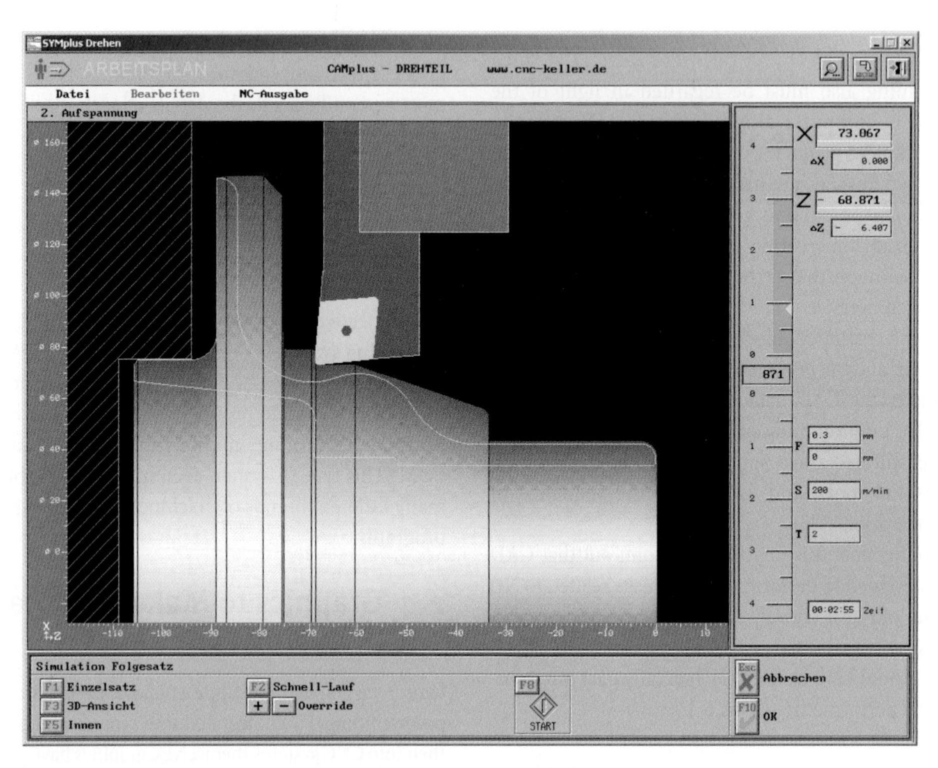

Figure 2.7: Real-time simulation of a turning operation (clamping second end). (CAMplus turning; image courtesy of Keller, www.cnc-keller.de.)

performed on a conventional machine, where the workpiece's compliance with the drawing can be checked at any time (although it is considerably more time consuming with the conventional method). This method provides the programmer with "guidance" and constant reassurance as programming proceeds. This allows large programs to be created in much less time.

The advantages of graphic support can be seen quite clearly in the **entry of complex contours and machining after reclamping**. This replaces the abstract thinking (in terms of axes and motion sequences) that otherwise would be required. **Auxiliary graphics** are also accessed several times during this input operation, for example, when tools are selected, safety zones are defined, drilling and milling cycles are programmed, and at other times.

When entry is complete, **simulation graphics** can be used to represent the machining operation, usually in several different views (→ *Figure 2.7*). The programmer can detect his or her programming errors either during or after programming depending on the system being used. If **real-time simulation** is provided, the machining time can be determined as the simulation is running. However, the overall process of real-time simulation takes too long to run and is therefore not always used. For this reason, machining also can be represented at time-lapse (or fast-forward) speed.

In **parallel simulation,** which runs simultaneously with the machining operation, the user can trace the machining of the workpiece in the machine at all times. This is advantageous when the coolant prevents direct monitoring, for example.

Almost all systems also show the individual tools in the correct scale in relation to the workpiece during the simulation of both internal and external machining. The observer can see how the shape of the part changes, whether the back end or shank of the tool might collide with the workpiece, or whether all points on the workpiece will be machined

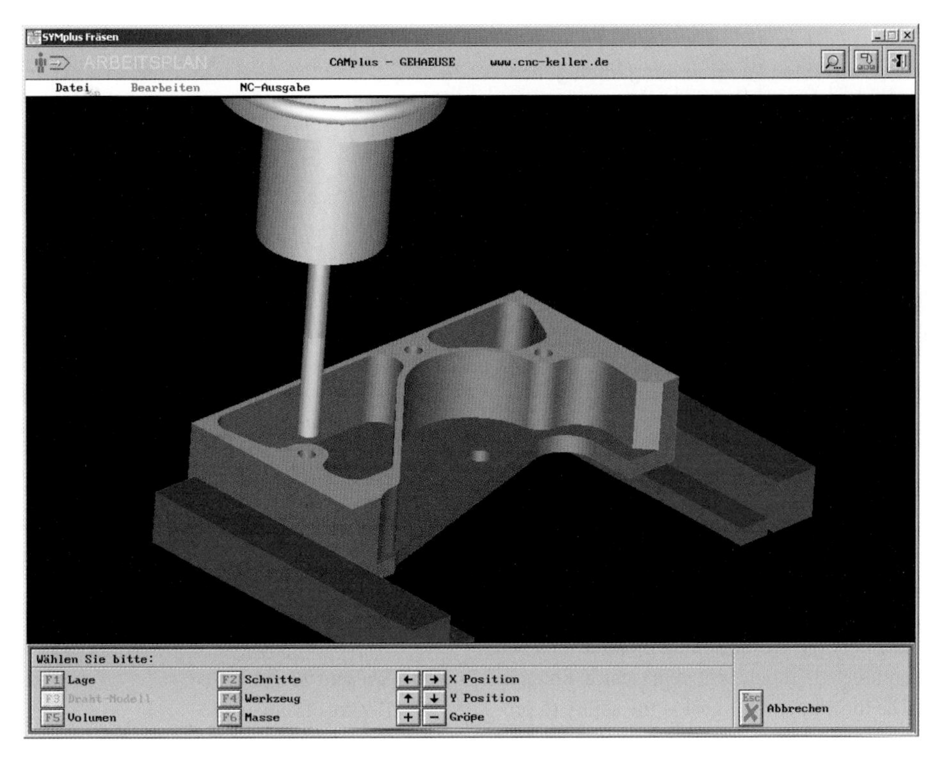

Figure 2.8: 3D view of a milling part during the simulation. (CAMplus milling; image courtesy of Keller, www.cnc-keller.de.)

satisfactorily. This requires that the user be allowed to enter and modify his or her own tools and all related data by graphic means.

Simulation graphics undoubtedly represent a significant step in the direction of greater safety. They make it easy to detect and correct collisions between the tool, the workpiece, and the clamping device, as well as geometric errors, before these potential problems can have disastrous consequences. This reduces the risks involved in machining unique, large, and expensive workpieces.

2.5 Distributed Intelligence *(Figure 2.9)*

Studies have proven that the degree of difficulty in NC programming is determined only by the complexity of the workpieces. It is relatively independent of the type of machine used, the machining method, and the numerical controller.

The more complicated the workpieces are with respect to shapes, transitions, surfaces, and finishing operations, the greater is the level of intelligence required for programming. This intelligence should be distributed among three areas in the proper ratio (→ *Figure 2.9*):

1. The programmer
2. The programming system
3. The CNC system, which today generally has an integrated programming system

When designed properly, each of these three areas offers certain advantages. However, none of these areas may be too weak; otherwise, NC programming will become unprofitable or even troublesome.

The **programmer** must be well trained and should be able to use his or her experience to concentrate on programming the workpiece geometries that need to be manufactured. This requires thorough familiarity with the programming system

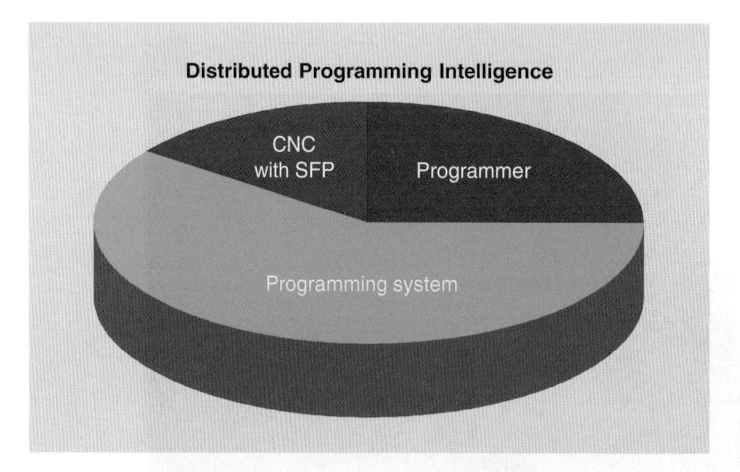

Figure 2.9: Principle of distributed intelligence among the three areas involved in program generation.

so that all the programming tools and aids provided by the system can be fully exploited.

The **programming system** should be selected carefully, based on its suitability for the machining methods that will require programming. It should not represent a financial compromise. The most important goal is the ability to provide error-free, ready-to-run programs to the machine right from the start. This can be verified largely on the basis of the machining simulation. Every system allows programs to be optimized for relatively large production quantities.

Finally, the **numerical control systems** also should possess certain capabilities. It should be no problem to insert small corrections in order to reduce the waiting time for modified programs.

2.6 Selecting a Suitable Programming System

Error-free part programs are an essential requirement for the profitable operation of NC machines. Furthermore, it must be possible to create the programs quickly and at the lowest possible expense. **The more flexibility is desired in the use of the NC machines, the more important a powerful programming system will be.**

The selection of an NC programming system from those available on the market is already complicated by the fact that each brand offers special advantages in some areas while requiring small or large compromises in other areas. Therefore, the following factors should be examined first as the **basis of decision making:**

- **Variety of workpieces**
  - Group technology
  - Size and weight
  - Degree of similarity
  - Geometry of unmachined parts
  - Complexity of workpiece geometry
  - Number of tools required
  - Expense of determining technologic data
- **Set of NC machines**
  - Number of NC machines
  - Machining methods
  - Variety of NC machines
  - Machine size
  - Variety of numerical controllers
  - Degree of automation
  - Workpiece changing and tool changing
- **Planning statistics**
  - Number of new NC programs per week/month/year
  - Number of archived programs
  - Frequency of repetition
  - Lot size
- **Organizational issues**
  - Available computers
  - Need for installation of new computers
  - Preexisting or planned data network
  - DNC system

- CAD/CAM integration
- Personnel qualifications
- Experience in numerical control and programming

Companies that are new to the field of numerical control usually bring some past experience to the selection of the NC machine itself. When it comes to programming, though, the company is faced with a wide selection of systems, each having unknown advantages and disadvantages. But one should not be held back by the fear of making the wrong purchase. Powerful programming systems running on PC platforms—including the necessary postprocessors—are quite affordable today and therefore should be included in planning and financing of an NC production system right from the start.

There are many factors that make a clear argument for computer-aided programming. These factors differ from case to case but may include the greater computational complexity that results from complicated workpiece geometry, a wide variety of machines and controllers, the high degree of automation of certain machines, a high annual number of new NC programs that will be repeated a relatively low number of times, a lack of NC experience, or a lack of programmers with NC experience.

Each company that sells programming systems is convinced that its own is the best for every application. For this reason, the buyer has to apply certain criteria to decide on a preferred system. The **utility-analysis method** is the best way to compare the buyer's own requirements with the performance features of the individual systems because priorities differ from one user to another. Finally, the prospective buyer should visit one or more users of the system under consideration as a last step in obtaining information about operation, evaluation, satisfaction, and service. This information cannot be expressed in numbers; at best, it may be inferred from cost figures.

In any event, the following features must be examined closely:

- the **computer hardware** and **operating system**
- Available **interfaces** to DNC and CAD systems
- **Geometric capabilities** allowing for simple and complex workpiece geometry, variable dimensioning, etc.
- The ability to make subsequent corrections to geometry and technology
- **Technological capabilities** pertaining to the selection of tools, cutting data, sequences of machining operations, and cycle call-up
- Graphic-dynamic **simulation** with **collision monitoring** between tools and the machine, clamping fixture, and workpiece
- **Universality,** that is, the programming of different machine types with two to five NC axes, 3D operation, turning machines with driven tools, and the available **postprocessors** (if necessary)
- **Ease of programming,** including user friendliness and convenient displays
- The available **training, documentation, and startup help**
- The **cost of implementing and operating** the system
- Anticipated **future enhancements**
- The degree to which the system is **known and widely used** in domestic and foreign markets

These criteria will determine whether you have chosen the right system. If your purchasing decision is based on promises made by the vendor, you definitely should get written confirmation of these promises. And above all, **allow the *programmer* (who will eventually have to work with the system) to share in the decision!**

Of course, it is also possible to use different programming software for each machining method, but this approach involves higher costs and additional problems.

2.7 Summary

NC programming has a decisive influence on the cost-effectiveness and profitability of NC manufacturing. For this reason, the company must approach the selection of the programming system and the training of its personnel at least as carefully as it approaches machine procurement. Later on, when the programming system is used on a daily basis, it will soon become clear that it is not enough to just "get the machine running."

Computer-aided programming systems are preferable for most applications. (There are a few exceptions—for example, applications involving hobbing machines or tube-bending machines, which can be programmed for several hours of operation by entering just a few parameter values.) Computer-aided programming systems spare the NC programmer a great deal of calculation work, make it possible to test the program on the screen—usually in a graphic/dynamic manner—and produce error-free NC programs in the shortest possible time. Furthermore, the stored source programs can be converted quickly and easily for use on replacement machines when necessary.

Powerful programming systems are indispensable when working with complicated workpiece shapes and surfaces; large, expensive, or complex machines; or expensive workpieces. However, these systems are no substitute for intensive training and hands-on orientation. According to the principle of distributed intelligence, the programmer, programming system, and CNC system, each must possess certain capabilities.

The **selection** of a programming system is essentially guided by the need for software that is suitable for the application at hand, readily available, and as universally applicable as possible. Custom solutions should be avoided because experience has shown that these will lead to unsolvable problems sooner or later.

The use of **two or more programming systems** can be quite practical and economical in some cases. This also applies to the use of CNC systems that are especially designed for shop-floor programming, combined with a central programming department. These two elements can complement one another quite nicely in practical, everyday operations.

The most important requirement is to be able to make optimal use of the productivity and flexibility of NC machines by means of trouble-free programming.

Programming of CNC Machines

Important points to remember:

1. In a general sense, the term **NC programming** refers to the creation of control data for machining workpieces on NC and CNC machines.
2. The term **manual programming** refers to writing and storing a part program for a particular machine/controller combination. The program is written in block format, as defined by German Standard DIN 66 025. The programmer must specify the required **tool motions** in a step-by-step manner.
3. In **manual programming,** manual calculations cannot be avoided. This requires certain knowledge of mathematics and trigonometry and a lot of time for auxiliary calculations. Programming errors are not detected until they reach the machine.
4. In addition to the geometric data, the programmer also must determine and enter the **technological data.** These include such data as rotational speed, feed rate, depth of cut, compensation values, and so on.
5. The term **computer-aided programming** refers to the creation of a part program with computer assistance and in a generally accepted output format (a cutter-line data [CLDATA] file). The **postprocessor** adapts the program to a specific machine-controller combination.
6. **Programming with graphical support** involves programming of the workpiece in its "before" and "after" states. For the most part, the computer program automatically determines the necessary sequence of tools, technological data, and motions that must be executed.
7. Almost all the **programming systems** that are available today offer color graphics monitors, interactive user guidance, graphic/dynamic simulation of the programmed machining operation, and the ability to be upgraded for DNC operation.
8. Uniform data interfaces are needed to transfer **workpiece geometry data from a CAD system.**
9. **SFP** can be performed by means of a programming system built into the CNC system, either at the machine **or** in an area close to the machine (e.g., on a PC).
10. The most important goal of NC programming is to create **error-free programs that are immediately ready to run.** In other words, these programs must be able to produce the desired workpiece without prolonged test runs or program corrections. Optimization for speed is of secondary importance. It becomes necessary when large production quantities are required.
11. The **profitability** of any NC manufacturing operation depends heavily on programming. The more powerful the programming system is, the quicker error-free programs will be supplied to the machine, and the more effective and flexible NC manufacturing will be.
12. **NC programming** should **not** be confused with laborious, time-consuming, and error-ridden efforts to solve complex trigonometric problems. This term actually represents the simple, systematic description of the workpiece in an interactive dialog with the computer, allowing the resulting program to be checked immediately by graphic methods.

3. CNC Programming Systems

Dipl.-Inform. (FH) Ralf Weissinger, Dipl.-Inform. Roland Aukschlat, Ing. Franz Imhof

Multifunctional machines and complex workpieces require the use of high-performance NC programming. Individual priorities will differ depending on the type of manufacturing. Often the data for the workpieces being produced come from different sources. For this reason, the user should examine the criteria very carefully before committing to a specific program.

3.1 Introduction *(Figure 3.1)*

The requirements placed on NC programming are subordinated to the demanding requirements of the manufacturing industry and thus are subject to constant changes. The production environment is characterized by a wide variety of products and ever-shorter product life cycles. The result is that components of ever-increasing complexity have to be produced in smaller and smaller lot sizes. In response, CNC machine tools are becoming more capable and more multifunctional. The required flexibility is reflected in the increased installation of "almost universal" machining systems.

A result of this is that the demands placed on CAM system have increased tremendously in recent years. This means that as the number and variety of machining processes contained in complete machining solutions increase, the more important it is to be able to manage them using the same CAM system.

3.2 Machining Processes Are Undergoing Major Changes

If one considers the various machining processes such as turning, milling, flame cutting, grinding, EDM, sheet forming, and even the machining of stone (*Figure 3.2*), it becomes clear that completely different machining strategies have to be chosen depending on the material (various metals, plastic, wood, glass, etc.).

Recently, more and more turning machines have been turned into universal machines that also can perform high-quality milling. And vice versa; there are now also milling machines that can be used for turning.

State-of-the-art programming systems therefore have to be able to cope easily with switching back and forth between these machining processes. What is demanded here is quite clear: to be able to program all processes in any desired mix using the same system and the same user interface (*Figures 3.3 and 3.4*).

This trend toward combined processes continues with a mix of 3D sculptured surface machining and 2½D milling/drilling. In the future, purely 3D machining will no longer be acceptable. There will be increasing demands to also be able to implement mixed programming for both requirements using a single system (*Figures 3.5 and 3.6*).

3.3 The Area of Application Sets the Priorities

The need for adaptability becomes obvious when one takes a closer look at the areas of application subdivided into prototype construction, one-off parts manufacturing, small series, and full-scale production. These make completely different demands both on the programming system itself and on the NC programs that are created. In prototype construction, the requirement is to quickly create reliable, collision-free programs. The programming time and the ability to implement changes quickly are the highest priorities. The run-in time on the machine is also important. Optimization of the process time is of only secondary importance. The reason for this is that there is really little benefit to be gained from optimizing the process time for just a few parts. Because running in may take place several times a day, this is where by far the greatest savings can be achieved. It is extremely important here to have programs that can be used to exclude the possibility of

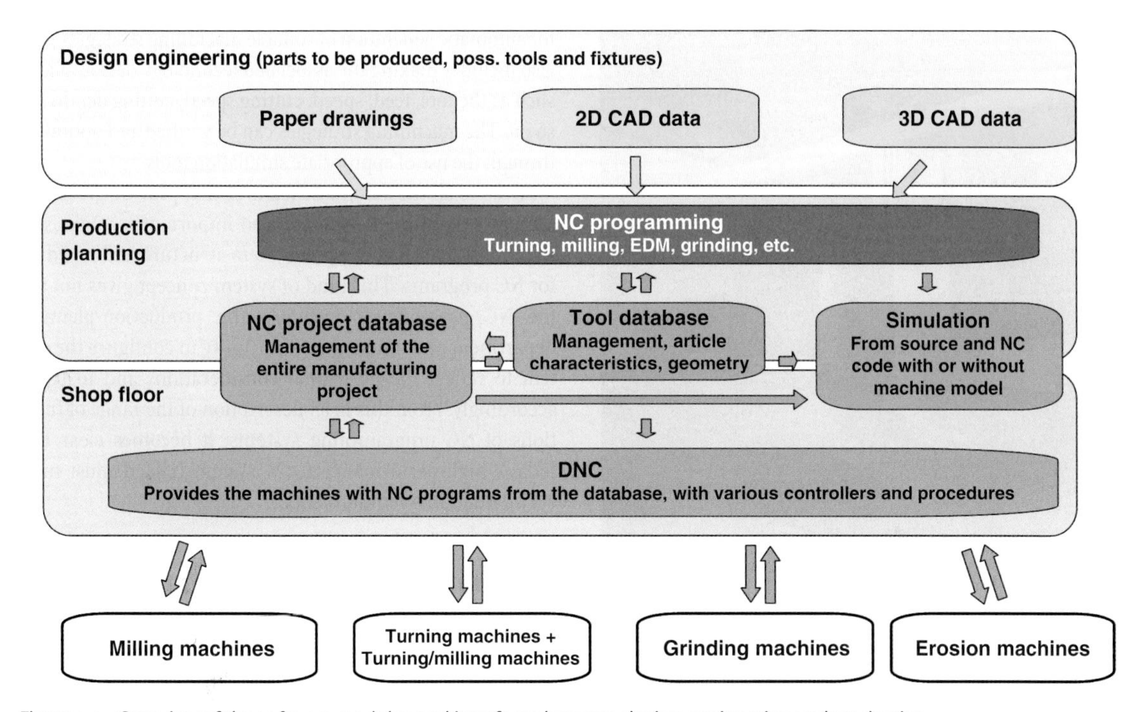

Figure 3.1: Overview of the software modules and interfaces between design engineering and production.

collisions from the very start, for example, by means of quality simulations.

In some fields, and especially in one-off production, the unmachined part is already very expensive and hard to obtain. In such cases, *zero defects* are essential even during running-in (*Figure 3.7*).

Series production involves completely different criteria. Here the process time itself is the most important point. The planning time is divided by the large number of units to be produced. It is thus advisable to look for ways to save even fractions of a second. This can be done most of all by optimizing the process sequence, optimizing tool paths and multiple fixtures, and using a very highly adapted postprocessor. All the same, simulation also must be used here to reliably avoid collisions.

3.4 Input Data from Various Sources

As the demands placed on NC programming systems have increased, so has the range of options for data input. While initially this involved only the geometry data of the parts being manufactured, it now also includes 3D data for the tools and clamping equipment, as well as the specific technology data for the individual machining processes. The standard here is generally 2D data in various interface formats and increasingly also 3D CAD data (usually the ISO 10303 series of standards [STEP]). These are supplemented by data about available tools, machines (for simulation), knowledge-based technology data, and recently, in some cases, feature data. Nevertheless, design drawings on paper also still exist. For such cases, CAM also has to provide a manual input option.

3.5 Capabilities of Modern NC Programming Systems (CAM)

The purpose of an NC programming system is to provide efficient support to the programmer. Its functions thus extend far beyond the creation of the NC program itself. They involve the

Figure 3.2: NC-controlled stone saw with 1.5-m saw blade. (Robert Schlatter GmbH.)

handling of geometric data for the part being produced, the handling and registering of tools, and making them available in the data pool. With internal system functions it is necessary to create a data model that can serve as an unambiguous basis for automatic generation of suitable machining strategies. This also includes making the associated technology data available, such as the tool, feed, speed, cutting speed, cutting depth, and so on. The machining strategies can be verified and optimized through the use of appropriate simulation tools.

Despite all the automatic features that provide great time savings on routine tasks, it is also important to make sure that the system has an open-system structure and an editor for NC programs. This kind of system concept gives not just the NC programmer but also the production-planning department and others the opportunity to configure the system to reflect their personal considerations and to use it accordingly. From this brief description of the range of functions of NC programming systems, it becomes clear that today's high-performance CAM system indeed must meet very demanding requirements.

3.6 Data Models with a Uniform High Standard

In order to obtain a precise data model in the system, the CAM system has to check the imported geometric data (2D or 3D CAD data) for completeness and consistency and to bring the data to a uniform "high standard." If the design drawing is available on paper, then the system has to have a 2D design tool so that the user can transfer the information on the drawing to an internal 2D depiction.

For the automatic transfer of 2D or 3D CAD data in various formats (STEP, initial graphics exchange specification [IGES], and CAD-specific formats), the important thing

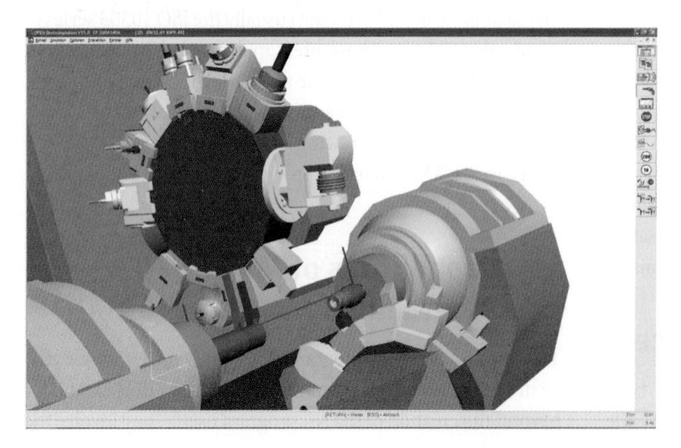

Figure 3.3: INDEX turning/milling center in a simulation.

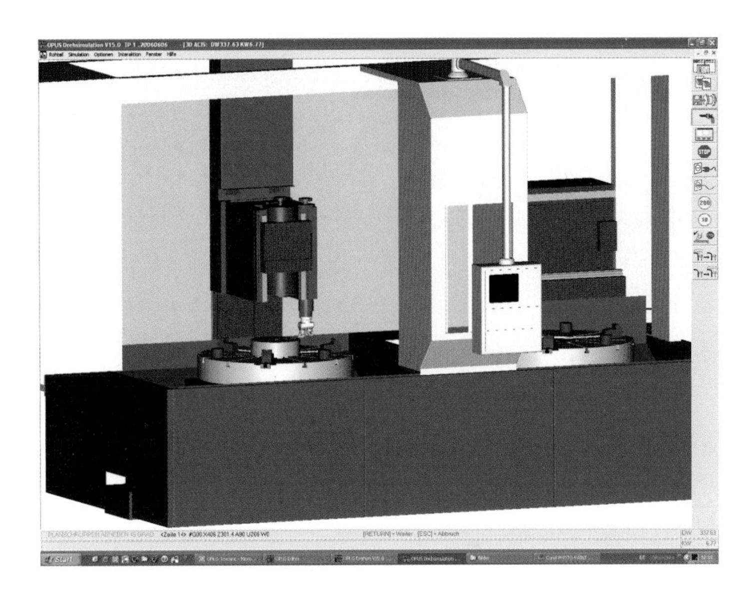

Figure 3.4: Turning/milling or milling/turning machine from Pittler.

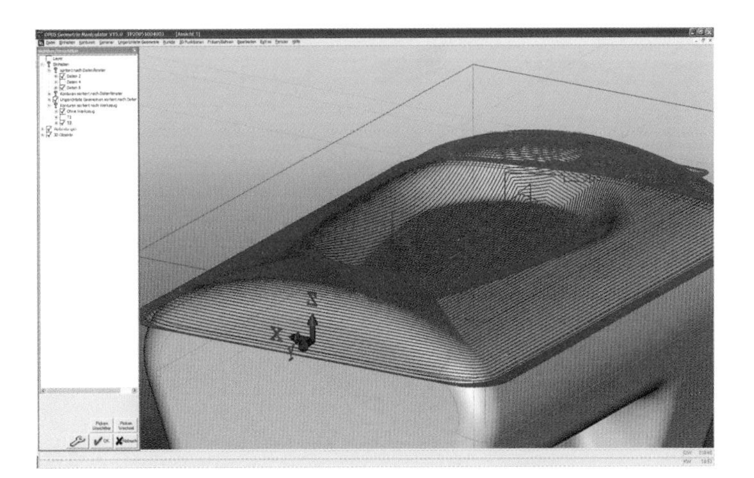

Figure 3.5: Depiction of sculptured surfaces (3D machining). The geometry is generated by simultaneous tool motion in all axes.

is to be able to read these formats and convert them into the internal format. Once this has been done, the geometry can be processed further in a uniform manner. CAD drawings contain large amounts of data that are completely irrelevant for programming. These include, for example, hatching, drawing frames, textual notes, dimension lines, and so on. To avoid extra work, it is important here to use an integrated data filter to select the right elements while the data are being read in.

3.7 CAM-Oriented Geometry Manipulation

If individual lines, circles, and circle segments are completely sufficient to depict a drawing, then contiguous contour segments are absolutely necessary for processing it into an NC program. Because data that have been transferred are normally not of high enough quality, the CAM system has to be able to sort the data, correct them, and

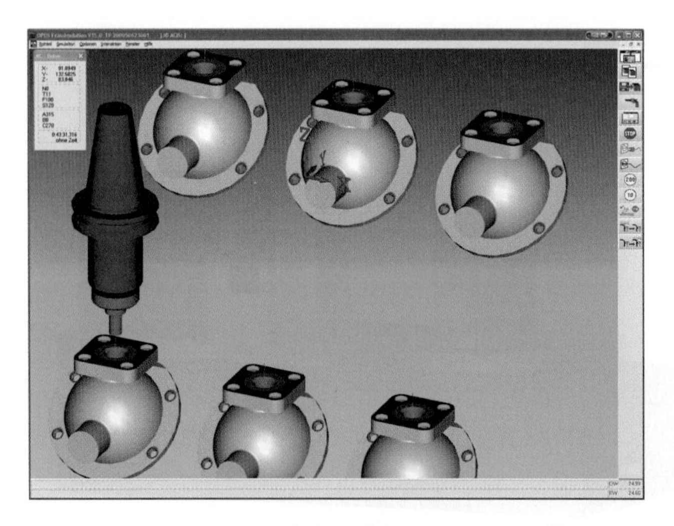

Figure 3.6: 2-1/2-D machining.
Form-reproducing tool generates resulting geometry through Z in-feed.

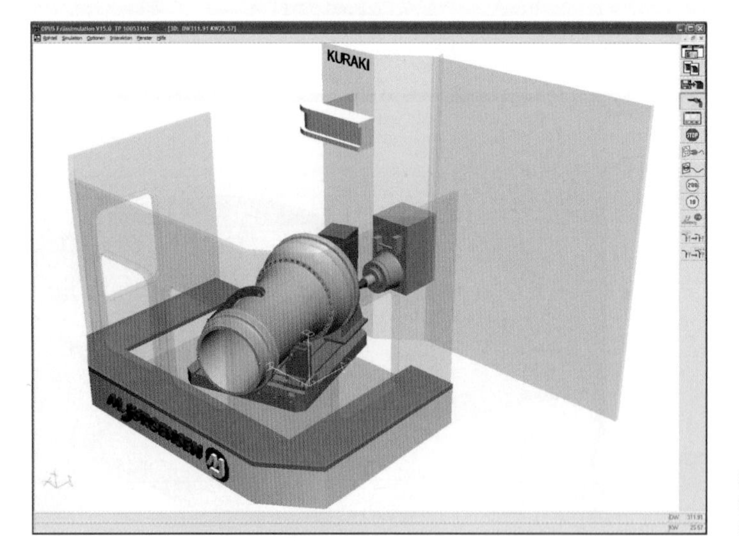

Figure 3.7: Cylinder liner (length ≈ 1,500 mm).
Example of a very expensive unmachined part in one-off parts manufacturing.

place them in the correct contour associations. This can include

- Extending or trimming contour elements
- Simplifying contours
- Splitting, connecting, and cutting
- Inserting and changing radii, chamfers, and recesses
- Shifting, rotating, stretching, copying, and mirror imaging
- Generating equidistant contours, erc.

It is also necessary to be able to correct elements whose tolerances are not centered, taking the connecting elements into account. In any case, it is essential to give the system information about the height of the body involved and the depth of any recess. If this cannot be done using the geometric data, then it will have to be done later via the depth setting of the tools. This is very time consuming.

The design attitude very seldom corresponds to the attitude in the NC programming or on the machine. It is therefore important to align the geometries in space as simply as possible. It must be possible to associate different sides with different zero points (on the machine).

With modern systems, if 3D models are available for the workpieces, then they also can be used to generate 2½D programs. Based on the precise 3D data in the system, it is possible to derive 2D geometries such as planes, cylindrical surfaces, sections through volumes, or projections of surfaces as a contour with supplementary information about the position in space.

If no CAD data are available, then the programmer must be able to generate all the geometry necessary to create the program using the functions available in the CAM system without any help from a CAD system.

Complex workpieces contain such a large amount of geometric data that in some cases it is hard for the programmer to maintain a clear overview. It is therefore absolutely essential to support the programmer by means of suitable tools and management functions. It is especially useful here to highlight elements, contours, solids, and surfaces through the use of different colors, transparencies, thicknesses, and types of dashed lines and other special ways of emphasizing edges. It is also advisable to assign names and layers, as well as ways to conveniently show and hide relevant and irrelevant objects. One excellent method, for example, is to automatically hide surfaces that have already been processed.

3.8 Only High-Performance Machining Strategies Count

The truly supreme aspect of CAM systems is the generation of machining strategies. The important thing here is to bring the geometric data of the part being produced together with the suitable tool, the associated technology data, and the individual machining steps to form a high-performance machining strategy.

For normal machining such as milling or turning, it is common to use standard strategies that simply can be called up, such as

- Milling
  - Moving over contours/contour areas, etc.
  - Broaching of recesses with and without islands
  - Thread milling
  - Slot milling
  - Engraving
  - Drilling cycles
- Turning
  - Rough machining
  - Roughing with cutting passes
  - Finishing
  - Recessing (*Figure 3.8*)

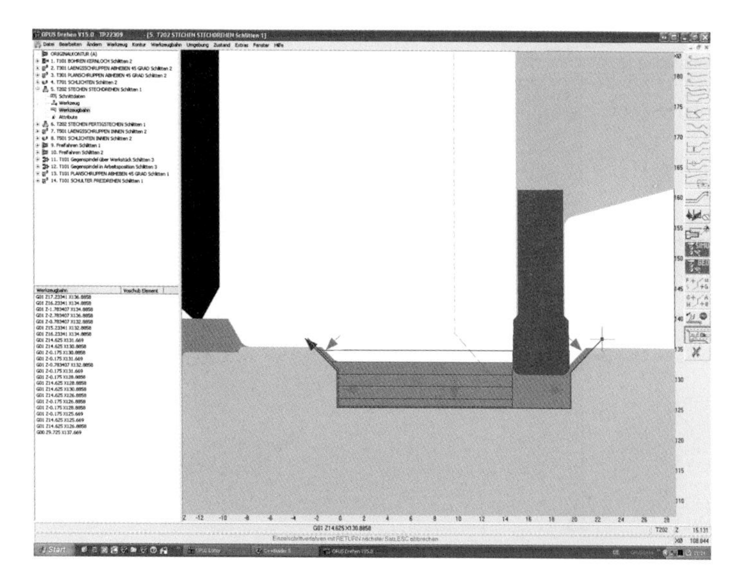

Figure 3.8: Path calculation for recess turning.

These types of machining strategy can be generated automatically, practically at the press of a button (*Figure 3.9*).

But the real challenges are as follows:

- To provide the user with a configurable, application-specific standard
- To provide for easy integration of nonstandard and new tools, technologies, and parameters
- To incorporate user-specific requests into the system, such as tool catalogs or special knowledge-based databases
- To transfer existing machining strategies to other geometries
- To create a direct association between the geometry and the machining so that in the event of a geometry change, the change is made automatically in the corresponding machining operations

3.9 3D Models Offer More

Already today CAM systems are appearing that can use 3D model data directly to generate 2½D programs. Transferring model data from various CAD systems no longer represents any problem. Previously, the interfaces often were problematic for the consistency and completeness of the data. Today, some interfaces are even capable of detecting errors in the consistency of the original model and may be able to repair them. As with drawings, looking good is not enough by itself. The model has to be logically correct, with no gaps in its edges, for example. In addition to these automatic correction steps, the user also has to have an option for manual simplification of the model. In the simplest case, this can involve hiding holes, chamfers, and recesses, perhaps in order generate a suitable model of the unmachined part. From the point of view of the programmer, it is important to be able to use his or her usual environment, for example, the CAM system, to manipulate the data as regards production. Even in the future, any direct, design-related changes still should be done only by the design engineer. Power over the product design and responsibility for it should remain there.

The 3D basis can be used to derive the appropriate edges and surfaces for motion paths (up to five axis paths) (*Figure 3.10*), thus unambiguously defining the positions and standards.

3.10 Innovation with Feature Technology

Definition According to VDI2218

Features are informational elements that represent areas of particular (technical) interest in individual or multiple products.

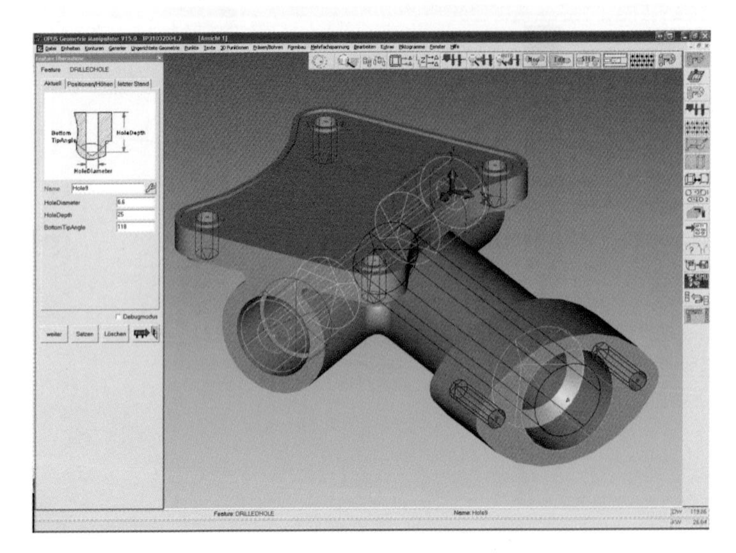

Figure 3.9: Transferred from CAD to CAM and automatically converted into machining sequences.

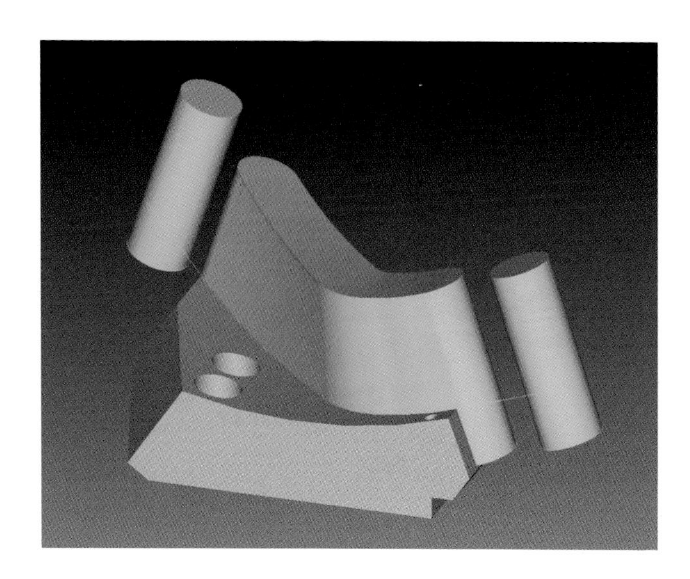

Figure 3.10: Five-axis path, derived from the surface and edge.

If the 3D CAD system that is being used is capable of attaching expanded information to geometric elements, then a modern CAM system can use this information to derive all the required tools and machining sequences. It should be noted here that only geometric and design-related information is needed in the 3D model.

Some systems try to achieve automatic program generation by attaching tool and technology information to CAD objects. This approach, of course, is completely wrong. The developer has to stick to his or her core competency—design engineering—and must not have to cope with production-related issues. What is more, it is very rare in the design phase for it to already be clear what production systems and which tools will be used later to manufacture the part. It is thus important to make a clear separation between the geometry and the manufacturing technology.

It becomes obvious here that model generation has to take into account more than just the external appearance. Information about tolerances, threads, surface quality, and much more has to be linked to the model in a consistent, logical manner. Whether it is integrated or used independently, the CAM system has to make use of the available information to derive optimal machining operations.

3.11 Automatic Object Detection

Unfortunately, the availability of such feature depends both on the CAD system being used (and its configuration) and on the individual developer and/or design engineer. It is therefore useful if the CAM system is also capable of deriving, for example, drilling data from the CAD model. In this case, it must be possible for the programmer to insert missing information manually, such as fit data. Once this information is included, it is possible to derive the same machining operations and tools as with fully fledged feature models (*Figures 3.11 and 3.12*).

3.12 Machining Database

Setting up a machining database can help to serve the dual purpose of greater transparency and faster processes. It is advisable to design this type of database with an eye to the user-specific infrastructure. The specific characteristics of use also should be taken into account; these are determined by specific properties, for example, frequently recurring machining strategies. On the one hand, common data structures are facilitated by data sets that are stored in a 1:1 relationship; that is, the geometry and the complete associated machining strategy are stored together. On the other hand, it

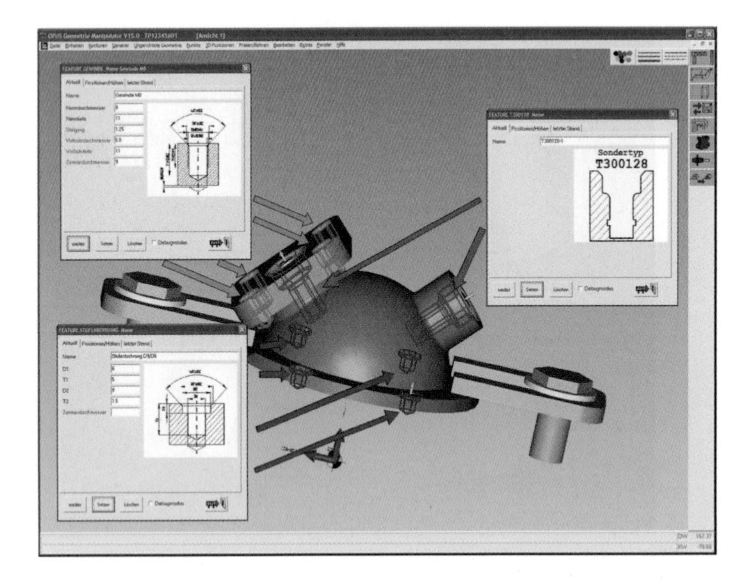

Figure 3.11: Part with detected features, including a special borehole with a special tool.

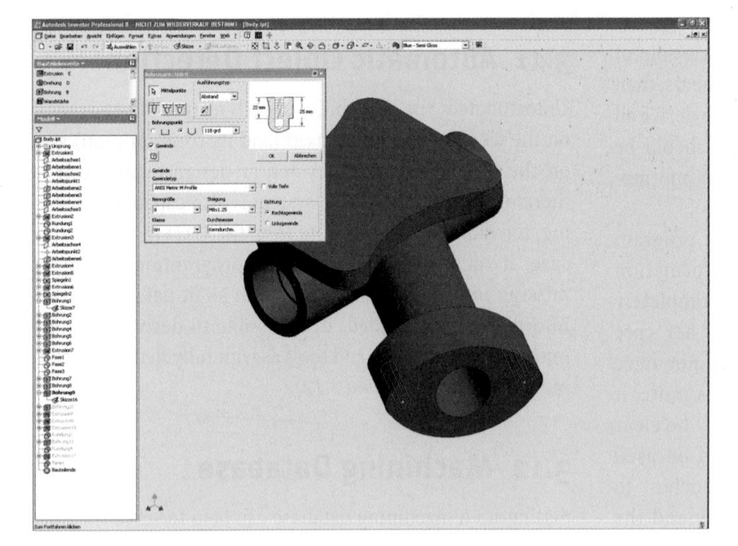

Figure 3.12: Features of the CAD system. (Autodesk Inventor.)

is also possible to assign predefined machining operations to other geometries.

3.13 Tools

In this process chain, the tools are of central importance. Neither automatic program generation nor quality simulations are at all possible without complete, consistent tool models. In manufacturing technology, tool management cannot be limited merely to the recording and saving of tool data. Far from it: Both the geometric and technological characteristics have to be available in such a way that the CAM system can access them as automatically as possible (*Figure 3.13*). In conjunction with the feature information—for example, for a

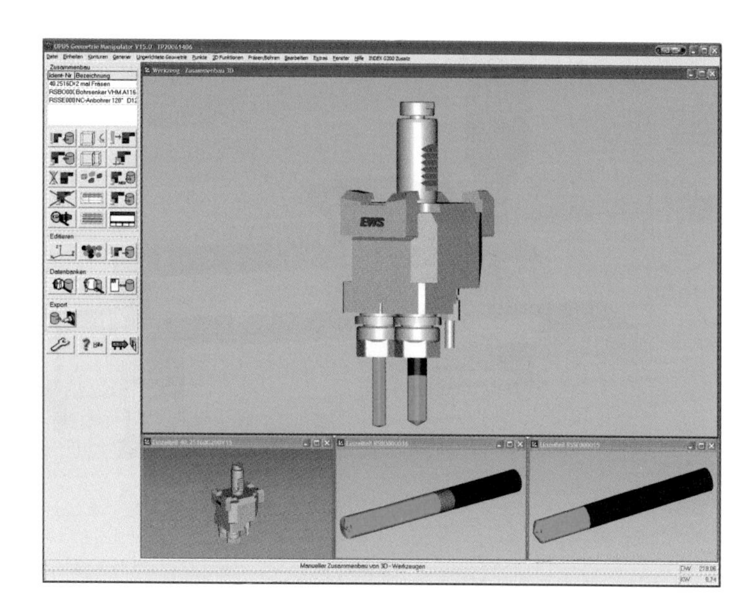

Figure 3.13: EWS tool for turning/milling centers; depiction of components and assembly.

stepped hole—this makes it possible to automatically determine the correct tools. Using this method, it is also possible to ensure that in the event of changes, for example, from an M8 tapped hole to an M10 one, both the machining operation and the tool selection will be integrated automatically. Entering new, technologically improved tools thus also will automatically affect future selections of the optimal tool.

3.14 Clamping Planning and Definition of the Sequence

Once programming is complete, the CAM system should be able to optimize the tool sequence and machining sequence automatically. For example, it should compile the spot drilling operations for all the holes and optimize the newly created motion paths. This also should take into account collision geometries on the model or resulting from the clamping operations. And, of course, it should be possible for the resulting sequence to be modified by the user. Here, too, it goes without saying that all actions should be visualized immediately for maximum user friendliness (*Figures 3.14 and 3.15*).

The planning of the machining operation initially should be as independent of the machine as possible. It is not absolutely necessary to manufacture a part in the position where it was programmed initially. Offsets and changes in position owing to clamping can be taken into account automatically by the postprocessors. This means that reclamping a part in a different setup/position is very easy. In series production, multiple setups contribute to a considerable reduction of the production time per piece. The machining operations can be defined on only *one* part. If this part uses a multiple setup, then these machining operations are "inherited" at the new position so that multiple parts can be processed in various positions in the process sequence. Here, too, sequence optimization can play a major role and can be used to recalculate the tool sequence and processing paths in relation to the overall scenario. For traversing motions, the basic objective is to achieve motions that are as short and quick as possible. This is generally achieved by reducing the retraction distances. The principal problem here is the danger of collisions between moving machine parts and the clamped part or the clamping equipment. In series production, it is not acceptable to always pull back to a position that is safe under all conditions.

3.15 The Importance of Simulation

In this context, simulation is becoming more and more important as an essential additional tool for NC programming. Its

Figure 3.14: Planning the machining of the same part from two sides.

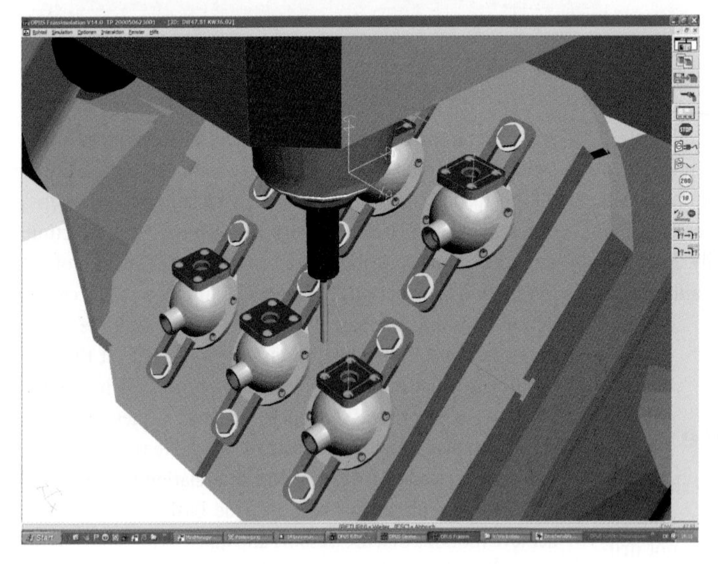

Figure 3.15: Simulation of a multiple setup.

scope covers everything from analysis of the tool paths on the workpiece to the overall scenario on the entire machine. This should involve source and NC code simulation, taking the specific machine model into account. One especially important aspect is the dynamic tracking of the machining progress so that the removal of material and the status of the part being machined become visible after every machining operation (updating of the blank). This makes it possible to have even the most complex machining sequences executed at the end of the programming to guarantee that there will be no collisions.

Therefore, before the program is transferred to the machine, there should already be a simulation of the entire scenario that reflects the features of the machine model as precisely as possible (→ *Chapter 4: Manufacturing Simulation*).

3.16 Postprocessors

Regardless of how convenient it has been to generate the data up to this point, the controller of the NC in any case will require its own special NC code. In the meantime, these codes deviate to a greater or lesser degree from the syntax defined in the DIN 66025 standard regardless of the type of controller. It is thus the task of the postprocessor (a program section belonging to the CAM system) to convert the data that have so far been generated into the precise syntax of the specific controller-machine combination. Good systems have a modular structure, which means that the postprocessor is not buried in the general system code. Nevertheless, the CAM systems supplier also should take care of this last step because the postprocessor has a decisive influence on the overall results of programming work. This is where it is determined whether control cycles and subprograms can be used and likewise whether full use can be made of the cutter radius and/or tool-tip edge-radius compensation. Last but not least, the postprocessor is also responsible for the generation of additional data such as tool lists, clamping plans, and auxiliary programs, for example, for the tool-presetting device. It also performs preassignment of milling operations on turning machines or distribution of the overall machining tasks among a number of individual machines.

3.17 Generated Data and Interfaces with the Machine Tools

When integrating the NC programming system into the existing machinery, the user's expectations should be taken into account if at all possible. This can include a number of additional production-specific requirements, including

- Assignment of NC programs to specific machines/groups
- Enabling/disabling of individual NC programs (including groups thereof)
- Sending back optimized programs
- Comparing the original program with the one that is sent back
- Logging of data transmission, service interventions, etc.

To implement this range of requirements, it is necessary to have an NC program-management system that is connected via DNC (→ *Part 6, Chapter 1: Direct Numerical Control*).

3.18 Summary

The profitability of a CNC manufacturing operation depends heavily on the performance of the NC programming system. Programming comprises not just the inputting of the workpiece dimensions but also has to provide appropriate specifications for the entire manufacturing process for each machine. All the essential factors have to be taken into account. Therefore, before committing to a particular manufacturer, it is absolutely essential for the purchaser to obtain extensive information and to compare the various systems. It necessary to take into account not only the various machining processes and CNC machines that have to be programmed but also the requirements as regards the variety of parts with 2½D and 3D machining operations, quality simulation of the finished NC program, the options for transferring data from the CAD system, feature programming as a way to make work easier, and last but not least, the ways and means for recording and calling up tool data and/or entire tool catalogs in a simple and easy manner. Also important is the automatic availability of the associated technology data, such as feed rates, speeds, cutting depths, and so on, for each of the materials being machined. A high priority is also placed on the ability to make changes or modifications quickly to finished programs without having to start again from the beginning. Finally, one of the supreme aspects is generation of an error-free machining strategy.

CNC Programming Systems

Important points to remember:

1. **CAM** is the abbreviation for "computer-aided manufacturing," also called *computer-aided programming*. This means programming with the aid of computers. This can involve systems integrated into CAD and independent CAM systems. All contemporary CAM systems provide graphics support.
2. The difference between **2½D** and **3D programming** does not refer to the nature of the graphics support. 2½D programs are also generated on the 3D model using high-performance systems.
3. In the case of **3D** programs, the tool is moved along the workpiece during the machining process with simultaneous participation of at least three axes of the CNC machine tool. The shape of the workpiece thus is created directly by the tool path (typical for mold construction).
4. **2½D** programs are needed for all processes in which a maximum of two axes move simultaneously during the machining operation with interpolation. This type of machining is typical for drilling operations and milling operations such as surface, contour, and pocket milling.
5. It should be possible for a CAM system to manage a **mix of processes,** for example, turning and drilling/milling without a module change.
6. With state-of-the-art **interfaces,** it is possible to transfer models from various CAD systems to external CAM systems with no loss of data.
7. The **tool database** is of central importance in the CAD/CAM process chain. Both in program generation and in simulation, it must be possible to access not only the numerical data but also realistic model data at run time.
8. Today, **simulation** is an indispensible aid in NC programming. This is essential in order to prevent collisions. Here, too, the better the model data, the better is the quality of the prediction from the simulation.
9. It is the **postprocessor (PP)** that generates the actual NC program for the specific controller-machine combination. It also can supply additional data such as tool lists, clamping plans, and programs for the tool-presetting device. A CAM system can be used only if the postprocessor is delivering the proper output.
10. It is advantageous if CAM systems have "feature" functions to be able to adopt programmed machining operations directly in the event of small geometric corrections.
11. During the design phase, it has generally not yet been determined which CNC machine will be used for the machining. Therefore, there should be a strict division between the CAM system geometry and the technology data.

4. Manufacturing Simulation

Dipl.-Ing. Karl-Josef Amthor, Ing. Franz Imhof, Dr.-Ing. Karsten Kreusch, Dipl.-Ing. Stefan Großmann

The simulation of technical systems and processes is considered to be one of the key technologies for computer-supported product-development and production technology. The goal of manufacturing simulation is to provide error-free manufacturing processes, optimization of machining times, and overall improved safety and profitability of the entire production process.

4.1 Introduction

In recent years, the simulation of CNC machines has become established more and more as the standard method for the verification of complex machining processes. In modern NC simulation systems, a real manufacturing system is reproduced true to life in a 3D graphical environment. Machining processes can be viewed and followed from any angle (*Figures 4.1 and 4.2*).

In contrast to this, in NC-controlled manufacturing systems, the visual observation of motions and sequences is possible only to a very limited extent. If the enclosure of an NC machine allows any observation at all, then it is generally from a very disadvantageous angle. The clamping situation and coolants also obstruct the view of the machining process and thus hinder the running in of new or modified programs.

4.2 Qualitative Classification of Systems *(Table 4.1)*

NC programming systems often include simulation components that show the motions between the tool and the workpiece on the basis of the NC source (before the postprocessor). This method is highly suitable for performing an initial check of the results of programming. For exact geometric investigations and analyses of the run time, systems based on the final NC codes are to be preferred for the following reasons:

- Errors in the postprocessor are not detected by the simulation system but only appear during run-in on the machine.
- It is only the postprocessing that determines the machine on which a program is to be processed; this also defines the workspace and the machine parameter set.

For programs that are created with the help of manual programming in the editor, that is, without the NC programming system, the "pre-postprocessor" simulation is eliminated from the beginning. This means that in this case it is particularly important to verify the programming result because a large number of errors appear especially with this programming method.

Truly definitive simulation results are only possible on the basis of the actual NC program; that is, the simulation uses exactly the same program text that is processed on the actual machine. This includes not just the program itself but all the subprograms, cycles, and parameter tables called up by the program, such as zero-point offsets, tool-compensation values, and so on.

We will now describe the state of the art for modern NC simulation systems. The functionalities described here correspond to the capabilities of the most powerful systems available at present. Not all suppliers may offer the same level of performance.

NC Simulation with Controller Emulation *(Table 4.1)*

For many years, the required computing power was a limiting factor for 3D depictions, collision checks, and machining simulations. In the meantime, even high-performance systems can run on standard commercial PCs and/or laptops.

Modern systems represent all motions in a **3D** machine model. Interfering elements in the workspace can be either depicted transparently or even hidden. They are nevertheless taken into account in the collision-detection process that is running in the background, thus ensuring that no collisions are "overlooked." The angle of vision navigation in

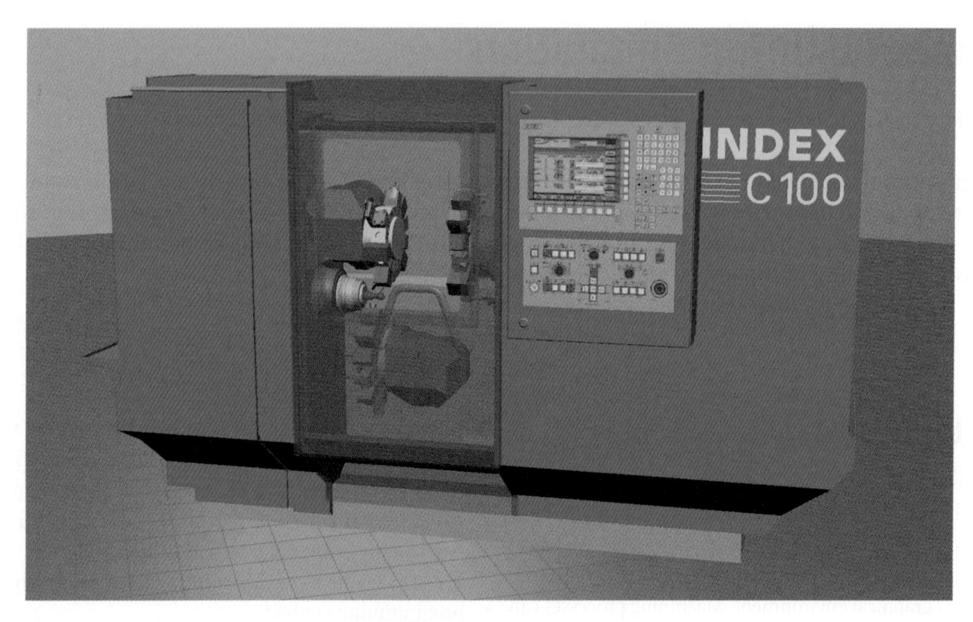

Figure 4.1: Machine model of an INDEX C100 simulation on the screen.

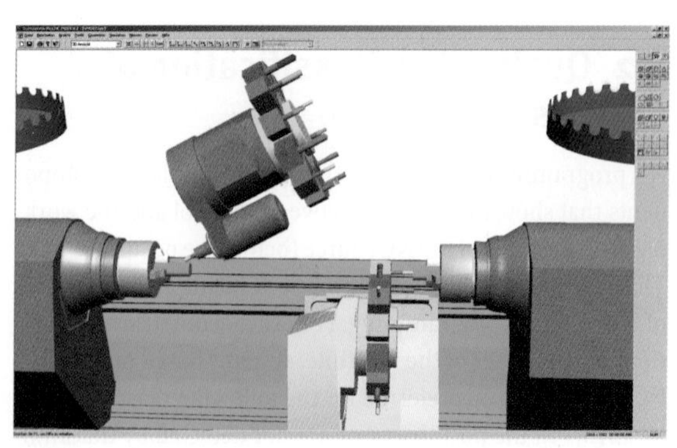

Figure 4.2: Comparison of the workspace of a turning machine in reality and in a simulation.

Simulation "Upstream of Postprocessor"	Simulation with Controller Emulation	Simulation with Virtual Machine
+ Possible at an early stage. + No need to specify a particular machine. – Kinematic parameters are not taken into account. – Postprocessor errors are not detected. – No simulation of free NC code.	+ A platform solution for many individual machines. + Controllers can be simulated for which there is no virtual variant. – Controller emulations are often inadequate for complex machines.	+ Full range of functions of the controller. + Operator control and programming are identical in the real and virtual machines. – Performance is generally worse than in controller emulations.

Table 4.1: Comparison of the Various Simulation Approaches

the simulation system allows very detailed examination of all machining situations. The purely 2D depiction of work-space components (broken-line graphics) that was previously common is hardly used in the meantime.

In addition to purely visual depictions, high-performance NC simulation systems provide functions such as

- Practically complete depiction of a wide variety of controllers
- Depiction of the material removal
- Automatic collision detection
- Syntax verification of the NC program
- Precise time analyses
- Depiction of system variables, axis values, and other process parameters
- Protocol functions for documentation of errors and collisions that occur

In order to make a distinction from the other systems available on the market, it is necessary to analyze the machining process being simulated. As a general rule, a distinction can be made between manufacturing processes with 3D machining and those with 2½D machining. In a 3D machining process, the feed motion of the tool takes place in at least three axes simultaneously. The shape of the workpiece is created directly by the tool path (e.g., five-axis milling).

In 2½D programs, two axes are moved with interpolation. These machining operations typically involve drilling or turning operations.

In modern turning/milling centers, it is even possible to execute both 2½D and 3D machining operations in parallel. Through the integration of various technologies in a machine tool, it is today no longer possible to make a strict distinction between turning and milling machines. A modern simulation system therefore not only has to support all kinds of machining but also has to be able to perform various machining operations on one and the same part.

Virtual Machine *(Table 4.1)*

Because the scope of functions and complexity of NC controllers have increased greatly in recent years, it has become much more difficult to reflect the functions in the form of a controller emulation. In the case of complex, multichannel machines, the amount of effort required for the programming of controller emulation has risen to such a degree that it is no longer possible to emulate the machine completely.

An escape from this situation has appeared in the form of **virtual NC systems.** Unlike simulation with controller emulation, the behavior of the controller is no longer emulated. Instead, the controller manufacturer provides a software component that contains the complete response of the controller. One certain advantage in favor of this development is the fact that today the functionality of an NC control system no longer resides in the hardware but rather in the software.

This has made it possible for controller manufacturers to provide a virtual controller for the simulation whose functionality is identical to that of the actual controller. A virtual NC control system is generally the software of an actual NC control system that has been encapsulated in such a way that it can run on a standard commercial PC and can communicate with simulation systems.

This virtual control system is put into operation using the data of an actual machine. This produces an exact copy of the control system of the actual machine. The word **emulation** is therefore no longer used in the simulation.

Use of the virtual NC controller then makes it possible to equip the simulation with the user interface of the actual machine. Because the user interface is also a software program, it can be operated with the help of the virtual NC in the same way as the actual machine (*Figure 4-3*).

The term **virtual machine** has become common on the market in recent times. Unlike NC simulation with controller emulation, the virtual machine has the following characteristics:

- Material removal, 3D depiction, and collision calculation according to the NC simulation with controller emulation
- Use of virtual NC controllers with a complete range of functions (syntax checking, run-time analysis, settings and parameters like on the actual machine)
- Use of the machine operating panel in the virtual environment
- The simulation should behave like the actual machine in as many areas as possible.

4.3 Components of a Simulation Scenario

Machine Model

A **machine model** refers to the true-to-original emulation of the physical machine on a computer. At the least, a machine model contains the following components:

- Geometric models of machine elements such as the frame, guides, and cover plates
- The kinematic structure
- The control model or virtual NC control system

At a minimum, the **geometric elements** of the machine model have to describe the workspace precisely. Additional machine elements such as parts of the enclosures may be relevant in terms of collisions but also have recognition value. For the purposes of simulation, they can be made transparent or hidden entirely (*Figure 4.1 and 4.2*).

The **kinematic structure** stores all the actual machine axes as virtual axes. The sequence of the axes and axis distances correspond to the actual conditions. It is not only NC

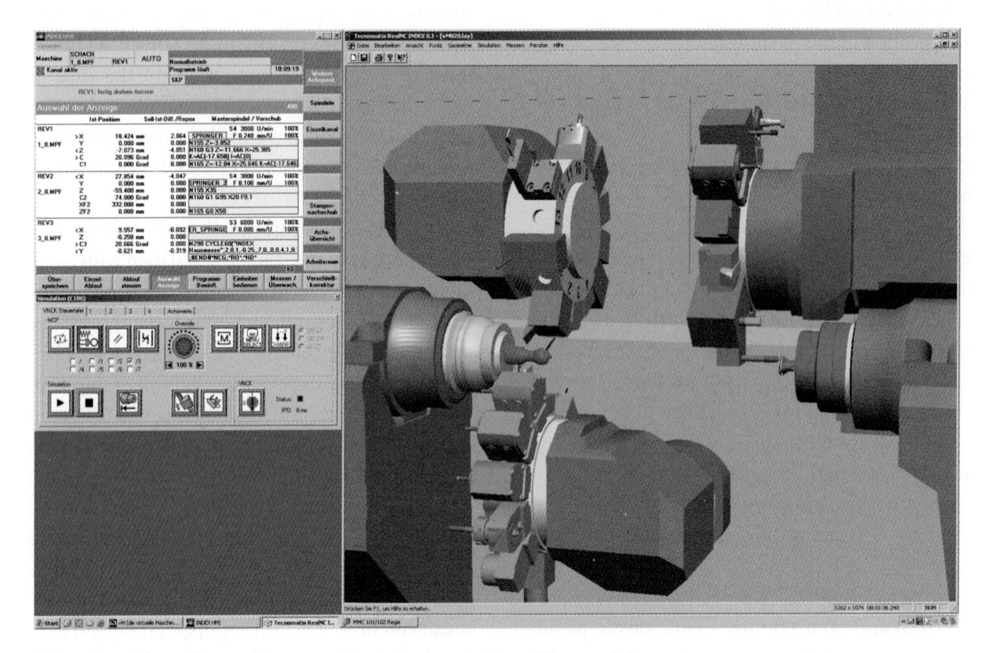

Figure 4.3: Virtual machine with machine operating panel and 3D machine model.

axes that are defined as axes—even PLC-controlled axes such as turret axes and hydraulically driven clamping elements can be a part of the kinematic structure.

The **control model** reflects the most important characteristics of the actual controller. This means that high-performance control models can execute almost the entire range of commands for the original controller. This is absolutely essential for functions such as motion planning, interpolation, milling cutter-radius correction, and so on. It also should be possible to support high-level language elements such as jump commands, conditional jumps, loops, and the programming of variables and parameters.

With a virtual machine, a **virtual CNC controller** is used instead of a control model. The virtual CNC system is put into operation using the data of the actual machine and thus is functionally identical to the actual machine. In this way, it is ensured that all the functions of the controller are exactly the same in both the virtual machine and the actual machine.

Virtual machines also make use of the **machine operating panel.** The user of the virtual machine has identical access to all the inputs and displays that are present on the actual machine. It is thus possible to use the virtual machine not only at the shop-floor level but also in a production-planning environment.

Workpiece Geometry and Clamping Devices

Today, workpieces generally are designed as unfinished and finished parts. The unfinished parts' geometry required for the simulation sequence is transferred to the simulation system via CAD interfaces of the CAD system or the simulation system. The same applies to the necessary clamping fixtures. If no CAD data are available, some simulation systems provide easy-to-use CAD functions that can be used to create even complex clamping kinematics from standard geometric figures such as cuboids and cylinders (*Figure 4.4*).

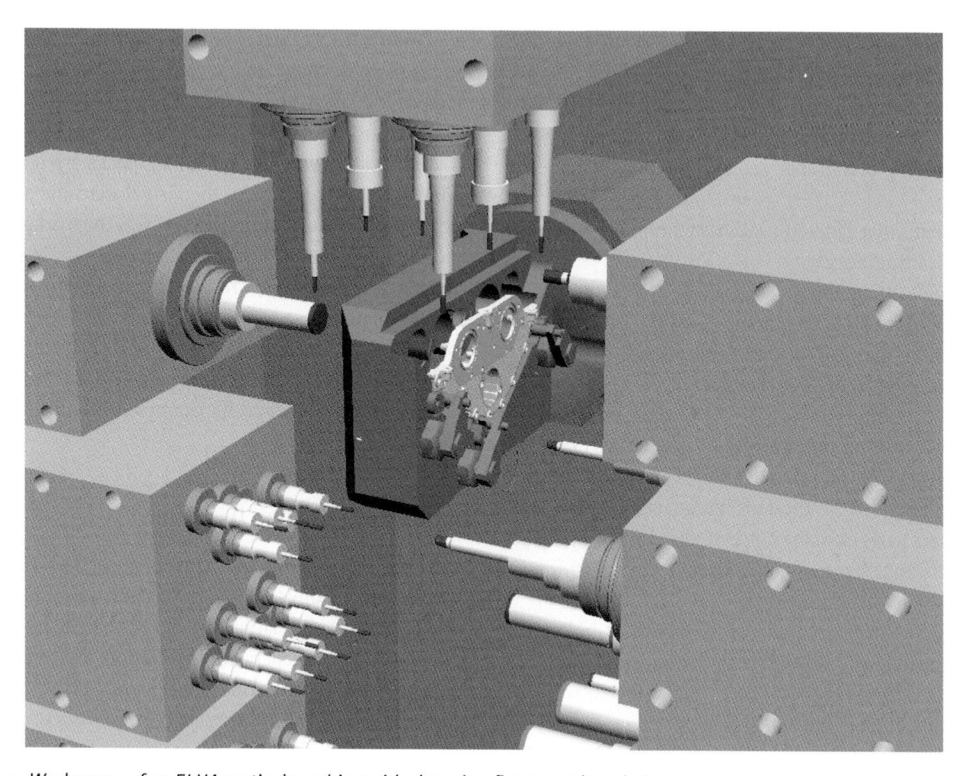

Figure 4.4: Workspace of an ELHA vertical machine with clamping fixture and workpiece.

High-performance systems support typical series-production methods such as multiple setups in machining centers and multiple-spindle machining on turning machines.

Tools

Tools are subdivided into the following categories:

- Rotationally symmetric tools (drilling and milling cutters)
- Non–rotationally symmetric tools (turning tools, single-blade tools)
- Complex tools (angle drilling heads, drilling heads with more than one tool)

Rotationally symmetric tools generally are easy to draw because they can be created via rotation from a simple 2D contour. Non–rotationally symmetric tools accordingly are more difficult to model.

There are two different approaches for generating tools for NC simulations. Tools can be either generated or compounded. With **generated tools,** the user defines the geometric characteristics of the tool using a number of defined parameters. The system then generates a complete 3D model consisting of a holder geometry and a cutting edge (*Figure 4.5*).

Tool systems that generate tools have the following advantages and disadvantages:

- Very simple operation
- No knowledge of CAD required
- Sufficient for many simulation applications
- Only predefined tools are possible.

In the case of **compound tools,** 3D geometries of individual workpiece parts are saved in a single database. The geometries are provided by the tool manufacturers. The user then can select parts from the range of stored 3D components and put them together to create complete tools (*Figures 4.6 and 4.7*).

Tool systems that are made up of groups of individual tools have the following advantages and disadvantages:

- Any desired 3D geometry
- Precise reflection of the tool geometry
- Effort required to maintain the database and to create tools

In some simulation systems, both types of tool creation are used. Thus the user can decide from case to case which approach is correct for the specific application.

Periphery of the Machine

Peripheral components such as pallet-changing devices, handling units, and magazines can be depicted and included in the simulation using the methods just described.

4.4 Procedure for NC Simulations

Graphical Depiction

During the simulation, the lowest level of detail depicts the workpiece and the tool. The next level shows in addition the clamping fixtures and the components that carry the tool (i.e., angle head, turret, and tool spindle). The complete graphical depiction covers the entire machine with all the cover plates, tool-changing devices, measuring devices, and so on that limit the workspace (*Figure 4.8*).

After the simulation is started, the NC program is executed like in an actual machine. The NC blocks are prepared and converted into motions of the virtual machine axes. Visual depiction of the machining progress on the workpiece is provided in the form of a material removal. The new surfaces created on the workpiece receive the color of the cutting edge of the tool being used.

As an option, the tool path is displayed during the machining process. In this manner, it is easy to reconstruct the machining sequence, for example, with drilling patterns (*Figure 4.9*).

Automatic Collision Detection

During the simulation, the components of the workspace are checked for inadmissible collisions and/or contact. If these occur, then the simulation sequence is halted and an appropriate message is generated. In addition, the colliding components are highlighted in color (*Figure 4.10*).

Some simulation systems also cover errors of a technological nature. If machining is performed with deactivated coolant feed or if the maximum permissible spindle speed

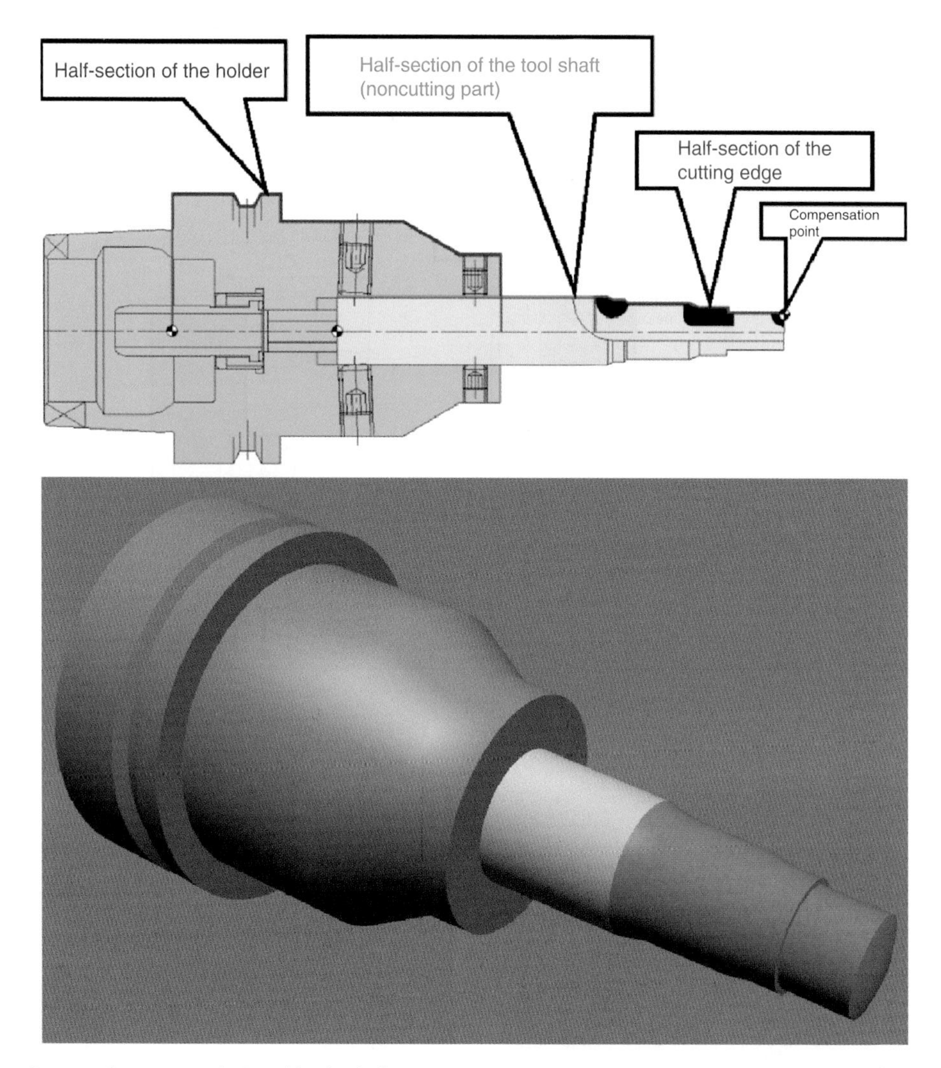

Figure 4.5: 2D external contour and 3D tool in simulation.

and feed rate are exceeded, then the appropriate warning messages will be issued.

Editing Options and Analysis Methods

While a simulation is running, the line of the NC program that is currently being executed or the subprogram that is loaded at the moment is always displayed. In some simulation systems, it is always possible to edit the program tests and to test the changes immediately.

In NC simulations with controller emulation, it is possible, as in a machine, to execute the program in single-block or automatic mode. Furthermore, it should be possible to simulate the sequences individually in each interpolation cycle. It is thus possible for the user to track, analyze, and optimize the motion sequences with accuracy to the nearest millimeter.

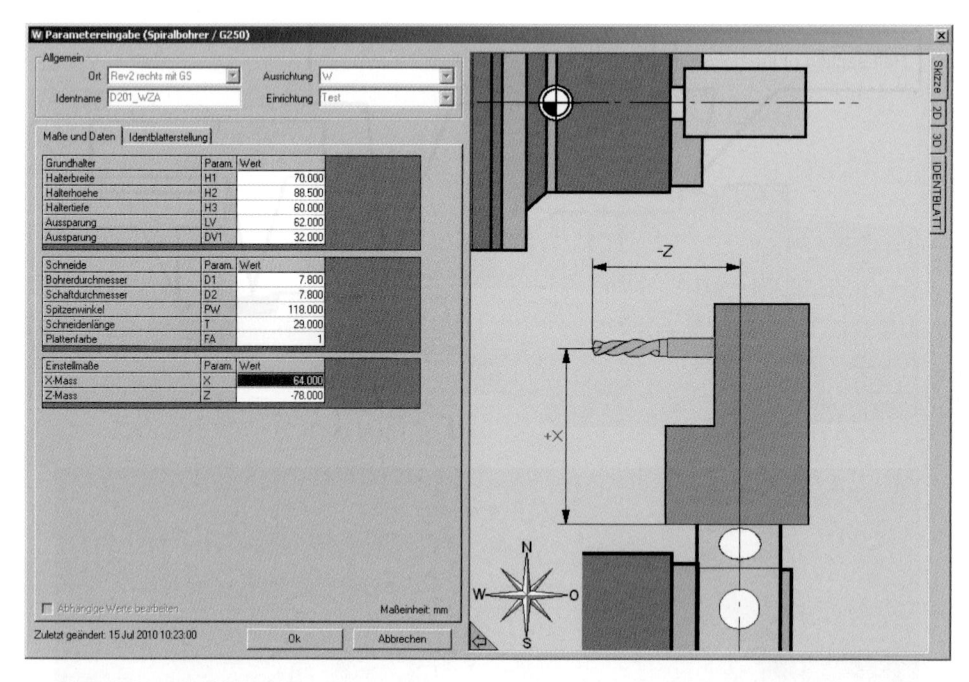

Figure 4.6: Tool wizard for generation of 3D tools.

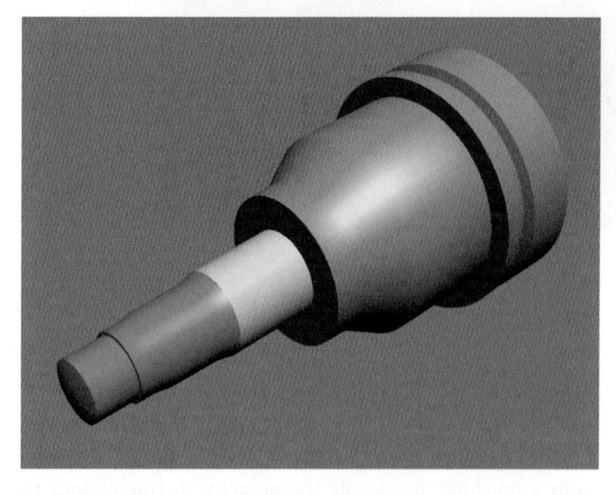

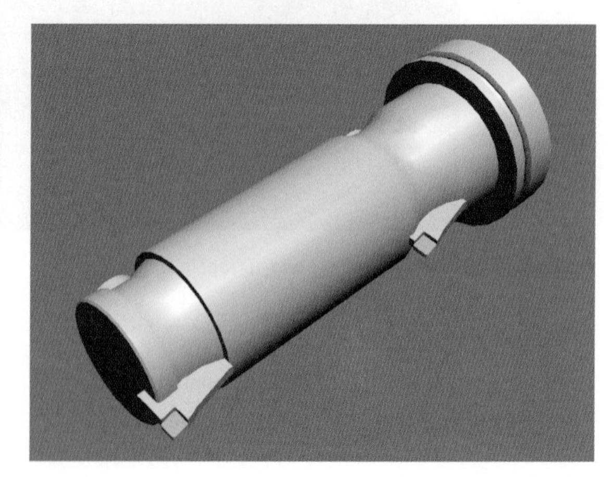

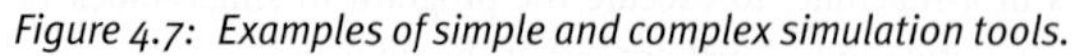

Figure 4.7: Examples of simple and complex simulation tools.

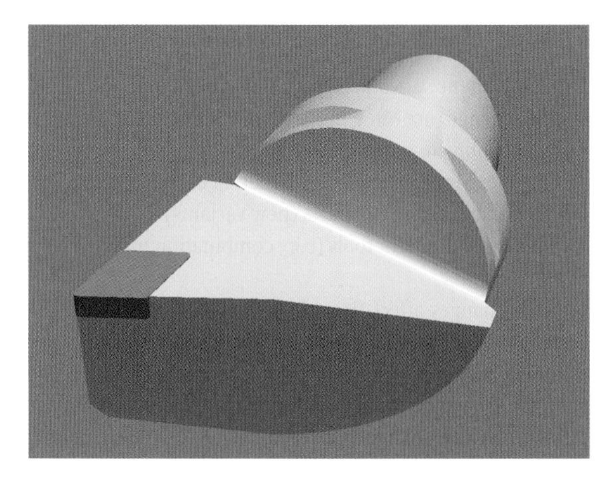

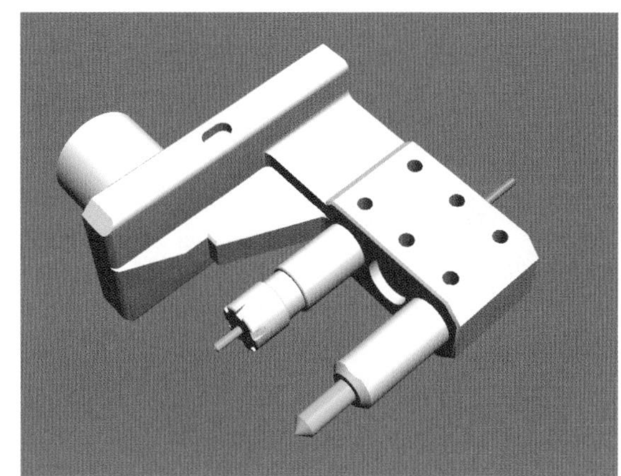

Figure 4.7: Examples of simple and complex simulation tools. (Continued)

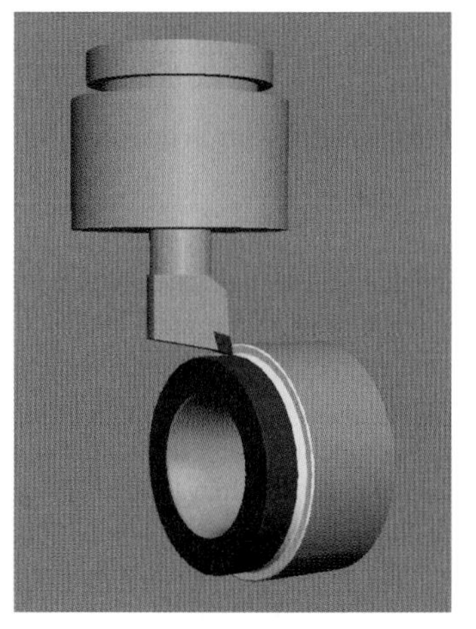

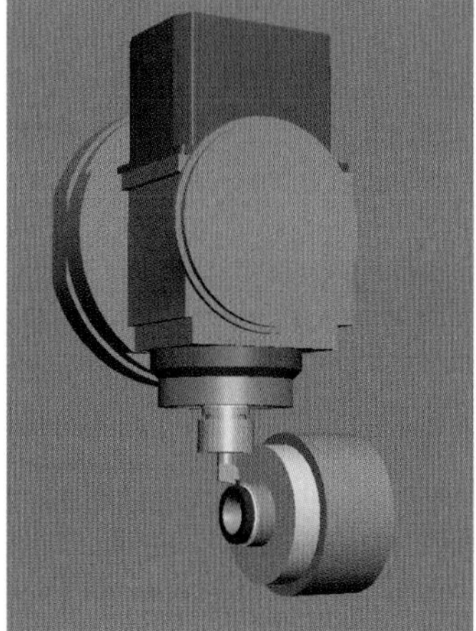

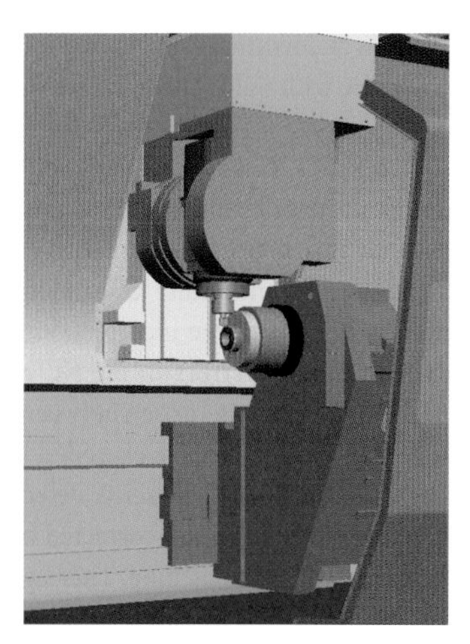

Figure 4.8: Levels of detail in the simulation.

4.5 Fields of Application

Running-in of New Programs

The use of NC simulation **before** the running-in of new programs has the following primary goals:

- Transferring the running-in processes from the expensive machine to the much less expensive computer workstation
- Parallelization of the setup process. The machine can continue production while the part is being run-in virtually.
- Reduction to a minimum of the run-in time on the actual machine (by up to 80 percent)
- Drastic reduction of the collision risk
- Reduction of the risk of reject production from faulty machining of expensive blanks (especially in one-off and small-series production)

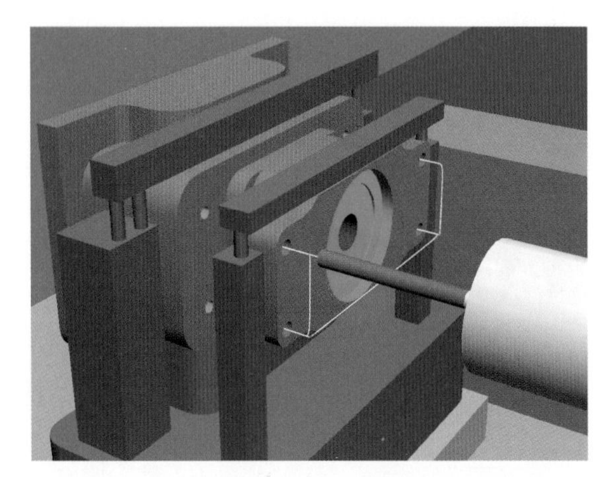

Figure 4.9: Material removal with adoption of the the tool cutting edge color and depiction of the tool path.

- Reduction of downtimes and costs for repairs
- Compliance with tight schedules thanks to shorter run-in times

Through the use of simulation, it is possible to detect and correct a wide variety of errors at an early stage. This includes both syntax errors and faulty coordinates, incorrect tool-correction switches, or even incorrect or missing zero-point offsets. In addition to errors in the NC program, it is also possible to analyze the interaction between clamping fixtures and tools. If the axis limits are exceeded by variation in the clamping position, then this fact is detected automatically. The errors thus described are corrected before the NC program is loaded into the machine. This significantly reduces the run-in time because troubleshooting on the machine is a very time-consuming process.

The actual potential savings naturally are heavily dependent on the conditions in the specific company in question. Before making a decision with regard to the use of an NC simulation, potential users should ask themselves the following questions:

- To what extent is it possible to reduce the programming time of new NC programs through the use of NC simulation? This question is especially relevant when no NC programming system is being used.
- What additional value creation can be generated from the machine capacity that is freed up?

Program Changes during Operation

Especially in the series-production phase of a manufacturing system, changes are often made to the NC program. This can be for the following reasons:

- Changes in the geometry and tolerances of the workpiece
- An expanded range of parts (new variants)
- The use of different tools (e.g., combination tools)

Such changes affect first of all the NC program and the tools; in some cases it is necessary to make changes in the clamping situation. Changes and/or expansions in the NC program are performed either during production planning or directly on the machine via an editor using the control panel.

Here, too the NC simulation offers the capability of making necessary changes on the PC ahead of time while the manufacturing system continues production. The modified NC program is transferred to the controller only after the necessary changes have been verified and accepted by means of simulation. This minimizes machine downtime.

Optimization of the Production Process

In series production, NC simulation is used not only for the verification of NC programs but also for cycle-time optimization. The exact geometric and temporal reproduction of the machining process in the simulation system makes it possible for the user to optimize motion sequences in the machine/machine periphery using methods that would be too risky or too complicated if NC simulation were not used. When one considers that a production process may have a service life of several years, even a single-digit percentage reduction in the cycle time may result in potential savings that far exceed the expense of achieving such optimizations.

The following methods may be used for optimization:

- Minimization of safety clearances
- Parallelization of motions (e.g., workpiece positioning parallel to table rotation)
- Avoiding tool changes through exchange of operations
- Optimization of use of linked systems

Planning Phase in Series Production

In series production, the system manufacturer is required to provide not just the physical components of the production system, that is, CNC machines, interlinking, and transport media, but

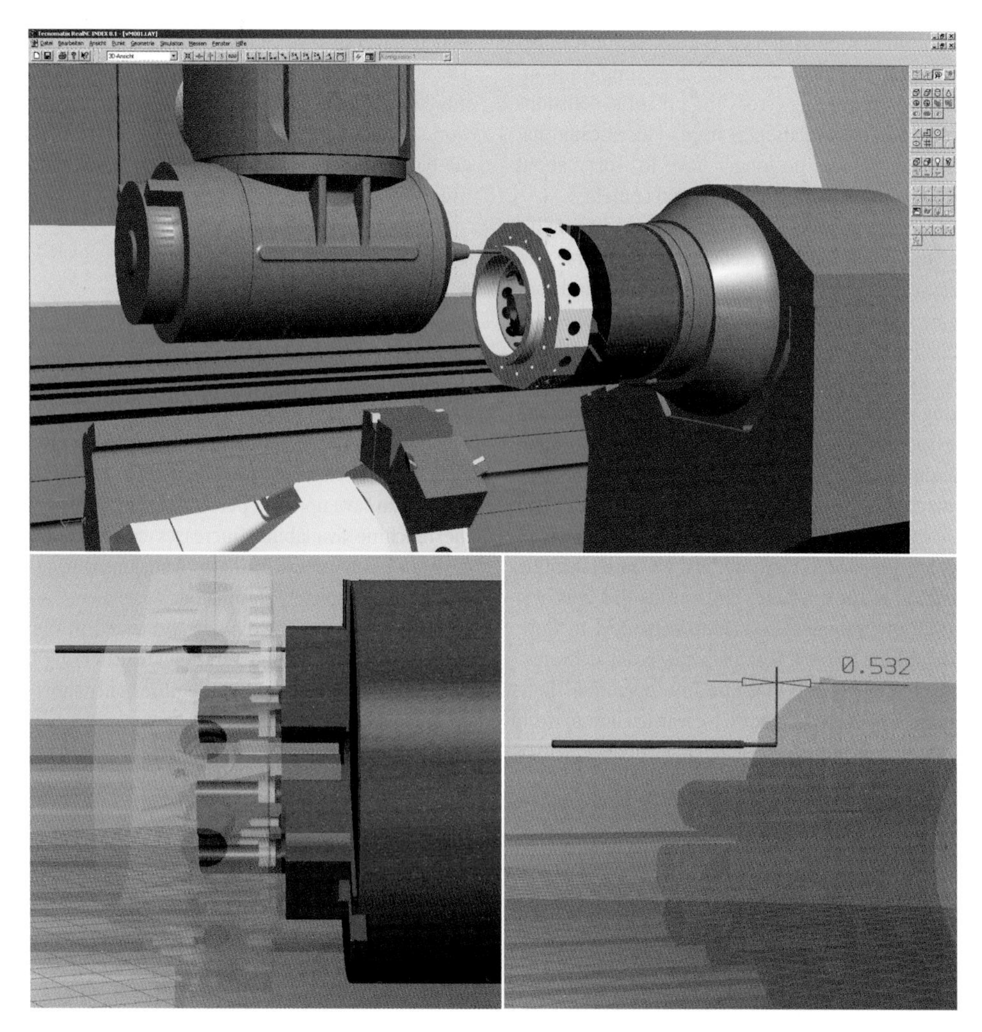

Figure 4.10: Depiction of a collision. The drill pierces the workpiece and drills into the chuck.

also implementation of the entire production process with all NC programs. Here the system manufacturer's task is, among other things, to put the system into operation successfully and to fulfill predefined acceptance criteria such as units produced per unit of time and compliance with product quality.

Today, many system manufacturers are already using NC simulation in the development and planning phase. The main emphases here are first on the development and verification of machine concepts and second on the validation and optimization of production processes before the machine is put into operation.

Training

Training departments generally do not have the resources to provide a wide range of modern NC machines for training purposes. If machines are available, then faulty operation often leads to long downtimes and high repair costs.

With an NC simulation system, it is possible for trainees to learn how to work with NC machines in as simple a manner as possible. Programming errors and potential collisions are no problem when simulation is used. This means that both trainers and trainees no longer have to worry about errors and thus can concentrate on the actual contents.

What is more, it is possible for each trainee to work on his or her "own" machine. Training becomes more diverse and realistic because it is possible to simulate machines and controllers that are generally not available in training mode.

NC simulation is also used in production as a means for user training. Especially when it comes to complex, abstruse manufacturing systems and transfer lines, system operators can obtain exact information on the processes in the system based on the simulation.

Virtual machines provide even greater potential. Because the virtual machine has use of the original operating panel of the machine, it is possible to teach not just NC programming but also operation of the machine. This means that it is also possible to train programmers and setup personnel to work with the new machine before it is actually physically delivered. It is then possible to put the machine into operation more quickly because the operators already have learned how to work with it and have already programmed the first workpieces.

4.6 Summary

Components of a Simulation Scenario

A simulation scenario represents the entire production facility. This includes

- The machine model with all geometric elements and axes
- The controller emulation or virtual controller with machine operating panel
- All workpieces and associated clamping fixtures
- Tools, drill heads, adaptor heads

Procedure for NC Simulations

All machine motions are depicted in three dimensions with maximum realism. The material-removal simulation shows how the workpieces change during the machining process. The automatic collision-detection process reports all impermissible contacts between the components located in the workspace. The simulated NC programs are executed in single-block or automatic mode and can be edited easily at any time.

Further options include

- Depiction of the tool trace
- Detection of technological errors
- Simulation in the interpolation cycle

Fields of Application and Potential Use

The majority of work when running in new programs takes place at cost-effective PC workstations. In this manner, it is possible to save up to 80 percent of the running-in time, and the machine availability increases correspondingly. Repair costs and downtimes owing to collisions during running in are almost completely eliminated.

Program changes are made and tested in the simulation system during ongoing operation. Transferring the changes to the production equipment requires only short downtimes on the machines.

In series production, NC simulation is used for cycle-time optimization, among other things. In the simulation scenario, safety clearances are minimized, feed motions are parallelized, and the machining sequence of the tools is exchanged. The prerequisite for this is the exact depiction of the action-control response with regard to motions and temporal response.

The main emphases in the planning phase are first on the development and verification of machine concepts and second on the validation and optimization of production processes before the machine is put into operation.

During vocational training, the ways of working with NC machines are imparted in a simple manner by means of NC simulation systems. Programming errors and collisions are no problem here. Both the trainers and the trainees can work without pressure and can concentrate on the actual contents of the instruction.

Manufacturing Simulation

Important points to remember:

1. **Simulation** refers to the reproduction of a system together with its dynamic processes in a model capable of experimentation in order to gain knowledge that can be transferred to reality (VDI Standard 3633).
2. **Manufacturing simulation** refers to the graphic dynamic reflection of an actual system using an model, for example, for
   - The design of components and their assembly
   - Work scheduling, handing, clamping, and production of workpieces
   - Structural analysis and response of components
   - Investigation of the kinematic and dynamic behavior of bodies (FEM)
   - The optimal design of flexible manufacturing systems (material flow, machine layout)
3. **Manufacturing simulation** comprises the following for manufacturing technicians:
   - The graphical-dynamic process simulation of NC programs on a specific machine
   - With depiction of the machine, workpieces, and tools in the correct relative size for analysis of the motion behavior during machining
   - In real time, time lapse, or slow motion with stopping if desired for detection of problems
   - Including execution of tool and workpiece changing
   - For the purpose of detecting and eliminating empty runs, safety reserves, collisions, programming errors, and so on
4. **Simulation of NC programs.** Goals:
   - Help in the time optimization of NC programs
   - Detection/avoidance of collisions
   - Reduction of the test phase for new programs
   - Increased machine service lives
   - Cycle-time optimization of FMS through coordination of machining times with multimachine processing
   - Check when special tools are used as regards the geometric dimensions and during machining
   - *As a general rule:* preventative damage control and increase in productivity
5. **A distinction is made between the following:**
   - Process simulation emulates complex production systems in such a way that it is possible to perform optimization with regard to the arrangement of the machines, the design of the overall system, and the sequences.
   - Graphical **3D kinematic simulation** for investigating the kinesic behavior of systems, such as robots, machines, or even humans.
   - **FEM simulation** for modeling and analysis of the physical response of materials and complex structures.
   - **Multibody simulation** of bodies that are connected to each other via joints in order to investigate the kinematic response, including physical effects of accelerations, static loads, and overall assessment of machine designs.

Part 6

Integrating Computer Numerical Control Technology into In-House Information Technology Systems

1. Direct Numerical Control or Distributed Numerical Control

Edgardo Mantovani

Computer networking now has become a general standard. This networking technology also can be used for numerical control (NC) machines. It offers many advantages that no manufacturing operation should fail to exploit.

1.1 Definition

The abbreviation *DNC* originally stood for **direct numerical control.** Recently, terms such as *distributed numerical control* and *distributive numerical control* also have been used. This term refers to a mode of operation in which multiple NC- or CNC-machines and other production equipment (e.g., tool-presetting devices, measuring machines, and robots) are connected to a computer by cables. The **direct transmission of data** has eliminated the data-storage media that previously were used commonly (e.g., punched tapes, magnetic tapes, and diskettes) as well as the devices required for reading and writing to these media. This brought several technical and economic benefits.

According to the VDI 3424 Standard, the essential **feature of direct numerical control** is the "management and timely distribution of control information to multiple NC machines, in which the computer can assume responsibility for the numerical control functions." The last phrase no longer applies to modern DNC systems, however. Machine control functions now remain within the purview of the CNC system.

The data networks and powerful DNC software allow all systems connected to the local-area network (LAN) to communicate with one another.

1.2 Functions of DNC

Although the specific technologies have changed greatly in recent years, the basic functions of DNC systems have remained the same even today. A DNC system has to fulfill two basic tasks:

- **Guaranteeing secure, timely data transfer from and to the CNC controllers**
- **Administering many thousands of NC programs**

The first task, guaranteeing secure data transfer, protects the company in question against the possibility of expensive damage to the machine and workpieces. NC program management, on the other hand, ensures the proper organization and storage of what are generally large amounts of data that represent a significant monetary investment. Both these tasks can be solved by means of a modern system, thus contributing significantly to increased production and quality assurance in the manufacturing process.

The VDE guideline that appeared in 1972 already made a distinction between basic functions and expanded functions of DNC systems, such as tool management and workpiece management in highly automated manufacturing systems. This will be discussed in greater detail below.

1.3 Application Criteria for DNC Systems

A company's requirement for introducing a DNC system can be based on a number of criteria. These include

Frequent program changes. The smaller the lot size, the greater is the problem of always having the correct NC program available at the right machine at the right time. In DNC operation, the NC program is available in the machine immediately after being called. In expanded systems, all the additional data, compensation values, and operator information are also available.

Number of NC and CNC machines. Profitable DNC operation can begin even with two to three NC machines, provided that a number of program changes are necessary every day. The arguments for DNC increase with each additional machine. In fact, some manufacturing processes assume that data will be provided via a DNC system.

Number of NC programs. The problems of administering several thousand NC programs with the associated modifications, updates, and changes can hardly be managed without the use of computers. A DNC system makes this work easier and minimizes the risk of human errors.

Program length. If programs are too large so that a number of portable data-storage media would be required, then there is a risk of confusing them, with potentially costly consequences. If the programs are larger than the memory capacity of the CNC machine, then a DNC system may be essential even for a single machine in order to be able to operate without interruption even for several hours.

Many new programs. If a company depends on many new programs or frequent program modifications, then the direct transfer of programs from the computer-aided design (CAD)/computer-aided manufacturing (CAM) system to the CNC system is essential. Especially with shop-floor programming, that is, programming directly at the machines by the machine operators, it is very important to be able to save the programs that are generated.

High transmission rates. Especially high speed cutting (HSC) machines and laser technologies require an extremely high data throughput. It is therefore necessary to transfer the NC program data to the CNC system very quickly so that the machining process does not stop owing to a lack of data. This requirement can only be satisfied with DNC systems.

Computer-aided tool management. Integrated management of tools and tool data can provide enormous cost savings. Better use can be made of tool lives, and unnecessary breakdown and assembly are avoided. The summarized transfer to the CNC systems of tool numbers with all the tool data reduces the time requirements and increases the safety.

Flexible manufacturing systems. Flexible manufacturing systems represent a special group that is categorized by that fact that both basic DNC functions and the recording, storage, and management of pallets, compensation values, measurement data, and so on are performed by a special **host computer.** This is generally provided by the supplier of the overall system and is equipped with a specially adapted, expanded software.

In this case, the DNC computer provides not just NC programs, subprograms, and cycles but also current tool data, zero-point offsets, and correction values. In the case of flexible manufacturing systems with complementary machining operations, it is necessary to provide each machine with a number of appropriate program sections for each workpiece.

With appropriate software expansions, it is possible for the flexible manufacturing system (FMS)/DNC computer to also make backups of the current-status data for each machine, such as the placement of the tools in the magazine, the interruption point for the machining operation, the compensation values, and other data. This provides substantial time savings in the event of data loss in the CNC memory if the machines have to be restarted.

As a general rule, DNC is a fundamental component of computer-integrated manufacturing (CIM) and should be included in the overall concept from the very start. DNC systems must be viewed as essential when one considers that the increased use of CAD/CAM systems, tool-management systems, and presetting devices means that increasingly large amounts of data have to be processed and made available in less and less time.

1.4 Data Communication with CNC Systems

Initial analyses were aimed at making NC path controls less expensive through the use of central computers and to provide the remaining **residual controllers** with precalculated data. Residual controllers were no longer able to function autonomously, however, which led to major problems during commissioning at the machine manufacturer's plant and in the event of computer failure. It also became clear that unexpected bottlenecks occurred already in the provision of data to a few controllers. The rapid drop in prices for CNC systems marked the final end of the DNC principle using residual controllers.

Before standardized serial interfaces became available, NC control systems were equipped with punched-tape readers, and the NC programs were loaded to the controllers by means of punched tapes. Later, after the RS232 (V.24) serial interface established itself as the standard, it became the most generally used interface because it was designed bidirectionally and could be used either for saving programs or for reading them out. The early DNC systems based on the minicomputers available at that time—which were very expensive in comparison with today's PCs—were already multitasking-capable and equipped with multiple serial interfaces. They were thus capable of handling several data transfers simultaneously and were operated from a single terminal. These minicomputers also were used to store the NC programs. All these systems operated with transfer technology; that is, the controller was set to the operating mode "read-in," after which it was possible to initiate data transfer to the machine from the terminal.

Because serial data transfer is very subject to interference—RS232 guarantees reliable data transfer only over 15 m and has only rudimentary data validation—the manufacturers of the so-called mold construction controllers (e.g., Bosch, Heidenhain, Fidia, etc.) soon realized that the data transfer had to be secured by means of a protocol. Therefore, the older controllers of these manufacturers transmit in a block-block fashion; that is, each data block is provided with a checksum that is recalculated by the controller and acknowledged as regards the computer. If a deviation is detected, then the controller automatically demands repetition of the packet. These protocols (FE1, FE2, LSV-2, etc.) do in fact result in a lower transmission speed but also ensure error-free transmission. They also make possible the reading in of excessively long programs that are larger than the memory capacity of the controller.

Thanks to the long service lives of NC machine tools in today's workshops, it is possible to encounter almost all the transfer modes that were used with older NC machines, such as

- BTR interface (BTR = behind tape reader)
- RS232 (V.24), serial interface for data input and output
- Ethernet interface (if the CNC controller is PC-based)

For CNC machines that are integrated into a network, that is, when the controllers have an Ethernet connection, the entire spectrum of operating systems and transfer protocols can be encountered. Thus there are controllers based on Unix, Linux, DOS, and Windows. While the majority of Japanese controller manufacturers decided in favor of FTP as the transfer protocol, European manufacturers use all the other protocols (Netbios, Netbeui, and NFS). This diversity should be taken into account when selecting a suitable DNC system.

1.5 Methods for Requesting Programs

While, previously, punched tapes, cassettes, or floppy discs were used, today it is possible to read NC programs into the controller directly without having to deal much with any data-storage media as long as the controller is set to the operating mode "reading." Two different methods are used today to load programs into the program memory of machines with a serial interface:

The transfer method. Here the programs are sent by the computer (or by the terminal that is assigned to the machine). In other words, the controller first has to be set to "read in," after which the program being loaded is sent to the machine by an external communication device.

The call-up method. This method is used in systems without terminals. Here, a calling program (also called a *dummy program* or *runner program*) requests the program that is to be loaded. The DNC computer makes available the program specified by a comment in the request program and then waits for the operator, who still has to set the controller to "read in." This method has now become established as the standard and is used widely.

Network machines with PC-based CNC systems can access the server directly via the network. It is thus possible to load and save programs directly from the controller with the aid of a few softkeys.

1.6 DNC Systems Currently Available

All current DNC systems connect the CNC machines to the computers via either a standard network or serial cables. There are fundamentally three different concepts, which we will now describe.

DNC Systems with Serial Cabling *(Figure 1.1)*

These systems are encountered primarily in small installations with only a few machines. Generally, a single PC is used as a central computer for saving programs and as a communication device. This computer often is equipped with a serial multiple interface card and connects the CNC controllers via RS232. If copper cables are used, the transfer rate is usually restricted in order to increase transfer reliability, thus increasing the service life of the machines. Therefore, fiberoptic cables are also used instead of copper cables to eliminate interference.

Fixed cabling is particularly disadvantageous when the DNC computer has to be moved to a different location. This type of DNC system is suitable for small companies with only a few machines, where the computer is located close to the machines (ideally less than 15 m). Both the transfer method and the call-up method can be encountered in such systems.

DNC Systems with Terminals *(Figure 1.2)*

These DNC systems, which were seen most commonly during the 1990s, use a terminal (IPC = industrial PC) for data communication between the server and the machines. Used for so-called paperless manufacturing, they bring all the manufacturing information to the machine. The operator can view all the manufacturing-related data and documents and can

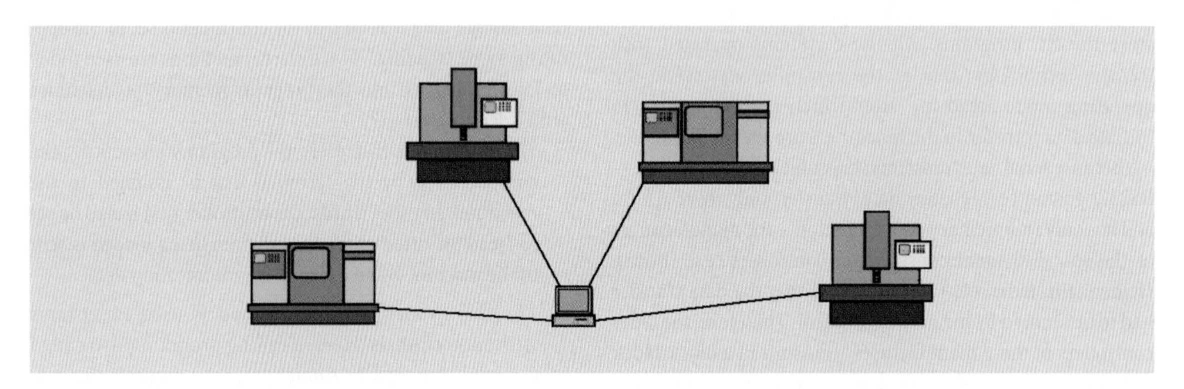

Figure 1.1: DNC system with serial cabling.

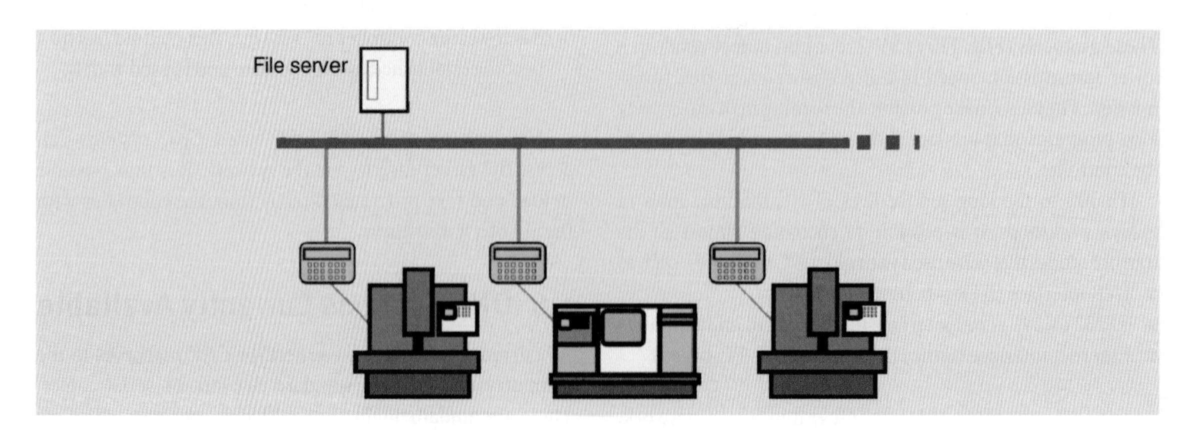

Figure 1.2: Terminal DNC system.

use the terminal to load the NC programs. Often these systems are also supplemented with manufacturing data acquisition (MDA)/production data acquisition (PDA) as well and can tap the relevant signals from the PLC while also enabling feedback of reasons for stoppages from the machine to the central computer.

For a long time, this type of DNC was the only one that avoided the disadvantages of long serial cables because it was based on network technology from the very beginning. These terminal DNC systems provide maximum convenience but require good personnel training and are, accordingly, expensive when first acquired.

Network DNC Systems *(Figure 1.3)*

These systems use network adapters for data transfer; that is, a controller is integrated into the shop's internal Ethernet with the serial interface by means of a **network adapter** (also called a *device server*, *com server*, or *terminal server*). This makes it possible to request the NC programs directly from the controller with no intervening terminals. These systems overcome the typical problem of transmitting data over long distances because correct data transmission is ensured by the network technology. Because they are relatively inexpensive, these systems are equally suitable for both small and large companies. A wireless LAN (WLAN) is also often used today instead of conventional Ethernet cabling. Modern PC operating systems (mostly Windows, rarely Linux or Unix) allow transfer to or from multiple machines simultaneously.

1.7 Network Technology for DNC

(Figure 1.4)

In standard networks (today practically exclusively Ethernet), protocols also secure error-free transmission of large quantities of data over long distances. The Transmission Control Protocol/Internet Protocol (TCP/IP) on which networks are based today ensures transmission even in industrial environments where there are all kinds of interference, guaranteeing absolutely error-free communication.

Today's DNC systems use almost exclusively Windows-based computers and standard networks (LANs). In order to connect CNC machines that have only a serial communication interface, network adapters (i.e., com servers and device servers) are used as media converters. These transform the data from serial to Ethernet and vice versa. Highly specialized network adapters contain the actual DNC functions and operate like separate computers with their own data-transfer program and implemented filter functions.

WLAN—Wireless Local-Area Network *(Figure 1.5)*

Another standard type of network today is WLAN. Wireless networks are Ethernet networks that check the data traffic using the TCP/IP. They operate in the gigahertz range and transfer data without any direct cable connection. This is done via an access point that forms the bridge between the cable network and the wireless network. The access point has transmitter and receiver functions and also can be used to connect two cable networks. On the receiver end, at the CNC machines,

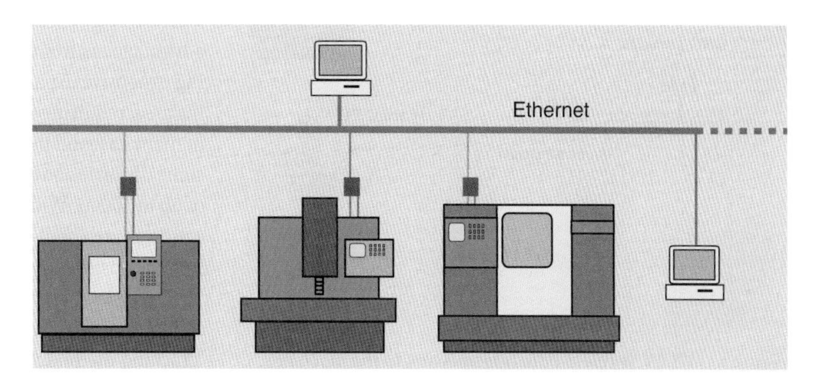

Figure 1.3: Network DNC system.

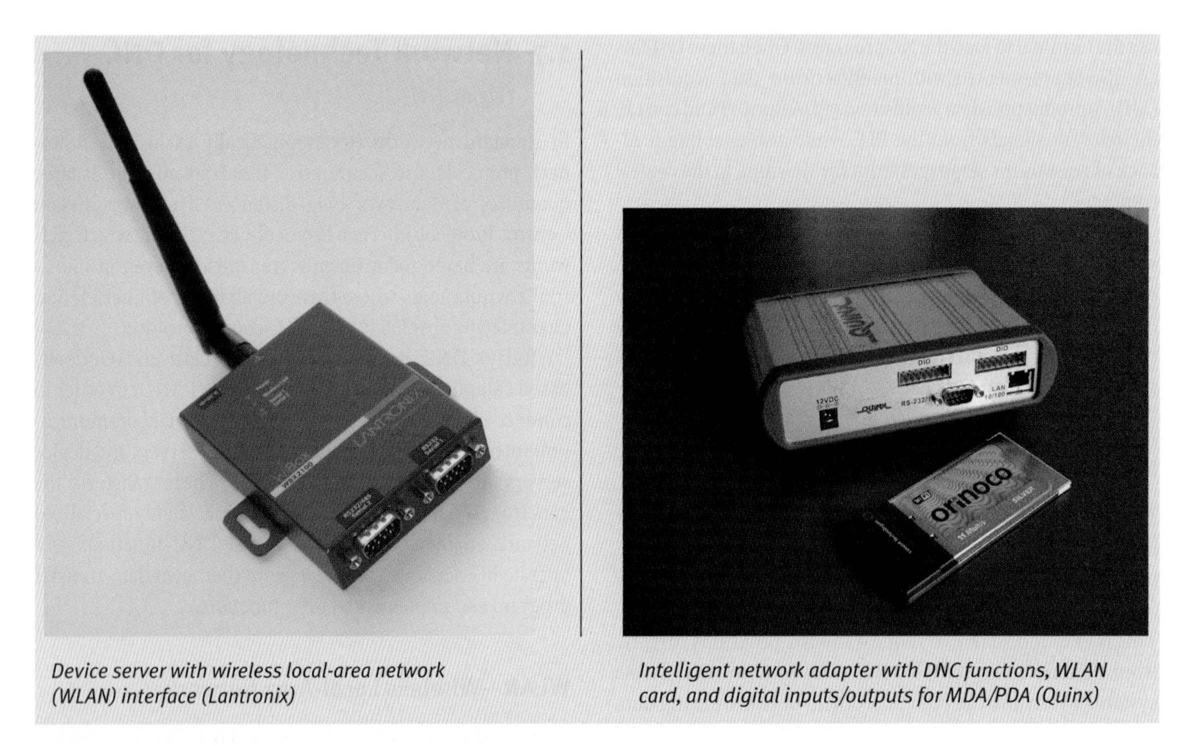

Device server with wireless local-area network (WLAN) interface (Lantronix)

Intelligent network adapter with DNC functions, WLAN card, and digital inputs/outputs for MDA/PDA (Quinx)

Figure 1.4: Modern devices for setting up a DNC system.

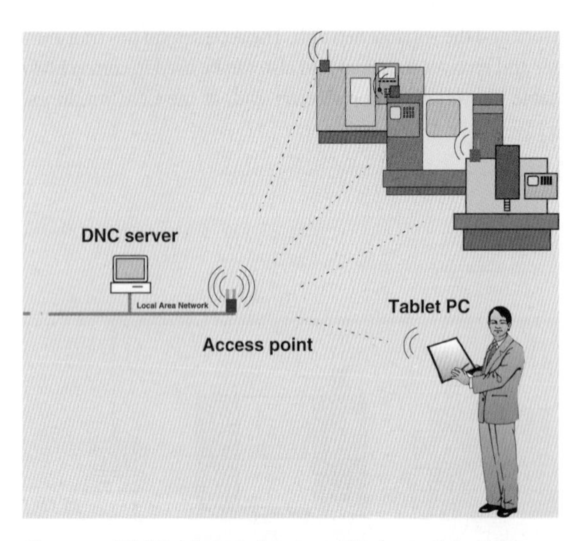

Figure 1.5: WLAN, wireless data transmission to the machines.

either network adapters with a built-in antenna are used or else conventional Ethernet adapters such as those described earlier. This is in addition to the WLAN-Ethernet bridges that convert from the WLAN to the cable-based Ethernet.

Benefits and Risks of WLANs

The use of a WLAN, that is, a wireless network, is especially advantageous from the point of view of cost. Instead of making expensive expansions to the network in the manufacturing area, requiring the laying of cables and the use of switches, in many cases all that is needed is an access point. Installed at some distance above the floor, receivers can operate at ranges of up to 100 m. With conventional Ethernet cabling, the average costs are about €300 to €600 per connection. In comparison, the costs for WLAN are only half as much.

The **risks** should not be underestimated, however. Because computer magazines are always keen to make public the latest "cracking methods" for various types of data encryption, we have to assume that no encryption will remain secure in the long run. This leaves the door wide open for industrial espionage. Moreover, in heavily built-up areas various WLANs may overlap and interfere with each other. This can lead to lost production if no emergency strategy is implemented. And finally, this factor could be used intentionally to disable a competitor's production at least temporarily.

DNC for Controllers with an Ethernet Connection

For machines with an Ethernet interface, the question is often asked whether DNC is even necessary at all because access to the computer for loading and saving programs is easy to implement. To be sure, quick and error-free data transfer is no longer a problem with this type of machine, but in practice, the following problems are encountered frequently:

- Because the data memory is very large, no regular data backups are made. A defective hard disk can result in an expensive loss of programs.
- The machine tender is granted full access to the program directories and can use the controller to delete or overwrite programs on the central computer or move them to locations where they are impossible to find. This can represent a security risk.
- As with all PCs in any company, PC-based controllers are vulnerable to computer viruses.
- Without a DNC system, transfers can no longer be traced. No logbook is kept.

While the first three problems can lead to expensive losses, the last point has to be taken into account for the International Standards Organization (ISO) 9001 Standard. There are therefore many companies that manage all their machine programs via a DNC system. Other companies avoid the risk of viruses by requiring connections via serial interface in all cases.

1.8 Advantages of Using Networks

The use of **standard networks** (Ethernet) offers many **advantages:**

- Absolutely error-free data transfer thanks to automatic error detection and correction
- Centralized management of all NC programs and production data
- An unlimited number of devices can be connected.
- Networking of machines and devices even over long distances
- Use of the maximum transfer rate of the controller's serial interface
- Trouble-free expansion of the system to suit the company's needs
- Direct communication between multiple computers, such as CAD/CAM systems, NC programming computers, production-planning system (PPS) computers, shop computers, DNC computers, and the CNC systems
- Simplified centralized management of all data and network devices in the company (computers, machines, etc.)

1.9 NC Program Management

Often too little attention is paid to data management of programs. Because older controllers only worked with four-digit NC program numbers, data management was previously adapted to the limited capabilities of the controllers, which meant that significant organizational work was required during production planning. Four-digit numbers generally do not allow the use of article numbers or part numbers, which generally have many more characters. The result of this was that comparison lists had to be used or complicated management systems had to be developed in order to assign the appropriate part program to the article. These old habits continue to exist in many places.

Modern DNC management systems overcome the limitation of four-digit NC program numbers and use a comments line within the program for unambiguous identification of the NC program independent of the NC program number. This makes it possible to use the information contained in every production order to request the required NC program for the part being manufactured without the operator having to know the NC program number.

Modern network-based DNC systems manage NC programs almost completely automatically. New programs that are sent from the machines or transferred from the CAM system are automatically placed in the database and are immediately ready for loading into the machine. Modified programs are automatically compared with the original and stored in a separate area. A click of the mouse will show all the changes so that the person responsible for the NC programs can enable

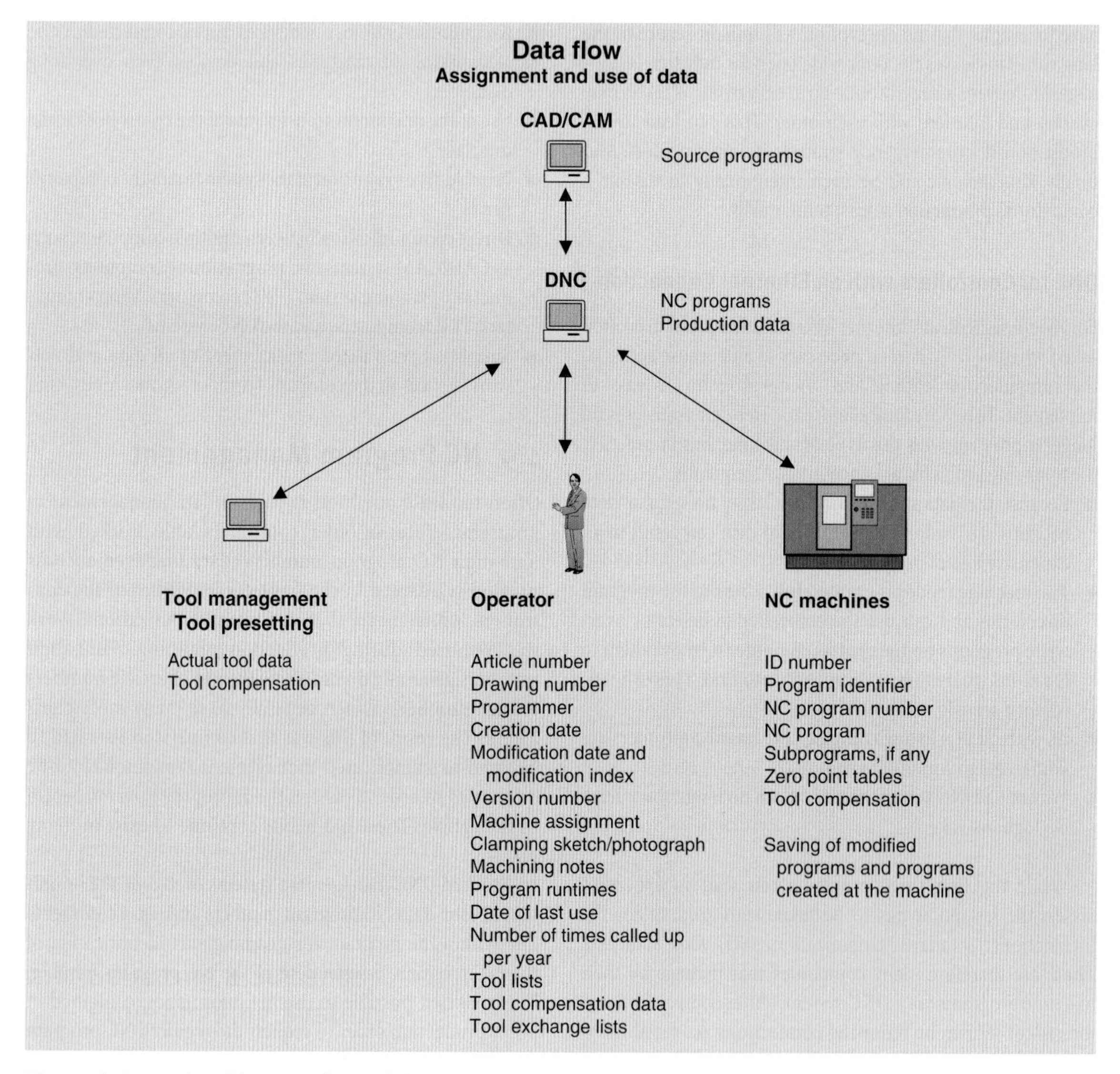

Figure 1.6: Categories of data according to their assignment and use in production.

the one that is desired. As long as a duplicate exists, the original program will remain disabled. Programs that have a missing or invalid identifier will be automatically detected and segregated by the DNC system.

Modern, efficiently designed DNC systems reduce the amount of handling required for NC programs by more than 90 percent compared with conventional work methods (based on research by Quinx AG). This means that these systems pay for themselves extremely quickly.

A detailed logbook listing all transactions is essential, especially when traceability is required, as in ISO 9001 and for medical technology (DIN EN ISO 3485:2003). In this manner it is possible to determine at any time what part was produced on what machine and using which program version.

Requirements and Functions *(Figure 1.6)*

Modern NC data-management systems

- Manage NC programs organized by machines or groups
- Manage manufacturing information (documents of all kinds that belong to the NC program)
- Manage NC programs according to a logical identifier in accordance with the conventions of the production-planning system (PPS)/enterprise resource planning (ERP) system
- Indicate ongoing transfers
- Automatically disable and enable programs that are in use (are being transferred or are open in the editor)
- Are designed for multiuser operation
- Allow integration of external editors (e.g., the CAM system)
- Show at the click of the mouse all the changes in a modified program (program comparison)
- Provide import and export functions for programs and documents
- Provide automated data import from CAM systems, tool-management systems, and presetting devices from a wide range of manufacturers
- Keep a logbook listing all transfers (actions of the PC users, loading and saving of programs from the machines, etc.)
- Automatically archive superseded program versions
- Maintain statistics concerning file transfers

Examples: *Figures 1.7 to 1.10*

1.10 Advantages of DNC Operation

Introduction of a DNC system provides the following advantages:

- Higher productivity thanks to shorter changeover times
- Absolute reliability of data transfer over long distances
- Programs read in at maximum speed
- Simplified data management
- Reduction of tedious, routine work to be performed by the person responsible for the programs
- Traceability with detailed transfer log (ISO 9001, DIN EN ISO 13485:2003)
- Absolutely safe from incorrect interchanging of data media
- Error-free data entry even at maximum transfer speeds, for example, during operation of high-speed cutting (HSC) machines
- Guaranteed use of the most current programs
- Simple, automatic, clearly structured program management
- Quicker availability of programs and compensation values
- No machine downtime owing to missing programs
- Trouble-free availability of tool data and compensation values
- Avoidance of extensive libraries of punch tapes or floppy disks, including the associated cabinets

For systems with an expanded scope of functions, advantages include

- Better utilization of tool-service lives thanks to a closed data circuit
- Minimization of tool changes when the program is changed
- Better transparency of information, especially with a view to interlinked production systems
- Operation of NC machines that is generally more flexible and more fully automated
- Greater use times of machines

1.11 Cost-Effectiveness of DNC

The total cost of a DNC system today ranges from about €1,000 to €5,000 per machine depending on the manufacturer, the CNC systems to be connected, and their interfaces.

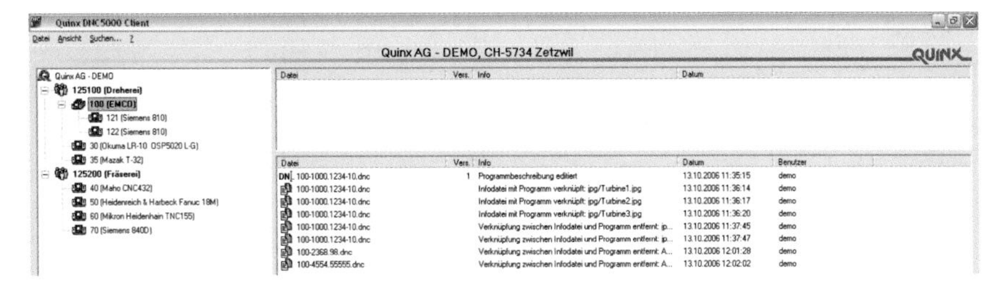

Figure 1.7: Machine overview (left); ongoing transfers (right).

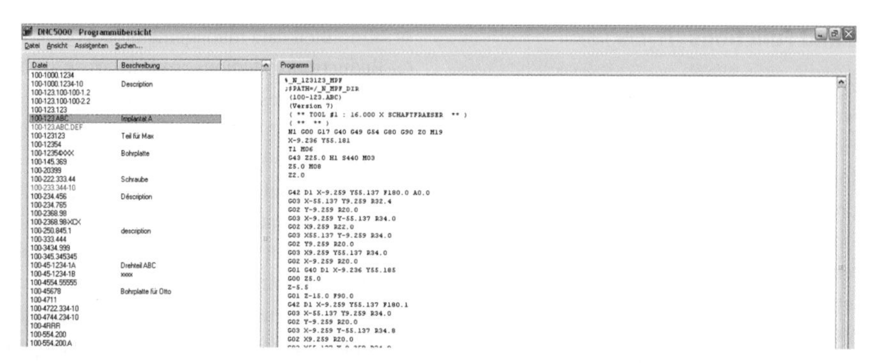

Figure 1.8: Machine directory and file details.

Figure 1.9: Production data for the highlighted NC program (e.g., digital photograph of a clamping situation).

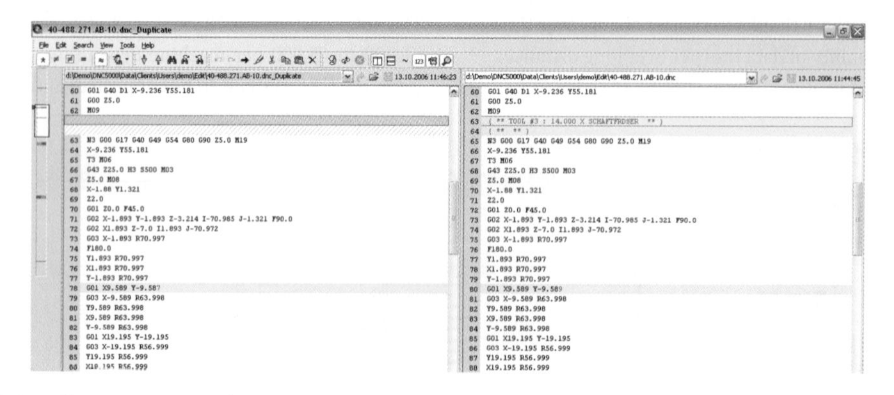

Figure 1.10: Automatic program comparison.

If the hourly cost of operating a machine is €150 and the increase in productivity is only 2 percent of 2,875 hours, then for a total investment of €15,000, the payback period is only two years in two-shift operation.

Each purchaser has to calculate the cost-effectiveness in accordance with his or her own situation. Inexperienced users should be aware of the fact that **entry-level DNCs** are available that are easy to expand if the experience with them is positive. The most important thing here is compatibility with the equipment that has already been bought and the portability of any special software that has been created to a follow-on system.

The following criteria can be used to **evaluate** a DNC system:

- **The DNC hardware:** Because not all manufacturers' products are able to cope with the tough conditions in production shops (e.g., interference resistance, temperature, physical shocks, continuous operation, atmosphere), the computer should be able to function using a standard operating system, and the electronic devices (i.e., network adapters, terminals, etc.) should be CE approved
- **The DNC software,** which should be modularly expandable, service-proven, and error-free and above all should fulfill the requirements as defined in writing
- **The transmission medium,** for example, coaxial cable, twisted pair, fiberoptic
- **The protocol that is used** (standardized if possible, such as Ethernet with TCP/IP)
- The manufacturer or supplier, which should have sufficient experience, its own development personnel, a well-thought-out product, a stock of spare parts, and adequate service support
- **An acceptable price** for good quality, which is a precondition for a short payback period

The **emergency strategies** in the event of failure of the DNC computer, the transmission line, or the connection terminal also must be subjected to a critical review.

1.12 Current State and Trends

Today, DNC systems are an integral component of industrial information technology (IT). They supply the NC machines and personnel with all the data related to production. They use high-performance industrial computers as a universal management and distribution system for all manufacturing data. Their scope of functions has been expanded considerably compared with earlier DNC systems. Reliability and speed also have increased. For use in flexible manufacturing systems, there is an option that allows access to all the manufacturing-related data.

The goal for the next few years is **digital manufacturing.** What this involves is that data generated with CAD are processed into NC programs with the aid of computers and can be passed on to the CNC with no errors via a DNC system. New technologies such as HSC, rapid prototyping, and laser applications require huge amounts of data and very high transfer rates. Punched tapes are completely inadequate for this. Electronic data media such as magnetic tapes and floppy disks are unsuitable for use in production shops and are not economical because of the reader devices that are required. The best solution is network-based DNC systems.

Upcoming DNC systems therefore include the latest developments in the field of communications (i.e., network technology, Internet, intranet, etc.), but they also meet the requirements of existing older machines and controllers. In almost all DNC installations it is necessary to connect at least some older NC machines.

The new CNC systems are equipped with universal data communication functions and standardized LAN interfaces. This means that as with other network devices (e.g., CAD, PPS, tool management, NC programming, etc.), it is no problem to integrate them into the flow of information within the company.

There still will be a market for DNC systems in the future because DNC systems still will be required for a long time, even if new machines come with their own network connections. DNC systems and the network are already provided for during the planning process for new installations.

In production shops where large numbers of NC machines are already present, the path to DNC is already clearly marked. The same applies for the installation of new technologies, even for only a few machines. DNC systems are absolutely indispensable for flexible manufacturing systems. Flexibility can only be achieved if fast changes in machining are possible because the NC programs, tools and tool data, compensation values, and zero-point tables are always at hand. Pictures of the workpieces in various clamping positions also help operating personnel to become familiar with new workpieces more quickly.

1.13 Summary

DNC systems have three main tasks:

- Management of NC programs for all the connected CNC machines
- Timely transmission of NC programs to the CNC machines
- Activation of data transmission:
  - Manually from a central location
  - Automatically or manually from the CNC system

This means that it should be possible to make the data request for transfer of an NC program to a specific CNC system either from the computer or from each CNC system. The DNC computer does not perform or replace any of the functions of the CNC system. Even in flexible manufacturing systems, in which all CNC machines are supplied with data via DNC, the CNC systems still retain all their control functions. Each CNC system has a DNC connection for automatic data exchange with the DNC computer. Today, this is generally an Ethernet interface. If more than one machine calls up programs from the DNC system simultaneously, then the transmission sequence has to take place according to specified priorities.

The high performance of DNC computers means that they have been given more and more additional tasks as time has gone by. For example, in addition to transmitting the NC programs, it is also necessary to transmit the programs for the workpiece-handling devices and any measuring programs that are necessary for postprocess measurements in the machine. On top of this are the current compensation values for the length and/or diameter of the tools, tool wear, necessary zero-point offsets, operator notices, and clamping plans.

As protection against incorrect program transfers, additional information is necessary, such as

- Checking whether the program number is enabled for this machine
- Operator, time, and date of call-up
- Enabling of the new program for activation and processing of the parts
- Comparison of the required tools with the tools present in the tool magazine and outputting a "Tool XXX missing" message to the operator

Another very important task is automatic transfer of the manufacturing data-acquisition (MDA)/production data-acquisition (PDA) data from the CNC system to the central computer. This comprises the following:

- The running times and downtimes of the machine
- Production quantities, reject messages
- Interruptions of production with causes
- Machine downtimes with causes
- Error messages generated for statistical analysis
- Maintenance notes and monitoring of execution

For purposes of operator control, it is also important whether the communication between the CNC system and the DNC computer takes place via the keyboard integrated into the CNC system or whether an additional DNC terminal is necessary. Today it is more economical and more common to use the operator panel of the CNC system, but this does require an interface adapter for each brand of CNC. This also may involve different operator control inputs. Using DNC terminals at the machines has the advantage that the operator controls remain largely identical.

Last but not least, the type and speed of data transfer are important. To prevent data transfer from becoming a bottleneck, the transmission speed should be as high as possible. Absolutely error-free transmission is also important. Today, both these requirements are fulfilled by Ethernet connections. Systems with wireless transmission are also available.

Direct Numerical Control or Distributed Numerical Control

Important points to remember:

1. **DNC** is an abbreviation for **direct numerical control** using computers. This version is now long obsolete.
   a. Today it is interpreted as **distributed numerical control,** meaning that the scope of functions of the DNC can be distributed among multiple computers that communicate via a LAN.
   b. Today's DNC systems have clearly defined **tasks**, such as
      (1) Supplying the connected machines and automation equipment with NC part programs and other manufacturing-related data in a timely manner
      (2) Reducing waiting times for NC programs and other manufacturing data
      (3) Introducing a system of data organization with the help of meaningful program identifiers
      (4) Replacing portable data media with all their disadvantages, such as storage and management, damage, loss of data, and cumbersome handling on the machines
      (5) Managing any required number of NC part programs and amounts of manufacturing information in the DNC computer
      (6) Returning corrected NC programs from the machines to the DNC system with indication of the changes
      (7) Better information for operators at the machines
2. There are two options available for **bidirectional data transfer:**
   a. Star connections from the computer to each device via a multiple-interface card
   b. A local-area network (LAN), with the advantages of absolute data reliability and high transmission speed
3. **Wireless transmission** to and from the machines is also possible.
4. The CNC systems connected to the DNC **do not require any data-reading devices.** This eliminates downtimes for maintenance work and the cost of keeping spares available.
5. A choice of various systems is available as the transmission medium. **LAN interfaces,** such as for Ethernet with the TCP/IP are recommended.
6. Today's CNC systems must in any case have an integrated LAN connection.
7. In the case of extra long NC programs, uninterrupted operation is possible through the **automatic loading** of multiple program sections.
8. The **master data** or **header data** of NC programs serve the following purposes:
   a. Better management, identification, and labeling
   b. Prevention of improper use
   c. Providing information to the operator
9. **MDA/PDA** are not DNC expansions but rather separate function modules that use the existing DNC hardware and software.
10. DNC systems have a clearly delineated range of functions and do not include any FMS process-control functions, for example.
11. The cost-effectiveness of DNC systems is indisputable.

2. LANs—Local-Area Networks

Information is becoming increasingly important in production shops and can be considered an essential production factor. This information must be available to connected users at the right time. This task is performed by the company's internal data network.

2.1 Introduction

The computers and computer numerical control (CNC) systems that are used in the various computer-aided technologies (e.g., CAD, CAM, CAQ, CAR, CAI, and CAE) are **data-generating** and **data-processing systems.** As more of these devices are installed in a plant, it becomes increasingly important for them to be capable of exchanging data with one another. This is the **company's internal data network** (*Figure 2.1*).

Broadband cables and networks transmit multiple types of information simultaneously between various *transmitters* and *receivers*. Broadband networks of this type are comparable with familiar cable television networks. These cables also carry multiple television and radio programs simultaneously, in addition to videotext. From this multitude of information, each receiver extracts only the program to which it is tuned. However, this requires a **cable tuner,** which converts the programs from the cable frequency to the standard input frequency of the devices.

Although the transmission occurs in **simplex operation** (i.e., in only one direction, from the transmitter to the receiver), it certainly would be possible from a technical standpoint to transmit the data back again. This is called **duplex operation.** In other words, it would be possible to send information from the cable television customer back to the central office during television programs.

There is also a third possibility: **half-duplex operation,** in which only one station can transmit at a time (*Figure 2.2*).

One critical difference between LANs and cable television networks, however, is that the individual customers on a cable television network cannot communicate with one another.

2.2 Local-Area Networks (LANs)

LANs are data networks that are limited to a particular area. LANs never expand beyond the boundaries of the company's building lot and thus are free of any regulation by outside authorities.

Figure 2.3 provides an overview of the range of communication options. A connection between computers located **far apart** is called a **wide-area network (WAN)** and requires publicly available facilities such as telephone cables, ISDN or DATEX-P, or direct subscriber line (DSL).

2.3 What Is Information?

Information can take the form of data, images, drawings, or control programs. Information processing involves the control and documentation of processes and relationships in various areas within a company. Information is used to regulate inventory management and production (in production-planning systems), to control NC machines and robots (DNC, CNC), to log production data and breakdowns (MDA/PDA), and to reduce unnecessary downtimes (CAQ, diagnosis, maintenance), thus increasing the profitability of a manufacturing operation. This is also the reason for the increasing interest in data networking within companies.

What tasks have to be carried out by companies that want to gather and distribute information more quickly and then use it to carefully control manufacturing?

- The most important requirement is the **use of computers and peripheral devices, databases, and suitable software tools** so that information can be generated, stored, and distributed.
- It is absolutely essential to build an **information network** that allows all users to access all the information they need. As automation increases, more and more locations in the production shop will use this information and feed back their results. These data networks therefore must be reliable, fast, expandable, and secure.
- The devices that will be connected to the network must be equipped with **data interfaces** that allow them to

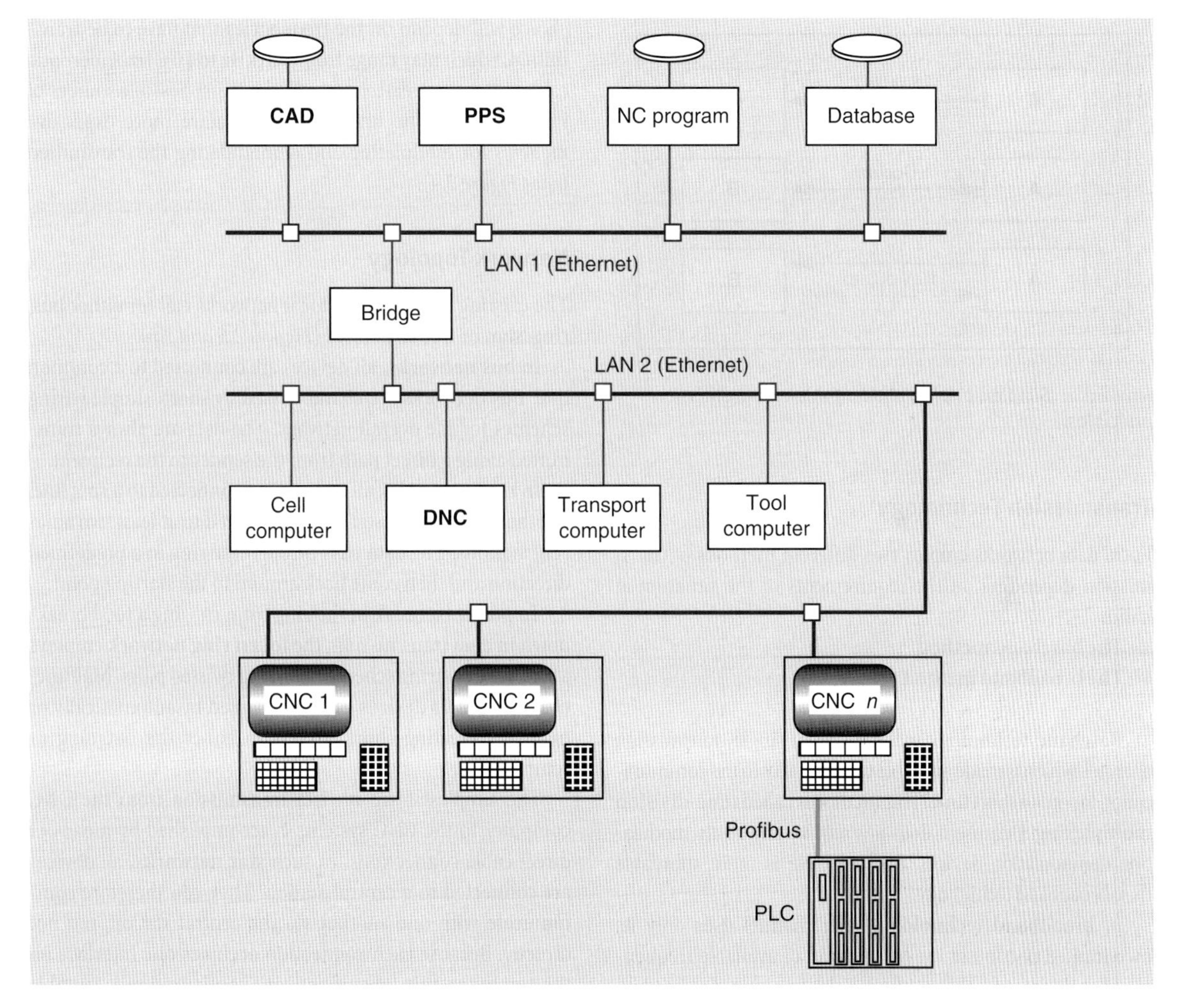

Figure 2.1: Communication between multiple computers and CNC systems via two identical Ethernet networks connected via a bridge. This enables free access and data exchange between selected devices on Network 1 and Network 2.

communicate with one another across the data network. The hardware and software involved in these interfaces must be adapted to the common data network.

2.4 Characteristics and Features of LANs

A number of different LAN systems exist today. Although their basic task is always the same, there are significant differences in features. These differences lie within the following seven areas:

- **Transmission technology**
- **Transmission media**
- **Network topology**
- **Access methods**
- **Protocols**
- **Transmission speed**
- **Maximum number of devices**

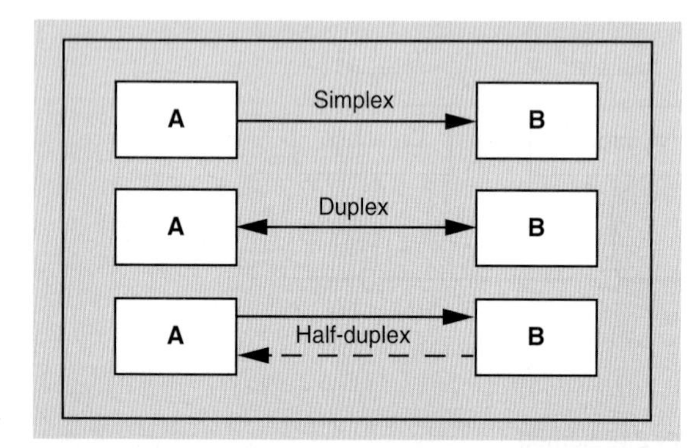

Figure 2.2: Simplex, duplex, and half-duplex transmission principles.

Transmission Technology

Local data networks employ two different methods of transmission depending on the requirements of the situation at hand:

- **The baseband method**
- **The broadband method**

Baseband technology makes use of a single transmission channel, which is made available to each party to the communication for only a short time. This method is called **time-division multiplexing.** Because it does not require any costly modulation/demodulation devices, this principle is more affordable than broadband technology.

In **broadband technology,** each channel uses only its own limited portion of the broad band of available frequencies. For this reason, this method is also known as **frequency-division multiplexing.** Each method has its own special advantages and disadvantages.

The information that is transmitted by either of these methods is electronically "compressed" to improve the integrity of the data transmission. The transmission occurs at such high speeds that the data cannot be read directly but must be made readable for the receiver by a so-called **converter.**

Transmission Media

Single-twisted pairs, shielded multitwisted pairs, coaxial cables, or fiberoptic cables may be used for this purpose. The choice will depend on the frequencies that have to be transmitted, which may range from 500 kHz to 10 GHz. Fiberoptic cables offer the greatest protection against possible sources of interference in the line, but they require more expensive *modems* for modulating and demodulating the transmitted data (*Figure 2.4*).

Network Topology

The existing standards for local networks call for either bus, ring, star, or tree structures (*Figures 2.5 and 2.6*).

In **bus networks,** all devices are connected to a common line. Bus networks are characterized by short, simple wiring schemes for the overall network. The data are always transmitted along a direct path from the sender to the recipient.

In **ring networks,** all stations are connected to a ring line. In other words, each station is connected to at least two adjacent stations. The data pass through the ring in a predefined direction and then come back around to the starting point.

To prevent a break in the ring from causing a total breakdown in data transmission, the **token ring network** employs a redundant wiring scheme involving twisted pairs. Malfunctioning lines or devices can be bypassed by automatically or manually creating short circuits in either the outgoing or return lines.

The simplest form of data transmission from the DNC computer to the CNC systems is accomplished by means of direct cable connections. In such **star networks,** all devices are connected to a central station. They can therefore communicate with one another via the central station, but not directly. Because each connection occupies one interface on the computer, this type of network configuration is suitable only for connecting a small number of devices.

Tree structures are a combination of the above-mentioned topologies.

Access Methods

The **access method** determines which station may send information and how the receiving station will recognize the message intended for it. A distinction is made here between methods that are collision-free and those which allow collisions to occur as the data network is accessed.

In the **token principle** (*Figure 2.7*), a special pattern of bits (the *token*) constantly circulates around the ring. It is

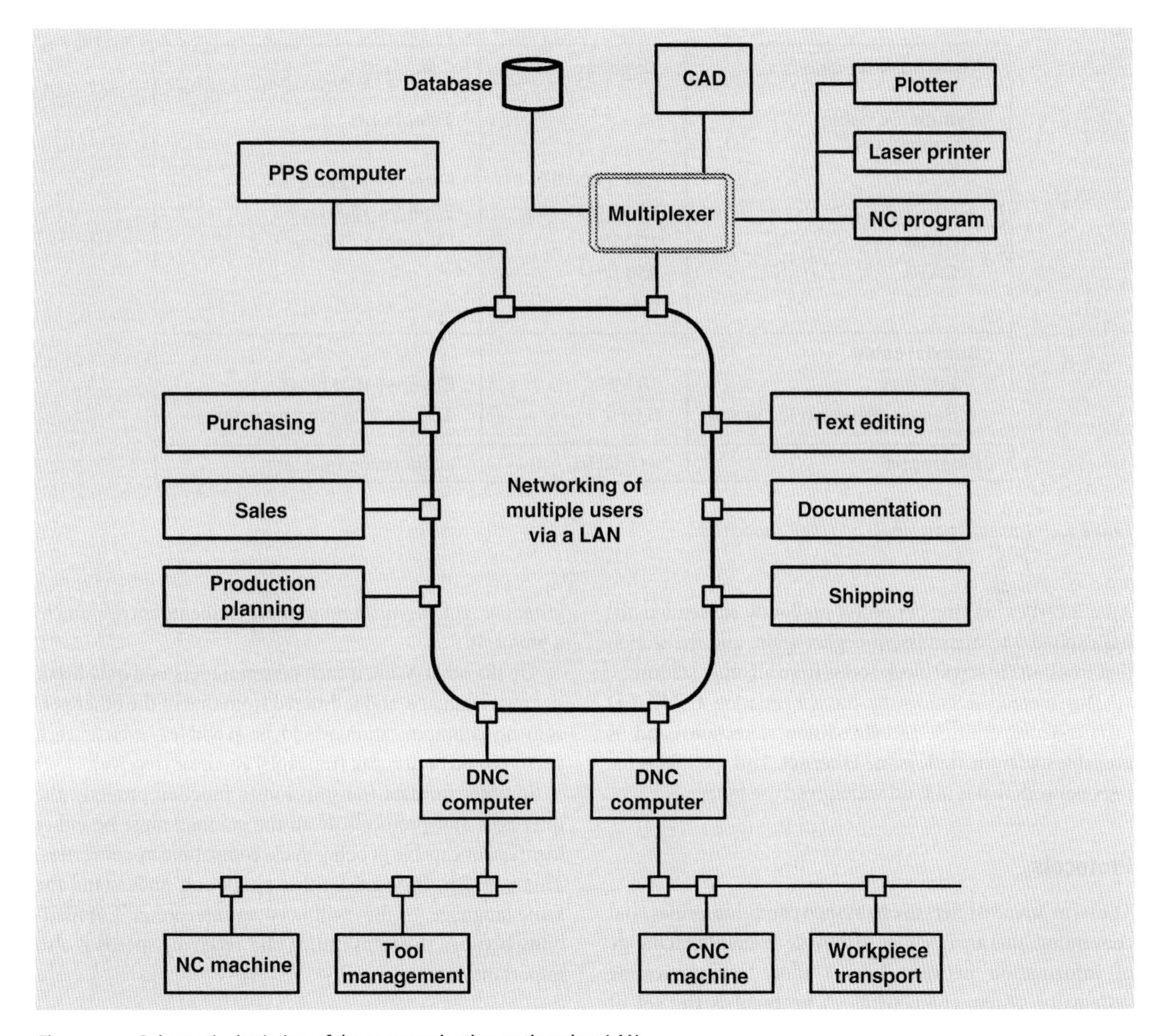

Figure 2.3: Schematic depiction of the communication options in a LAN.

passed like a relay baton from one active device to the next. In order to transmit, a station must possess the token. To do this, the station that is ready to transmit must wait until the arriving token indicates that it is "free." The station then sets the status of the token to "busy" as it flies past, so to speak, adds the source and destination addresses, and appends the data to be transmitted. The destination station recognizes its address, copies the data into its input buffer, and sets the "copied" bit as a flag. The entire data stream eventually comes back around to the transmitting station. Just to be sure, the transmitting station compares these data with the data it had transmitted, then removes these data, and sets the status of the token back to "free" before passing it along the ring. This prevents data collisions.

In methods using **access that is susceptible to collisions,** any station can transmit at any time. Data collisions are accepted as a potential risk, and when such a collision occurs, it immediately causes the transmission to be discontinued.

Transmission medium	Capacity	Comment
Symmetric cable (twisted pair)		**Telephone**
– <CAT3	upto 500 MHz	Ethernet 10 Base T
– CAT3	10 MHz	Ethernet 100 Base T
– CAT5	100 MHz	Ethernet 1 G Base T
– CAT6	250 MHz	
– CAT7	600 MHz	
Coaxial cable		
– Ethernet	10 MHz	Ethernet 10 Base 2
– Radio + television technology	2 GHz	e.g., satellite receiver system
Fiberoptic	>10 GHz	Interference resistant

Figure 2.4: Transmission media and their capacities.

After a certain waiting period (of randomly chosen length) has expired, the transmission begins again, and the station that transmits first will block access from all other stations.

This method is known by the abbreviation **CSMA/CD** (carrier-sense multiple-access/collision detection) and is available under the trade name **Ethernet**. This method is the LAN network that is in most widespread use today.

Protocols

The term *protocol* refers to established conditions, rules, and conventions that are intended to provide a **reliable exchange of information** between two or more communicating systems or system components. A **protocol is the set of rules** governing the communications in a data-transmission system. The protocol establishes the code, the method of transmission, the direction of transmission, the format of the transmission, the establishment of the connection (*call setup*), and the termination of the connection.

Example

Human Language and Its Rules. People who converse in the same language can engage in a two-way exchange of information by using their mouths and ears (which correspond to an *interface*). As a medium of communication, language is defined by *vocabulary* (the words and their meanings), *grammar*, and *pronunciation* (these elements correspond to a *protocol*).

On the other hand, if each person understood only his or her own language, and it differed from that of the other person, no communication would be possible—even if each person could both speak (transmit) and hear (receive).

In order for data transmission to function properly, the interfaces and protocols of all the stations must be either identical or capable of being made compatible by converters. This is to say, they must either speak and understand the same language, or they will need an interpreter. Compatibility between the transmitting and receiving speeds is also important.

Transmission Speed *(Table 2.1)*

The transmission speed in a data network is defined in terms of *bits transmitted per second* (1 b/s = 1 baud). This speed must be as high as possible in order to transmit as much data as possible each second and to avoid waiting time. On the other hand, the speed at which data can be read into the CNC controllers is usually limited to values of around 1,000 characters per second, which roughly corresponds to 8,000 b/s. This is another reason why a *converter* needs to be inserted between the LAN and the CNC system. As the data arrive from the LAN at a high speed, the converter puts them into a

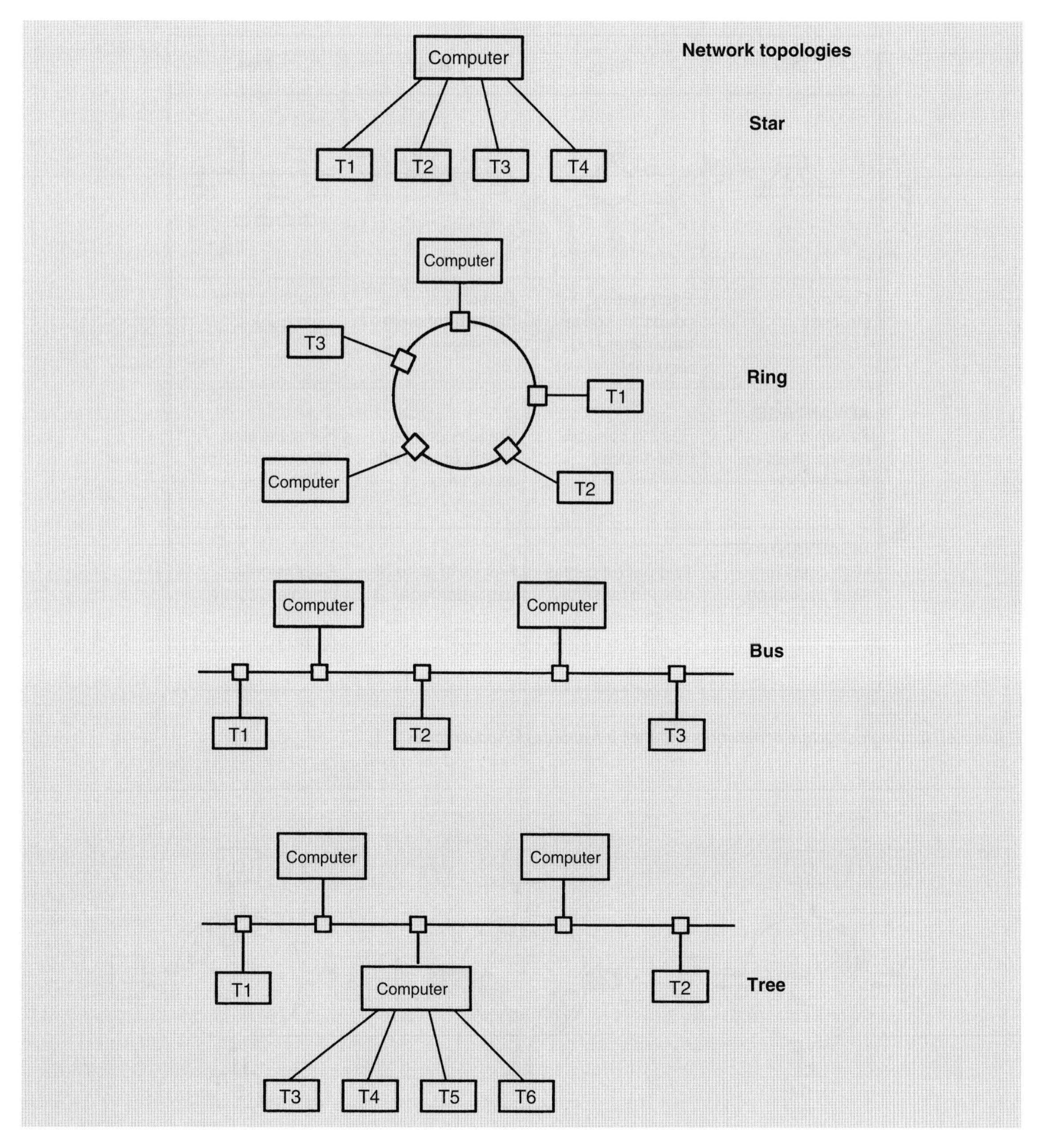

Figure 2.5: Star, ring, bus, and tree structures of local networks (T1, . . . , T6 = devices 1 to 6).

Star	Ring	Bus	Tree
	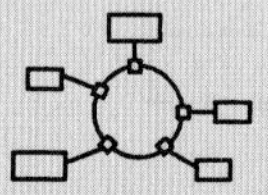	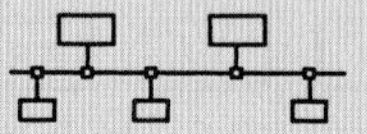	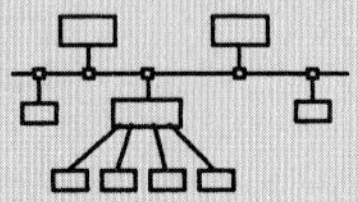
Common structure	Each station needs an active transmitter and receiver	Coaxial cable as passive medium	Branches off of main trunk
ADVANTAGE: Failure of a unit has no effect on the network	Easy to connect new devices	Stations can be switch on and off as desired	For separated networks or later expansions
DISADVANTAGE: High costs for lines over large distances	Two paired cables needed for safety if a station fails	Failure in case of cable damage	Expansion is complex, costly for different networks

Figure 2.6: Topologic structure of networks and their advantages/disadvantages.

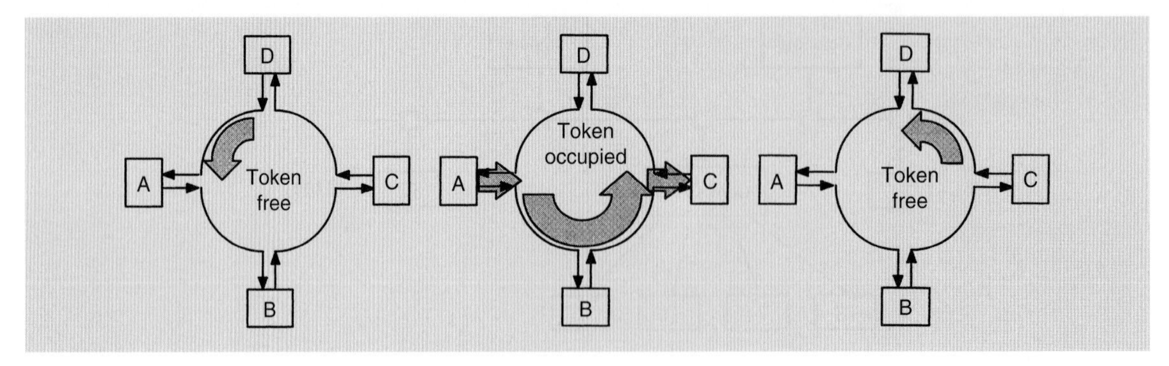

Figure 2.7: Token ring principle (transmitting from A to C).

buffer and then issues them to the CNC controller at the proper reading speed.

Most data-processing devices can be adjusted to standard transmission speeds: 110, 150, 300, 600, 1,200, 2,400, 4,800, 9,600, 19,200, 38,400, 57,600, and 115,200 baud are the specified values.

Although the data constantly flow past all the stations, they are only supposed to be delivered to the intended recipient, as specified by the destination address. To this end, the converter also assumes the tasks of filtering out and unpacking the data.

Maximum Number of Devices

A maximum of 1,024 stations can be connected to an Ethernet network, and the length of the network can be up to 2,500 m, not including gateways or routers. Standard Ethernet has a transmission speed of 10 Mb/s, Fast Ethernet 100 Mb/s, and Gigabit Ethernet 1,000 Mb/s. Fast Ethernet and Gigabit Ethernet require twisted-pair or fiberoptic cables because coaxial cables do not allow such high frequencies.

Because the CAN, InterBus-S, and PROFIBUS field buses are physically based on RS485 serial interfaces, their performance data are also practically identical. Differences exist, however, with regard to the bus access procedures, the security mechanisms, and the transfer protocols.

The maximum length of CAN is 1 km at a transfer rate of 50 kb/s. The maximum number of devices is limited to 64 (or 128 with restrictions). InterBus-S can transmit 500 kb/s to up to 256 devices at ranges up to 40 m. Thirty-two devices are possible with PROFIBUS, and 127 devices with repeaters. The transfer rate is 500 kb/s at ranges up to 200 m or only 93 kb/s at up to 1,200 m.

2.5 Gateways and Bridges

The purpose of data communications is to provide information at the desired location, no matter where that may be. Local networks are actually incompatible with this objective when they are restricted to a certain building or department, as mentioned earlier. Thus it is possible for multiple LANs to exist within a single company. This gives rise to a demand for devices that can carry information from a LAN to another network. These devices are called **bridges** or **gateways** (*Figure 2.8*).

The term **bridge** refers to a device—usually a computer with accompanying software—that makes it possible to **connect similar LANs** and provide communication between the stations of these similar networks. By definition, the connected networks are identical in type. Therefore, the bridge does not provide conversion between protocols. However, the bridge does have to recognize the address of each data packet as it arrives and determine whether its intended recipient resides in the other network. Only then will the bridge direct this packet to the other network. This prevents unnecessary overloading of the networks.

Television	5 Mbit/s	corresponds to	5 MBaud
Eyes	>5 Mbit/s		5 MBaud
Ears	20 Kbit/s		20 KBaud
Telephone (ISDN)	64 Kbit/s		64 KBaud
Teletext	ca. 96 Kbit/s		96 KBaud
LANs	10–1000 Mbit/s		10–1000 MBaud
V.24	1–115 Kbit/s		1–115 KBaud
20 mA	4,8 Kbit/s		4,8 KBaud
VMEbus	ca. 30 Mbit/s		30 MBaud

Table 2.1: Examples of Information Transfer Speeds

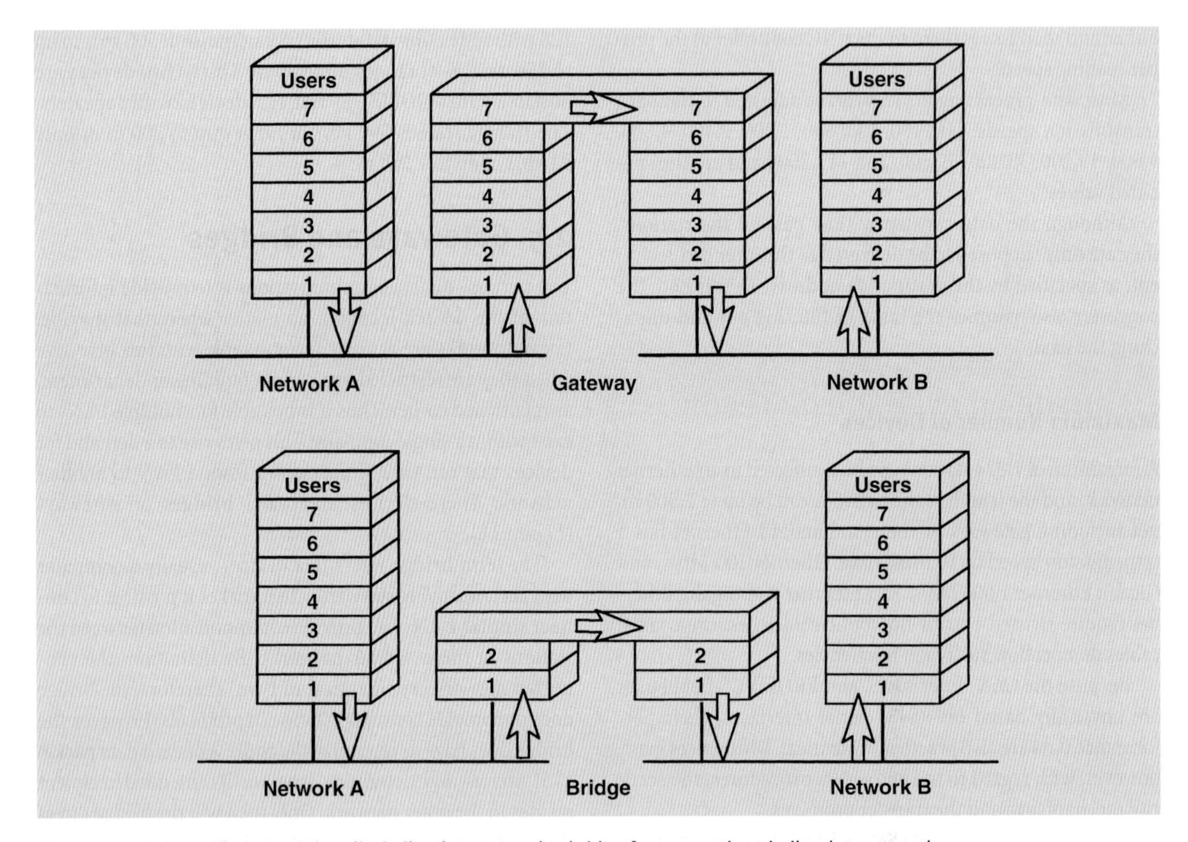

Figure 2.8: Gateway for connecting dissimilar data networks; bridge for connecting similar data networks.

Bridges also can be used as repeaters to extend the otherwise restricted length of a network.

Gateways, on the other hand, serve to connect **dissimilar networks.** Gateways must assume a wide variety of additional tasks, such as protocol conversion, formatting, and matching. This makes it easy to understand why the structure of a gateway can become relatively complex.

2.6 Criteria for Selecting a Suitable LAN

When selecting a LAN, the buyer must consider 10 essential criteria:

1. The maximum **transmission speed** in baud
2. The maximum **quantity of data expected** to be transmitted
3. The maximum **number of devices that can be connected** without problems
4. The issue of **transmitting data** to all devices in simplex, duplex, or half-duplex operation
5. **Protection against failure** of a single device or node (This must not lead to a breakdown of data transmission.)
6. The maximum allowable **line length** without intermediate repeaters
7. The **number of conductors** in the cable and the **type of cable** (shielded, twisted, coaxial, or fiberoptic)
8. The **minimum bending radius** of the cable (important for purposes of installing cable in ducts)
9. **Installation conditions**, for example, whether cables are installed together with power cables, in environments "contaminated" with electromagnetic radiation, or where severe network interference is present

10. **Price**, comprising (a) the base price for the LAN and (b) the price per device connection

2.7 Interfaces

The transmitted data must be capable of being entered into any of the connected systems. This is the purpose of the device interface.

An **interface** is the precisely defined boundary between two hardware systems (e.g., computers, printers, or CNC systems) or two software programs within a computer. It also can be defined in a broader sense as the "point where responsibilities are transferred," for example, from a cable television network to the service line of a particular house.

An interface can serve to connect two similar or dissimilar systems, for example, two identical computers, two different computers, a computer and a printer, a computer and a CNC system, a DNC system and a CNC system, or a person and a computer (in which case the keyboard and display screen form the interface).

This discussion will be restricted to interfaces that pertain directly to the control of NC machines (computer-aided manufacturing [CAM]). This primarily includes the interfaces that are used to transmit data to the NC machines, robots, transport systems, tool-management systems, measuring machines, and other devices. *Figure 2.9* shows how interfaces are classified.

Hardware Interfaces

This term refers to the hardware-related specification of a device interface. For example, the hardware interface defines how many wires are used for transmitting, receiving, control functions, messages, the clock, and so on, as well as the

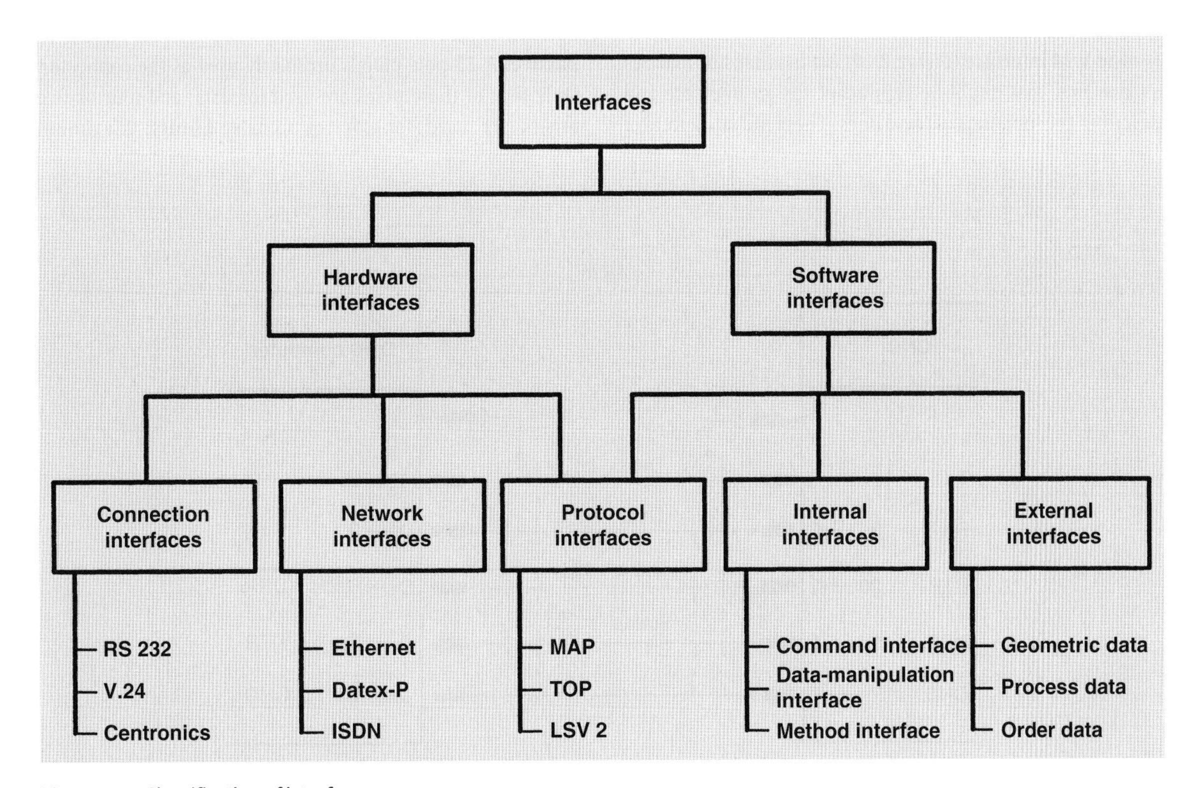

Figure 2.9: Classification of interfaces.

appearance of the plug being used and its pin assignments. Hardware interfaces also can be described as *device-connection interfaces* through which all information must pass when flowing into or out of a device. A distinction is drawn here between **bit-serial** and **bit-parallel interfaces.**

In the case of bit-serial interfaces, the information is transmitted in a certain chronologic sequence on a single line. For this reason, this type of transmission is slower than bit-parallel transmission, in which the individual bits of a given character are carried simultaneously on multiple lines.

Parallel interfaces have some serious **drawbacks,** however:

- At least nine lines are required for 8-bit characters.
- These interfaces are technically complex.
- The line length is limited to 1 to 3 m.
- **Examples:** Centronics, IEC-bus.

Because of the general preference for the use of **bit-serial interfaces** in data transmission, the most commonly used examples are examined in greater detail below.

The **V.24 interface** (*Figure 2.10*) is a list that summarizes all the lines that might be significant for an interface and describes the functions of these lines. These include a total of 25 lines for transmitting, receiving, control, messages, and so on. Only seven of these lines are actually used, however.

The **RS-232C** standard provides a much more precise interface definition than V.24 insofar as it defines the mechanical and electrical properties as well as the pin assignments for the 25-pole plug. Only 4, 9, or 15 of these 25 poles are assigned in actual practice. The electrical properties of the RS-232C interface are defined as follows:

+3 to +15 V = ZERO

−3 to −15 V = ONE

In most cases, the transmission speed is approximately 4.8 to 9.6 kbaud. The limit is approximately 100 kbaud, and the maximum allowable line length is 20 m.

The **20-mA interface** employs two wires for transmitting and two wires for receiving. A current of 20 mA corresponds to a value of ONE, whereas an absence of current is interpreted as ZERO. There is also a permanent quiescent current of 20 mA that has to be supplied by one of the devices. This device is considered to be **active,** and the other is regarded as **passive.** A 25-pole plug is ordinarily used as the connector. Only four of these poles are used for data, and four or five others are used for control signals. **Limits:** Maximum

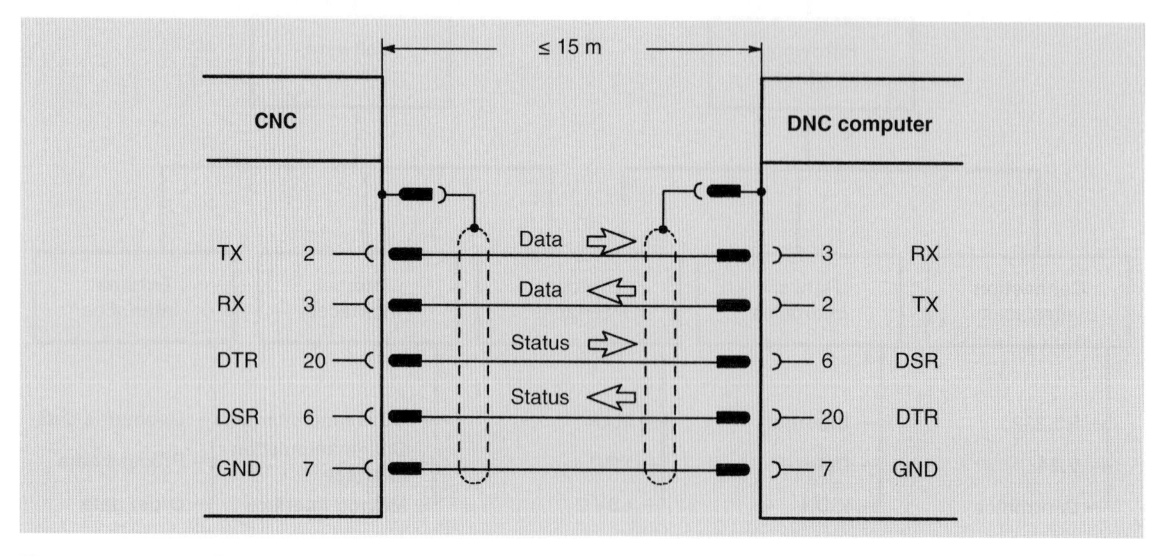

Figure 2.10: V.24 interface.

transmission speed of 9.6 kbaud, maximum line length of approximately 1,000 m.

The **RS-422** standard was specially developed for symmetric data transmission at baud rates greater than 100 kbaud and a line length of 1,200 m or at 10 Mbaud and a line length of 12 m. The information is interpreted as the difference between the signals of the two lines. Interfering signals and ground currents between different ground points can be prevented from causing interference in the transmission signals.

This standard is not intended for bus operation. In bus operation, multiple drive outputs are carried on the same line. At any given time, only one driver is free to "drive" the line. The outputs of the other drivers then must exhibit very high resistance so that they do not short-circuit the transmission signal. A modification of this type is provided in the **RS-485** standard.

Handshaking

This is the procedure for controlling data transmission between two devices. In start/stop operation, handshaking controls data transmission by allowing the receiving station to accept the transmitted data. If this becomes temporarily impossible, the receiver will issue a predefined signal to stop the transmitter until it is ready to receive again. A distinction is drawn here between **software handshaking** and **hardware handshaking** (*Figure 2.11*).

The principles of these two types of handshaking functions differ only as to

- The number of lines required
- The assignment of data lines and control lines

In hardware handshaking, two separate lines are defined for data signals, and two lines are also defined for control signals. Software handshaking requires only two data lines. The function of the respective data lines varies according to the direction of transmission.

Software Interfaces

Software interface is a predefined point at which data are transferred from one software package to another (e.g., within a computer). The purpose and design of the interface

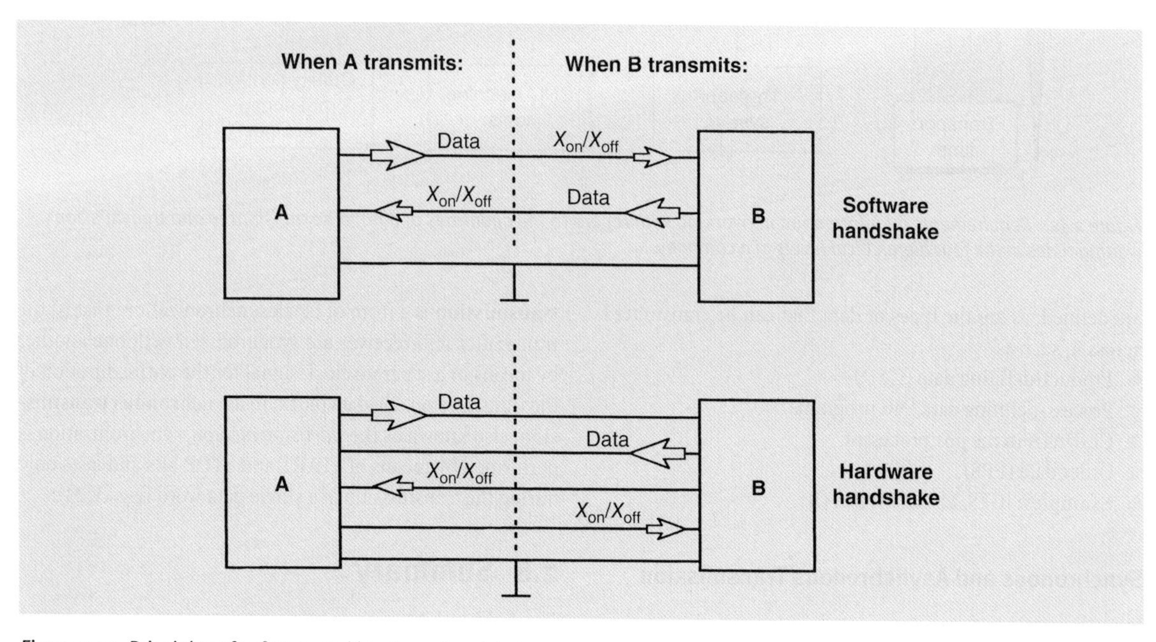

Figure 2.11: Principles of software and hardware handshakes.

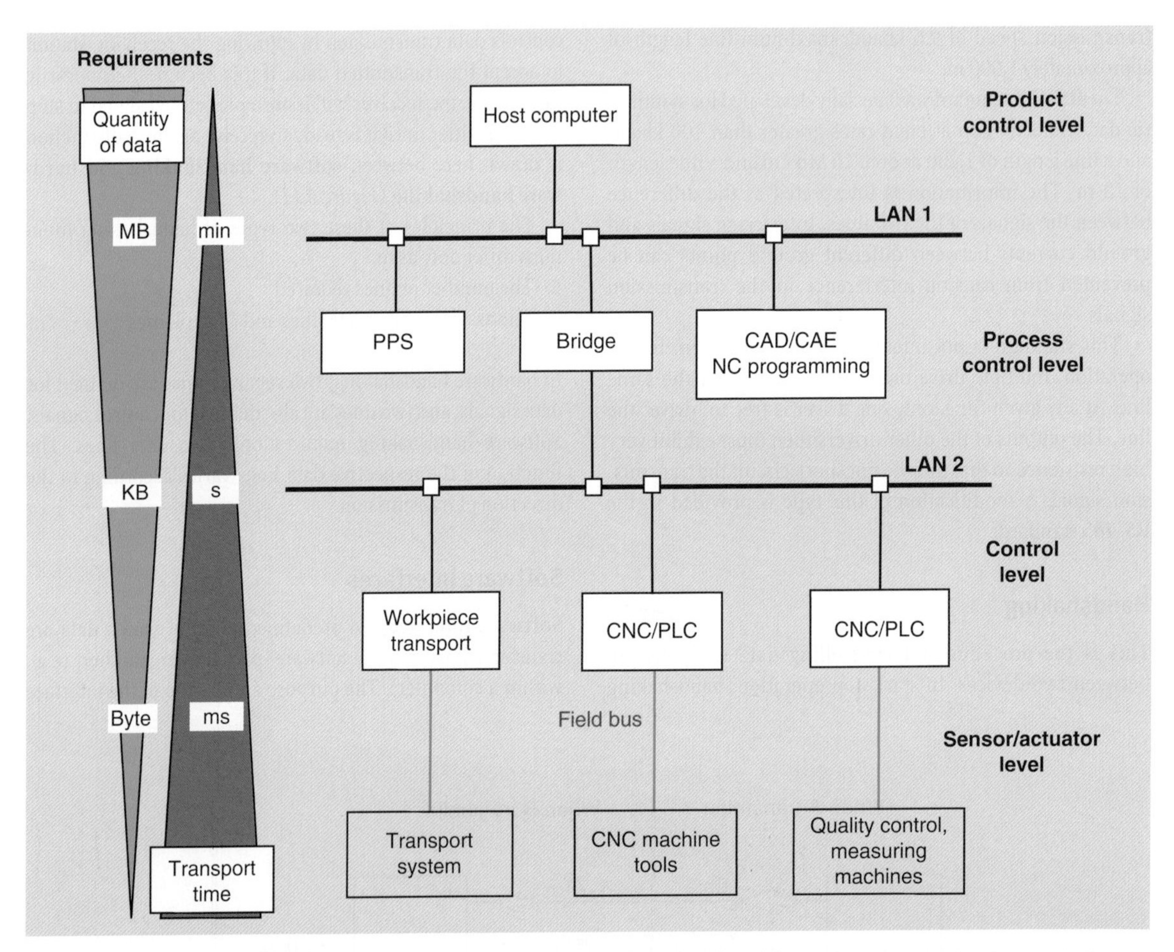

Figure 2.12: Requirements for information networking with regard to the quantities of data to be transmitted and transmission waiting times in the four hierarchical levels of a company.

are defined, as are the types of data that can be transmitted across it, such as

- Product-defining data (CAD)
- Process-defining data (NC programs)
- CL-DATA in the postprocessor
- Order data (PPS)
- **Examples:** IGES, LSV 2, MAP, TOP

Synchronous and Asynchronous Transmission

Data transmission requires coordination of the timing between the transmitting and receiving stations. **Synchronous transmission** is a form of block synchronization; that is, the transmitter and receiver are synchronized with one another by means of a separate clock signal for the entire duration of the transmission of a data block. In **asynchronous transmission,** also known as the *start/stop method*, synchronization is performed by means of START and STOP bits and lasts only during the transmission of a single data word (e.g., V.24).

2.8 Summary

The field of LANs, protocols, and interfaces is very broad and complex. It is typically the province of specialists.

A company that expects to be faced with the problem of networked data transmission in the future therefore should begin training its own skilled workers today. It is also advisable to seek the advice of experienced specialists until the company has built up its own know-how. The purchaser should at least be able to specify his or her requirements in technical terms in order to prevent unpleasant surprises and additional costs later. The installation of LANs is also a matter for specialists.

There is a widespread belief that a company only needs to commit itself to a single type of LAN that is fully integrated from the top down (e.g., Manufacturing Automation Protocol [MAP]), and then all its future data-transmission problems will be solved. This is false for the following reasons:

- LANs are subject to a variety of different requirements with respect to the quantity of data being transmitted and the necessary data rates (*Figure 2.12*).
- MAP interfaces are still not provided by every manufacturer.
- Most manufacturers provide Ethernet interfaces as standard.
- Aside from the technology, the costs of an overall system play an important role.
- Several of the available data-transmission systems are worth recommending because they can already be implemented without problems. In this area, the purchaser should let the supplier take responsibility for recommending a suitable choice and then accept the supplier's advice.
- The use of separate LANs will not necessarily be an obstacle to implementing a computer-integrated manufacturing (CIM) approach in the future.
- Another general trend in future networking could be to construct several small, self-contained, easily managed networks and then connect them by means of bridges or gateways. But this approach should be adopted only when it is really necessary.
- The expanding field of local networks is a scene of ongoing development. At this point in time, **Ethernet** appears to be the most widely used standard with the greatest acceptance and the largest body of experience.

The organization for the promotion of Ethernet in automation, abbreviated IAONA, provides a product database for Ethernet-based automation equipment on its website. At www.iaona-eu.com it is possible to select products from various suppliers in six categories and subcategories.

LANs—Local-Area Networks

Important points to remember:

1. The term *LAN* is an abbreviation for local-area network. This term refers to **standardized local data-transmission networks** that allow communications between different data processing devices within a company.
2. **Various brands of LANs** are available, each specially oriented to emphasize different types of transmission and applications. The purchaser must choose between these brands.
3. Although the basic task of a LAN is always the same, these networks **differ** in several respects, such as
   - Transmission technology
   - Transmission medium
   - Network topology
   - Access methods
   - Protocols
   - Maximum number of devices
   - Transmission speed
4. **Bridges** are used to connect similar LANs so that they can communicate with one another, and **gateways** provide a similar connection between dissimilar networks.
5. Only a limited **number of devices can be connected** to a given LAN. In situations involving many different devices in different hierarchical structures, it is advisable to select the most suitable LAN for each of these organizational structures and to connect the networks only to the degree that is absolutely necessary.
6. **Ethernet is the most commonly used type of LAN.**
7. The **goal** of choosing a single LAN from the brands that are most powerful in theory and then using only this system to connect all the users in the entire company can be ill-advised and very costly.
8. Before a company decides in favor of a particular LAN, the issue of **available interfaces on the devices** also should be explored.
9. The use of separate and dissimilar LANs will not necessarily block a company from implementing **computer-integrated manufacturing (CIM) in the future.**
10. A company should begin to consider the installation of a LAN either when the **management computers** are networked or when the initial **DNC system** is installed.
11. A basic distinction is drawn between LANs that function by the baseband method (transmission of signals without modulation) and those which function by the broadband method (transmission with modulated carrier frequencies).
12. Access control in a LAN is usually accomplished by the CSMA/CD method or the token-passing method.
13. The term *protocol* refers to the rules for exchanging information between computers or users in communication networks.
14. *Field bus systems* have the task of transmitting small quantities of data in a rapid sequence between the controller and the sensors/actuators, for example, CAN, InterBus-S, and PROFIBUS-DP.

3. Digital Product Development and Manufacturing: From CAD and CAM to PLM

Dipl.-Ing. Niels Göttsch, Dr. Thomas Tosse

Here we will discuss what is possible today in product development and manufacturing using computer-aided technologies and what has already been implemented successfully in some segments of large-scale industry and their subsuppliers. The automotive industry plays a pioneering role here because of its high number of units produced. In practice, however, the most important carrier of information between development and manufacturing remains the technical drawing, which is generally derived from three-dimensional (3D) models in 3D computer-aided design (CAD) systems.

3.1 Introduction

Industry, including the predominantly metal-working segments, is under pressure from globalization. This means, on the one hand, many new competitors in open markets with different conditions and, on the other hand, new consumers with their product requirements and purchasing habits. To be successful, companies have to take all these factors into account. New product variants, an increasing pace of innovation in all technological fields, and continuous measures to increase productivity are forcing industrial firms to take extreme efforts to improve their processes, to automate their production, and to make use of worldwide communication (*Figure 3.1*).

In these conditions, the fields of development, design engineering, production planning, and production itself all have undergone a transformation: The previous practice of developing innovations in seclusion so that they were first described and then produced now has given way to the management of ideas, requirements, variants, manufacturing capabilities, and market opportunities. Design engineering of components has been subsumed by **digital product definition.** Downstream company departments such as testing, analysis, production planning, and production have to be brought in at as early a point as possible. Development and manufacturing can remain successful only if they maintain an exchange with all areas of the company (*Figure 3.2*).

Today, this exchange takes place electronically on the basis of Internet technologies. E-mail programs and browsers have almost completely replaced earlier **CAD/CAM** concepts (see below) and **CIM** (computer-integrated manufacturing) concepts. Digital 3D product models have served as a catalyst, offering a wide range of potential uses in simulation and visualization, production planning and testing, production and logistics, spare-parts management, and service. Software systems for **product-definition management (PDM)** ensure seamless management, distribution, and use of this information. These systems store knowledge, a type of working capital that is kept up to date in **product life-cycle management (PLM)** concepts and put to use in all areas of the company.

3.2 Terminology and History

(→ Figure 3.2)

CAD (Computer-Aided Drafting)

This initially meant the computer-based creation of design drawings. This saved design engineers from time-consuming work in applying templates, hatching, filling in drawing headers, or creating various views and modifications. Later the term evolved into **computer-aided design.** The first surface modelers defined irregular surface characteristics that could no longer be calculated from points, lines, angles, and primitives with the aid of two-dimensional (2D) CAD. They represent the roots of 3D CAD. The most important ones, however, developed with solid modelers for definition of components in 3D space. They contained all the geometric information with regard to intersections and penetration of bodies that

Figure 3.1: Product life-cycle management integrates digital product development and production into a total concept for all areas of a company.

are relevant in subsequent process steps such as visualization, simulation, depiction of assemblies, prototyping, and manufacturing. **Parametrics** is an additional state of development that makes it possible to influence these models using numerical values (*parameters*), thus making it easier to implement change processes, creation of variants, and repeated use of similar models. **Hybrid modelers** combine functions for the design of complex surfaces with those for volumes. This is the current state of the art. **Object-oriented programming methods** allow automatic processes for the creation of components, **wizards,** and other **artificial intelligence** and **knowledge-based design** applications. The history of the creation of the component logged by the CAD systems provided additional options for influencing the model, including **associative links:** These define how changes in one object (i.e., feature, component, or module) affect another one. This property can be used for work with **assemblies** as well as for CAD/CAM integration (*Figures 3.3 and 3.4*).

CAM (Computer-Aided Manufacturing)

This comprises the creation of programs for computer numerical control (CNC) machines, whether with or without the use of CAD data from 2D or 3D systems. Today, an important role is played by the graphical **simulation** on the screen of the machining process defined in the numerical control (NC) programs. On top of this comes the creation of

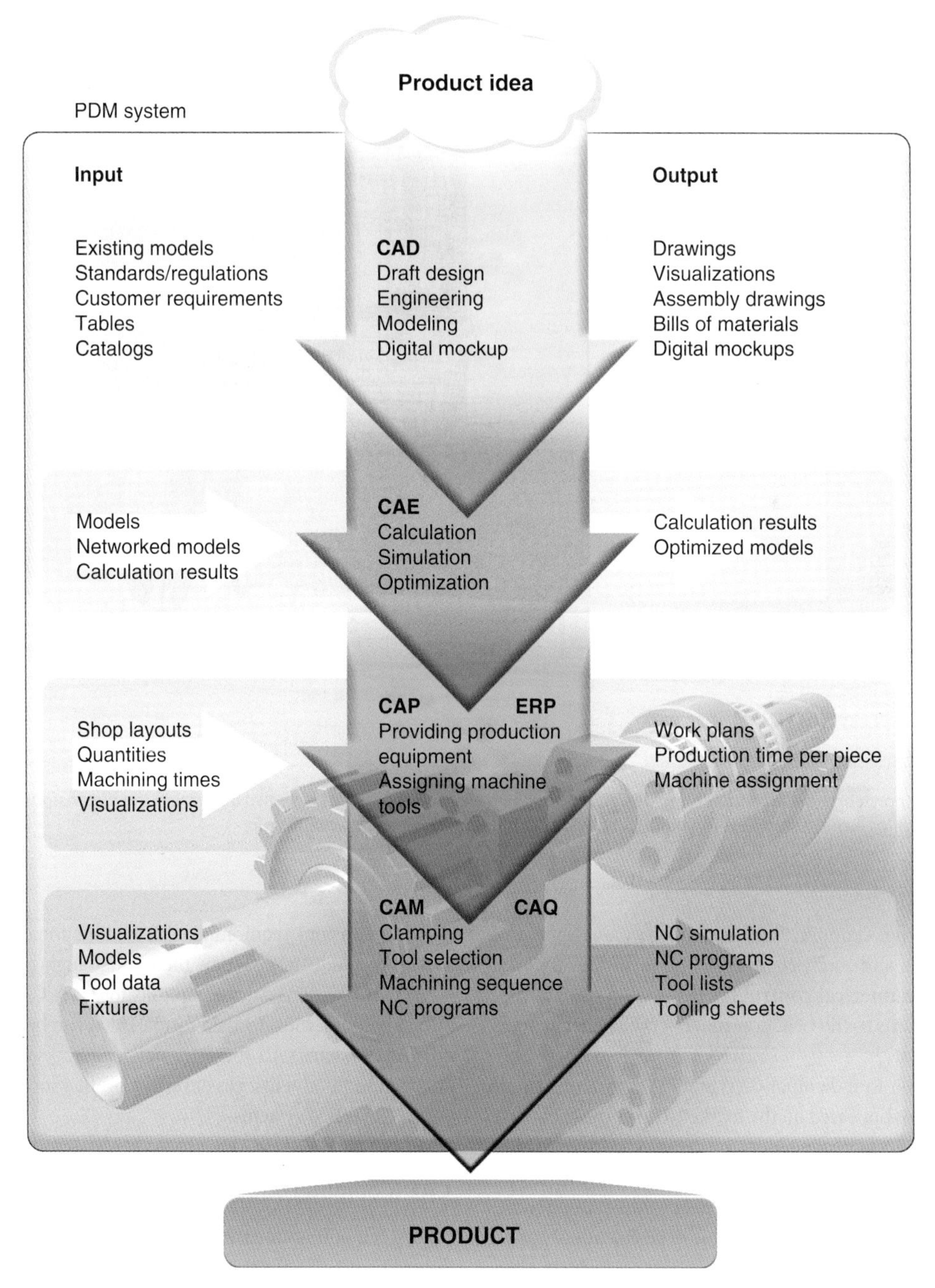

Figure 3.2: Computer-aided (CA) technologies in the information network. This depiction shows the principal interactions between networked CA systems.

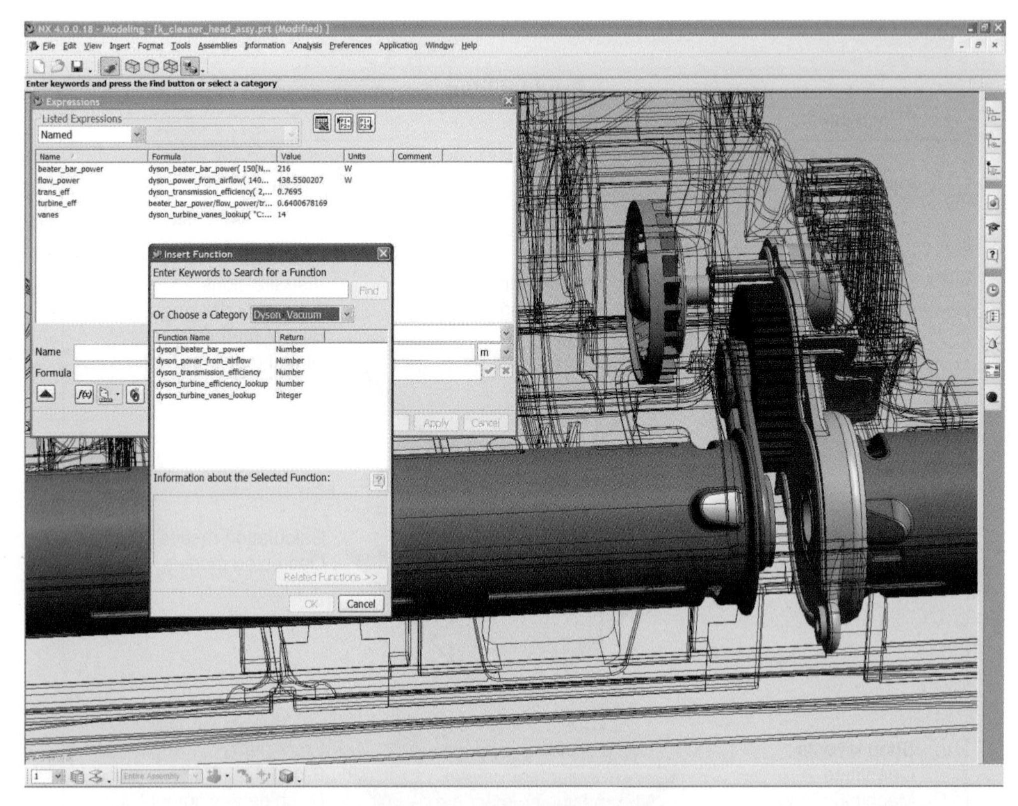

Figure 3.3: Knowledge-based engineering design for a vacuum cleaner. The formulas shown determine the dimensioning for entire assemblies—the stored development knowledge is applied quickly and without error.

the machining, clamping, and tool plans; management of the tool data and tools; clamping equipment and fixtures; and **direct numerical control (DNC)**—the direct transfer of NC programs to the CNC controllers of the machine tools via a network.

In addition to individual CAD and CAM systems, **CAD/CAM systems** appeared on the market. The purpose of these was to use internal connections of the two functional areas on the basis of a common data model in order to ensure that the data generated with CAD systems could be used for NC programming as directly as possible. The intention was to eliminate the cumbersome use of **interfaces,** which often involve additional work and data repairs, thus enabling seamless sequences and error-free processes. In practice, these goals could only be achieved to a very limited extent. Often the various requirements of the functional areas were not met adequately. Today there exist high-performance CAD/CAM systems, various options for CAD/CAM integration, and numerous 3D data formats that can be used easily in CAM systems. All these formats can cope with a wide range of requirements. Organization of the processes is performed by **PDM systems.**

CAE (Computer-Aided Engineering)

After computing-intensive beginnings with calculations according to the **finite-element method** (FEM), this has become more and more practical. After transfer of a 3D model and **meshing,** components and assemblies can be investigated in a **solver** with regard to numerous mechanical, acoustical, thermal, and other characteristics. Feedback to the CAD model is used to perform numerous optimization loops in a short time—long before the first component is manufactured or the first prototype is created. Originally reserved for specialists and appropriately dimensioned mainframe computers, today the tasks of digital simulation and checking are often associated with everyday activities in development departments.

Today, the term **digital product definition** provides a more appropriate summarization of the three historically conditioned areas of application (CAD, CAM, and CAE): The computer is no longer "additional equipment" but rather a workstation that is taken for granted. The areas of application have grown together and now overlap. Development and manufacturing processes can no longer be understood linearly—each of them runs in concentric circles that influence each other mutually.

CAP (Computer-Aided Process Planning)

(→ Figure 3.4)

This brings together the **manufacturing-oriented planning tasks** for new production tasks in manufacturing. Already during the investment preparation process the CAD model is used as the basis for creating machining plans for the component, simulating NC program sequences, and determining the optimal work processes, setups, and tools. Flexible manufacturing cells can be changed over for new

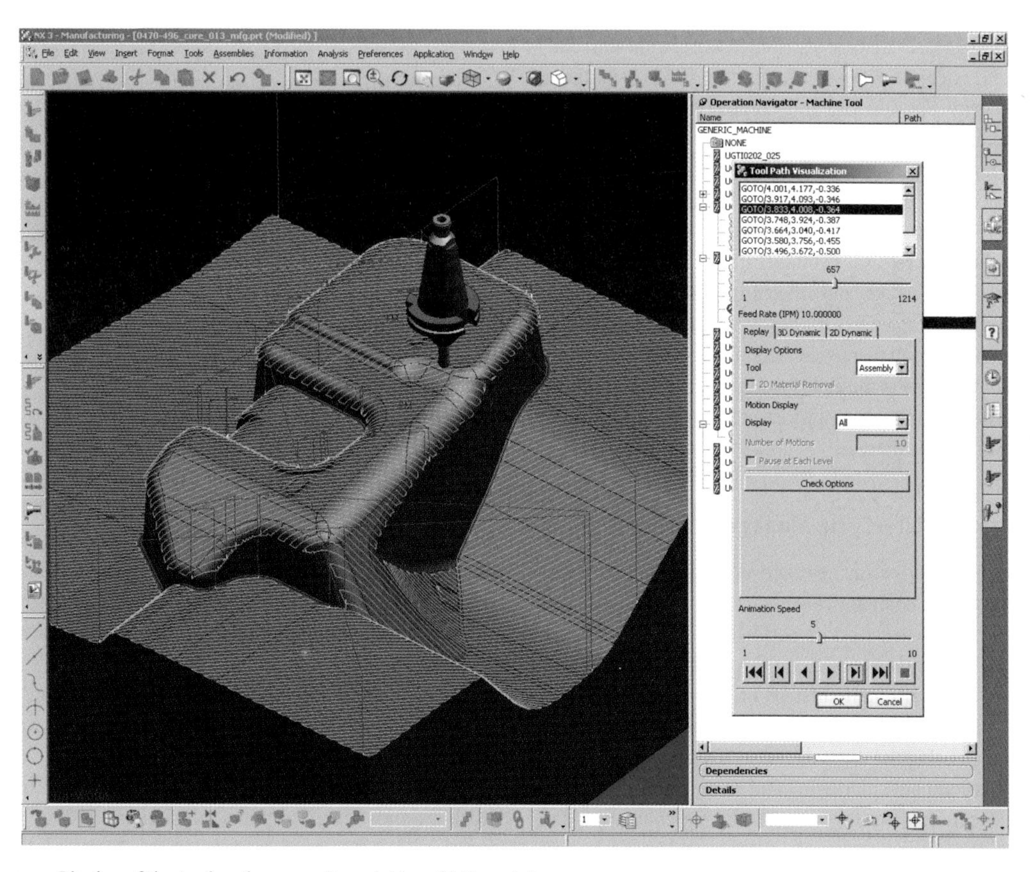

Figure 3.4: Display of the tool paths on a 3D model in a CAM module.

components more quickly, and complete processes in interlinked production lines can be planned minutely—an important step in the direction of the digital factory. The **material flow** is also simulated. For current daily detailed manufacturing planning, **manufacturing execution systems (MES) systems** are the right tool, replacing the earlier production-planning systems (PPSs). They are used to plan deadlines, machine utilization, production equipment and staff, and exchange data with **enterprise resource planning (ERP) systems.**

PDM (Product-Definition Management)

Today this extends far beyond its origins in **drawing management** or product management and **engineering data management (EDM).** Using modern database systems, all product-related data from the various departments are stored and managed. Databases that are distributed worldwide with daily replication, extensive provisions for granting access rights and other security precautions, and company- or role-specific menu systems in browsers offer not just 24-hour access to all relevant manufacturing data, but they also contain service-proven processes typical for specific branches of industry, covering the basic routings of all manufacturing companies. Modification and release, manufacturing release, and reuse can be organized quickly and efficiently. Modern work methods thus can be organized flawlessly, including **concurrent engineering** (simultaneous development of different but related components at multiple locations) and **simultaneous engineering** (simultaneous processing of the same components at various locations).

PLM (Product Life-Cycle Management)

This does not describe any new software but rather the maximum utility that a company can extract from the technologies mentioned here when work is performed in service-proven sequences, efficiently and with no errors, from the product-emergence phase through all steps of the process chains in development, production, installation, and customer service—and all using the same database. This encompasses all areas of the company: marketing and sales, training and service. PLM concepts are how manufacturing companies can cope with the current challenges of the world economy that were described earlier.

3.3 Digital Product Development

Draft Designs

Although 2D CAD systems are still in use in industry, they are being supplanted by 3D modeling. 2D functions are also expected and required in 3D CAD systems, where they are used for 2D drafting. Above all in early drafting phases, high-performance sketchers provide new ways to quickly make a record of new ideas. Existing 3D objects can be placed and modified (*Figure 3.5*).

Component Design

Previously, new components could be designed by selecting primitives that were placed in planes, given dimensions, mirror-imaged, and rotated, copied, and cut or joined using surfaces or lines. Today, however, high-performance 3D CAD systems cover a much wider scope. The selection of standardized parts from catalogs, parameter-controlled modification of existing components, automatic geometry generation based on specified parameters, and many other techniques provide an efficient method for any task. And in addition to the purely geometric description of a component, the models are enhanced with a great deal of further information that is stored in assigned data sets: Surface quality, material, weight, tolerances, mounting orientation, and many other characteristics can be appended here. This information will be needed in later process steps (*Figure 3.6*).

Assemblies

The capabilities of systems today generally are sufficient to edit even large assemblies on the screen. At the least, it is possible to design subsystems of products in relation to the assembly and to relocate component parts virtually. This makes it possible to immediately verify the accuracy of fit, tolerances, assembly options, or interference geometries (*Figure 3.7*).

One problematic aspect remains the machining of large assemblies, for example, complete machine units (e.g., complex machine tools). On most modern (standard commercial) computers, these can be edited only through the use of work-arounds (i.e., simplified depictions, hiding internal features, etc.). Besides faster processors, this problem can be dealt with especially by eliminating the limitations on the

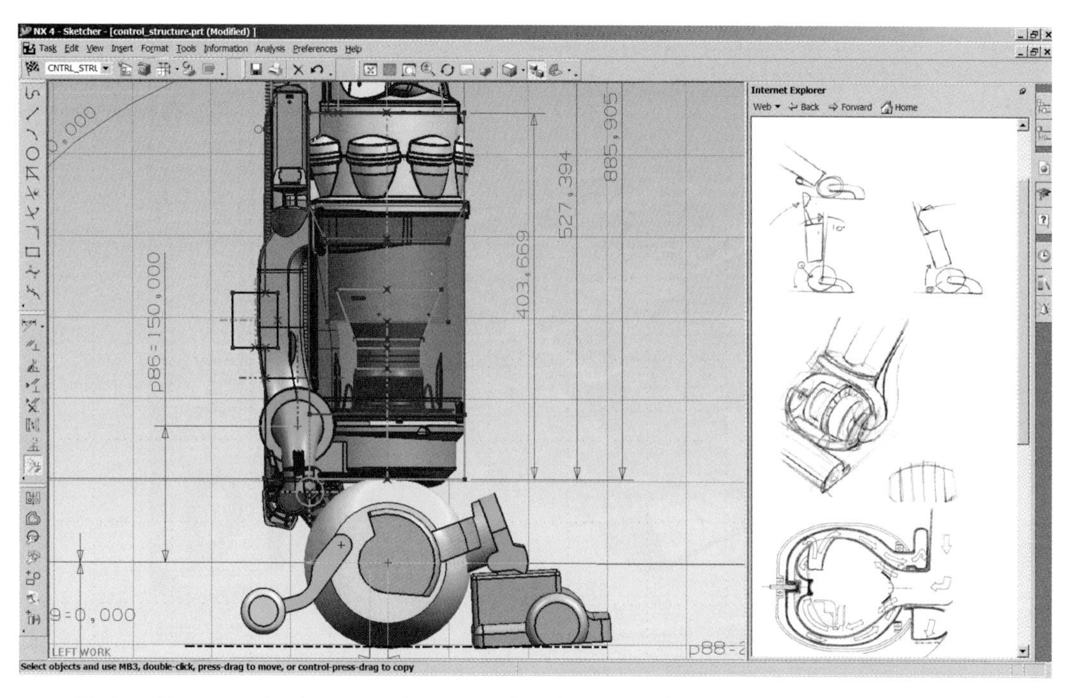

Figure 3.5: Modern 3D systems already support the concept phase by means of high-performance sketchers that take existing 3D models into account.

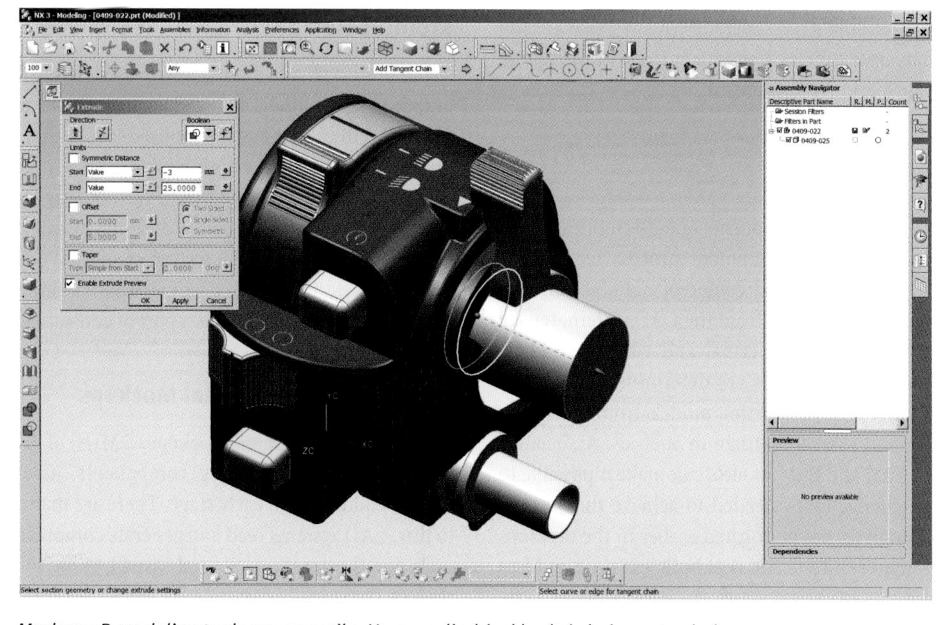

Figure 3.6: Modern 3D modeling tools are versatile. Here a cylindrical body is being extruded.

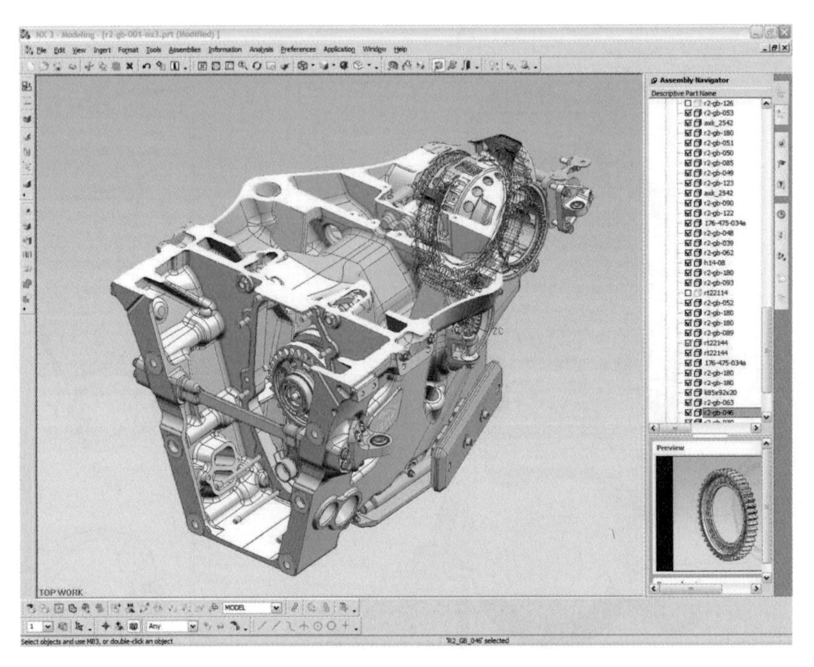

Figure 3.7: Modern product development in the context of all assemblies (Jaguar gearbox).

main memory (random-access memory [RAM]) that are still common today. Operating systems are currently undergoing a major transformation in this regard.

Simulation

Many properties of both components and assemblies can be checked directly in the development process using FEM analyses. Although weight and center of mass are already part of the model information from the CAD system, tensile strength, carrying capacity, thermal and vibrational characteristics, and acoustic effects are determined using CAE software. This is done by exporting models from CAD, networking them, and subjecting them to analysis. Associative links between CAD and FEM models can make it possible to abbreviate the change loops needed to achieve the optimal model. These analyses are performed earlier in the development process than real tests and cost both less time and less money. The results of development are verified at an early stage, allowing other areas to build on them.

Prototyping

Rapid prototyping technologies are used here: 3D models are output in stereolithography (STL) format, which divides the solid into individual superimposed layers that can be produced using various processes. Functional models made of plastic thus can be created right in the design department. Models made of wood, paper, and other materials are also possible. These enable the testing of specific characteristics: The visual design, surface properties, handling, and functions are even investigated as a part of consumer research.

Visualization and Digital Mockups

Visualization and digital mockups (DMUs) of 3D models are also methods for checking components, assemblies, and entire products at an early stage. There are many approaches to this. CAD systems read and generate compressed-data formats for the Internet. Special visualization software compresses data, reads common formats, and provides review functions: Marking and making comments probably are the ones used most commonly. These functionalities are also offered by PDM systems. An important role in the depiction and checking of 3D data from various CAD systems is played by the data format JT. It provides the component information from many CAD systems in a standardized manner and features a very lean data format even for accuracies that allow exact checking of subassemblies and installation spaces, collision checks, tolerance checks, and other investigations in the context of a complete DMU depiction of geometric data. The most spectacular use of DMUs is in the automotive industry: Long before the first model is produced, a so-called cage can be used to compress various projections to form a 3D model that can be viewed from all sides. With virtual vehicles, it is not just the visual attractiveness of the external shell that is checked: The functions of doors, trunk lids, and internal installations are depicted so clearly that one can form a realistic overall impression.

2D Drafting

With the increased use of 3D CAD, the digital product model is supplanting the drawing as the central carrier of development information, at least for mass-produced items. Drawings are still not superfluous, however. Sets of drawings for production are still standard alongside setup and tool information and NC programs—even when they are supplemented with 3D visualizations in many places. 2D drafting is an important and efficient function of 3D CAD systems, and it can be automated to a large degree. Of course, it is still possible to create cutting passes, fill in drawing headers, and select views manually. But often they are generated automatically, whenever necessary, together with an assembly drawing, transferred in the form of TIFF files, and archived for reasons of product liability and warranty. The task of feeding drawing data to the specifically intended uses with minimal effort is performed by PDM systems (*Figure 3.8*).

PDM (Product-Data Management)

Once the number of CAD workstations reaches five or more, it is no longer possible to cope with engineering data without the use of intelligent management functions that go beyond the capabilities of the directory structure of computer operating systems. But this is no longer the main reason for the rapid expansion of PDM systems on the market: The capabilities of even simple PDM solutions today extend far beyond classical functions for structure data storage, assignment of number ranges, compilation of bills of materials, or drawing management. For a long time now they have provided not only intelligent referencing and search functions but also the opportunity to define and permanently fix processes and access rights. Professional work methods are especially predicated on the management of modification numbers, internal company processes such as development releases and production releases, including support for the compilation and

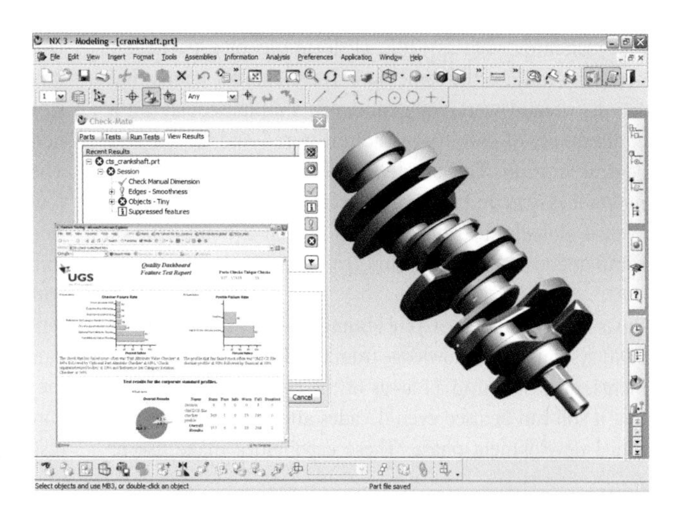

Figure 3.8: Reports concerning specific characteristics also can be generated based on 3D models.

transfer of all the necessary data to the downstream departments and processes.

PDM as an Integration Platform for CAD Systems

Anyone who develops complex products with numerous subassemblies, who works with multiple suppliers, or who has to serve various customers with semifinished products will not be able to maintain a single CAD standard, desirable though that may be from a cost standpoint. It is precisely then that PDM takes on a central role: Regardless of the source of the data, it is essential to establish uniform access rules, processes, and higher-level management functions. PDM solutions also provide these functions with multi-CAD/PDM. System-specific CAD managers in the menu system of the specific unfamiliar CAD solution give the user their familiar work environment. The user interface is the same, but PDM functions are added. The original file management of the CAD application is replaced by centralized database operation.

The management of subassemblies from various sources in a common overall production likewise is facilitated. Attributes, relations, and documents of various kinds can be assigned to the geometry data.

PDM as a Key to Information Distribution

Just as the use of 3D CAD systems has turned the digital model into a catalyst for product development, in like manner, Internet technology has allowed PDM solutions to become communication platforms in all areas of product development. Thanks to the use of Internet technologies and standards such as J2EE, all locations of a particular company can use the same data. Development teams that are distributed worldwide can work simultaneously on a common set of data that is administered and stored at a central location. This single sourcing of the product data makes it easy to find the most current development version, even if globally distributed teams exploit the potential of simultaneous cooperative work (collaboration).

Basis for Product Life-Cycle Management (PLM)

PDM systems store knowledge from the development departments, distribute it to all areas of the company, and ensure that it still can be used even decades after conclusion of the actual development phase. These capabilities make use of concepts from product life-cycle management (PLM) in order to achieve concrete economic goals: shorter development times, higher product quality, shorter innovation cycles, production at locations distributed all over the world. They thus make an important contribution to industry's ability to deal with the challenges it faces today.

3.4 Digital Manufacturing

In the interaction between CAD and CAM systems, three principal combinations have developed in recent years:

- Stand-alone CAM systems independent of CAD
- CAM solutions that optionally can be integrated into the user interfaces of CAD systems
- Complete CAD/CAM integration

Each of these solutions is oriented toward a particular method of work organization in industrial firms and toward the range of parts being produced. Component suppliers without their own design department will need different systems than manufacturers that perform their own product development (*Figure 3.9*).

2-1/2D Programming Systems

For many years, CAM systems that were independent of CAD were typical for component suppliers and series-production manufacturers that had to create a spectrum of products in the 2½D range, that is, components comprised of ruled surfaces and primitives. Data transfer from 2D CAD systems via DXF, initial graphics exchange specification (IGES), and other standard interfaces was in practice generally more labor-intensive and costly than creating the desired components using the capabilities of the NC programming system. The component and a fictitious blank were subtracted from each other based on the manufacturing drawings, dimensions, and material properties. The utility of 2½D programming systems was based on the difficulty of communication with the CNC controllers of the machine tools. CNC machines for turning, milling, EDM, and laser cutting require different data formats, and the scope of their commands goes beyond the standardized language defined in the DIN 66025 Standard, which only covers a basic vocabulary of necessary work instructions.

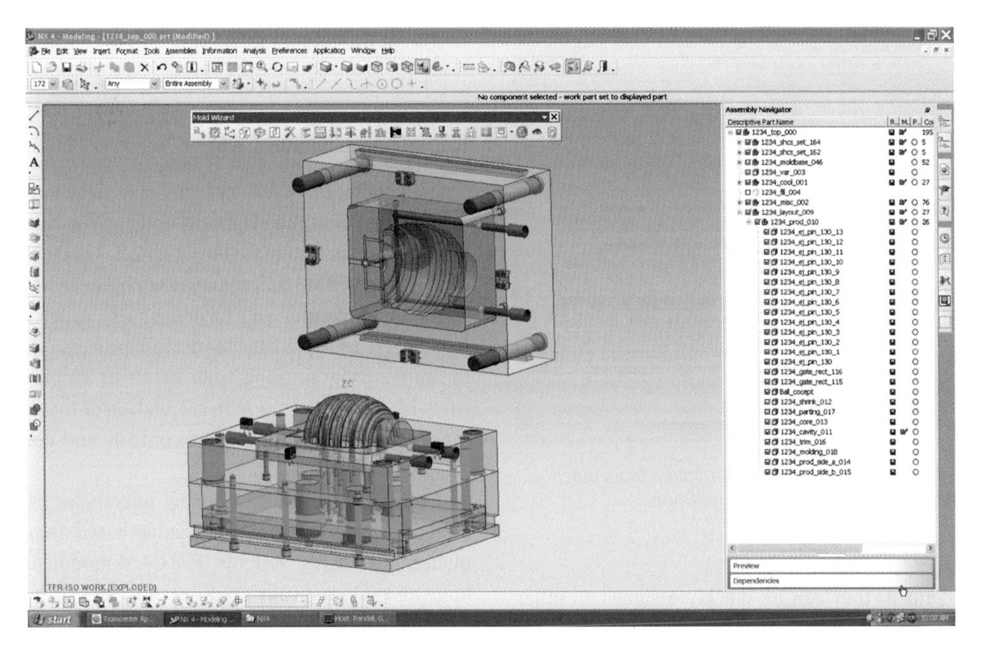

Figure 3.9: The design of the mold structure has been largely automated using knowledge-based design technology.

3D CAM Systems

In addition to the main aspect already described, 3D CAM systems also exist for individual operation at the shop-floor level, especially for demanding manufacturing tasks in one-off and small-series production, for example, in tool and die making. These systems naturally can receive data in the common 3D formats, allow the definition of blank contours, and then propose predefined strategies for roughing, finishing, pocket milling, and many other specialized machining options. They issue very complex NC programs for various machining sequences that often depend on each other, such as 3D milling and die-sinking EDM. However, as a rule, the results of this process cannot be returned to the 3D CAD system. On data transfer, it is often seen that the models that have been provided do not meet the requirements of 3D milling. Surfaces have to be repaired, and chamfers and rounded areas have to be reworked or patched (*Figure 3.10*).

3D CAM Systems as a Plug-in

CAM solutions can be acquired as an independent programming system or as a plug-in for one or more 3D CAD systems. The advantages of this variant, which is closely linked with midrange 3D CAM systems, lie in their uniform menu system and optimized data transfer. Training and familiarization are less costly, and data consistency is automatically improved. This is so because the geometries that are used are based on one of the common kernels and do not lose most of their information when passed through a "translator." The calculated tool paths are stored in a component file together with the 3D model. All manufacturing information is automatically taken from the solid model, thus largely eliminating input errors. After the component has been placed in the clamping position, it is easy to assign individual features—the number of mouse clicks and keyboard entries is reduced. Complete associativity between the solid model and the machining process allows quick modifications and avoids errors when the model is changed. In addition, once machining strategies

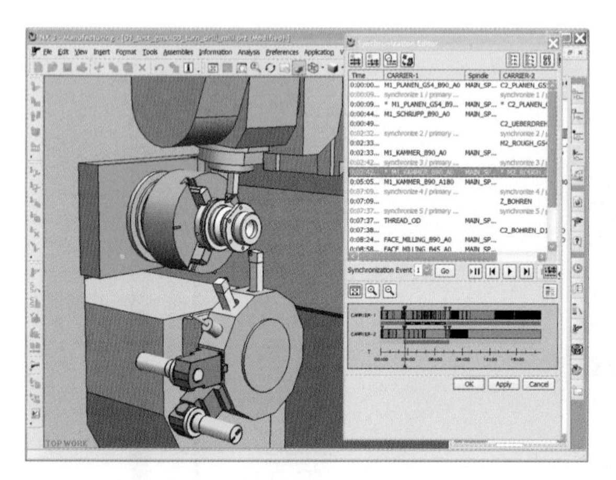

Figure 3.10: Modern CNC machines with driven tools place high demands on programming and simulation.

have been defined, they can be quickly assigned to other members of groups of related parts that have been generated parametrically.

Integrated CAD/CAM Systems

As already described, improved design engineering methods represent only a small part of the utility of solid modeling. The capital value of 3D data is only truly exploited through the superimposition in time of all product-emergence processes, collaboration of distributed teams on the same component, and the complete documentation of all process steps. This basic concept is best embodied in the form of seamlessly integrated CAD/CAM solutions.

Then even the CAM module itself provides integrated access to all product-development and manufacturing functions. This is helpful not only in data transfer and repair of models from external systems, as well as the development of clamping fixtures and new tools, but also for optimization of geometries that otherwise would be impossible to mill.

The mutually associative linking of all design engineering and manufacturing data brings the functional areas close together. Subsequent changes to the 3D model take effect immediately in the CAM area, resulting in quick updating of the NC program. Each step remains fully trackable; the tool selection, the clamping fixture, and the documentation are also updated right away for the worker. Subsequent changes to the 3D model take effect immediately in the CAM area, resulting in quick updating of the NC program.

Knowledge-Based CAM

Latest-generation CAM systems can detect and analyze form elements in the 3D model and can use this geometric information for automated production. This **feature-based machining (FBM)** technology uses engineering knowledge in the form of rules and formulas (also called *knowledge-based engineering* [KBE]) to increase process reliability and to reduce programming time by up to 90 percent. From rules-based tool selection to calculation of the manufacturing strategy, the CAM system controls and generates the NC paths.

Especially the field of boring operations can be automated completely in a knowledge-based manner using modern CAD/CAM systems. The CAM module detects drill holes and automatically proposes processes such as countersinking, drilling, and subsequent thread cutting; defines the tool paths; and provides support in the selection of tools. This scope of functions is enhanced greatly by special knowledge-based modules. For example, for tool and die makers (*Figure 3.11*).

Simulation of the Machining Process

(Figures 3.10 and 3.12)

The graphical simulation of the machining process on the screen generally represents the conclusion of NC programming in the office. Undesirable collisions and damage to the workpiece contour, the tools, holders, and spindles are detected at an early stage by suitable simulation solutions. More user-friendly systems can simulate the tool paths in real time within the programming environment. Here the user selects the desired details—from block-by-block machining up to a complete 3D depiction of the machining process with all the tools in the interior of the machine. Although effective collision monitoring has been moved forward to the program-creation phase, this offers no guarantee of safety in the event of errors in the NC program!

Another stage of development is **controller-specific simulation** (→ *Figure 3.12*). Here CAM systems or special simulation programs access a special version of the original CNC control software that has been supplemented with a

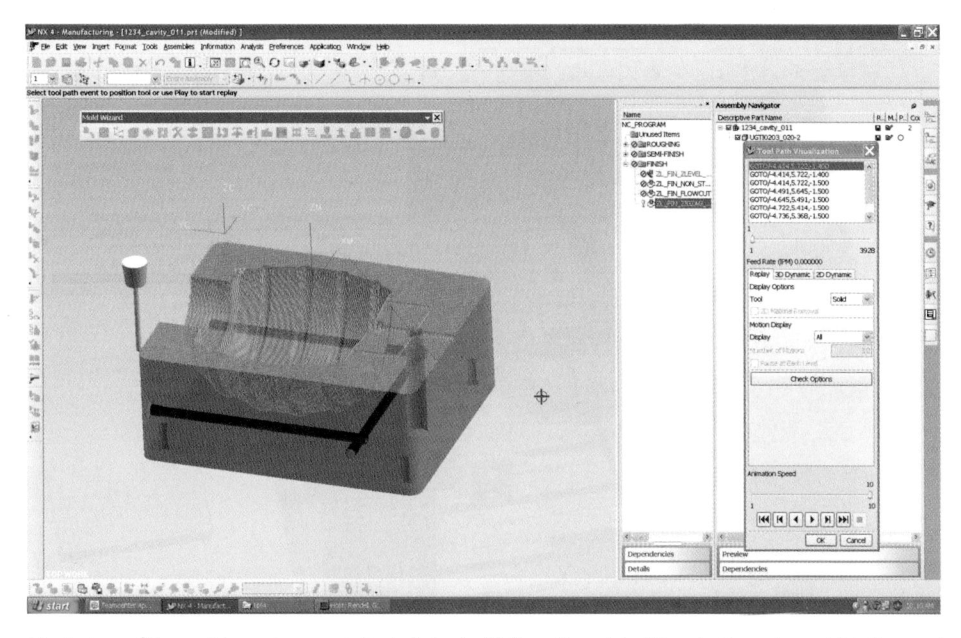

Figure 3.11: The halves of the mold remain associatively linked with the 3D model of the plastic part on which they are based. This means that last-minute changes can be transferred automatically in the CAM area.

module containing important parameters of common machine tools. These can include mass moments of inertia, axis information, and spindle forces. The result of the simulation *may* eliminate or shorten the verification of programs on the machine. Calculations of the run time of NC programs on the machine are sufficiently accurate to meet the requirements for large-series production, with certain limitations and tolerances.

NC program simulations are also expected to produce some additional results. These include first of all realistic, true-to-scale depictions of the machine, tool, and fixture; dynamic processes during tool changing or rotation/tilting of the workpiece with indication of collisions; and the type of finished boreholes (with or without thread), etc. (→ *Part 5, Chapter 4: Manufacturing Simulation*).

3.5 Summary

Today, the actual process chain of a manufacturing company from draft design to development, design engineering and verification, and production and shipping is effectively supported by a digital "backbone" that also encompasses customers, service providers, and suppliers. It must be possible for other companies to tap into this backbone using different software solutions. As for other software components, the main requirement for CAD and CAM solutions therefore is openness. There has been much improvement in this area in recent years. Interfaces and standards, the open exchange of data, and cooperative agreements between system suppliers are the right approach to providing greater utility for the user.

The modularization and specialization of the individual software solutions along the process chain have eliminated many of the problems that existed before and have led to great progress. For this reason, there are hardly any CAD systems appearing on the market that have significant new design engineering functions. Today the greatest potential for improved efficiency in companies lies in their processes.

Product life-cycle management (PLM) is a very promising approach toward using modern software solutions to integrate all areas of a company in lean, efficient processes. Employees in marketing and sales, service, and installation

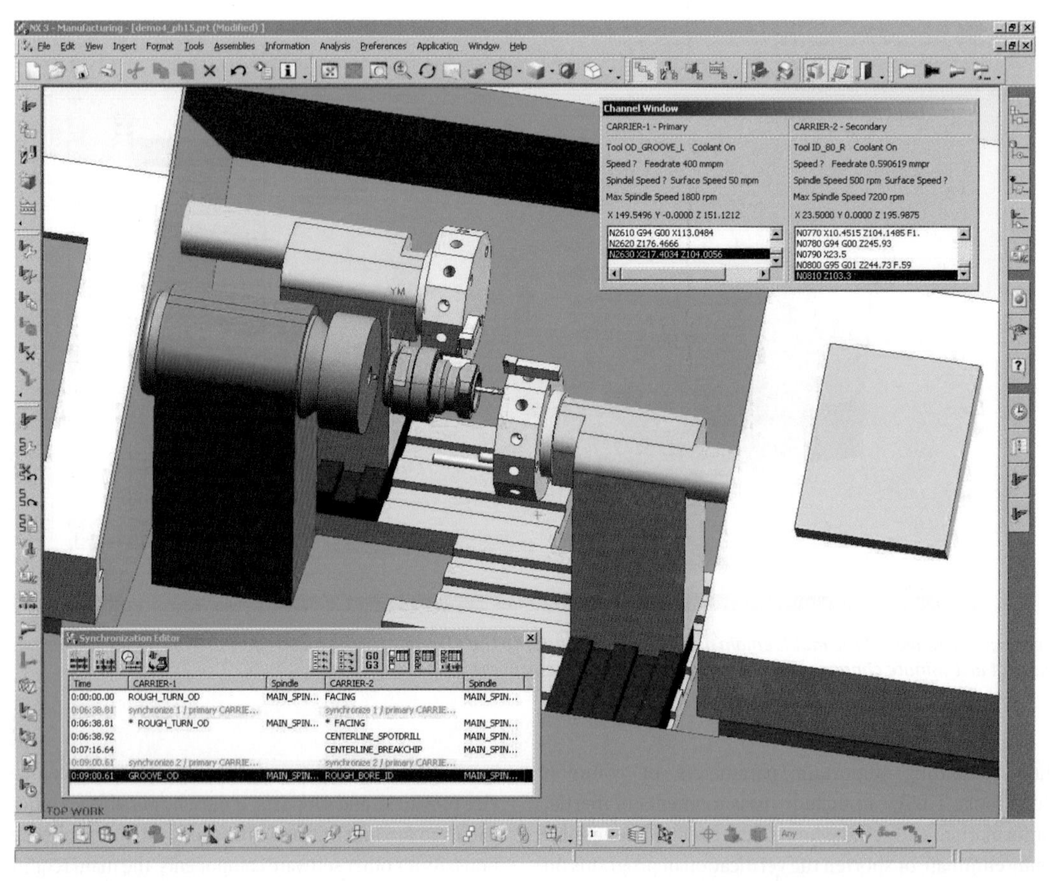

Figure 3.12: The monitoring functions on the screen should be as comprehensive as possible and must realistically depict the entire interior of the machine.

are networked with development and production. A spectrum of digital information from a wide variety of sources is available in a structured form whenever it is required. This is one reason why engineering "on which the sun never sets," international collaboration, and production facilities located all over the world have become an option for medium-sized companies.

These new work methods also place new requirements on employees. Openness and a sense of responsibility are just as important here as the ability to work in a team, to understand other cultures, and to communicate clearly. Despite all these changes, or perhaps even because of them, product development and manufacturing will continue to present interesting career challenges in the future.

Digital Product Development and Manufacturing

Important points to remember:

1. A distinction is made between 2D and 3D CAD systems depending on their methods of operation. 3D CAD systems are able to describe 3D bodies precisely and comprehensively. They also have 2D drafting functions.
2. Utility can be derived from 3D CAD systems not just during design engineering work but also in all subsequent steps of the process chain because more information is linked to the model.
3. CAE software can be used to check components at an early stage of the development process, thus saving development time and the cost of errors in product development. More and more design engineers are performing analyses on the results of their own work.
4. CAM should remain as tightly linked to product development as possible. A high degree of data consistency saves duplication of work, makes it easier to make last-minute changes, and makes manufacturing more flexible.
5. Every time an NC program is created, the last step should be a 3D simulation of the machining process on the screen. Simulated machining can provide absolute safety only if the conditions are realistic, that is, taking into account the machine, clamping fixture, tools, and all motions.
6. CAD, CAE, and CAM each describe subsections of the digital product definition. It is useful to view and organize them as a continuous process.
7. One good way of integrating these processes is PDM systems. Once the number of CAD workstations reaches five or more, it is no longer possible to adequately reflect product development data using directory structures.
8. PDM systems organize the management, distribution, and use of all product information in modern database structures. They also specify established procedures for many processes in the company.
9. Even today, drawings remain an important source of information. They primarily play a role in transferring information to the production shop but also for product liability and warranty purposes. When working with 3D CAD systems, drawings are 2D drafts of the 3D model. Whether the information is correct and up to date depends on the 3D model.
10. Digital manufacturing encompasses all the preparatory measures for efficient, productive work in the production shop. Besides CAM solutions, CAP solutions and software for material flow and production-system simulation also play an important role when changing over from one product to another in large-series production.
11. Product development and production must exchange information constantly with all other areas of the company. The best way to organize this is with product life-cycle management (PLM) concepts. Greater productivity, capacity for innovation, and shorter development times provide economic advantages. PLM is a way to achieve this.

Glossary

Rapid advances in technology mean that the technical terminology is also constantly being adapted, expanded, and modified. The following terms and explanations refer primarily to the field of NC technology.

Absolute Measuring System Position-measuring system for NC axes in which all measured values are relative to a specified zero point. Each point of the measuring distance is identified by means of a unique measurement signal.

The measuring systems used most often are encoded rulers, rotary encoders.

Ant.: Incremental Measuring System.

Access Time Time needed to call up a certain quantity of data from memory.

Accuracy, Precision In NC machines, a distinction is made between static and dynamic accuracy.

Static accuracy refers to absolute, repeatable positional accuracy. It is affected by systematic and random errors. → *VDI/DGQ 3441.*

Dynamic accuracy takes into account inaccuracies arising from the feed rate and acceleration. → *VDI 3427.*

The accuracy that can be achieved on the workpiece is always lower than the machine accuracy because it is also influenced by other factors, such as the stiffness of the machine and clamping fixture, thermal affects, tool wear, tool weight, and the machining operation.

Actual Value Value of a controlled variable, for example, spindle speed, feed rate, or position of the NC axis, returned by a measuring system at a particular moment.

Actuators Actuating drives in a control system that which transform control signals into mechanical motions. Actuator types currently in use include electric motors, hydraulic and pneumatic cylinders, piezoelectric actuators, and ultrasonic motors. Linear technology makes use of electromechanical lifting and positioning systems. Actuators can be controlled using open-loop or closed-loop control. Muscles can be viewed as the biological counterparts of actuator technologies.

Adaptive Control (AC) Control system for automatically adapting the cutting conditions to optimize a specific variable, such as maximum stock removal rate or optimal tool utilization. These specifications can be achieved, for example, by altering the feed rate during milling. This requires special measuring sensors for spindle flexion, motor output, torque, motor heating, and chattering. These sensors continuously report the current cutting and output data to the AC controller. These data are used to calculate the command variables for the feed rate and spindle speed, which are then output to the actuators.

Adaptive control is seldom implemented owing to its great technical complexity and expense.

Address Within an NC program, a letter that defines the assignment of the numerical values that follow.

Example: X27. 845 means the dimension 27. 845 mm in the *X* axis, F125 means 125 mm/min feed, etc.

Additional Functions

→ *Auxiliary Functions.*

Additive Manufacturing Processes An overall term for the process that was previously often referred to as *rapid prototyping*, a quick and cost-effective production of models, specimens, prototypes, tools, and products. Fabrication is performed layer by layer based on internal CAD data models using amorphous materials (fluids, powders, etc.) or materials with a neutral form (strips, wires, paper, or film) and using chemical and/or physical processes. Examples of these processes include →*Stereolithography*, →*Selective Laser Sintering*, →*Fused-Deposition Modeling*, →*Laminated Object*, modeling, and 3D printing. They can be used economically to manufacture parts with a high level of geometric complexity.

AGV (Automatic Guided Vehicle) Automatic transport system, primarily used to transport parts and tools in FMSs.

Driverless tracked vehicles using various guiding technologies: **inductively,** with wires embedded in the floor; **optically,** with markings on the floor; **laser-guided,** with reflectors in the surrounding area, or **RFID**, with position sensors in the floor or via satellite using **GPS.**

Alphanumeric Code A code for NC programs using a combination of letters, digits, and special symbols.

Example: N123 G01 X475,5 Y-235,445 F 250 T7 M02 $.

Alternate Tool Also called *replacement tool.* A tool of similar type in the tool magazine that can be used when the original tool is broken or has reached the end of its service life.

Amplification Factor Factor for amplification in the closed-loop control or the lag (following error) of an NC axis depending on the feed rate (v).

$$kv = v\text{: following error (m/s), mm}$$

The greater the amplification factor, the stronger is the dynamic behavior of the control circuit.

Amplifier An electronic unit for increasing the power of an input signal to output signal. In NC systems, this is generally a servo amplifier that supplies the power for the servo drives.

Analog In the sense of analogous, a physical variable is analogous to another physical variable if it changes in a certain way depending on the first variable. This relationship may be linear but does not have to be so.

The key feature: An analog signal can have any value between the limit values (theoretically infinite resolution).

Examples:

A) Sundials, thermometers.

B) Representation of a physical variable in the form of an electrical voltage, for example, speed as tachometer voltage, angle of rotation as potentiometer voltage, and temperature as the voltage of a thermocouple.

C) Resolvers. These supply two output voltages that change proportionally to the sine and cosine of the mechanical angle of rotation of the resolver axis.

Ant.: Digital.

Analog-to-Digital Converter Generally, an electronic device that converts an analog input signal into a digital output signal.

Artificial Intelligence Abbreviated AI. Branch of computer science research concerned with the development of computers that can replicate human intellectual performance. This term is difficult to define because no exact definition of *intelligence* exists.

Examples: Pattern recognition, ability to learn, ability to carry on a conversation.

ASIC Abbreviation for *application-specific integrated circuit.* A customer-specific electronic circuit implemented as an integrated circuit (IC) chip. ASICs are manufacturer assembly language.

ASCII Abbreviation for *American Standard Code for Information Interchange.*

A standardized code for data storage and transfer; also used for punched tape. Characters are depicted using seven bits for information; the eighth bit is a parity bit for even-parity-checking number per character. With 7 bits, a maximum of 128 characters can be coded.

ASCII Keyboard Alphanumeric keyboard for entering data in the form of letters, numerals, symbols and auxiliary commands in ASCII format. Widely used keyboard with personal computers and CNC systems.

Assembly Language, Assembler

1. Low-level symbolic programming language for a computer.
2. Translation program that converts programs written in an assembly language and converts them to machine instructions.

Asynchronous Axes Auxiliary axes that can be programmed and controlled independent of the main axes of an NC machine, for example, for a loading robot.

Asynchronous Mode Mode of operation or transmission that is not time-dependent and functions independent of other processes.

Asynchronous Motor Electric motor for three-phase current in which, depending on the load, the rotor always runs slower than the rotating field generated in the stator via the windings. This is also called *slip.*

The rotor of the **squirrel-cage motor** consists of copper or cast-aluminum bars short circuited to both ends of the rotor.

This type of motor is very widely used. Torque is created by the induction currents generated by the rotating field.

In the case of **slip-ring rotors,** the rotor is designed with three windings connected via three slip rings and three brushes to enable high-inertia starts using external electrical resistors.

Both motor types can be connected directly to the three-phase network. Depending on the stator winding (2-, 4-, 6- or 8-pole), the rated speed is 3,000, 1,500, 1,000, or 750 rpm minus the slip. In normal operation, asynchronous motors are not designed for and are not suitable for speed control.

In order to allow speed control all the same, a frequency converter has to be connected upstream of each motor. In this combination, squirrel-cage motors are used primarily as **speed-controlled main spindle drives.** Speed is changed by changing the voltage and frequency. Contemporary asynchronous servomotors are specially designed to allow a large wide speed range.

Automatic Guided Vehicle (AGV) Trackless, unmanned, computer-controlled transport vehicle for workpieces and tools.

Automatic Mode NC operating mode in which a part program is automatically processed without interruption.

Automation Automatic sequence of a number of successive manufacturing processes in order to release humans from the need to perform repetitive mental or manual functions and to free them from dependence on the machine cycle. Unlike mechanization, in which the entire work process is repeated without changes, in automated systems, work is performed according to a program that is specified from the outside and that can be changed. The entire process is monitored and deviations are corrected in a self-regulating manner, that is, automatically.

Auxiliary Functions, Miscellaneous Functions Commands for controlling switching functions of an NC machine, such as spindle ON/OFF, coolant ON/OFF, tool changing, workpiece changing, or end of program.

Axis The principal direction in which the motion of the tool relative to the workpiece takes place. In milling machines, these are the three linear axes *X*, *Y*, and *Z*, which lie perpendicular to each other. Additional parallel axes are designated *U*, *V*, and *W*. Additional rotational axes (rotary and swivel axes) are designated *A*, *B*, and *C* (→ *DIN 66217, ISO 841*).

When defining the positive axis direction (+*X*, +*Y*, +*C*, etc.), it is assumed that the tool is moving and the workpiece is stationary. If the workpiece moves, such as on coordinate tables and rotary tables, then the axis is marked with an apostrophe: *X*′ or *Y*′ or *C*′. This makes it possible to carry out the NC programming independently of the structural design of the machine.

Axis Calibration Also called *axis error compensation*. A CNC function to compensate systematic position measuring errors in the NC axis, thus achieving greater accuracy. This can be done, for example, by measuring the NC axis precisely with a laser interferometer and saving the dimensional deviations of the path-measuring system as *compensation values*. In subsequent operation of the machine, these values are added to the measured values depending on the position and direction (positive or negative).

Axis Change Switching the *Y* and *Z* axes in the NC in order to process NC programs on machines with different axis arrangements and angular heads placed in front.

Axis Lag Also called *following error*. Dynamic distance of an NC axis between the calculated position and the actual position during axis motion. The lag value depends on the control-loop amplification (*kv* factor) of the drive and the feed rate.

Backlash A relative movement delay between mechanical parts resulting from looseness.

Backlash Compensation Also called *axis reversal-error compensation*. A CNC function for electronic compensation of the backlash in the mechanical drive chain of an NC axis with rotary encoders.

Ball Screw Thread spindle with low friction between the spindle and recirculating ball nut; used for force transfer in NC machine slides. Additional advantages are high pitch accuracy, minimized backlash between spindle and nut, and a high efficiency of approximately 98%.

Barcode This code consists of a band of bars of various widths that represent alphanumeric characters. There are a number of different encoding systems. These are hard to decipher visually, which is why Arabic numerals are generally also printed. Barcodes are used for labeling and automatic identification of tools and workpieces.

Instead of barcodes, increasing use is being made of **datamatrix codes,** which require less space (2 × 2 cm) and can contain up to 200 characters. The general trend is toward the replacement of optically readable encoding systems by radiofrequency identification (RFID) systems.

Baseband The unmodulated frequency range of a signal, for example, in a LAN. Only one transmission channel is available, which is used by the connected stations one after the other.

Ant.: Broadband.

Baud Data communication term that defines the transmission speed as the number of bits transmitted per second.

BCD Code Abbreviation for *binary-coded decimal code.* A method for encoding decimal numbers using several binary numbers. Each digit is represented by its own binary number.

Example: 0001 0010 0011 0100 1000 1001 = 123.489.

The BCD code makes it easy to represent any numerical value because the number of digits is unlimited, and conversion is very simple.

Binary Two mutually exclusive states, for example, yes/no, correct/incorrect, on/off, 1/0. These are very easy to depict digitally, for example, with hole versus no hole in a punched tape.

Binary Code Code that uses only two different elements to represent data, that is, 0 and 1. Especially suitable to represent data that has to be transmitted in digital form or processed further in computers. n binary signals can be used to define 2^n states.

→ *Binary Numeral System, BCD Code.*

Binary Numeral System Base-2 number system; that is, all numbers are represented as powers of 2.

For example

$77 = 2^6 + 2^3 + 2^2 + 2^0 = 64 + 8 + 4 + 1$

The binary representation is:

1001101 = 77; that is, $\mathbf{1} \times 2^6 + \mathbf{0} \times 2^5 + \mathbf{0} \times 2^4 + \mathbf{1} \times 2^3 + \mathbf{1} \times 2^2 + \mathbf{0} \times 2^1 + \mathbf{1} \times 2^0$

→ *BCD.*

Bit Abbreviation for *binary digit.* The smallest representation unit in a binary number: 0 or 1. A hole (or no hole) in a punched tape corresponds to one bit.

Block In an NC program group of words containing all instructions for one operation. A complete block consist of the following "words": N (number); G (prep function); X, Y, Z (coordinates); F (feed); S (speed); T (tool number); and M (miscellaneous functions) and ends with the "end of block" ($) or "line feed" character. One block generally comprises one line in the NC program printout.

Block Cycle Time Indication of how quickly an NC system prepares and presents sequential NC blocks for processing. This can be used to calculate the following:

1) The maximum feed rate for polygonal lines (linear interpolation) produced by very short displacement intervals, or
2) The minimum permitted displacement intervals per block for a specified feed rate.

Block Delete An NC function for selectively deleting blocks designated for deletion in an NC program. Such blocks are generally indicated by a slash in front of the block address (/N478).

Block Number Ordinal number for numbering the blocks of an NC program, programmed under the N-address. Primarily used to inform the operator about the status of the machining.

Block or Sequence Search An NC function for reading an NC program over to a specific point, with all machine functions being disabled. In order to start the program at the required block number, it is also necessary for the tool, speed, feed, and compensation values to be activated.

Bridge Electronic device that joins two data networks of similar type.

→ *LAN.*

Broadband Transmission Denotes a data network that has a number of frequency bands for simultaneous, selective data transmission between numbers of stations. The best example is cable television.

BTR Input Abbreviation for *behind-tape-reader input.* Data input for an NC "downstream of a tape reader" to input data to controllers that do not have a DNC interface. No data output is possible.

Bus Originally named for *bus lines.* A group of conducting wires used for a group of electronic devices that are connected in parallel in order to exchange binary signals. A bus is a bidirectional data link between parts of an electronic system. A distinction is made between address bus, data bus, and control bus.

Byte A group of eight binary numbers stored and processed as a unit.

Example: A letter, a digit, or a symbol on punched tape.

CA Abbreviation for *computer-aided,* with the following additions:

A for assembly,
D for design,
E for engineering,
M for manufacturing,
P for planning or production planning,
Q for quality assurance, and
T for testing

CAD/CAM Abbreviation for *computer-aided design/computer-aided manufacturing.* The integration of CAD and CAM, CAP and CAE techniques through standard communication interfaces to ensure access to a common CAD database by all departments.

CAN Bus Abbreviation for *controller-area network,* developed by BOSCH for applications in the automotive industry. Together with Interbus and Profibus, CAN bus is one of the three most important *field buses,* also called *sensor/actuator buses.* Suitable for transmitting small quantities of data in a rapid sequence between the controller and the sensors/actuators via a short communications path.

CANopen, DeviceNet CANopen and DeviceNet are CAN-based layer 7 communication protocols that are primarily used in automation technology. CANopen is mostly used in Europe. It was initiated by small and medium-sized German companies and developed in the framework of an ESPRIT project led by Bosch. It has been managed by CiA since 1995, and now it has been standardized in the form of European Standard EN 50325-4.

DeviceNet, on the other hand, is more common in America.

Cardrack Mechanical holder for a number of electronic modules for controllers. The modules are generally of plug-in design.

Cartesian Coordinates Perpendicular coordinate system with axis designations *XYZ* for determining the position of a point in a plane or in space.

CATIA Abbreviation for *computer aided three-dimensional interactive applications.* Programming language for 3D surfaces and volumes developed by the French company Dassault and marketed by IBM.

Cell, Manufacturing Cell

→ *Flexible Manufacturing Cell.*

Channel Communication channel; a path for the transmission of signals, data, control information, etc.

Channel Structure Capability of a CNC system to subdivide all NC axes if necessary into synchronous, that is, mutually interpolating main axes, and asynchronous, that is, auxiliary or secondary axes that function in a time-independent manner from the main axes.

CIM Abbreviation for *computer-integrated manufacturing.* Information network including all of the departments involved in production (CAD, CAP, CAM, CAQ, PPC) with the goal of being able to respond more flexible to changes and corrections. In additional to technical functions, it is also possible to integrate organizational functions (i.e., sales, purchasing, cost accounting, etc.) into a CIM system in order to act more quickly when it comes to tenders, promised delivery dates, and pricing.

Circular Interpolation Calculation within the NC system of the points on a circle between the programmed start and end points. Circular interpolation is generally only possible only in the planes *XY*, *YZ*, and *XZ* and not diagonally in space.

CISC Abbreviation for *complex instruction set computer*.

CLDATA Abbreviation for *cutter-line data*, that is, the preliminary results for the tool path calculated by the computer program in computer-aided programming systems. These data are only converted to the block format of the specific NC machine by the postprocessor.

Closed-Loop Control, Servo Loop Definition according to DIN 19226: "The characteristic feature of closed-loop control is the closed action flow in which the controlled variable continually influences itself in the action path of the control loop." In other words, an electronic comparator compares the feedback signal with the command variable and generates an error signal as input for an amplifier to minimizing the deviation.

NC machines have a number of closed-loop systems, such as for the spindle speed, the feed rate and the axis position.

CNC Abbreviation for *computer numerical control*. In CNC systems containing one or more microprocessors. The identifying characteristics are the color display, the keyboard, and the ability to store programs, correct them, and to read them in and out automatically.

Because all current NC systems have at least one microprocessor, the terms *NC* and *CNC* can now be regarded as synonyms.

Code Rules that have been agreed on for converting data from one representation to another, for example, the assignment of specific hole combinations in a punched tape to specific numerals, symbols, or letters or combinations of bars in a barcode.

In NC systems, data encoding is used in order to store data on data carriers and to read data in and out automatically.

Code Checking Control function of the NC or programming system to detect incorrect symbols during data transmission. This is done by means of a parity check of all bytes in the ISO code and checking the meaning versus the information permitted in the programming rules. Information with the wrong values *cannot be detected by code checking.* Hereto → *Error-Detecting Code*

Code Converter This electronic device converts digital input signals in one code into another code, for example, binary code into BCD code.

Command Position Programmed dimensional data for an NC axis (machine slide).

Comparator A functional unit that, for example, compares the command value for an axis with the actual position value and in the event of a deviation generates a correction signal for the closed-loop control to minimize this difference.

Compatibility Two systems (hardware or software) are compatible if they can work with each other without additional equipment or modifications or if they can be exchanged for each other.

Compiler A computer program, which translates a programs written in a higher-level programming language into the machine language or operation code of a specific computer.

Computer Program-controlled electronic computing machine used to solve mathematically defined tasks. A distinction is made between analog and digital computers; CNC technology uses digital computers exclusively.

Computer-Aided Programming The generation of NC programs using a computer-based NC programming system. These systems provide interactive programs with graphical support for the programmer. The output is a CLDATA file (cutter location data) which has to be translated by a postprocessor for use with any suitable NC machine.

Concurrent Engineering (CE) Simultaneous development of a number of assemblies for a product in different locations, preferably using a uniform CAD system.

Continuous Path Control, Contouring Control Numerical control system in which the relative motion of the tool and workpiece is controlled continuously along a programmed path. This is achieved through the coordinated simultaneous motion of two or more machine axes. In addition, continuous path control includes so-called interpolators, which for CNC systems come in the form of software. When the start and end points of a defined path element (straight line, circle, parabola, etc.) are entered, the interpolator calculates the exact path and velocity of each axis between these two points. In each NC block, all involved axis start simultaneously and arrive at the programmed end position simultaneously.

Contour Segment Programming Programming function for inputting a number of contiguous geometrical contour segments that are not dimensioned individually but which, when taken together, produce a clear pattern. The system automatically calculates the individual points of intersection, transition radii, and tangential transitions and generates the appropriate NC program.

Controller, Machine Control Unit Electrical or electronic device for controlling the programmed functions of a machine. In NC technology, all of the control functions are divided between the NC system and the PLC.

Converter Electronic device which changes signals, codes, or numerical data from one form to another. In connection with NC: code converters, digital-to-analog converters, serial-to-parallel converters.

Coordinates Numerical values used to define a point in space. NC technology is generally based on Cartesian coordinate systems and polar coordinates.

Coordinate System Mathematical system in which the position of a point in a plane or in space can be determined using numbers. Possible types are Cartesian, cylindrical, joint, polar, and spherical coordinates.

Coordinate Transformation Mathematical term for the conversion of space coordinates into the axis coordinates of an NC machine with swivel or rotary axes or into the axis coordinates of a robot with a nonlinear kinematic system. This makes programming of such systems easier because programming can be performed in space coordinates.

Corner Deceleration In order to prevent overloading of a milling cutter when milling inside corners, a special G-function reduces the feed automatically to previously programmed values and sets it back to 100 percent behind the corner.

Corner Rounding Automatic insertion of transition radii at erratic transitions in order to achieve rounded transitions.

CPU Abbreviation for *central processing unit*. The part of the computer in which programs are interpreted and executed. The CPU contains the main memory, the arithmetic logic unit, the control unit, and the input/output interfaces.

CRT Abbreviation for *cathode-ray tube*. Simply called *screen* in normal usage. In today's CNCs and NC programming systems, flat screens are used to display data and graphics.

CSMA/CD (Carrier-Sense Multiple Access with Collision Detection) Access procedure for transmission media (buses). A station that wants to transmit senses the transmission medium to check whether the network is free of other traffic.

Cutter Compensation → *Cutter Radius Compensation, Tool-Tip Compensation.*

Cutter Radius Compensation A control function to compensate for the difference between programmed and actual cutter diameters. For each cutter diameter, the NC system calculates the correct equidistant center point path from the workpiece contour, as well as the points of intersection and transition radii at corners.

The NC system uses the functions G41 = cutter left and G42 = cutter right of the contour to calculate the compensation value.

For turning machines, this function is known as → *Tool-Tip Compensation.*

Cycle A fixed sequence of a number of individual steps that are stored in the NC system and can be called up using a programmed M- or G-function. Cycles are supplemented with specific parameter values (withdrawal level, drilling depth, or cutting passes) in order to adapt them to the task at hand. Cycles make programming easier and reduce the length of programs considerably. NC programs with cycles are much easier to modify than those without.

Examples: Cycles for tapping, deep drilling, rough machining, tool changing, pallet changing, and measurement operations.

Cycle Time Minimum time required by the NC system to prepare sequential program blocks and make them available for execution. If the time needed to execute a block is shorter than the cycle time, then the machine stops briefly until the next block is released for execution. To prevent this, the feed rate has to be reduced.

D/A Converter Digital-to-analog converter, an electronic unit that converts a digital input signal into an analog output signal.

Data Numerical values intended primarily for automated processing.

→ *Information.*

Database Data stored in a computer that have been introduced from various sources, can be sorted according to various criteria, and which can be accessed by multiple users.

Data Carrier, Data-Storage Medium Portable data memory that can be used to save and transport data so that they can be read out and used again later. Today almost exclusively electronic memory cards, such as CD-ROMs, DVDs, USB sticks, and memory cards.

Data Interface Connection point between a CNC system and external systems for automatic transfer of data and control information. Examples: DNC interface, drive interface, computer interface.

→ *Interface.*

Data Memory Sorting of data in an electronic medium from which it can be used later.

Data Processing Performing calculations or other logical operations according to certain rules in order to obtain new information, in order to put it into a specific form, or to use it to control devices, for example, NC machines.

Datum Point, Zero Point

1) The origin of a coordinate system, or
2) The starting point (zero point) of a measurement system.

Decade Group of 10 units or the interval between two numerals with a ratio of 1:10.

Decimal-Point Programming Programming and input of displacements using the decimal point instead of leading or trailing zeroes.

Examples:

- 417 instead of 417000 with trailing zeroes
- .75 instead of 750 with trailing zeroes
- .001 instead of 000001 with leading zeroes

Decimal System Base-10 decadic number system. This uses the numerals 0 to 9, where the positional values of adjacent numerals are integer powers of 10.

Dedicated Computer Computer that is used exclusively for a specific task, such as a DNC computer for a limited group of machines or a CAD computer as a workstation for the mechanical design and compute machine parts.

Degree of Automation Ratio of automated work processes to the total number of work processes. As the degree of automation increases, the costs also increase progressively, thus also increasing the production costs.

Deviation For NC machines, the deviation of the actual path from the programmed command path. Also called *dynamic deviation.* This has an effect on the contour accuracy of the workpiece.

To keep the deviation as small as possible, the CNC should be equipped with special features (e.g., look ahead, stiff feed drives, etc.).

→ *VDI 3427, Part 1.*

Diagnosis Special functions of computers or CNC systems that make it possible to find and locate error sources using the screen. For CNC systems, for example, this includes the software functions Logic Analyzer, PLC Monitor, multichannel storage oscilloscopes, logbooks, graphical display of measured values, etc.

Dialog Mode Data-input method where the user is "guided" by means of screen masks. This makes it easier to input data and helps to prevent input errors.

Digital Measuring System Path- or position-measuring system that uses discrete individual steps to measure either the path that has been traveled (incremental) or the specific position (absolute).

Digital Readout, Numerical Display Values displayed directly as decimal numbers. In NC systems, for example, the axis positions are shown in mm or inches, the feed rate in mm/minute, and the speed in rpm.

Digitizing Capturing a physical model or a mathematically undefined curve from a drawing in the form of individual, sequential coordinates. The digitizer systems for 3D surfaces use a spindle-mounted probe or a laser system.

Dimensional Data In an NC program, all of the command-value specifications for the NC axes under the addresses X Y Z I J K R U V W.

Direct Measurement A linear position-measuring system in an NC machine is called direct if the system is directly coupled to the machine slide without conversion into a rotational motion for position measurement via a rotary encoder. This means that the measuring accuracy is not compromised by any inaccuracy in the spindle, measuring rack, or rotary transducer.

DNC Abbreviation for *direct numerical control* or *distributed numerical control.* A system in which one or more computers are used to save and manage all NC programs and to transmit them to the connected NC machines via cable or network connections (LAN) when they are called up. To do this, the **CNC** requires a DNC or LAN interface (e.g., Ethernet) for bidirectional data exchange.

DRAM Abbreviation for *dynamic random-access memory.* Dynamic read/write memory in which stored information has to be refreshed periodically in order for it not to be lost.

Drilling Cycles (G80–G89) Work processes for drilling, reaming, countersinking, tapping, etc. that recur frequently and are stored as subprograms. When they are called up (G81–G89), they are supplemented with parameter values (reference plane, drilling depth). Subsequently, the appropriate machining is performed at each *X/Y* position.

→ *Cycle.*

Duplex Mode Simultaneous data transmission in two directions.

Dwell, Dwell Time Waiting time after a block end, for example, to cut the tool free. Can be programmed in seconds or spindle revolutions.

EIA Code Eight-track code for NC machines, standardized by the U.S. Electronic Industries Alliance. EIA-358B corresponds to the ISO code, where the number of holes is even for each character.

EIA-232C, RS-232C Standardized interface for serial data transmission in NC systems primarily used as a DNC interface. Because its transmission speed is generally inadequate, it is replaced with faster LAN interfaces such as Ethernet.

Electronic Handwheel Electronic insert for the absent mechanical handwheels on NC machines. Small handwheels built into the NC control panel or on the machine. It consists of a handwheel connected to an electronic pulse generator and can be used by the operator to adjust each axis manually in the operating mode "SET-UP." The adjustment motion can be changed from 1 to 1000 increments per step.

Encoder → *Feedback Device.*

Encoding In general, encoding of information, observing certain defined rules. In NC, converting the control data into automatically readable coded symbols and saving them on data carriers or any other memories.

End-of-Block Character A defined character at the end of the block ($ or "line feed" code) that separates the individual machining information from each other.

End of Program (EOP) Programmable miscellaneous function (M00, M02, M30) for stopping an NC machine once machining is completed. It generates the commands for "Spindle off," "Coolant off," "Tool back into magazine," and "Move all axes to starting position."

Energy Efficiency Current CNC systems include special programs for recording, analyzing, and reducing energy consumption for each individual machine.

Engraving Cycles Programming directly at the machine of any text or numbers that need to be engraved on the workpiece, such as the date, time, quantity, or serial number. These are *executed on* the workpiece using laser beams or special milling cutters.

EPROM, EEPROM, FEPROM Electronic memory chip whose contents can be erased and reprogrammed by means of ultraviolet (UV) light or electrical pulses (FEPROM).

Error-Detecting Code

1) Simple method for data validation by checking each character for an even number of bits per character (ISO code).

   → *Parity Check.*

2) Higher-level data check by comparing the total of digits of each block with the value programmed under an additional E-address on the end of each block in the NC program (only possible with computer-assisted NC programming).

Ethernet A widely used data bus that uses various transmission media (coaxial cables, twisted-pair cables, fiberoptic cables). Thanks to the availability of various low-cost speed levels from 10 Mb/s to 10 Gb/s, the Ethernet standard has become the clearly predominant network protocol. This is true not only for local networks used for office communication but also increasingly for networks in industrial automation. In the medium term, Ethernet also may become established in regional networks metropolitan area network (MAN) because a fully integrated standard without protocol conversion is simpler and more economical than a large number of different systems with the necessary converters (from www. tecchannel.de).

Executive Program System software for a CNC system containing the range of functions of the controller and the special functions for a specific type of machine.

→ *Open-Ended Control.*

Expert System A knowledge-based computer program that uses previous experience and available knowledge for problem solving within a limited technical field and also applies the necessary rules, methodology, and procedures.

External Memory Data memory (memory expansion) outside the central element of a computer, generally designed as a mass storage device (hard disk) and often connected by USB interface.

Feed, Feed Rate The distance a tool travels per minute or per revolution. In NC programs, programmed under the address F and defined using a specific G-code (G94 = mm/min or G95 = mm/revolution).

Feedback Transmission of a signal from a later stage to an earlier stage within a closed-loop system, for example, the actual speed or actual position. Using the feedback signal, a controller performs adjustments automatically when there is a deviation from the command value.

Feedback Device Includes sensors, that is, devices, that measure a physical variable and convert the measured values into output signals that can be evaluated electrically. Various feedback devices are used in NC technology, for example, to measure axis positions, velocity, speed, torque, current, and temperature.

Field Bus Globally standardized (IEC 61158) industrial communication systems that link a large number of field devices (such as measuring probes, actuators, and drives) with a control device. Field-bus systems were developed to replace analog signal transmission and the costly parallel wiring of binary signals. Instead, digital transmission technology is used. A number of field-bus systems with various characteristics are available:

BitBus, PROFIBUS, InterBus, ControlNet, CAN, etc.

File Defined memory area in a computer or CNC system where certain data are stored for subsequent reuse, such as tool data, materials data, cutting data, or undercut data.

Firmware Software permanently saved by the manufacturer of an electronic device on a programmable chip, generally flash memory, EPROM, or EEPROM. The manufacturer's intention in doing this is to prevent tampering by third parties.

Fixed-Zero Dimension Also called *absolute coordinates*. Dimensions relative to the coordinate zero point.

Ant.: Incremental Dimension, Relative Dimension.

Fixture, Clamping Fixture, Clamping Device Mechanical clamping equipment that is used to hold workpieces of the same type in a precisely defined position on the machine table in order to machine them with high repeatable accuracy.

Flexible Manufacturing Cell Highly automated, autonomous production unit consisting of an NC machine with tool- and workpiece-changing equipment, additional monitoring equipment, and a DNC connection.

Flexible Manufacturing Island A clearly delineated shop area with a number of machines and devices necessary to be able to perform all of the required operations on a limited selection of workpieces. The people who work there carry out planning, decision making, and monitoring of the operations themselves.

Flexible Manufacturing System (FMS) Grouping of a number of machining centers and/or flexible manufacturing centers to enable fully automated complete machining of groups of related parts in any desired lot size and desired sequence and without any manual intervention because the elements are linked by means of a common automated workpiece-transport and workpiece-changing system. Generally, the entire system is linked to a host computer.

Following Error →*Axis Lag.*

Frame Commonly used term for a calculation rule, such as coordinate translation or rotation.

Fused Depositing Modeling (FDM) RPD (rapid prototyping) process for layer-by-layer creation of precision plastic parts.

Fuzzy Logic Logic using "degrees of truth" used, for example, to optimize or correct programmed machining processes within certain limits by a relatively simple control system. Application examples: Wire-cut electrical-discharge machining, die-sinking electrical-discharge machining.

G-Functions Also called *preparatory functions*. Consists of the word address G and a two-digit coded number. These commands for NC machines specify, for example, how the programmed end point should be moved to: in a straight line, a clockwise or counterclockwise circular path, or in combination with a particular cycle (G80–G89).

Gantry-Type Machine Machine tool with a movable gantry. Used mainly in the aerospace industry to machine multiple surfaces and long parts simultaneously using multiple spindles.

Advantage: Requires less floor space as compared to table milling machines.

Gateway Electronic device, generally a computer, used to connect dissimilar networks with different protocols by providing protocol translation.

Geometric Data Complete information concerning the geometry and the dimensions of a workpiece as they appear on the workpiece drawing.

Graphic Assistance Using the screen of a CNC or a programming system for multiple graphical depictions, such as input graphics for programming that show the workpiece contour, simulation graphics for testing the program sequence with dynamic depiction of the tool paths, auxiliary graphics to inform the operator quickly about problems that arise, diagnostic graphics for troubleshooting, etc.

Graphical Programming Hardware and software, which helps the programmer to produce, modify, or simulate the part program, for example, graphic terminals, editors, simulators, etc.

Group Technology, Groups Of Related Parts Groups of geometrically and technologically similar workpieces that can be processed using the same machines and tools without any significant changeovers.

Handling Unit Another term for a robot used for loading/unloading of a machine, for tool changing, or for assembling components.

Handshake In data transmission, a process that serves to coordinate the transmission and prevent transmission errors. The individual data blocks are transmitted only if the recipient has acknowledged error-free reception of the previous block.

Hard Disk Encapsulated magnetic system for data storage with a significant storage capacity.

Hardware All devices and components comprising a computer or controller system.

Ant.: Software.

Hard-Wired NC An NC in which all functions and commands are processed in fixed-wired circuits and components. Control functions can be altered only by changing the wiring and possibly exchanging or adding modules.

Ant.: Soft-Wired NC = CNC.

Helical Interpolation, Helix Interpolation In addition to circular interpolation in a single plane (X, Y), linear interpolation is performed in a third axis (Z), perpendicular to this plane. It is used to create internal and external threads with form cutters and to mill lubrication grooves (thread milling).

Hertz, Kilohertz, Megahertz, Gigahertz Abbreviated Hz, kHz, MHz, GHz.

SI unit for frequency, that is, *number of cycles per second.*

$$1 \text{ kHz} = 10^3 \text{ Hz};\ 1 \text{ MHz} = 10^6 \text{ Hz};\ 1 \text{ GHz} = 10^9 \text{ Hz}$$

Hexadecimal Base-16 number system, that is, with 16 numerals. Mainly used by computers; 0 to 9 are used for the first 10 digits, and the first six uppercase letters of the Latin alphabet (A to F) are used for the remaining six digits.

Decimal	Binary	Hexadecimal
0	0000	0
1	0001	1
3	0011	3
7	0111	7
9	1001	9
10	1010	A
11	1011	B
12	1100	C
13	1101	D
14	1110	E
15	1111	F

Hexapod Also called a *Stewart platform.* Kinematic structure of a machine or a robot with the characteristic that linear and rotational motions of a platform can be executed using six *struts* (axes) with adjustable lengths (i.e., jacks). This provides six degrees of freedom. Each position corresponds to a defined combination of the six axes. Via simultaneous control of all six struts, it is possible to generate any desired spatial motion sequence in the platform and the spindles and tools mounted on it. The working space is not cubic but rather hemispherical.

→ *Parallel-Kinematic Machines.*

High Tech Abbreviation for *high technology.* The state of the art achieved based on the latest results of research and development; cutting-edge technology. It will be outmoded by further developments in the near future, at which point it will become the normal/ordinary state of the art.

Examples: Microprocessors, bus systems, CDs, data memory.

Host Computer A central computer in a computer network that provides services to terminals or satellite computers. In a flexible manufacturing system, the host performs process-control functions for data distribution, transportation control, tool management, materials management, and error monitoring and collects feedback messages to generate management reports.

HSC Machine Abbreviation for *high-speed cutting machine*, that is, milling machines with extremely high speeds (up to 100,000 rpm) and feed rates (up to 60 m/min). This places particularly demanding requirements on the machine and NC system, such as high stiffness, low mass, short block cycle time, zero lag, look-ahead, etc.

Hub Telecommunication devices used to connect a number of computers in a star configuration; also used to describe *multiport repeaters*. These are used to connect network nodes or additional hubs to each other, for example, via Ethernet.

HMI Abbreviation for *human-machine interface*, another term for the operator control and display facilities of a machine.

IGES Abbreviation for *Initial Graphics Exchange Specification*. A non-manufacturer-specific standardized data format for transferring geometric data between different CAD systems. It is also used to transfer data from CAD systems to CAM systems.

→ *STEP, VDA-FS.*

Increment Growth of a variable in individual steps of the same size.

Incremental Dimensioning All dimensions are relative to the previous position.

Incremental Jog A manual control function that allows the operator to move an NC axis in defined increments, for example, 1 μm, 10 μm, 0. 1 mm, 1 mm, etc.

Incremental Measuring System A measuring system in which each displacement is measured by adding up displacement increments (e.g., 0.001 mm) in an electronic counter or in the CNC system. The counter value thus is a measure of the actual position.

Incremental Programming NC programming in which the coordinate values are specified as an increment relative to the previous position (G91).

Indirect Measurement Position-sensing method in which a rotational measuring system (e.g., rotary encoder, pulse generator) is driven by the lead screw or measuring rack and pinion. The inaccuracies of the transmission elements have a negative effect on the measuring accuracy. Modern CNCs can compensate for systematic measuring errors.

Ant.: Direct Measurement.

Industrial Robot (IR) A mechanical device that is freely programmable in several axes (degrees of freedom) that is equipped with grippers or tools and which can perform handling and/or manufacturing tasks (e.g., workpiece or tool changing, welding, laser machining, painting, and assembly).

They can be categorized according to

A) Their **kinematics**, that is, Cartesian, cylindrical, spherical, or joint-coordinate robots
B) Their **programming,** with teach-in/play-back method or with external data input
C) Their **control**, with pick-and-place or NC continuous path control
D) Their **type of drives**, that is, hydraulic, pneumatic, or electric
E) Their **use** as universal, special-purpose or movable robots (gantry or Cartesian gantry robots
F) Their **load capacity and carrying capacity**

Information Data in a clearly organized, easy-to-understand summary, primarily intended as information for humans.

→ *Data.*

Interface

A) Electrical interface*:* In an NC system, a special hardware and/or software link between the NC and the machine-controller for data transfer and control signals.
Examples: IGES, VDAFS, SERCOS, V.24, RS 232, etc.
B) Human-machine interface or man-machine interface: Also called HMI or MMI, consisting of the control panel with displays and input elements.

International Standards Organization (ISO) Code Standardized 8-bit code with 7 information bits and 1 parity bit in track 8. It is the standard code used for NC.

Internet Global computer network via which users can communicate with each other and exchange data. Various protocols are available for this (TCP/IP). For CNC machines, the Internet is also used for remote diagnosis and correction of errors and intervention to correct CNC executive programs.

Interpolation Calculation of intermediate points between specified start and end points to produce a smoothed curve. If the connecting segments are straight lines, then it is called *linear interpolation*; if they are arcs or parabolas, then it is called *circular* or *parabolic interpolation*. Modern systems also offer the option of *spline interpolation*.

Interrupt The temporary or lasting interruption of a running program at a point that is not intended as an end of program.

Intranet Private networks that exist within a company or other entity and use Internet protocols. Their main purpose is to make internal data and information available to employees while preventing access to these data by outsiders. If access to the external Internet is required, this can be implemented via gateway computers.

Just-In-Time Production A manufacturing strategy aimed at reducing stocks and work in-process by scheduling the delivery of the right amount of parts at the required time.

KB Abbreviation for *kilobyte*. Used to define the memory capacity of a computer or CNC system. It is written with a uppercase K and specified using the binary numeral system:

$$1 \text{ KB} = 1 \times 2^{10} \text{ bytes} = 1{,}024 \text{ bytes}$$

$$8 \text{ KB} = 8 \times 2^{10} \text{ bytes} = 8{,}192 \text{ bytes}$$

Kernel Elementary component of an operating system, with the following functions:

- Interface to application programs (i.e., starting, stopping, input/output, memory access)
- Controlling access to the processor, devices, and memory
- Allotting resources, such as allotting processor time to the application programs
- Monitoring access rights to files and devices in multiuser systems; etc

Kinematic Configuration Kinematics generally is a branch of classical mechanics. In machines and robots, the kinematic configuration describes their structure as regards motions, that is, the motions they can perform in Cartesian, cylindrical, spherical, or joint coordinates.

LAN Abbreviation for *local-area network*, that is, a data network that is limited with regard to its scope and range and which is not subject to any regulation by postal authorities. It is used to connect a number of computers and peripheral devices within a limited area and enables direct communication between the devices.

Data transmission mostly takes place via broadband technology, that is, using frequency-modulated carrier frequencies.

LaserCusing This rapid prototyping laser process can be used to build up high-density components layer-by-layer by melting single-component metallic powders.

Lead-Error Compensation NC function for programmable correction of measured lead errors in a ball screw or a measuring rack and pinion.

LED Abbreviation for *light-emitting diode*, that is, illuminating diode or luminescent diode. Semiconductor that emits colored light. Used, for example, to replace incandescent lamps while using less energy.

Linear Interpolation Calculation within the CNC system of all points on a straight line between the programmed start and end points. A distinction is made here between simple 2D interpolation, interpolation with plane switching (2½D), and interpolation in space (3D).

Linear Motors Electrical drives for linear motions of machine axes without additional mechanical gearing. With linear direct-drive technology, it is possible to avoid elasticity, backlash, and friction effects, as well as oscillations in the drive mechanic. Linear motors enable maximum dynamics and high precision.

Logistics Organization, planning, and control for the systematic availability and targeted employment of production factors (workers, equipment, and materials) in order to achieve the corporate goals, warehousing, and transport.

LOM Abbreviation for *laminated-object manufacturing*, an additive CAD/CAM-based rapid manufacturing process. The workpiece geometry is generated by gluing together successive layers of paper and then cutting them to shape by an NC-controlled laser. The result is a wood-like 3D model.

→ *RPD.*

Look-Ahead Function Automated function of CNC systems to look ahead to a tool path a number of blocks in advance in order to reduce the feed rate at critical contour transitions (corners, radii) in accordance with the machine kinematics. This makes it possible to maintain contour accuracy on the workpiece. It is also possible to detect and avoid impending contour violations, for example, if the tool diameter is greater than the workpiece contour during tool entry.

M-Function Also called *on/off command* or *switching functions*. Abbreviation for *miscellaneous functions* or *auxiliary functions*.

Programmable commands that control the switching functions of a machine tool, for example, switching the coolant (M08/M09) or spindle (M03/M05) on and off, or activating the tool changing process (M06).

M Functions Switching functions of the machine, programmed in the NC program behind the M address.

Machine-Control Interface Electrical or electronic control for adapting a CNC to a machine tool. It has the following functions:

- Decoding, saving, and amplifying the coded signals output by the CNC system and forwarding them to the actuators.
- Linking the signals with limit-switch feedback signals from the machine.
- Locking out commands to prevent impermissible on/off commands.

  Most often a function of the

→ *PLC.*

Machine Data Acquisition (MDA) Automatic acquisition and storage of essential machine data during the machining phase, supplemented with additional manually entered information. Allows greater transparency of the production equipment and facilitates quick analysis of technical and organizational weak points.

Examples of the data acquired include the machine operating time, downtime, and outage time and the reasons for them; fault messages and their causes; manual interventions in the automatic process; and correction-value inputs; etc.

MDA is a subcomponent of a comprehensive → *Manufacturing Data-Collection (MDC)* system.

Machine Tool Machines that use different tools to perform cutting or noncutting machining of workpieces, made of metal, wood, plastic, or other materials.

Examples: Turning, milling, drilling, electrical-discharge, and grinding machines; nibbling; shearing, punching, pressing, and rolling. Newer types of machine tools include water-jet cutting machines and laser-beam machines for welding, cutting, stock removal, or forming. Machine tools are operated either manually or automatically, with the latter being significantly faster and more precise. In the case of **NC machine tools**, a freely programmable sequence of machining operations can be executed, thus allowing different workpieces to be machined automatically in any desired order.

A distinction is made between **single-purpose machines** or **production machines** for one or more operations and **universal machine tools** for a variety of different operations and subsequent processing steps. → *Machining Centers.*

In series production, a number of machining units or machine tools are often grouped together and linked to each other in such a way that different machining operations take place one after the other (automatic rotary indexing or transfer units, assembly lines, and flexible manufacturing systems).

Robots, painting machines, measuring machines, welding equipment, and many other types of production equipment are **not** considered to be machine tools.

Machine Zero Point Defined zero position of an NC axis, generally the coordinate zero point that can be reproduced precisely by the position-measuring system.

Machining Center These are CNC machines with a high degree of automation for fully automated complete machining of components.

Machining centers, also called *manufacturing centers*, are CNC machine tools equipped for automated operation. Automation functions can be expanded by providing additional peripheral devices, such as a tool magazine with workpiece changer or pallet changer. In modern machining centers, the tool-changing times and chip-to-chip times are sometimes under 3 seconds, resulting in significantly shorter cycle times.

Machining centers are categorized according to the orientation of their main spindle (horizontal machining centers and vertical machining centers). Another distinction involves the number of NC axes, mainly 3 to 5.

Machining centers can have one or two additional NC axes for rotating and swiveling tables, some machines even allow the execution of sophisticated turning operations on rotating tables (turning/milling centers, hybrid machines).

The cost-effectiveness of this type of machine has led to similar upgrading and redesign of other kinds of machines, such as turning centers, grinding centers, sheet-metal working centers, and combined laser/water-jet machines. The goal is the fully automated numerically controlled machining of different parts without manual intervention.

(*Comment:* There is no standardized definition of a *machining center.*)

Machining Cycles Processing cycles for standard geometries, circular pocket milling, thread undercuts, engraving cycles, deep-hole drilling, etc. in a plane or also on the end face or lateral surface of turned workpieces or on swiveled workpieces.

Machining Time Also called *cutting time* or *production time.* For machine tools, the sum of all machining times during which the feed is active.

Macro A group of prestored instructions (control data) that can be called up as a unit and which reduce the programming effort required for repetitive tasks.

Mainframe Computer A computer with high computational power, a large byte width, and very fast, simultaneous processing. Generally the central computer of an enterprise, with a number of terminals or separate computers connected to it from various departments.

Manual Data-Input Control A CNC with an integrated programming system that enables programming of complete machining processes directly at the machine.

→ *Dialog Mode.*

Manual Part Programming The preparation of NC programs for a specific machine/controller combination without the use of a computer-based programming system.

MAP Abbreviation for *Manufacturing Automation Protocol.* Bus-based broadband network using the token-passing principle with seven defined hierarchical levels.

Standardized protocol developed by General Motors for interconnection of individual, independent manufacturing islands and computers in order to enable a seamless flow of information. The objective was for all data and information relating to proposal preparation, order management, work scheduling, materials management, production control, quality assurance, and accounting to be exchangeable via standardized, non-manufacturer-specific interfaces and communication protocols.

MB Abbreviation for *megabyte* (1 million bytes), but actually 1,048,576 bytes = (2^{20}).

→ *Byte.*

Mb Abbreviation for *megabit* (1 million bits), but actually 1,048,576 bits = (2^{20}).

→ *Bit.*

MDC Abbreviation for *manufacturing data collection,* such as production quantities, rejects, interventions, times, employees, etc. MDC allows greater transparency in the operating sequence and quicker analysis of weak points.

Measuring Gear High-precision gear unit that is used between an indirect position-measuring encoder and the mechanical measure material (rack and pinion or ball screw and nut).

Measuring System

→ *Feedback Device, Path-Measuring System.*

Mechatronics An interdisciplinary field of engineering based on mechanical engineering, electrical engineering, and computer science. The emphasis is on supplementing and expanding mechanical systems with sensors and microprocessors in order to implement semi-intelligent products and systems.

Memory Electronic functional unit that can receive, save, and then output data. In NC technology, various memory elements are used, including RAM, ROM, EPROM, and FEPROM.

Menu Selection of options offered to the user on the screen to perform a necessary task.

Method, Process A planned, deliberate, goal-oriented process to achieve a specific goal.

Microcomputer Functional unit consisting of a microprocessor, program memory, main memory, and an input/output unit. Minimum configuration for functional computer hardware.

Microprocessor A large-scale integration module containing the central element (CPU) of a computer. It essentially comprises the arithmetic unit, various working registers, and the sequential control. A microprocessor is still not functional in this configuration, however.

→ *Microcomputer.*

Microsystem Technology (MST) Also called *microelectromechanical systems (MEMS)*. Production of extremely small precision components weighing approximately 1 mg and up, with a correspondingly small size.

Minicomputer A medium-performance computer occupying an intermediate position between mainframe and microcomputers.

MIPS Abbreviation for *million instructions per second*, a unit of measurement for the processing speed of a computer.

Mirror-Image Operation By changing the direction of an NC axis (swapping of + and –), it is possible to manufacture two mirror-imaged workpieces using the same NC program.

Examples: Milling left-hand and right-hand doors, drilling housings and lids.

Modal Function Commands of an NC program that remain in effect until they are deleted or overwritten by another command.

Examples: G90/G91, G80–G89, or F, S, and T words.

In contrast, nonmodal functions only take effect in the block where they are programmed, such as G04 or M06.

Modem Electronic device for modulation and demodulation of data in data-transmission systems. It converts data from one form to another, for example, characters from 8-track code into bit-serial pulses for transmission via telephone lines.

Modular Design Assembling principle for complex, extensive electronic controllers from a number of simple modules.

Monitor Screen for textual or graphical display of data, operations, processes, dynamic simulations, and results from electronic devices and computers.

Motor Spindles A main spindle with an integrated, coaxially arranged motor.

The advantages are high control dynamics; a stiff, compact design; extremely high power density; and lower overall mass and frictional force. The direct acquisition of the measured values for the spindle position and speed allows high production accuracy when operated as *C* axis.

NC Abbreviation for → *Numerical Control.*

NC Axis Numerically controlled machine axis whose positions and motions are programmed by entering the dimensional values directly and are controlled using an NC system. To do this, each NC axis requires a path-measuring system and a drive with closed-loop control.

NC Machine Tool Machine tool equipped with a numerical control system.

NC Program Control program for machining a workpiece on an NC machine. It contains all of the necessary data and control commands in the correct sequence and is processed by the NC machine step by step.

NC Programming Also called *part programming*. Creating control programs for the machining of workpieces on NC machines. A distinction is made here between manual NC programming, computer-aided NC programming (CAD/CAM), and shop-floor programming (SFP).

Network Protocol Also called *transmission protocol.* A precise convention for data exchange between computers and processes that are linked via a network. It consists of a set of rules and formats (syntax) that defines the data traffic (semantics) between the communicating computers.

Nonproductive Time In machine tools, the sum of all times that the machine is not in production, for example, when it is in rapid traverse, times for tool changing, workpiece changing, measurement operations, etc.

Numerical Control A control system that "understands numbers"; that is, all commands are input in numerical form. In the case of machine tools, this means in particular the numerical values that control the relative motion between the tool and the workpiece (dimensional or geometrical data). In addition, there are the numerical values for speed, feed rate, tool number, and various miscellaneous functions (switching information). The control information can be input either via a keyboard, a data-storage device, or via a direct cable connection (DNC). The typical characteristic of an NC system is the ability to make quick program changes without manual intervention at limit switches or cams.

NURBS Abbreviation for *nonuniform rational-basis splines.* Method for mathematical modeling of ruled and free-form surfaces such as cylinders, spheres, or tori by means of points and parameters. NURBS allow more efficient editing of such curves and surfaces than is possible with point models. As opposed to other splines, these have the advantage of being able to represent cleanly all kinds of geometries, even sharp corners and edges. Newer CAD/CAM systems can process the NURBS output by the CAD system directly in the CNC system. Advantages: Reduced amount of data, higher accuracy and speed, even motion of the machine, and longer service life of the machine and tool.

Off-Line Not connected to the computer, that is, an operating mode of a computer system in which the peripheral devices operate separately and independently of the central computer. This is done by saving the data generated by the computer temporarily and only processing them later.

Ant.: Online.

Offset Compensation Electronic compensation for clamping tolerances in the workpiece or tools in order to eliminate the need for mechanical alignment or adjustment.

→ *Zero Offset.*

Offset Path Path with constant distance from the programmed workpiece contour.

Example: Cutter-radius compensation that is calculated automatically by the NC system.

On/Off Control Simple control circuit in which the feedback signal (actual value) is used only for switching a process on and off when a specified set point is reached. Also called a *two-position controller.*

Examples: Refrigerator temperature, filling-level control, and hot-water tanks.

Online Connected to the computer. An operating mode of a computer system in which the peripheral devices are controlled directly by the central computer. This is done by connecting them to the computer via a data cable and executing the data generated by the computer immediately.

Open-Ended Control A CNC system containing an industrial PC and using a PC operating system (Windows or UNIX). *Open* means that the user has the ability to intervene in the executive program.

Thus the user himself can introduce customer-specific modifications and machine-specific functions. It is not possible to execute older NC part programs.

Open-Loop Control A control system in which proper execution of a control command is not monitored and regulated by means of a feedback signal.

→*On/Off Control. Ant.:* Closed-Loop Control.

Operating System Software for automatic control and monitoring of program processes in a computer, for example, UNIX, Windows, or Linux.

OS Abbreviation for *operating system*, that is, the system software of a computer.

Override A manual function for NC machines that enables the operator to modify the programmed feed rate or spindle speed temporarily to adapt it to the machining circumstances.

Pallet A transportable workpiece-clamping table that makes it possible to clamp/unclamp workpieces outside of the machine and which can be exchanged into the machine automatically for machining. This reduces machine downtime and enables automated transport of the clamped workpieces by an automatic transport system to multiple machines.

→ *Flexible Manufacturing System.*

Parallel Axes

1) Two mechanically coupled NC axes, for example, *Y1/Y2* of gantry-type machines, that have to be activated simultaneously to avoid skewing.
2) Two NC axes acting in the same direction, such as the machine column and the spindle of a boring machine or tables and spindles with the same axis directions.
3) Two mutually independent NC axes, for example, main spindles *Z1* and *Z2* of two different headstocks of a vertical milling machine that can work simultaneously on two identical workpieces.

Parallel Data Transmission In parallel data transmission, a number of bits are transmitted simultaneously (in parallel), that is, on a number of parallel lines or via a number of logical channels at the same time. The number of data lines is not defined, but a multiple of 8 is generally used, thus allowing the transmission of full bytes (e.g., 16 lines mean 16 bits = 2 bytes). Often additional lines are used to transmit a checksum (parity bit) or a clock signal.

Ant.: Serial Data Transmission.

Parallel Programming, Parallel Input Mode Programming a new workpiece using manual data input controls while a different workpiece is still being machined.

Parametric Programming NC programming by inputting the parameter values instead of numeric values in single words describing the process. For rings of holes, this could be, for example, the diameter, the number of holes, the start angle and angular step, and the desired drilling cycle. With just these few input values, the system calculates the individual positions and machining sequences.

Parity Check A method for checking binary data for single-bit errors by adding an additional *parity bit* on track 8 to make the sum of bits per character always odd. The parity check detects incorrect characters during data transmission

Example: An uneven number of bits for a character in an ISO code.

Part-Program Zero Generally identical to the program zero point defined by the programmer in each coordinate to which all the dimensions of the NC program refer. The relationship to the machine zero point is taken into account via the zero offset.

Performance This term has various meanings depending on the specific field. In a technical context, it means the **sum of all features.**

Peripheral Units Collective term for additional devices connected to a computer.

Examples: Printers, DVD players/recorders, plotters, and monitors.

Photoelectric Line Tracer A non-NC-controller with a photoelectric read head that traces the lines of a special template in 1:1 scale. Primarily used with older flame-cutting machines, but declining in popularity due to the large 1:1 templates.

PKM Abbreviation for *parallel kinematic machine*. Collective term for machines that generate all motions via kinematic struts.

Examples: Tripods, hexapods.

Playback Programming method mostly used with robots in which the robot is guided manually, and the NC system saves the entire motion sequence simultaneously (teaching). The stored motion sequence can be automatically repeated at a different (higher) velocity.

PLC Abbreviation for *programmable-logic controller*. In their simplest configurations, these electronic controllers replace earlier relay controls for interlocks and linking of switching commands and functions (bit processing).

High-performance PLC systems are special process computers with a number of inputs and outputs for continuous monitoring and control of the process with data feedback (word processing).

Plotter Also called *coordinatograph*. A computer-controlled drawing machine which provides a graphical representation on paper.

Pocket Milling, Pocketing NC function for removing metal from an enclosed surface up to a certain depth using a single instruction instead of programming each single step.

Point-to-Point Control NC system that moves to all programmed positions over an uncontrolled path. No tools are engaged during this. The machining starts after positioning is complete. Used for drilling, punching, and spot welding.

Polar Axis Term for swivel axes, jointed axes, and rotary tables of NC machines.

Polar Coordinates Mathematical system for determining the position of a point in a plane using the length of its radius vector and the angle of this vector relative to the datum line.

Polynomial Interpolation Interpolation method in which the NC axes follow the function:

$$f(p) = a_0 + a_1p + a_2p^2 + a_3p^3 \text{ (pol., max. 3rd degree)}$$

This can be used, for example, to generate straight lines, parabolas, or power functions.

Position Display Visual display of the absolute position of a machine slide returned by the path-measuring system, measured from the axis zero point or the program zero point.

Position-Feedback Control, Closed-Loop Control Closed-loop control system that continually compares the command values with the actual values and, in the case of deviations, outputs a correction signal until the difference between the two values has been corrected and the desired position has been reached.

Position-Measuring System Measuring devices with signals that can be evaluated electrically to measure the axis motions and positions of an NC machine. A number of different types of measuring systems and measuring processes are available for this purpose: Linear scales and rotary encoders, absolute and relative systems, analog and digital systems, and pseudo-absolute systems.

Postprocessor (PP) Software program required for computer-aided programming; converts the standard data for the tool motions (CLDATA) calculated by the computer into a machine-specific NC program. A special postprocessor is required for each machine/CNC combination.

PPC Abbreviation for *production planning and control*. Integrated use of IT in production for organizational planning, control, and deadline monitoring for production processes from preparation of proposals to shipping. The main functions are forward planning of machine utilization, production deadlines, checking stocks of materials and assembly times.

Preparatory G-Functions In an NC program, the G-function (G00–G99) defines how the tool should move to the programmed position: in a straight line, in a circular path, in rapid traverse, or with a drilling cycle. Likewise, what to do at the programmed positions: drilling holes, thread cutting, tool-diameter compensation right or left, etc.

Standardized in ISO 6983 and DIN 66025.

Preset NC function for defining a specific position of the machine coordinates without axis motion. Merely a new position value is set for the current axis position.

Probing Cycles Subprograms stored in the NC system for automatic measurement of boreholes, pockets, or surfaces using a sensing probe and to calculate positions, accuracies, tolerances, centers of circles, center distances, or inclinations.

Procedure The process and/or way in which operations for obtaining, manufacturing, or disposing of products is performed.

Processor

1) Special electronic hardware in a computer that performs certain functions such as data processing, arithmetic or logic functions, etc.

2) Computer software for converting a part program that has been written in a problem-oriented language and processed by the computer into a general NC system independent form (CLDATA). This is converted into an NC program by the postprocessor.

PDM Abbreviation for *product data management.* A concept for storing and managing product-defining data and documents, and to make these data available in downstream phases of the product life cycle. It is based on an integrated product model. PDM largely arose from problems in the management of CAD drawings that resulted from dramatic increases in the quantities of product data associated with the introduction of CAD systems. The ISO 10303 series of standards (STEP) has assumed a dominating position for data exchange between participating systems and for describing product models.

PDM can essentially be viewed as an outgrowth of CAD systems. The purpose of PDM is to increase the quality of product development and to reduce the time and expense of product development. By achieving a fully integrated flow of information, it should be possible to pass these benefits on to downstream points that are tied in with the product life cycle.

PDM systems are generally specific to a particular industry and company.

PLM Abbreviation for *product life-cycle management.* Designates an IT solution system in which all data that are generated during creation, warehousing, and marketing of a product are stored, administered, and called up in a uniform manner. This means that all departments and systems access a common database: planning (PPC/ERP), design (CAD), production (CAM, CAQ), controlling, sales, and service.

Because of its complexity, PLM should not be viewed as a product that can be purchased but rather as a strategy that has to be implemented on a company-specific basis by means of suitable technical and organizational measures.

PROFIBUS DP (Decentralized Peripherals) A data bus used mainly in manufacturing technology to connect sensors and actuators with a centralized controller. Additional areas of application are connecting *distributed intelligence*, that is, the mutual networking of multiple controllers (similar to PROFIBUS FMS). Data rates of up to 12 Mb/s are possible using twisted-pair cables and/or fiberoptic cables. (*From the German Wikipedia.*)

Program → *NC Program.*

Program Edit Correcting NC programs by inserting, deleting, or modifying digits, words, or blocks in NC programs.

Program Format Rules and definitions for arranging data on a data carrier, for example, to design an NC program consisting of addresses, characters, words, and blocks with a variable block length.

Program Interrupt Programmed STOP command (M00) that interrupts the machining process and allows the operator to perform checks, make measurements, change tools, or reclamp the workpiece. Machining is resumed with a START command.

Program Word Basic unit of an NC block, consisting of an address and a numeric value.

→ *NC Programming.*

Programming Language Artificial language for symbolic description of computer instructions using mnemonic code words. In the context of NC programming, this means a problem-oriented language that can be used to create a source program. This is the input into the computer, compiled by the language processor, and then processed into a generally valid NC program (CLDATA). After that, it is converted by the postprocessor for the particular NC machine converts it.

Examples for problem-oriented languages are APT, EXAPT, and RADU.

Programming System Equipment for programming NC machines, consisting of a computer with keyboard and screen, programming software, and the corresponding peripheral devices.

PROM Abbreviation for *programmable read-only memory*, an electronic memory that can be programmed only once, after which its data are retained permanently.

Protocol Rules for the exchange of data between computers and/or other electronic devices.

Pseudo-Absolute Measuring

1) The use of two cyclic absolute encoders (resolvers) in connection with a special measuring gear and an electronic evaluation system. The fact that the angular positions of the two encoders differ by approximately one degree per rotation creates a phase shift that the electronics can use to calculate the absolute position. It is possible to measure a limited distance absolutely depending on the resolution, for example, approximately 5–8 m at 0.001-mm resolution (23 bits = 838,860,800 increments).
2) Pulse scale that is provided with distance-coded reference marks and special evaluation electronics at intervals of 20 mm. The absolute displacement is available once two reference marks have been moved over.
3) Rotary pulse generator with a backup battery for the encoder and counterelectronics so as to detect every motion of the machine even when switched off. The absolute axis positions are available immediately after switch-on.

Pulse Generator, Digitizer A rotary measurement device that generates a defined number of pulses per rotation with very high angular accuracy. In NC machines, digitizers are used as incremental encoders for measuring displacements or the spindle position of turning machines.

→ *Feedback Device.*

Quadrant

1) One quarter of a circle or the area of a quarter of a circle.
2) One of the four parts of a plane divided by two perpendicular coordinates.

Quality Assurance Generic term denoting quality engineering, quality control, and quality audit.

Radiofrequency Identification (RFID) This is a method for reading and writing data in data memory (in this case, a transponder) without physical or visual contact. This transponder can be attached to objects to allow them to be identified quickly and automatically based on the data saved there. The term *RFID* is used as a blanket term to describe the entire technical infrastructure. An RFID system contains the following:

- A **transponder** (also called an *RFID label, RFID chip, RFID tag,* or *radio tag*)
- A **send/receive unit** (also called a *reader* or *interrogator*)
- Integration with servers, services, and other systems such as cash-register systems or inventory-control systems.

Data transmission between the transponder and reader/receiver unit takes place by means of electromagnetic waves. At low frequencies, this occurs inductively via a near field and at higher frequencies via an electromagnetic far field. The distance over which an RFID transponder can be read can vary between a few centimeters and more than a kilometer depending on its design (passive/active), the frequency band being used, the transmission power, and environmental influences. (*From the German Wikipedia.*)

RAM Abbreviation for *random-access memory*. Electronic memory (read/write memory) in which each storage location can be addressed, written to, or read out directly. A distinction is made between DRAM = dynamic RAM with very short read times and SRAM = static RAM.

Random Tool Access A method for identifying tools in the tool magazine. Tools are placed in any free magazine pocket when the magazine is loaded. The tool location within the magazine is variable and changes with every tool change. The CNC performs the logical management of the tools and pocket numbers.

Advantages: Use of uncoded tools and programming of tool numbers in the NC program. The programmed tool is searched via the shortest distance.

Rapid Product Development (RPD) Additive manufacturing process in which workpieces with very complex internal and external contours can be produced directly from 3D CAD data in a single process using layer-by-layer deposition. The materials used include films, thermoplastics, or photopolymers, which can be processed using lasers, lamps, blades, extruders, or printers.

Examples: Stereolithography, laser sintering/melting, layer (laminate) process, and 3D printing.

Rapid Prototyping (RP) Computer assisted additive manufacturing processes to generate prototype workpieces by use of special machines.

→ *Additive Manufacturing Processes.*

Rapid Tooling Creating tools that are used as permanent molds for plastic injection molding or metal injection molding using → *Rapid Prototyping (RP) Technologies.*

Rationalization Technical and organizational measures for increasing efficiency, such as increased productivity or decreased resource requirements or costs. Rationalization saves means of production (raw materials, capital, and work) as well as time. In a production environment, rationalization is generally associated with automated processes and personnel reductions.

Real-Time Processing, Online Processing Computer operating mode in which calculated data are immediately processed. The numerical control of machine tools is a typical example.

Reference Point Defined position of an NC axis that has a particular relationship to the axis zero point. This is moved to after the machine is switched on in order to zero the machine coordinate system unambiguously. With absolute encoders, there is no need to move to the reference point.

Repeatability Also called *repetitive accuracy*. Accuracy achieved when a machine slide is moved multiple times to the same position under the same conditions. The deviations result from random errors and not from systematic errors.

Reproducibility Essentially identical to repeatability but measured over a longer period.

Reset Command to set a device or control system to a predefined initial state.

Resolution Also called *measurement step*.

1) The smallest increment that can be detected by a measuring system in order to distinguish between two discrete positions. In NC machines, generally 0. 001 mm.
2) Distance between pixels on a screen.

Resolver A rotary analog position-measuring system comprising a rotor and a stator with two windings spaced at 90° to each other. It converts rotary motion into electrical sine and cosine output signals (voltage) that are proportional to the angular position of the rotor and provides cyclic-absolute position measurements.

Restricted Area Also called → *Software Limit Switch*. Programmable NC function that temporarily prevents the tool from moving out of a programmed area or entering a specific area. The programmable limit switches prevent collisions between the tool and workpiece based on incorrect dimensional data input.

Retrofitting Converting an older NC machine to a newer, higher-performance CNC machine. Comes generally in conjunction with modernization of the drives and the measuring systems. In most cases, the electrical control systems will be replaced by a PLC. Cost-effective only for large, expensive machines that are in good condition.

RFID → *Radiofrequency Identification Device.*

Robot A program-controlled device that can perform complex motion sequences and is equipped with grippers or tools. Used primarily for handling tools or workpieces, or for assembly.

Examples: Coating, welding, deburring, polishing, and tool changing.

ROM Abbreviation for *read-only memory*. Electronic memory whose contents can only be read and cannot be changed.

Rotary Encoder, Shaft Encoder A measurement device that converts angular positions into digitally encoded data. This is done using an encoded disk that is subdivided into a certain number of discrete positions. These are scanned by photocells; each track of the coded disk is assigned to its own measuring cell. The position values are output as binary code. The Gray code is the most suitable.

Rotary Milling Economical alternative for turning operations on large workpieces using rotating tools (milling cutters). A distinction is made between rotary milling parallel to the axis and orthogonal rotary milling.

Not to be confused with turning machines that also can be used for milling or milling machines that also can be used for turning (turning/milling centers).

Rotary Table Rotating clamping and worktable to allow machining of cubic workpieces from several sides. In machining centers, the ***B'*** axis, that is, rotation about the *Y* axis.

→ *Tilting or Swiveling Table.*

Safety Clearance Minimum distance of the *Z* axis from the workpiece for automatic tool changing without any risk of collision.

Safety Functions Additional safety functions are used, for example, to fulfill the requirements for Safety Integrity Level (SIL) 2 of IEC 61508 and Performance Level (PL) EN ISO 13849 for adjustment and test operation when a safety gate is open. This makes it possible to implement the primary requirements for functional safety in a simple, cost-effective manner:
- Monitoring of velocity and standstill
- Reliable delimitation of workspaces and protected spaces
- Safety-related signals and their internal logical links

Scaling This CNC function can be used to produce workpieces with different scales using the same NC program. This is done by specifying a scaling factor for each axis, thus modifying the programmed dimensions accordingly.

Scanner In NC technology, a device for digitizing the coordinate values of a workpiece and saving them to a data carrier. In principle, a measuring machine that uses a measuring probe to scan the workpiece line by line and stores the measurement data.

SCARA Robot Abbreviation for *selective compliance assembly robot arm* or *selective compliant articulated robot arm.*

Sculptured Surface Complex surface, generally with multiple curves, that cannot be defined mathematically using simple basic geometrical forms such as straight lines, circles, and conic sections.

Selective Laser Sintering (SLS) Rapid prototyping process for heavy-duty prototypes based on CAD data with layer-by-layer melting of powdered materials using focused laser beams. With special molding sand, it is possible to create molds and cores for metal casting.

Semiconductor Components Electronic switching elements with special characteristics to trigger, generate, rectify, switch and control electrical currents.

Examples: Photoconductive cells, diodes, transistors, microprocessors, RAM, ROM, etc.

Sensing Probe, Touch Probe, Measuring Probe Precision gage with high switching accuracy and reproducibility of the switching point. These devices are mounted like a tool in the main spindle of an NC machine and used to measure the tool length or workpiece position or for monitoring the machining accuracy. The CNC system has to have special software programs in order to save the measured position values and to use them to calculate compensation values, centers of circles, or tolerances, for example.

→ *Probing Cycles.*

Sensors Electrical feedback devices for nonelectrical variables such as lengths, angles, pressures, torque, forces, and temperatures. When the measured variable changes, the sensor's output signal has to react as quickly as possible.

SERCOS Abbreviation for ***SER****ial Real-Time* ***Co****mmunication* ***S****ystem.* Comprehensive specification for digital interfaces between CNC systems and drives. Its purpose is to combine CNC systems and drives from various manufacturers and to synchronize them with greater precision than with analog control circuits.

Serial Data Transmission Transmission of information one bit at a time via a single data channel.

Ant.: Parallel Data Transmission.

Server A computer in a LAN that uses special software to control the connection to other connected devices (computers, drives, printers, etc).

→ *LAN.*

Servo Control A closed-loop control circuit whose controlled variable is a mechanical motion. In NC machines, this can be the position feedback control of the NC axes.

→ *Closed-Loop Control.*

Servo Cycle Time Time in microseconds for how often the actual position of an NC axis is scanned electronically and fed back to the closed-loop position control. This is an important factor for the dynamic accuracy of an NC machine.

SFP (Shop-Floor Programming) Easy-to-use programming procedure suitable for use by the machine operator on the shop floor by interactive menus and graphically based input functions.

A defining characteristic of SFP is the strict separation of geometry and technology inputs; that is, inputting the workpiece geometry does not specify the subsequent machining process. The workpiece contours are programmed and not the tool paths.

Simulation Computer-generated reproduction of a complicated technical process in an as realistic manner as possible, which is then presented on a screen. Can be used to check the subsequent process with regard to costs and time requirements.

Examples: Graphical-dynamic simulation of NC machining or robot motions for the purpose of error detection.

Simulation of a planned flexible manufacturing system in order to check bottlenecks, capacities, options for expansion, expansion variants, or scheduling problems even before the start of installation.

Single-Block Mode Also called *block-by-block mode*. Operating mode of an NC system in which the operator must initiate the execution of each block.

Slope Programmable or settable soft acceleration and deceleration of NC axes in order to prevent jerking motions and prevent wear on the mechanical elements.

SMD Abbreviation for *surface-mounted devices*, a special method for mounting microelectronic components on circuit boards.

Softkeys Freely assignable function keys, also called *multifunctional keys*. Generally, 5 to 10 mechanical or electronic touch keys with varying functions arranged around the screen of a CNC system. A number of various functions are assigned to these keys in sequence via software for operation of the CNC system. Softkeys replace many hardware keys with discrete individual functions.

Software In general, programs that are necessary for operation of a computer or a computer-based system. The software of a CNC comprises the basic software for the microprocessors, interface software, and application software for the machine tool. Should not be confused with the user's NC programs.

Software Limit Switch Programmable axis limitation for NC machines intended to prevent undesirable exceeding of the limits. Used as a replacement for mechanical axis limit switches or for temporary limitation of the working space in order to protect the machine and workpiece from damage caused by incorrect dimensional data input.

→ *Working Area Limit Switch.*

Source Program An NC program written in a problem-oriented programming language.

Spindle Orientation Programmable NC function for main spindles with position feedback for stopping the spindle in a defined or programmable angular position (0–359°). Necessary when withdrawing a boring tool with a one-sided cutting edge out of a borehole or for tool changing in the case of tool holders with gripping and fixing positions or for drilling/milling operations on lathes.

Spline Function A mathematical function for approximation of curves. This produces smooth curves that connect the specified interpolation points with a smooth, constant progression. A distinction is made between *A*, *B*, and *C* splines.

→ *NURBS.*

Spline Interpolation Concatenation of third- or higher-degree polynomials that exhibit a transient behavior with minimal bending. Used to calculate free-form curves based on a small number of interpolation points.

SRAM Abbreviation for *static RAM*. Electronic read/write memory whose contents are retained without periodic refreshing.

STEP Abbreviation for Standard for the Exchange of Product Model Data. International standard to allow trouble-free exchange and processing of CAD data (ISO 10303).

Stepping Motor An electric motor in which the rotor moves in small, discrete angular steps (e.g., 400 steps/

revolution). The displacement angle of the rotor corresponds to the number of pulses generated by the control device, and the rotational speed corresponds to the pulse frequency.

There is no feedback and therefore no closed-loop control. Only relatively low speeds and torques can be implemented. Hydraulic amplifiers are connected downstream to achieve greater torques.

Stereolithography CAD/CAM process for creating prototypes without a casting mold and without tools. The basis material is a bath of liquid plastic that is subjected to a numerically controlled laser or ultraviolet beam, which cures the plastic layer by layer.

→ *RPD.*

Straight-Cut Control Numerical control system that can only move the tool parallel to the axes (*X*, *Y*, *Z* in sequence) at the feed rate.

Subprogram, Subroutine, Macro Frequently-used program sections stored in the memory that can be called up by the main program. After that, the program sequence jumps back to the main program.

Synchronous Motor Electric motor for three-phase current in which, regardless of the load, the rotor always runs synchronously with the rotating field generated in the stator. The stator winding is identical to that of an asynchronous motor.

The **rotor** is equipped with permanent magnets or separately excited magnets, with one or more pairs of poles depending on the design.

Synchronous motors cannot be run up directly by switching on the three-phase current; to reach their rated speed, they also require a startup device, generally in the form of a squirrel cage.

Speed Control of synchronous motors is performed by changing the feed voltage and frequency using a frequency converter. The speed-control range extends from standstill (with standstill torque) up to the maximum permitted speed. This is motor-dependent and can be from 2,000 > 9,000 rpm.

The advantages of the speed-controlled synchronous motor are such that it is the preferred **axis drive** for CNC machine tools. Contemporary synchronous servomotors are special designs that allow a large control range and dynamic speed response.

Syntax

1) In grammar, study of sentence structure and the placement and functions of individual words and phrases, main and subordinate clauses, etc.
2) In NC technology, defined rules for structuring instructions in characters, words, and blocks.

System An integral combination of things, processes, and components; may have been created by humans.

Examples: The periodic table of the chemical elements, planetary systems, systems of measurement, cybernetic systems, programming systems, and manufacturing systems.

Systematics Subdivision according to objective, logical relationships.

TCP/IP Abbreviation for ***T****ransmission* ***C****ontrol* ***P****rotocol/* ***I****nternet* ***P****rotocol.* A network protocol that due to its great importance for the Internet is also simply called the *Internet Protocol.*

Teach-In Mode, Teaching Mode, Teach Mode Programming by moving NC axis to various positions in a sequence. Used primarily for robots by moving the robot arm step by step to the desired positions in setup mode. The CNC stores these values when a "Store" button is pressed.

After that, the individual positions are moved to automatically by "Playback" mode.

Technological Data As a supplement to the geometrical data, all of the information in an NC program for selecting the technological functions such as the speed, feed rate, coolant, and tools.

Terminal Devices for data input and display, generally consisting of an ASCII keyboard and screen.

Thread Milling → *Helical Interpolation, Helix Interpolation.*

Thyristor, Silicon-Controlled Rectifier (SCR) Semiconductor component whose transition from the conducting state to the blocking state (and vice versa) can be controlled. Has a large area of application in power electronics for speed and frequency controllers.

Tilting or Swiveling Table Clamping table or worktable at machine tools that can be tilted or swiveled in order to machine diagonal surfaces and holes and sculptured surfaces on cubic workpieces.

In machining centers, the ***A′*** axis, that is, swiveling about the *X* axis.

→ *Rotary Table.*

Time Sharing Computer operating mode in which multiple users can use the computer simultaneously for various tasks. This results in cost-effective utilization of the computer without noticeable waiting times for the individual users.

Token Ring More precisely, token-access protocol for local networks (LANs). Controls access of the individual participants to the bus. The *token* (a specific bit pattern that grants transmission rights) can only be transmitted from one participant to another. This ensures that only one participant is transmitting at any given time and that the data are transmitted without collisions.

Tool Changer Mechanical device on NC machines for automatically exchanging tools from the magazine in the working spindle and vice versa. The changing process is performed by single or double grippers or directly from the tool magazine to the spindle.

Tool Compensation Compensation values stored in the NC system in order to compensate for deviations in the tool length, different tool radii, the tool position, or tool wear.

Tool Data The tool data describe a tool, such as diameter, length, and service life; in some cases supplemented with cutting parameters, weight, shape, type, and other data concerning the tool.

Tool Function A programmable function that identifies a tool in an NC-program by the T-address and a multidigit number. In most CNCs, an additional D-word activates the related tool-offset file.

Tool-Length Compensation Value stored in the CNC system to compensate for the actual tool length compared with the programmed tool length, for example, for drills, countersinks, or taps.

Tool Management

1) Within machines, an NC function for management of the tools located in the magazine according to their tool numbers, location numbers, service life, compensation values, breakage, and wear and tear.
2) Outside machines, a central tool computer that carries out all of the management tasks that take place outside of the NC system, such as tool numbers, tool data, compensation and adjustment values, availability, remaining service life, etc.

Data exchange between the computer and the NC system is either via →*DNC* or special read/write capable data memory chips in the tool holder.

Seamless management and permanent updating of tool data is very important for cost-effectiveness of machine tools.

Tool Monitoring, Tool-Life Monitoring CNC function for monitoring the theoretical service lives (periods of use) for each individual tool in the magazine of an NC machine. The NC system does this by adding the individual usage times for the tools and comparing them with the specified theoretical service life. At the end of it service life, the tool is locked out and an alternate tool may be called up.

Tool Path The path calculated by the NC system on which the tool center point moves relative to the workpiece in order to create the programmed contour.

Tool Presetting Precise measurement and adjustment of tools to defined values as regards length (drills, mills) and diameter (boring tools). Special presetting devices are available for this purpose. These are equipped with dial gauges, microscopes, profile projectors, and computers for measuring or setting the values exactly. These tool data are stored for subsequent transfer to an NC machine.

Tool-Tip Compensation Equidistant path compensation value for turning machines to compensate for various tool-nose radii of cutting tools. It is different from cutter-radius compensation.

TOP Abbreviation for *Technical and Office Protocol.* Communication concept (LAN) developed by Boeing for administrative networks.

Topology, Structure In computer networks, topology designates the structure of the interconnections between a number of devices in order to ensure the mutual exchange of data. A distinction is made between star, ring, bus, tree, meshed network, and cell topology. The topology of a network has a decisive effect on its reliability. When individual links fail, functional capability is retained only when alternative paths exist between the nodes. In this case, there are one or more replacement paths (or detours) in addition to the normal path.

Torque Motor Torque motors are gearless direct drives with very high torque (more than 8,000 Nm) and relatively low speed. They are used for quick, precise movement and positioning tasks. Thanks to their compact construction and limited number of components, they require very little space. They are suitable for use with rotary tables, swivel and rotary axes, spindle machines, dynamic tool magazines, and spindles in milling machines.

Torque motors can be manufactured with internal or external rotors. With external rotors, they have greater torque with the same external dimensions.

Touch Screen Screen with touch sensors to activate menu options by touching them with the finger (replacement for softkeys).

Transfer Line Grouping of a number of machine tools in a manufacturing line, where all parts pass through the individual stations in a specific sequence and are processed using subsequent supplemental programs. The machining processes can be changed only within certain limits. This means that transfer lines are ideal for series production without major product variations.

Turning Center NC turning machine with automatic tool- and workpiece-changing and additional equipment providing expanded machining options, such as eccentric drilling and milling, reclamping of parts, milling of surfaces, and possibly even grinding, measuring, and hardening.

Ultrasonic Technology A new technology for the cost-effective machining of ceramics, glass, carbides, silicon, and similar materials. This is done by superposing a mechanical oscillation of approximately 20,000 Hz on the milling tool in the direction of the *Z* axis using a special *ultrasound spindle* at speeds of 3,000 to 40,000 rpm. The vibrating diamond tool knocks very small powdery particles off the workpiece, thus creating a surface with extremely high quality.

Undershoot Undesirable rounding of corners and other erratic transitions on the workpiece made by the cutting tool, caused by the axis lag (following error) of the NC axes. This can be avoided by programming "exact stop" (G60, G61), a high kV factor or "corner braking".

→ *Corner Rounding.*

Upward Compatibility The capability of a newer CNC to execute NC programs written for an older NC/CNC.

USB Stick Portable semiconductor memory, generally flash ROM. Unlike the main memory of a PC, these memory chips retain their contents even when no operating voltage is present. USB is the ideal connection technology for this because it is now in widespread use, brings the required power supply along with it, and specifically supports plugging in and unplugging during operation.

V. 24 Interface Data interface recommended and standardized by CCITT that largely corresponds to the EIA-232-C interfaces. It is used in NC systems for automated data input/output.

Variable-Block Format NC program format in which the length of each block can vary depending on number of addresses and numeric values.

VDAFS Abbreviation for *Verband der Automobilindustrie – Flächen-Schnittstelle*, a surface-translation format from the German Association of the Automotive Industry (VDA). Became a DIN standard (DIN 66301) in 1986. A purely geometric translation format especially for exchanging three-dimensional (free-form) curve and surface data, for example between CAD systems. It is characterized by having only a few basic elements, a simple data format and simple syntax.

Vector Federate Feedrate that results when a tool moves along the workpiece contour. The axes involved change their velocities in such a way that the resulting vector feedrate of the tool corresponds to the programmed value and remains constant.

Virtual Product Realistic, computer-based representation of a product as a solid on the screen, with all of the required functions. Realistic assessments can be made without creating a physical model.

VLSI (Very Large-Scale Integration) Designates the scale of integration of electronic semiconductor modules. VLSI processors have between 100,000 and 3 billion transistors, for example.

Volatile Memory A data-storage device that loses its stored information when power is turned off.

Example: RAM.

WAN Abbreviation for *wide-area network*, that is, a data connection between computers over long distances using public facilities such as telephone lines or ISDN.

Wizard A software tool used, for example, in CAD and PDM systems to automate processes and significantly reduce machining times.

Word In a NC a word consists of an address character followed by a number, for example, N123, G01.

Working Area Limit Programmable limitation of the permitted working space of an NC machine by entering the upper and lower limit values for each axis. The machine switches off immediately if a command position outside these limits is entered.

→ *Software Limit Switch.*

Workpiece Changing The programmable, automatically executed loading/uploading process in an NC machine with the aid of a pallet changer or a robot.

Workstation A terminal for communication with a computer, generally equipped with its own CPU.

***X, Y, Z* Axes** Addresses for the three main linear NC axes, which are generally perpendicular to each other. *X* is the horizontal longitudinal axis, *Y* is a transverse motion, and *Z* is the motion in the direction of the spindle axis.

Zero Lag Also called *feed forward control.* This NC function corrects the expected → *Deviation*, that is, the NC axes move directly to the desired contour without any → *Following Error.*

Zero Offset An NC function for shifting the program zero point as desired, manually or by means of a program. For this purpose, CNC systems have separate memory areas for a number of zero offsets that can be called up by the NC programs.

Zooming, Zoom Function Continuously adjustable enlargement or reduction of a graphical depiction on a screen to allow better identification of the details.

Index

An "f" follows page numbers referencing figures; a "t", tables.

A

B

C

D

G

N

O

P

S

T

U

V

W

X

Y

Z